Birkhäuser

Modeling and Simulation in Science, Engineering and Technology

More information about this series at http://www.springer.com/series/4960

Yuri Bazilevs • Kenji Takizawa
Editors

Advances in Computational Fluid-Structure Interaction and Flow Simulation

New Methods and Challenging Computations

Birkhäuser

Editors
Yuri Bazilevs
Department of Structural Engineering
University of California, San Diego
La Jolla, CA, USA

Kenji Takizawa
Department of Modern Mechanical Engineering
Waseda University
Tokyo, Japan

ISSN 2164-3679 ISSN 2164-3725 (electronic)
Modeling and Simulation in Science, Engineering and Technology
ISBN 978-3-319-40825-5 ISBN 978-3-319-40827-9 (eBook)
DOI 10.1007/978-3-319-40827-9

Library of Congress Control Number: 2016952091

Mathematics Subject Classification (2010): 65N30, 76D05, 76Z05, 74S05, 74F10

Printed on acid-free paper

This book is published under the trade name Birkhäuser, www.birkhauser-science.com
The registered company is Springer International Publishing AG
The registered company address is: Gewerbestrasse 11, 6330 Cham, Switzerland

Dedicated to Prof. Tayfun E. Tezduyar on the occasion of his 60th *birthday*

Preface

This edited book contains chapters written by invited speakers of Advances in Computational Fluid-Structure Interaction and Flow Simulation (AFSI) 2014 – a conference that celebrated the 60th birthday of Prof. Tayfun E. Tezduyar. The conference was held on March 19–21, 2014, in the Green Computing Systems Research and Development Center at Waseda University, Tokyo, Japan. Over 70 people attended the birthday celebration conference, and over 60 presentations were delivered as part of the conference technical program. This edited volume, much like the AFSI 2014 conference, is dedicated to Tayfun Tezduyar and celebrates his exemplary research achievements in computational fluid-structure interaction (FSI) and flow modeling and simulation, and his impact in the computational mechanics community at large.

This book has seven distinct parts arranged by thematic topics. The first three parts are devoted to contributions in computational fluid dynamics (CFD) and cover basic methods and applications, flows with moving boundaries and interfaces, and phase-field modeling, the latter being a relatively new research direction in CFD. The remaining four parts are focused on computer science and high-performance computing (HPC) aspects of flow simulation, mathematical methods, biomedical applications, and, last but not least, FSI. The book opens with a humorous yet impressive account of Tayfun Tezduyar's career, which is a printed version of the laudation that Prof. Thomas J.R. Hughes delivered at the AFSI 2014 conference banquet.

The editors would like to thank all the authors for choosing to submit their contributions to this edited volume, for meeting the editor- and publisher-set deadlines, and for being patient with the editors throughout this process.

La Jolla, CA, USA — Yuri Bazilevs
Tokyo, Japan — Kenji Takizawa

Contents

Laudation at the AFSI 2014 Conference Banquet Celebrating Tayfun Tezduyar's 60th Birthday, Tokyo, Japan, March 2014

Thomas J. R. Hughes

1 Laudation

Thanks to Kenji Takazawa and Yuri Bazilevs for organizing this wonderful conference, and thank you to all the attendees who are here to honor Tayfun.

Let us begin with a little history.

1954 was an incredible year in world history:

Marilyn Monroe married Joe Dimaggio.
The first issue of *Sports Illustrated* was published in the USA.
Bill Haley and the Comets recorded "Rock Around the Clock" and Rock and Roll was born.
Roger Bannister, in Oxford, England, was the first to run a mile in under four minutes.

Famous people were born in 1954:

Oprah Winfrey, American actress, talk show host, publisher, producer, and entrepreneur.
Socrates, Brazilian footballer (attacking midfielder, doctorate in medicine), medical doctor, intellectual, and journalist.
François Hollande, President of France.
Shinzo Abe, Prime Minister of Japan.

T.J.R. Hughes (✉)
Institute for Computational Engineering and Sciences, University of Texas at Austin, 201 East 24th St. Stop C0200, Austin, TX 78712, USA

Aerospace Engineering and Engineering Mechanics, University of Texas at Austin, Austin, TX, USA
e-mail: hughes@ices.utexas.edu

Y. Bazilevs, K. Takizawa (eds.), *Advances in Computational Fluid-Structure Interaction and Flow Simulation*, Modeling and Simulation in Science, Engineering and Technology, DOI 10.1007/978-3-319-40827-9_1

Last but not least, the most famous of them all:

Tayfun Ersin Tezduyar, gentleman, scholar, engineer, researcher, author, editor, etc., etc., etc.

Not much happened in the world until 1977 when Tayfun arrived at Caltech from Turkey and shortly thereafter became my PhD student. This is where I met Tayfun. (Tayfun came to Caltech with no undergraduate degree. His professors in Turkey indicated this would just be a waste of his time as he was ready to pursue the PhD.) I do not remember our first meeting, but I do remember one of the Mechanical Engineering professors at Caltech, Alan Acosta, telling me "He wants to work with you" and "He might really do something"—prophetic words. Tayfun had almost immediately established a reputation at Caltech as the brightest among many bright students. He had impressed many of his professors. As we began to work together I noted how thoroughly he would analyze every idea, his neat handwritten derivations, and his meticulous organization. It was a pleasure to work with him and discuss ideas. Everyone here knows about his accomplishments in fluids and fluid–structure interaction, but what you might not be aware of was that, very early on, we co-authored an important paper on a finite element for plate bending. It was the first four-node quadrilateral element that did not lock but at the same time attained full rank. It was immediately implemented in several commercial nonlinear finite element codes (e.g., MARC, Abaqus). When I left Caltech for Stanford in 1980, Tayfun came with me and worked on his PhD thesis there. I obtained support from NASA Ames Research Center and NASA Langley Research Center to work on finite elements for compressible flows. This was the topic of Tayfun's PhD thesis and he became the first person ever to solve the compressible Euler equations with finite elements. The first time I gave a talk about this landmark accomplishment was in Tokyo in 1982 at the Finite Element in Fluids conference organized by Professor Tad Kawai, among others. My presentation created quite a stir and drew the attention of French researchers from Dassault Aviation, namely Jacques Periaux and Pierre Perrier whom soon after visited me at Stanford and began supporting my research, which lasted for many years. After completing his thesis, Tayfun continued working with me as a post-doc and then joined the faculty of the University of Houston, subsequently moving to the University of Minnesota, and, finally, Rice University where he is today. At the University of Minnesota he was named the Distinguished McNight University Professor and became Principal Investigator and Director of the Army High Performance Computing Research Center, and led it through a period of intense activity, great accomplishment, and international recognition. He left Minnesota to become James F. Barbour Professor of Mechanical Engineering and Materials at Rice, and he chaired the department for six years.

Tayfun is famous for his groundbreaking research in computational fluid dynamics and his fundamental contributions to the solution of complex fluid–object and fluid–structure interaction problems, for which he has received a multitude of honors. Some of the simulations he has performed of fluid–structure interaction are among the most impressive calculations ever performed in any discipline. If I had to select one, as truly awe-inspiring, I would say it is the work he has performed for the NASA Johnson Space Center on multiple interacting space parachutes.

Over the years Tayfun and I have become very good personal friends, and we have interacted professionally in a variety of capacities. Tayfun maintains very high professional standards, but is also a kind and generous individual and a loyal friend. His strong support for young people has been mentioned, but I am personally happy to say he also strongly supports old people ☺. Tayfun has had a brilliant academic career and achieved great things and I personally feel it is especially nice that we are here to celebrate his 60th birthday in a country that he loves and where he has many dear friends and admirers, and where his wife comes from.

60: The number 60 has mathematical and metaphysical significance. It is divisible by 1, 2, 3, 4, 5, 6, 10, 12, 15, 20, and 30. It is the smallest number divisible by 1, 2, 3, 4, 5, and 6. The base-60 number system (i.e., "sexagesimal") was originated by the Sumerians in the third millennium BC and passed down to the Babylonians. The legacy of this system remains with us today. For example, the angles of an equilateral triangle are all 60°, hours consist of 60 minutes, minutes of 60 seconds, etc. Ptolemy's *Almagest*, a treatise on mathematical astronomy, used the base 60 system to express fractions. Book VIII of Plato's Republic involves an allegory of marriage centered on the number 60^4. The first fullerene molecule to be discovered was the Buckminsterfullerene, C_{60}, consisting of 60 carbon atoms, known to some of us as a soccer ball.

Tayfun Tezduyar will be 60 in August of this year. What is the significance of becoming 60 years old? It is a time to look back and take stock of one's life, and in the best of circumstances it is also a time to look ahead to the future and continue to build upon one's accomplishments. In the USA, retirement is an abstract concept. Mandatory retirement is illegal. Discrimination because of age is likewise illegal. One can retire if one chooses, or work as long as one is willing and able. I hope Tayfun will continue with his work for a long time, because his work is scientifically elucidating and of great importance to engineering. He is the true modern master at developing technologies for solving the most complex engineering problems.

But most of all I hope to have him for my dear friend for many years to come.

Part I
CFD: Methods and Applications

Numerical Comparison of the Particle Finite Element Method Against an Eulerian Formulation

Juan M. Gimenez, Pedro Morin, Norberto Nigro, and Sergio Idelsohn

Abstract The main goal of this paper is to validate experimentally the principal conclusions previously published in [17]. Two manufactured test cases were considered with their respective analytic solutions. First, a scalar transport equation is considered written in such a way that several parameters are included to stress the limiting situation where the Eulerian and the Lagrangian approaches behave better. The results show conditions to be fulfilled in order to choose between both formulations, according to the problem parameters. A brief discussion about the projection needed for PFEM-2 method is included, specially due to its impact on the error convergence rate. Lately, an extension to Navier-Stokes equations is introduced using also a manufactured case to verify again the same conclusions. This paper intends to establish the first steps towards a mathematical error analysis for the particle finite element method which supports the preliminary theoretical and experimental results presented here.

J.M. Gimenez • N. Nigro
Centro de Investigación de Métodos Computacionales, CIMEC-UNL/CONICET, Predio CONICET Santa Fe, Colectora Ruta Nac 168, Km 472, Paraje El Pozo, 3000 Santa Fe, Argentina
e-mail: jmarcelogimenez@gmail.com
http://www.cimec.org.ar

Facultad de Ingeniería y Ciencias Hídricas - Universidad Nacional del Litoral, Ciudad Universitaria, Paraje "El Pozo", Santa Fe, Argentina
http://www.fich.unl.edu.ar

P. Morin
Instituto de Matemática del Litoral, IMAL-UNL/CONICET, Predio CONICET Santa Fe, Colectora Ruta Nac 168, Km 472, Paraje El Pozo, 3000 Santa Fe, Argentina
http://www.imal.santafe-conicet.gov.ar

S. Idelsohn (✉)
International Center for Numerical Methods in Engineering. CIMNE Edificio C1, Campus Norte UPC, C/ Gran Capitán S/N, 08034 Barcelona, Spain
http://www.cimne.com

ICREA, Institució Catalana de Recerca i Estudis Avançats (ICREA), Barcelona, Spain
e-mail: sergio@cimne.upc.edu

Y. Bazilevs, K. Takizawa (eds.), *Advances in Computational Fluid-Structure Interaction and Flow Simulation*, Modeling and Simulation in Science, Engineering and Technology, DOI 10.1007/978-3-319-40827-9_2

1 Introduction

Over the last decades, computer simulation of incompressible fluid flow has been mainly based on the Eulerian formulation of the fluid mechanics equations on fixed domains [3]. During this period, hardware has evolved considerably increasing the speed performance of computations and allowing better facilities for data entry and the display of results. However in these decades there have been no substantial improvements on the numerical methods used concerning the efficiency of the algorithm. In most practical engineering problems, very fine mesh and very small time-steps are needed to reach acceptable results. This handicap exceeds most time the efficiency of current powerful computers.

More recently, particle-based methods in which each particle is followed in a Lagrangian manner have been used for fluid flow problems. Monaghan [21] proposed the first ideas for the treatment of astrophysical hydrodynamic problems with the so-called Smoothed Particle Hydrodynamics Method (SPH), which was later generalized to fluid mechanics problems [6, 7, 21]. Koshizuka and coworkers [18, 19] developed a similar method to SPH, named Moving Particle Simulation (MPS). SPH and MPS belong to the family of the so-called meshless methods, as well as the Finite Point Method [24–26]. Lately, the meshless ideas were generalized to take into account the finite element type approximations in order to obtain more accurate solutions [9]. This method was called the Meshless Finite Element Method (MFEM) and uses the extended Delaunay tessellation [8] to build the mesh in a computing time, which is linear with the number of nodal points. A natural evolution of the last work was the Particle Finite Element Method (PFEM) [10]. The PFEM combines the particle precept with the Finite Element Method (FEM) shape functions using a background finite element mesh. This mesh may be quickly rebuilt at each time-step (PFEM with moving mesh) or may be a fixed mesh (PFEM with fixed mesh). In the last case, the results from the Lagrangian particles are projected on a fixed mesh at each time-step. The idea of combining fixed meshes with moving particles is not new. It was introduced for convection-diffusion problems in [22] and was used in the so-called Particle in Cell method (PIC) [2] and later in its extension called the Material Point Method (MPM) [29]. All these methods use a Finite Element (FE) background mesh. Despite that both the PFEM and the MPM use a fixed FE mesh and a set of Lagrangian particles, there are important differences in the way the particles are employed: thus, while in the MPM all computations are performed on the mesh, in the PFEM the aim is to calculate as much as possible on the particles, leaving small corrections to be performed on the mesh. However, the most important difference is that in the PFEM the particles do not represent a fixed amount of mass, but rather material points that transport only intrinsic properties of the fluid. This allows to use a variable number of particles and therefore simplifying refinement.

The PFEM has been successfully used to solve the Navier-Stokes equations [1, 20] and fluid–structure interaction problems [11, 12, 28] as well as solid mechanics problems [23]. The advantages of the PFEM concerning the tracking of internal interfaces have also been explored and used to solve fluid mechanics problems including multi-fluid flows [13].

The possibility to use the PFEM to solve non-linear problems with large time-steps in order to obtain an accurate and fast solution was successfully explored by the authors for the solution of the homogeneous incompressible Navier-Stokes equations [5, 14, 15] and multi-fluid problems [4, 16]. This new strategy was named PFEM-2.

In [17] the first trial of an error analysis for Lagrangian based methods like PFEM and its comparison with an Eulerian formulation was presented in order to demonstrate why the former is more accurate than the latter in certain particular cases when large time-step and/or coarse meshes were used. There the authors claim that nowadays, the best way to improve the efficiency of the algorithms in order to take advantages of the increasing computer power is using this particle-based method.

The goal of this paper is to demonstrate numerically the validity of that first error analysis attempt for scalar transport problems and also for homogeneous fluid flow problems. During this analysis the projection error arises as one of the main limitations to reach that goal and a new proposal of projection algorithm is presented. This projection shows to have several collateral benefits like having good mathematical properties as the existence of reciprocity in the operation of going back and forth between mesh and particles and also smoothing properties in the computation of fluid flow forces.

The layout of the paper is the following. After a review of the main results presented in [17] a numerical validation of these results for scalar transport problems is shown highlighting the conditions under which a Lagrangian approximation is preferred, in particular when using large time-steps. Next, the new projection scheme is presented putting in evidence the possibility and the requirement of reaching a second order approximation in space and time. Finally, an extension to fluid flow problems is presented with a manufactured example that serves to demonstrate the conclusions written in [17]. Some conclusions and future trends are highlighted at the end.

2 Error Analysis Applied to a Scalar Transport Equation

Many problems in engineering may be mathematically expressed by transport equations written in Lagrangian or Eulerian reference frames. For such mathematical models there is a strong division between transporting scalar fields or transporting vector or tensor fields. Also the linear and non-linear approaches become another important feature normally included in the modeling. Moreover, advection and diffusion effects present very different behaviors from the physical, mathematical, and numerical point of view. With such a complexity it is very difficult to find analytic solutions to be used as reference for formal error analysis. In order to circumvent such a drawback in this paper manufactured test cases are developed to be used as good candidates for getting exact solutions to our test problems.

Several control parameters for the exact solution are used to place the problem in various conditions from some beneficial to Lagrangian schemes to other more beneficial to an Eulerian one.

To begin, in this section a scalar transport equation is solved. The differential equation associated to solve this problem may be written in an Eulerian (fixed) frame as

$$\frac{\partial T}{\partial t} + \mathbf{v} \cdot \nabla T = Q(\mathbf{x}) \tag{1}$$

where $T = T(\mathbf{x}, t) = T^t(\mathbf{x})$ is the scalar unknown, $\mathbf{v}$ is the velocity field, and $Q(\mathbf{x})$ includes the source term $q(\mathbf{x})$, the diffusion term $\nabla \cdot (k\nabla T)$, and the linearized version of a reactive source term cT, etc., being

$$Q(\mathbf{x}) = q(\mathbf{x}) + \nabla \cdot (k\nabla T) - cT + \ldots \tag{2}$$

To write Eq. 1 in the Lagrangian (mobile) frame, it is necessary to use the material derivative D/Dt, which condenses the temporal derivative and the convective term into a unique term. However, as it is well known for Lagrangian formulations not only the respective field should be computed, but also the particle trajectories, obtaining

$$\begin{cases} \dfrac{DT}{Dt} = Q(\mathbf{x}_p) \\[2ex] \dfrac{D\mathbf{x}_p}{Dt} = \mathbf{v} \end{cases} \tag{3}$$

where $T = T(\mathbf{x}_p^t, t) = T^t(\mathbf{x}_p^t)$ is the scalar unknown. Here the superscript t means the time dependency and the subscript p represents the particle itself.

Eqns. (1) and (3) are integrated in time in an Eulerian frame using a linear method named θ-method:

$$T^{n+1}(\mathbf{x}) = T^n(\mathbf{x}) + \int_n^{n+1} \left(Q^t - \mathbf{v}^t\nabla T^t\right)\, dt \approx T^n(\mathbf{x}) + (Q - \mathbf{v}\nabla T)^{n+\theta}\ \Delta t \tag{4}$$

where $f^{n+\theta} = (1-\theta)f^n + \theta f^{n+1}$ is the linear in time interpolation for any function f. The time integration error of a function f^t for $\theta = 1/2$ is proportional to the second derivative of the function and Δt^2, i.e.,

$$\int_{n}^{n+1} f^t \, dt \approx f^{n+\frac{1}{2}} \Delta t \pm \epsilon \;\text{ with }\; \epsilon = \mathscr{O}\langle \frac{\partial^2 f}{\partial t^2} \Delta t^2 \rangle \Delta t. \tag{5}$$

In the case of time integration in the Lagrangian frame, there are several options to evaluate that integral [17]. An option is to perform the integration with the moving particles following the streamlines (taking into account intermediate positions), splitting the integration into an explicit part and an implicit part, or decoupling the trajectory and temperature integrations employing two different values for θ, i.e.,

$$\begin{cases} T^{n+1}(\mathbf{x}_p^{n+1}) \approx T^n(\mathbf{x}_p^n) + (1-\theta_1) \displaystyle\int_{n}^{n+1} Q^n(\mathbf{x}_p^t)\, dt + \theta_1 Q^{n+1}(\mathbf{x}_p^{n+1}) \Delta t \\ \mathbf{x}_p^{n+1} \approx \mathbf{x}_p^n + (1-\theta_2) \displaystyle\int_{n}^{n+1} \mathbf{v}^n(\mathbf{x}_p^t)\, dt + \theta_2 \mathbf{v}^{n+1}(\mathbf{x}_p^{n+1}) \Delta t \end{cases} \tag{6}$$

The case of performing the unknown integration following streamlines with $\theta_1 = 1/2$, and the trajectory, also following streamlines, but with $\theta_2 = 0$, has been named X-IVAS+implicit correction [15]. On the other hand, if the choice for the unknown integration is $\theta_1 = 1$ the method is named X-IVS, which was presented in [14]. The latter integration will be used in all the examples presented here when using a Lagrangian frame. However, in the present study for the error analysis, the standard θ-method will be considered:

$$\begin{cases} T^{n+1}(\mathbf{x}_p^{n+1}) \approx T^n(\mathbf{x}_p^n) + Q^{n+\theta}(\mathbf{x}_p^{n+\theta}) \Delta t \\ \mathbf{x}_p^{n+1} \approx \mathbf{x}_p^n + \mathbf{v}^{n+\theta}(\mathbf{x}_p^{n+\theta})\, \Delta t \end{cases} \tag{7}$$

This assumption is based on the fact that the X-IVAS integration improves the results and decreases the integration errors. The evaluation of this difference is out of the scope of this work.

2.1 Eulerian Errors

As presented in Equation (5), the linear θ-method introduces a minimum error for $\theta = 1/2$ which is proportional to the second derivative of the integrated function and the square of the time-step, i.e.,

$$T^{n+1}(\mathbf{x}) = T^n(\mathbf{x}) + \int_{n}^{n+1} \left(Q^t - \mathbf{v}^t \nabla T^t\right) dt \approx T^n(\mathbf{x}) + \left[(Q - \mathbf{v}\nabla T)^{n+\frac{1}{2}}\right] \Delta t \pm \epsilon_t \tag{8}$$

with the time integration error

$$\epsilon_t = \mathcal{O}\langle \frac{\partial^2}{\partial t^2} [Q - \mathbf{v}\nabla T] \Delta t^2 \rangle \Delta t = \mathcal{O}\langle [Q - \mathbf{v}\nabla T]^{\cdot\cdot} \Delta t^2 \rangle \Delta t. \tag{9}$$

On the other hand, the FEM approximation of the functions and the space derivatives introduces spatial errors. As it was mentioned, this analysis only considers linear finite element approximations of the unknown, therefore the spatial errors are proportional to the second derivative of the functions and the square of the mesh-size, i.e.,

$$T^{n+1}(\mathbf{x}) = T^n(\mathbf{x}) + \int_n^{n+1} \left(Q^t - \mathbf{v}^t \nabla T^t \pm \epsilon_x\right) dt \approx T^n(\mathbf{x}) + \left[(Q - \mathbf{v}\nabla T)^{n+\frac{1}{2}} \Delta t \pm \epsilon_x\right] \Delta t \pm \epsilon_t \tag{10}$$

with the spatial discretization error

$$\epsilon_x = \mathcal{O}\langle \frac{\partial^2}{\partial \mathbf{x}^2} [Q - \mathbf{v}\nabla T] \Delta x^2 \rangle \Delta t. \tag{11}$$

Finally, avoiding higher order terms, the unknown function after a time-step using the Eulerian framework is

$$T^{n+1}(\mathbf{x}) = T^n(\mathbf{x}) + (Q - \mathbf{v}\nabla T)^{n+\frac{1}{2}} \Delta t \pm \epsilon_x + \epsilon_t = T^n(\mathbf{x}) + (Q - \mathbf{v}\nabla T)^{n+\frac{1}{2}} \Delta t \pm \epsilon^E \tag{12}$$

with $\epsilon^E = \epsilon_x + \epsilon_t$.

2.2 *Lagrangian Errors*

In this section first a general analysis taking into account the whole problem involved in a Lagrangian formulation, i.e., solving not only for the scalar field, but also for the particle trajectories, is done. Afterwards a specific analysis of projection errors for PFEM-2 scheme due to the mapping of the field between particles and the mesh is presented considering the importance of this projection stage in the global error of the method.

2.2.1 General Analysis

In the case of the Lagrangian frame, the errors in the numerical evaluation of the unknown function and the particle position are

$$\begin{cases} T^{n+1}(\mathbf{x}_p^{n+1}) \approx T^n(\mathbf{x}_p^n) + Q^{n+\frac{1}{2}}(\mathbf{x}_p^{n+\frac{1}{2}})\Delta t \pm \mathcal{O}\langle Q''\Delta x^2\rangle\Delta t \pm \mathcal{O}\langle Q^{\cdot\cdot}\Delta t^2\rangle\Delta t \\ \mathbf{x}_p^{n+1} \approx \mathbf{x}_p^n + \mathbf{v}^{n+\frac{1}{2}}(\mathbf{x}_p^{n+\frac{1}{2}})\,\Delta t \pm \mathcal{O}\langle \mathbf{v}''\Delta x^2\rangle\Delta t \pm \mathcal{O}\langle \mathbf{v}^{\cdot\cdot}\Delta t^2\rangle\Delta t \end{cases} \tag{13}$$

The error in the evaluation of the particle position $\epsilon_x^p = \mathcal{O}\langle \mathbf{v}''\Delta x^2\rangle\Delta t \pm \mathcal{O}\langle \mathbf{v}^{\cdot\cdot}\,\Delta t^2\rangle\Delta t$ introduces also an error in the evaluation of the unknown function. Performing a series expansion around $\mathbf{x}_p^{n+1}$, i.e.,

$$T^{n+1}(\mathbf{x}_p^{n+1} + \epsilon_x^p) = T^{n+1}(\mathbf{x}_p^{n+1}) + \nabla T\epsilon_x^p + \mathcal{O}\langle T''(\epsilon_x^p)^2\rangle \tag{14}$$

and replacing in the first equation of (13), it is

$$T^{n+1}(\mathbf{x}_p^{n+1}) \approx T^n(\mathbf{x}_p^n) + Q^{n+\frac{1}{2}}(\mathbf{x}_p^{n+\frac{1}{2}})\Delta t \pm \mathcal{O}\langle Q''\Delta x^2\rangle\Delta t \pm \mathcal{O}\langle Q^{\cdot\cdot}\Delta t^2\rangle\Delta t \pm \nabla T\epsilon_x^p \tag{15}$$

the expression for the Lagrangian error ϵ^L can be found

$$\epsilon^L = \pm\mathcal{O}\langle\big(Q'' \pm \mathbf{v}''\nabla T\big)\,\Delta x^2\rangle\Delta t \pm \mathcal{O}\langle(Q^{\cdot\cdot} \pm \mathbf{v}^{\cdot\cdot}\nabla T)\,\Delta t^2\rangle\Delta t \tag{16}$$

Comparing the Eulerian error expression in 12 with the Lagrangian one in 16, the main differences are in the following terms:

$$\underbrace{[\mathbf{v}\nabla T]''\,\Delta x^2 \pm [\mathbf{v}\nabla T]^{\cdot\cdot}\Delta t^2}_{Eulerian} \neq \underbrace{\mathbf{v}''\nabla T\Delta x^2 \pm \mathbf{v}^{\cdot\cdot}\nabla T\Delta t^2}_{Lagrangian} \tag{17}$$

The difference presented in Equation (17) leads to a big advantage of the Lagrangian framework against its Eulerian counterpart for some problems. For instance, in the standard convection-diffusion problem of a non-constant unknown, where the convective field is known and has a constant or nearly constant velocity, Equation (17) reads

$$\underbrace{[\mathbf{v}\nabla T]''\,\Delta x^2 \pm [\mathbf{v}\nabla T]^{\cdot\cdot}\Delta t^2 \neq 0}_{Eulerian} \;\;;\;\; \underbrace{\mathbf{v}''\nabla T\Delta x^2 \pm \mathbf{v}^{\cdot\cdot}\nabla T\Delta t^2 = 0}_{Lagrangian} \tag{18}$$

that is, the Lagrangian integration does not have any error because $\mathbf{v}'' = \mathbf{v}^{\cdot\cdot} = 0$, but the Eulerian framework presents an error due to $[\mathbf{v}\nabla T]'' \pm [\mathbf{v}\nabla T]^{\cdot\cdot} = \mathbf{v}\,[\nabla T]'' \pm \mathbf{v}\,[\nabla T]^{\cdot\cdot} \neq 0$.

2.2.2 Projection Errors

In order to complete the error analysis for the particular Lagrangian approximation employed by PFEM-2, the projection errors should be taken into account. In [15] PFEM-2 had been introduced as a hybrid method using particles and mesh in order to exploit the advantages of Lagrangian and Eulerian methods. This duality between mesh and particles requires that data should be sent back and forth along the whole computation several times. If this projection is not designed properly a huge source of inaccuracies is introduced and the error degrades a lot. Moreover, the projection is normally responsible of getting noisy values represented by the mesh, like forces in a fluid–structure interaction problems. In [17] Idelsohn et al. conclude that these errors depend on the distance between particles which is denoted by h. Therefore, the importance to use a great amount of particles with $h << \Delta x$ becomes obvious for decreasing projection errors. However this assumption relies on an ideal projection strategy which consists in generating a mesh with the particles positions guaranteeing that the projection error depends only on particle mesh-size h.

Even being an ideal strategy, the triangulation of the particle positions at each time-step may be unaffordable because of the computational cost involved when particles tend to grow in number.

Typical strategies employed by PFEM-2 [4] to project a given field ϕ between nodes using subindices (j) and particles using subindices (p) are of the following form:

$$\phi_j = \Pi_{p\to j}\phi_p = \frac{\sum_P \phi_p W_j(\mathbf{x}_p)}{\sum_P W_j(\mathbf{x}_p)} \tag{19}$$

where the function W_j, associated with the node j, can be either the typical kernel functions used in particle methods such as SPH[7] or the linear shape functions raised to a power $\alpha > 0$ (it is $W_j(\mathbf{x}) = N_j(\mathbf{x})^\alpha$), while $\mathbf{x}_p$ is the position of the particle p with state ϕ_p and P is the number of particles in a region around the node j.

This projection algorithm does not preserve the global second order error introducing a degradation of the solution accuracy proportional to the number of projections done. This last variable depends on the time-step with a smaller impact when the time step is large and a bigger one when the time-step tends to reduce. This fact is observed experimentally in the next section.

On the other hand this family of projections algorithms is in general not commutative, i.e., the projection does not satisfy the inverse property of operators:

$$\Pi_{p\to j}\left(\Pi_{p\to j}\right)^{-1} \neq \mathbf{I} \tag{20}$$

This last statement strongly endangers the solution accuracy and may also be responsible for excessive diffusion of the primal variables or for noisy secondary (dual) fields computed on the mesh, like forces, heat fluxes, etc.

In order to circumvent this drawback in this paper a new projection is presented overcoming most of the above cited weakness. Here only its definition and the main idea behind this development are presented. The mathematical demonstration of its incidence on the error analysis is still in elaboration.

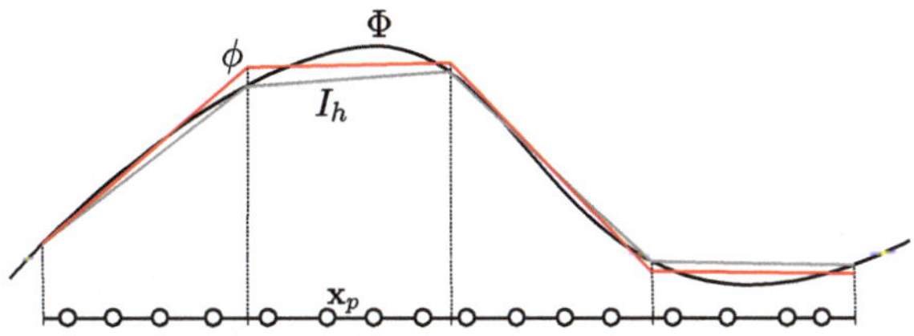

Fig. 1 Graphical example of the target function Φ and its two possible FEM approximations ϕ and I_h

The idea is based on using least squares approximations to find the nodal values from particles states.

More precisely, the idea is to find ϕ_j that realizes the following minimum:

$$\min\left[\frac{1}{2}\sum_p\left(\phi_p - \sum_j \phi_j N_j(\mathbf{x}_p)\right)^2\right] \tag{21}$$

Solving the Equation (21) leads to an equation system of J unknowns $\mathbf{M}\boldsymbol{\phi} = \mathbf{f}$, where J is the number of nodes, $\mathbf{M}_{ij} = \sum_p N_i(\mathbf{x}_p)N_j(\mathbf{x}_p)$ is a consistent mass matrix, and $\mathbf{f}_i = \sum_p N_i(\mathbf{x}_p)\phi_p$.

Figure 1 shows a graphic representation of the approximation of a target function Φ where the particles look like quadrature points where the FEM solution is evaluated. The function ϕ is obtained minimizing (21), while the function I_h is the Lagrange interpolator which takes the values of the function Φ at nodal positions. Because an approximation as I_h, which belongs to the same discretization space than ϕ, has an error proportional with the square of the mesh-size, i.e., $\epsilon \approx C\Delta x^2$, and ϕ has the lowest error among the functions of this space, then it follows that ϕ has also an error proportional with the square of the mesh.

Finally the projection error is introduced at each projection step independently of the time-step size. Then, in a total period of time $(t_f - t_0)$ the Lagrangian error can be extended to

$$\begin{aligned}\epsilon^L = &\pm\mathscr{O}\langle(Q'' \pm \mathbf{v}''\nabla T)\,\Delta x^2\rangle(t_f - t_0) \pm \mathscr{O}\langle(\ddot{Q} \pm \ddot{\mathbf{v}}\nabla T)\,\Delta t^2\rangle(t_f - t_0)\pm \\ &\pm \mathscr{O}\langle T''\Delta x^2\rangle\frac{(t_f - t_0)}{\Delta t}\end{aligned} \tag{22}$$

2.3 Validation Test

The differences presented in Equation (17) should be analyzed in depth. In the Eulerian case the error formula includes the spatial and temporal derivative of the product between the velocity and the scalar gradient, while in the Lagrangian formula they only affect the velocity. It has important theoretical consequences, which can be enumerated:

1. Eulerian frames are better for diffusion dominant problems. In these cases, the errors between the Lagrangian and Eulerian approaches are of the same order, but in the Lagrangian frames the projection errors must be added.
2. Lagrangian frames are better for convective dominant problems when the convective flow is constant or nearly constant in time $\mathbf{v}^{\cdot\cdot} \approx \mathbf{v}'' \approx 0$. The remaining cases, when the convective flow presents high variations the Lagrangian or the Eulerian frame will be better or worse depending on the projection errors.
3. Eulerian frames are better for stationary problems. In these cases, $[\nabla T]^{\cdot\cdot} = 0$ and there are no projection errors.

The aim of this section is to verify these facts experimentally, with numerical simulations employing both frames. The modification of some parameters in these tests will increase or decrease the terms which introduce errors in each formula, being the objective to match the experimental and theoretical results. In order to do this a particular 2D problem was tailored with an analytical solution to compare the results.

The proposed domain is $[-1, -1] \times [1, 1]$. The velocity field $\mathbf{v} = u\hat{\mathbf{i}} + v\hat{\mathbf{j}}$ has variations both in time and space: a rigid rotation with a periodic temporal variation of its angular velocity which is modulated in space by a parabola. The proposed velocity field is

$$\begin{cases} u(x,y,t) = -\omega_1 y(1-x^2)(1-y^2)(1+C\sin(\omega_2 t)) \\ v(x,y,t) \ = \omega_1 x(1-x^2)(1-y^2)(1+C\sin(\omega_2 t)) \end{cases} \tag{23}$$

where ω_1 is the mean angular velocity of the rotation field, ω_2 is the frequency at which the field increases or decreases its rotation. The parameter C allows to control the amplitude of the variation of the rotation direction, such as if $|C| > 1$ the rotation is inverted for certain time, if $C = 1$ the movement vanishes at some instant, and if $|C| < 1$ the field does not change its rotation direction. A source term q was included such that the unknown function becomes

$$T(x,y,t) = \sin(\omega_3 t)\sin(\pi x)\sin(\pi y) \tag{24}$$

which is identically zero at boundaries and has four hills at points $[\pm 0.5, \pm 0.5]$. A snapshot of the proposed scalar field is presented in Figure 2. An initial value of $T = 0$ and Dirichlet boundary conditions $T = 0$ were used. The diffusivity k and the oscillation frequency ω_3 can take independent values allowing to analyze different situations.

The reference grid used has 50×50 nodes conforming 4802 triangles, which gives a mesh-size $\Delta x = H = 0.04$. In the Eulerian case, Crank-Nicholson as time discretization scheme was employed. For Lagrangian simulations four particles per element were evenly seeded. The particle grid size h is defined as $h = \sqrt{2A/N_p}$, where A is the area of the element containing N_p particles. With this definition, the area of the element is divided by the number of particles contained.

Fig. 2 Snapshot for the manufactured scalar field solution

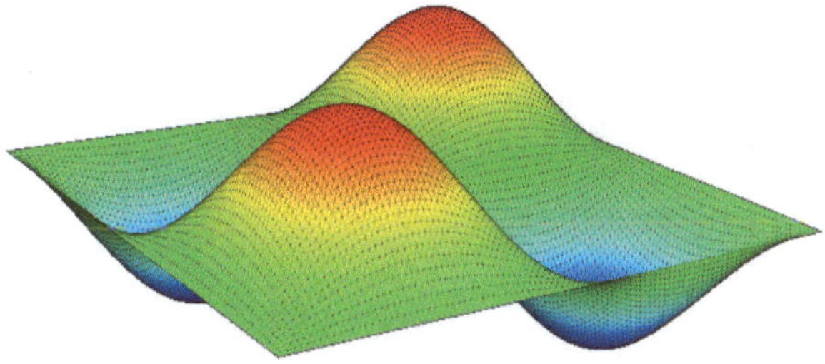

Table 1 Parameters for the manufactured 2D scalar transport case

Case	C	ω_1	ω_2	ω_3	k
1	1	2π	0	2π	0
2	1	2π	2π	2π	0
3	1	2π	2π	3π	0
4	1/2	2π	2π	2π	0

Four cases were designed to cover a wide range of problem types. Table 1 presents the parameters employed in each case. Two time-steps are employed in order to simulate with small $\text{CFL}_{max} \approx 1$ and large $\text{CFL}_{max} \approx 10$, respectively, being $\text{CFL}= |\mathbf{v}|\Delta t/\Delta x$ the Courant-Friedrich-Levy number.

In the following cases the root-mean-square (RMS) as an error measure is employed, defined as

$$RMS(t) = \sqrt{\frac{1}{N}\sum_{j=1}^{N}\left(T_j^{ex}(t) - T_j^{ap}(t)\right)^2}$$

where N is the number of nodes on the mesh, T^{ex} is the analytic reference solution, and T^{ap} is the numerical solution.

The results are presented in Figure 3. In the first case, which has a steady velocity field ($\mathbf{v}^{\cdot\cdot} = 0$) and no diffusion, as anticipated in previous section, the Lagrangian framework has better results than the Eulerian one, being specially remarkable when large Courant numbers are employed. In the mentioned case, the PFEM-2 error does not depend on the time-step size showing a periodic variation due the spatial error which is proportional to the temporal harmonic function ∇T. On the other hand, FEM shows large errors when the time-step is increased as it is expected.

In the second case, the velocity field is unsteady having a harmonic variation in its amplitude. Here $\mathbf{v}^{\cdot\cdot} \neq 0$, then the Lagrangian solution shows similar problems to the Eulerian one. However, if the oscillation frequency of the solution is larger than the amplitude variation of the velocity ($\omega_3 > \omega_2$), as happens in the third Case, the temporal error term in the Eulerian case ($[\mathbf{v}\nabla T]^{\cdot\cdot}\Delta t^2$) increases, while the Lagrangian one ($\mathbf{v}\,[\nabla T]^{\cdot\cdot}$) remains almost equal. In this way, the second and third case also prove experimentally the theoretical error formula presented in Equation (17).

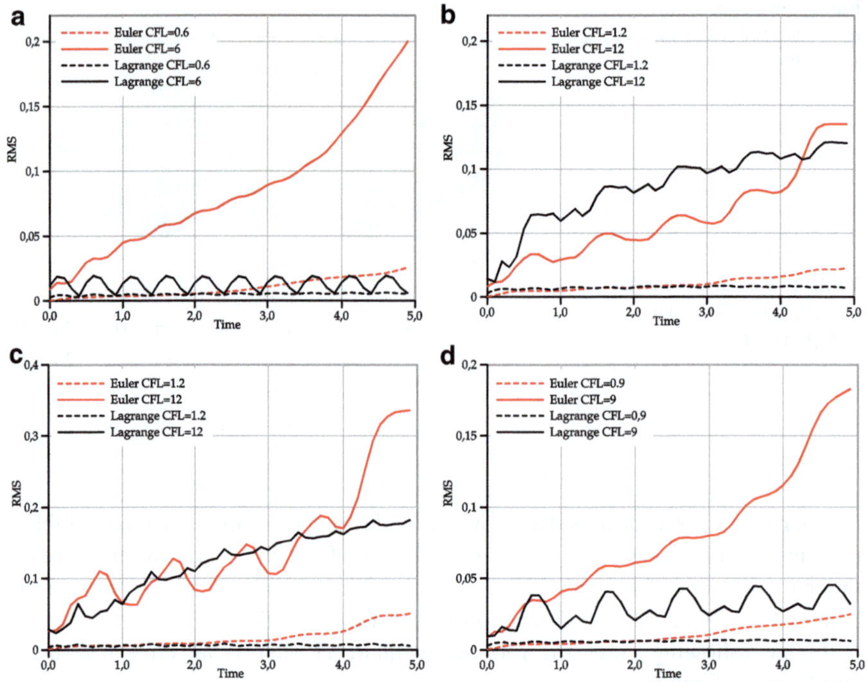

Fig. 3 Case Four hills without diffusion. RMS error evolution for Eulerian and Lagrangian simulations with steady (Figure 3(a)) and unsteady (Figures3(b), 3(c), and 3(d))) velocity fields. The Lagrangian error is presented with black lines and the Eulerian with red lines

The second and the third cases use a velocity field which vanishes every $T = 1/\omega_2$, this gives to the Eulerian formulation some advantages due the spatial error term ($[\mathbf{v}\nabla T]'' \, \Delta x^2$) also vanishes. That advantage is, in some cases, recovered by the Lagrangian approach when the time-step is increased. The fourth case overcomes this fact using another value for the constant C which avoids that $\mathbf{v}(t) = 0$ for any t. Figure 3(d) shows that the error differences are more visible, proving again the theoretical error formula.

As a preliminary conclusion, the results presented in Figure 3 confirm the proposed error formula in [17], showing that the Lagrangian error using large time-steps is affected by the unsteadiness of both, the velocity field and the source term. As a footnote based on experimental facts, it can be mentioned that an X-IVAS calculation of the unsteady source is mandatory to obtain accurate results, which is not possible with the time-step selected (it only samples seven points inside the source's period). Regarding to Eulerian simulations, as expected, they show larger errors when the CFL grows. Also, this error increases when high $\nabla T''$ are employed.

On the other hand, Figure 4 presents a convergence analysis for Case A. Three mesh refinements were used $\Delta x = 2H$, H, and $H/2$, modifying the Δt in order to keep the CFL$_{max}$ in a constant value. The RMS error presented for each case is the

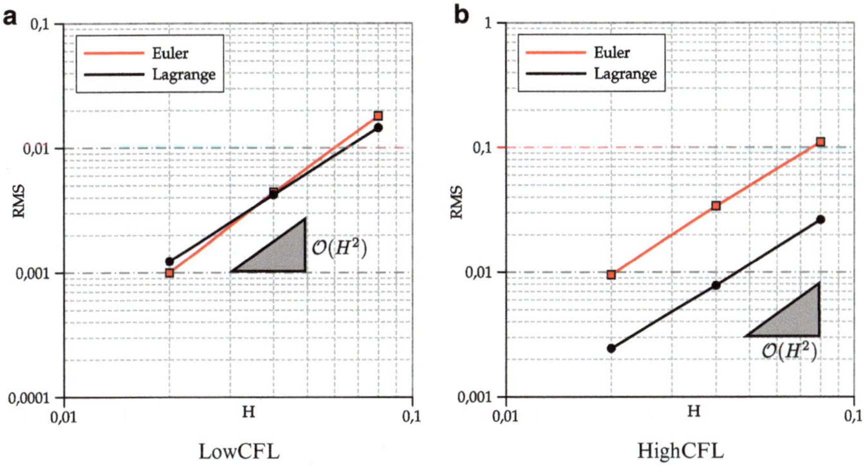

Fig. 4 Convergence of the RMS Error. Case A. $CFL_{max} = 0.6$ in 4(a) and $CFL_{max} = 6$ in 4(b)

average value of RMS from $T = 0$ to $T = 2[s]$ of simulation time. As presented in Figure 3(a), using $\text{CFL}_{max} \approx 0.6$ the Lagrangian and Eulerian errors are almost the same. However, when the dimensionless number is increased to $\text{CFL}_{max} \approx 6$, the Lagrangian solution obtains better accuracy than Eulerian even using both time-step and mesh-size twice larger. This fact represents a notable advantage of the Lagrangian approach over the Eulerian in this type of problems, showing that PFEM-2 simulation can obtain the same precision than FEM even solving a problem eight times smaller.

Regarding to convergence orders, both Lagrangian and Eulerian solutions show concordance with theoretical formula, presenting a second order convergence in every case. In the Lagrangian approach this behavior is reached if the projection operator fulfills this requirement.

3 An Extension to Viscous Incompressible Fluid Flow Simulations

In this case the model to be solved is represented by the Navier-Stokes equations added with the mass conservation that imposes a constraint on the velocity field to be divergence free. This condition is enforced by the pressure acting as a Lagrange multiplier. Inspecting the momentum equation and comparing this with the scalar transport equation just analyzed, here the same unknown variable acts as the velocity field that drives the convection term. This fact produces a non-linear term that is responsible for the chaotic nature of the model that pretends to emulate the physical effects produced by the turbulence.

3.1 General Comments

In this case the error equation looks like the same [17], but the conclusions are fundamentally different.

Generalizing Equation (17), the difference between the errors in the Eulerian and Lagrangian frames for convective dominant problems (high Reynolds number) is

$$\underbrace{[\mathbf{v}\nabla\mathbf{v}]''\,\Delta x^2 \pm [\mathbf{v}\nabla\mathbf{v}]^{\cdot\cdot}\Delta t^2}_{Eulerian} \neq \underbrace{\mathbf{v}''\nabla\mathbf{v}\Delta x^2 \pm \mathbf{v}^{\cdot\cdot}\nabla\mathbf{v}\Delta t^2}_{Lagrangian} \quad (25)$$

In this case, it is not anymore possible to separate the case in which the convective field is constant from the case in which the gradient of the transported field ($\nabla\mathbf{v}$) is not constant. Nevertheless, some particular cases may be analyzed in order to draw some conclusions:

- Eulerian frames are better for low Reynolds number. In these cases, the errors between the Lagrangian and the Eulerian approaches are of the same order, but on Lagrangian frames the projection errors must be added.
- Lagrangian frames are better for convective dominant problems when the velocity has a smooth variation in time but the gradient of the velocity has high spatial variations. This case is very common in fluid mechanics problems, such as in shock waves. The remaining cases are better or worse depending on the projection errors.
- Lagrangian frames are better for multi-fluid flows. This is because Eulerian frames need to solve a level set equation to know the position of the interface. The level set equation [27] is a convection equation that requires small time-steps to yield accurate results due to the considerations concluded in the previous section, i.e.,

$$\mathbf{v}^{\cdot\cdot}\nabla\mathbf{v} < 1 \quad \text{and} \quad [\mathbf{v}\nabla\mathbf{v}]^{\cdot\cdot} \approx \mathbf{v}\,[\nabla\mathbf{v}]^{\cdot\cdot} \gg 1$$

3.2 Validation Test

Following the same criteria as the validation test in Section 2.3, a manufactured case is employed where a solution is proposed and the external force $\mathbf{f}$ must be adapted in order to satisfy the differential equation.

The case is adapted from the book of Donea & Huerta [3] where the authors solved a stationary Stokes flow. In this work, the problem includes also a convective term, leading to recalculation of the external force needed to satisfy the Navier-Stokes equations. A two-dimensional problem in the square domain $[0, 0] \times [1, 1]$ is considered, which possesses a closed-form analytical solution. The problem consists in determining the velocity field $\mathbf{v}$ and the pressure p such that

$$\begin{cases} \frac{\partial \mathbf{v}}{\partial t} + \mathbf{v} \cdot \nabla \mathbf{v} - \nabla \cdot (\nu \nabla \mathbf{v}) + \nabla p = \mathbf{b} & in\ \Omega \\ \nabla \cdot \mathbf{v} = 0 & in\ \Omega \\ \mathbf{v} = 0 & in\ \Gamma \end{cases} \tag{26}$$

where the fluid viscosity ν can be changed in order to simulate for different Reynolds numbers, being $Re = |\mathbf{v}|L/\nu$ with $L = 1$. In order to solve the previous equation system, an analytical solution is proposed:

$$\begin{cases} u(x, y, t) = x^2(1-x)^2(2y - 6y^2 + 4y^3)(1 + 0.5\sin(\omega t)) \\ v(x, y, t) = -y^2(1-y)^2(2x - 6x^2 + 4x^3)(1 + 0.5\sin(\omega t)) \\ p(x, y, t) = x(1-x) \end{cases} \tag{27}$$

which allows to find the expression for $\mathbf{b}$. Varying ν and ω is possible to modify the Reynolds number and the unsteadiness of the solution, respectively. Cases were solved using a 50 × 50 Cartesian mesh split into triangles and setting the time-step such as $\text{CFL}_{max} \approx 10$. Figure 5 shows the shape of the proposed manufactured flow for two different times.

Table 2 shows the configuration and the RMS errors at $T = 1000[s]$ of the simulated cases.

Figure 6 presents graphically the evolution of the *RMS* for Cases 1 and 2. In the first case, a low Reynolds number was selected, therefore the Eulerian solution obtains better results although simulating with large CFL. This is an equivalent

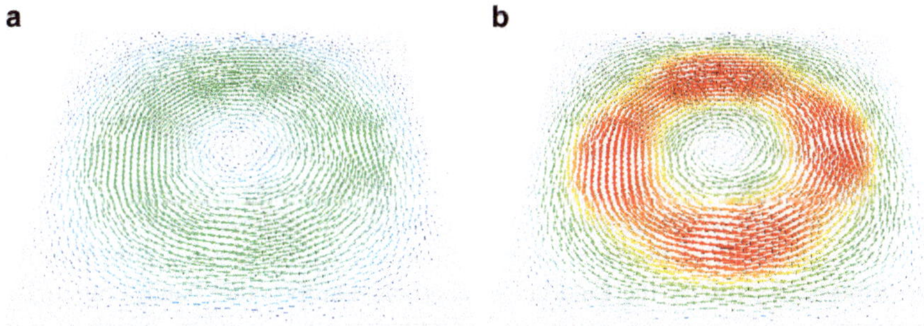

Fig. 5 Manufactured velocity field solution of the Navier-Stokes equations. Snapshots for two different times. Arrows indicating direction of the velocity field are colored by its magnitude

Table 2 Configuration and RMS × 10^3 errors. Every case was run with $\text{CFL}_{max} \approx 10$

			Eulerian			Lagrangian		
Case	ω	*Re*	RMS $\mathbf{v}_x$	RMS $\mathbf{v}_y$	RMS p	RMS $\mathbf{v}_x$	RMS $\mathbf{v}_y$	RMS p
1	$\pi/5000$	10	0.75	0.73	0.036	2	1.9	0.058
2	$\pi/5000$	1000	0.012	0.011	0.014	0.0071	0.0062	0.013
3	$\pi/50$	10	0.8	0.82	0.026	0.83	0.85	0.023
4	$\pi/50$	1000	$\gg$ **1**	$\gg$ **1**	$\gg$ **1**	1.3	1.5	0.06

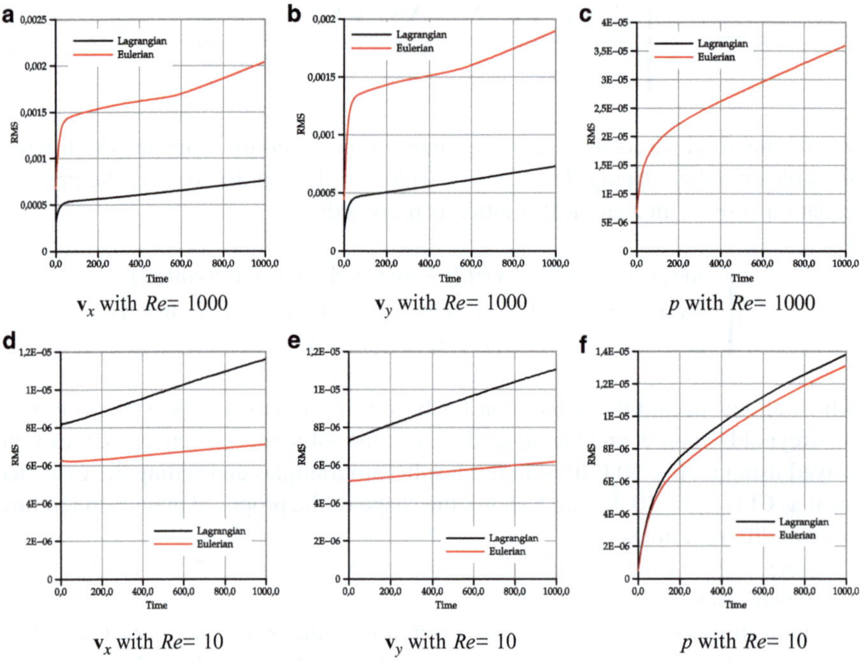

Fig. 6 RMS measured for a Navier-Stokes analytical solution using Lagrangian and Eulerian schemes. Cases with $\omega = \pi/5000$

problem to that diffusion dominated scalar transport problem. However, when the Reynolds number is increased, the Lagrangian framework recovers its advantage. In the third and fourth cases, the unsteadiness of the solution is increased, leading to an increment of the temporal derivatives. In the Lagrangian approach the error enlargement is lower than the Eulerian case, mainly because $\mathbf{v}^{\cdot\cdot}\nabla\mathbf{v} < [\mathbf{v}\nabla\mathbf{v}]^{\cdot\cdot}$. This fact is of extreme importance in the fourth case where the Eulerian solution diverges, concluding that beyond the increasing of the Reynolds number, the growth of the unsteadiness also affects the Eulerian simulations. On the other hand, Lagrangian framework is not adequate in the case of low Reynolds number, which is also consistent with the theoretical formula.

4 Conclusions

From the theoretical results published in [17] several conclusions were achieved allowing to decide in which situation an Eulerian framework is preferred against a Lagrangian one, and vice versa. In this paper the focus was on the experimental validation of those statements with the main goal of beginning to write a formal error analysis based on mathematical tools.

After confirming the validity of those conclusions through the results presented here not only for scalar problems, but also for vector systems like Navier-Stokes, the next step is establishing the main baseline for writing some a priori error analysis including not only the spatial and temporal approximations, but also the projection error for the particle finite element method.

As a conclusion, when advective dominant flows were analyzed, the PFEM-2 methodology has shown several advantages over pure Eulerian strategies. Beyond the possibility of enlarging time-steps, the Lagrangian method also allows to get as accurate solutions as Eulerian ones even using coarser meshes. Last means a significant saving of computational cost which is an invaluable feature in order to solve the challenging problems of next decade

Acknowledgements This work was partially supported by the European Research Council under the Advanced Grant: ERC-2009-AdG Real Time Computational Mechanics Techniques for Multi-Fluid Problems. Norberto Nigro and Juan Gimenez want to thank to CONICET, Universidad Nacional del Litoral and ANPCyT for their financial support (grants PICT 0830 (2013), PIP-2012 GI 11220110100331 and CAI+D 2011 501 201101 00435 LI.

References

1. Aubry, R., Idelsohn, S., Oñate, E.: Particle finite element method in fluid mechanics including thermal convection-diffusion. Comput. Struct. **83**(17–18), 1459–1475 (2005)
2. Brackbill, J., Ruppel, H.: Flip: a method for adaptively zoned, particle-in-cell calculations of fluid flows in two dimensions. J. Comput. Phys. **65**(2), 314–343 (1986)
3. Donea, J., Huerta, A.: Finite Element Method for Flow Problems. Wiley, Chichester (1983)
4. Gimenez, J., González, L.: An extended validation of the last generation of particle finite element method for free surface flows. J. Comput. Phys. **284**, 186–205(2015)
5. Gimenez, J., Nigro, N., Idelsohn, S.: Evaluating the performance of the particle finite element method in parallel architectures. J. Comput. Part. Mech. **1**, 103–116 (2014)
6. Gingold, R., Monaghan, J.: Kernel estimates as a basis for general particle methods in hydrodynamics. J. Comput. Phys. **46**(3), 429–453 (1982)
7. Gingold, R.A., Monaghan, J.J.: Smoothed particle hydrodynamics, theory and application to non-spherical stars. R. Astron. Soc. **181**, 375–389 (1977)
8. Idelsohn, S., Calvo, N., Oñate, E.: Polyhedrization of an arbitrary 3d point set. Comput. Methods Appl. Mech. Eng. **192**, 2649–2667 (2003)
9. Idelsohn, S., Oñate, E., Calvo, N., Del Pin, F.: The meshless finite element method. Int. J. Numer. Methods Eng. **58**(6), 893–912 (2003)
10. Idelsohn, S., Oñate, E., Del Pin, F.: The particle finite element method a powerful tool to solve incompressible flows with free-surfaces and breaking waves. Int. J. Numer. Methods **61** 964–989 (2004)
11. Idelsohn, S., Oñate, E., Pin, F.D., Calvo, N.: Fluid-structure interaction using the particle finite element method. Comput. Methods Appl. Mech. Eng. **195**, 2100–2113 (2006)
12. Idelsohn, S., Marti, J., Limache, A., Oñate, E.: Unified Lagrangian formulation for elastic solids and incompressible fluids: application to fluid-structure interaction problems via the {PFEM}. Comput. Methods Appl. Mech. Eng. **197**(19–20), 1762–1776 (2008). Computational Methods in Fluid-Structure Interaction
13. Idelsohn, S., Mier-Torrecilla, M., Oñate, E.: Multi-fluid flows with the particle finite element method. Comput. Methods Appl. Mech. Eng. **198**, 2750–2767 (2009)

14. Idelsohn, S., Nigro, N., Limache, A., Oñate, E.: Large time-step explicit integration method for solving problems with dominant convection. Comput. Methods Appl. Mech. Eng. **217–220**, 168–185 (2012)
15. Idelsohn, S., Nigro, N., Gimenez, J., Rossi, R., Marti, J.: A fast and accurate method to solve the incompressible Navier–Stokes equations. Eng. Comput. **30**(2), 197–222 (2013)
16. Idelsohn, S., Marti, J., Becker, P., Oñate, E.: Analysis of multifluid flows with large time steps using the particle finite element method. Int. J. Numer. Methods Fluids **75**(9), 621–644 (2014)
17. Idelsohn, S., Oñate, E., Nigro, N., Becker, P., Gimenez, J.: Lagrangian versus Eulerian integration errors. Comput. Methods Appl. Mech. Eng. **293**, 191–206 (2015)
18. Koshizuka, S., Oka, Y.: Moving-particle semi-implicit method for fragmentation of incompressible fluid. Nucl. Sci. Eng. **123**(3), 421–434 (1996). Cited By 646
19. Koshizuka, S., Tamako, H., Oka, Y.: A particle method for incompressible viscous flow with fluid fragmentation. Comput. Fluid Mech. J. **113**, 134–147 (1995)
20. Larese, A., Rossi, R., Oñate, E., Idelsohn, S.: Validation of the particle finite element method (PFEM) for simulation of the free-surface flows. Eng. Comput. **25**(4), 385–425 (2008)
21. Monaghan, J.: An introduction to SPH. Comput. Phys. Commun. **48**, 89–96 (1988)
22. Neuman, S.P.: Adaptive Eulerian-Lagrangian finite element method for advection-dispersion. Int. J. Numer. Methods Eng. **20**(2), 321–337 (1984)
23. Oliver, J., Cante, J., Weyler, R., González, C., Hernandez, J.: Particle finite element methods in solid mechanics problems. In: Computational Plasticity, pp. 87–103. Springer, Berlin (2007)
24. Oñate, E., Idelsohn, S., Zienkiewicz, O., Taylor, R.: A finite point method in computational mechanics. applications to convective transport and fluid flow. Int. J. Numer. Methods Eng. **39**(22), 3839–3866 (1996)
25. Oñate, E., Idelsohn, S., Zienkiewicz, O., Taylor, R., Sacco, C.: A stabilized finite point method for analysis of fluid mechanics problems. Comput. Methods Appl. Mech. Eng. **139**(1–4), 315–346 (1996)
26. Oñate, E., Sacco, C., Idelsohn, S.: A finite point method for incompressible flow problems. Comput. Vis. Sci. **3**(1–2), 67–75 (2000)
27. Osher, S., Fedkiw, R.: Level set methods: an overview and some recent results. J. Comput. Phys. **169**, 463–502 (2001)
28. Pin, F.D., Idelsohn, S., Oñate, E., Aubry, R.: The ALE/Lagrangian particle finite element method: a new approach to computation of free-surface flows and fluid-object interactions. Comput. Fluids **36**(1), 27–38 (2007). Challenges and Advances in Flow Simulation and Modeling
29. Wieckowski, Z.: The material point method in large strain engineering problems. Comput. Methods Appl. Mech. Eng. **193**(39–41), 4417–4438 (2004). The Arbitrary Lagrangian-Eulerian Formulation

An Implicit Gradient Meshfree Formulation for Convection-Dominated Problems

M. Hillman and J.S. Chen

Abstract Meshfree approximations are ideal for the gradient-type stabilized Petrov–Galerkin methods used for solving Eulerian conservation laws due to their ability to achieve arbitrary smoothness, however, the gradient terms are computationally demanding for meshfree methods. To address this issue, a stabilization technique that avoids high order differentiation of meshfree shape functions is introduced by employing implicit gradients under the reproducing kernel approximation framework. The modification to the standard approximation introduces virtually no additional computational cost, and its implementation is simple. The effectiveness of the proposed method is demonstrated in several benchmark problems.

1 Introduction

While Galerkin methods have proven successful in a variety of problems, the application of standard versions of these methods can yield disastrous results for non-self-adjoint problems, such as the Eulerian descriptions of conservation laws with strong convection. In particular, when boundary layers are present, these methods yield large oscillations that destroy the solution. A class of stabilized Petrov–Galerkin methods [1–3] has been developed that provide stable solutions for these problems.

Stabilized methods have been analyzed mathematically [2–4], can be justified by the variational multiscale framework [5, 6], and can be constructed by static condensation of bubble functions [7–9]. In these methods, portions of the differential operator, or the entire operator, are included in the test function. The streamline upwind Petrov–Galerkin (SU/PG) method [1] was motivated by performing

M. Hillman • J.S. Chen (✉)
Department of Structural Engineering, University of California, San Diego, 9500 Gilman Drive, La Jolla, CA 92093-0085, USA
e-mail: js-chen@ucsd.edu

Y. Bazilevs, K. Takizawa (eds.), *Advances in Computational Fluid-Structure Interaction and Flow Simulation*, Modeling and Simulation in Science, Engineering and Technology, DOI 10.1007/978-3-319-40827-9_3

stabilization in a consistent and streamline manner, and also put to rest notions of artificial diffusion. The Galerkin/least squares (G/LS) method [2] gave a more general framework to achieve stability by employing a weighted least squares of the residual. The subgrid scale (SGS) method gave improved stability for higher order elements [3] and reaction-dominated problems [7].

Though originally developed for finite elements, these approaches have been applied to meshfree methods as well. The reproducing kernel particle method (RKPM) with the standard stabilization approach has been applied to flow problems [10]. Suitable stabilization parameters for meshfree methods have been discussed [11], and stabilized meshfree and finite element coupled schemes have also been proposed [12] to achieve a combination of the computational efficiency of finite elements with the flexibility of handling difficult topological changes of the domain such as moving obstacles. A higher order accurate time integration scheme [13] has also been developed for meshfree methods for convection-dominated problems. Although the higher order derivatives involved in the gradient-type stabilization techniques can be calculated straightforwardly by taking advantage of the arbitrary smoothness in the meshfree approximation functions, they are computationally demanding.

The issue of computational efficiency of meshfree methods with gradient-type stabilization for convection-dominated problems is the cost of constructing derivatives of meshfree approximations. One remedy is to introduce implicit gradients [14], which originated from the synchronized convergence approximation [15, 16], where the completeness properties of approximation derivatives are imposed directly. Implicit gradients have been utilized for easing the computational cost of meshfree collocation methods [14], which require higher order derivatives. Implicit gradients have also been used to achieve gradient-type regularization for strain localization problems [17] and avoid the issue of boundary conditions associated with these methods.

In this work, the implicit gradient reproducing kernel particle method (IG-RKPM) is introduced for convection-dominated problems. A gradient reproducing condition is employed to construct the stabilizing gradient terms. This allows three standard stabilization techniques to be performed under a unified framework without the explicit computation of higher order derivatives of the shape functions. The proposed technique is computationally efficient, and it also simplifies stabilization procedures.

The remainder of this text is organized as follows. Section 2 reviews the numerical difficulties associated with convection-dominated problems, and common methods for stabilization. The implicit gradient reproducing kernel particle method is then introduced in Section 3, and the selection of a suitable stabilization parameter for meshfree methods is discussed. Several benchmark problems are solved in Section 4 to demonstrate the effectiveness of the proposed method. Concluding remarks are then given in Section 5.

2 Stabilization for Convection-Dominated Problems

2.1 Advection–Diffusion Equation

The advection–diffusion equation is considered herein as a model problem for convection-dominated problems. The strong form asks to find u such that

$$\begin{aligned} \mathcal{L}u &= s \quad \text{in } \Omega \\ u &= g \quad \text{on } \partial\Omega_g \\ \mathcal{B}u &= h \quad \text{on } \partial\Omega_h \end{aligned} \tag{1}$$

where s is a source term, $\partial\Omega_g$ and $\partial\Omega_h$ are the essential and natural boundaries of the domain, respectively, the flux boundary conditions $\mathcal{B}u = k\nabla u\cdot \mathbf{n}$ are considered herein, and the operator $\mathscr{L}$ in (1) is

$$\mathcal{L}u = -\nabla\cdot(\mathbf{K}\cdot\nabla u - \mathbf{a}u) \tag{2}$$

where $\mathbf{a}$ is the advection velocity and $\mathbf{K}$ is the diffusivity tensor. Without loss of generality, we consider a divergence-free advection field and the case of isotropic diffusion with $\mathbf{K} = \mathbf{I}k$, where $\mathbf{I}$ is the identity tensor, and k is a constant scalar, for which $\mathcal{L}u = -\left(k\nabla^2 u - \mathbf{a}\cdot\nabla u\right)$.

The weak form of (1) is to find $u \in U$ such that

$$B(v,u) = L(v) \qquad \forall v \in V \tag{3}$$

where

$$B(v,u) = (k\nabla v, \nabla u)_\Omega + (v, \mathbf{a}\cdot\nabla u)_\Omega, \tag{4}$$

$$L(v) = (v,s)_\Omega + (v,h)_{\Gamma_h}. \tag{5}$$

Here $U = \left\{u \in H^1(\Omega)\middle|\, u = g \text{ on } \Gamma_g\right\}$, $V = \left\{v \in H^1(\Omega)\middle|\, v = 0 \text{ on } \Gamma_g\right\}$, and $(\cdot,\cdot)_\Omega$ and $(\cdot,\cdot)_{\Gamma_h}$ denote the L^2 inner product on the domain and natural boundary, respectively. The Galerkin method for solving (3) is to find $u^h \in U^h$ that satisfies

$$B\left(v^h, u^h\right) = L\left(v^h\right), \qquad \forall v^h \in V^h \tag{6}$$

where $U^h \subset U$ and $V^h \subset V$ are suitable finite-dimensional subspaces.

For discrete solutions of the advection–diffusion equation with grid dimension or node spacing h, the critical value defining how the numerical solution will behave is the grid Péclet number $Pe_h = \|\mathbf{a}\|\, h/2k$. When the grid Péclet number is greater than unity, standard Bubnov–Galerkin methods lose coercivity and become unstable, and the instability manifests as large oscillations in the presence of fine-scale features such as boundary layers which can appear in the solution of the model problem (1).

2.2 Stabilized Methods

The stabilized methods SU/PG, G/LS, and SGS can be phrased in terms of the Galerkin form of the original problem with a modified test function expressed in a unified fashion as:

$$\tilde{v}^h = v^h + \tau \mathbb{L} v^h \tag{7}$$

where τ is a stabilization parameter, and $\mathbb{L}$ is a differential operator that varies depending on the method:

$$\begin{aligned} &\mathbb{L} = \mathcal{L}_{\text{adv}} && \text{SU/PG} \quad [1] \\ &\mathbb{L} = \mathcal{L} && \text{G/LS} \quad [2] \\ &\mathbb{L} = -\mathcal{L}^* && \text{SGS} \quad [3] \end{aligned} \tag{8}$$

In the above, $\mathscr{L}_{\text{adv}}$ is the advective portion of $\mathscr{L}$, and $\mathcal{L}^*$ is the adjoint of $\mathscr{L}$:

$$\mathcal{L}_{\text{adv}} = \mathbf{a} \cdot \nabla, \tag{9}$$

$$\mathcal{L}^* = -\left(k\nabla^2 + \mathbf{a} \cdot \nabla\right). \tag{10}$$

Stabilized methods are well justified [2–4], and can also be viewed as approximate variational multiscale methods [4, 6]. Stability estimates for SU/PG and SGS require invoking inverse estimates for the shape functions, while for G/LS stability follows directly. The stabilized form of the problem (6) can be stated as to find $u^h \in U^h \subset U$ such that

$$B\left(\tilde{v}^h, u^h\right) = L\left(\tilde{v}^h\right), \qquad \forall \tilde{v}^h \in V^h \tag{11}$$

where $V^h \subset V$, and U and V are adequate Sobolev spaces.

3 Implicit Gradient RKPM for Stabilization of Convection-Dominated Problems

3.1 Reproducing Kernel Approximation

The reproducing kernel (RK) approximation $u^h(\mathbf{x})$ of a function $u(\mathbf{x})$ is constructed by the product of a kernel function $\Phi_\rho(\mathbf{x} - \mathbf{x}_I)$ and a correction function [18, 19]:

$$\Psi_I(\mathbf{x}) = \mathbf{b}^{\mathrm{T}}(\mathbf{x})\, \mathbf{H}(\mathbf{x} - \mathbf{x}_I)\, \Phi_\rho(\mathbf{x} - \mathbf{x}_I) \tag{12}$$

where $\Psi_I(\mathbf{x})$ is the RK shape function, $\mathbf{H}(\mathbf{x}-\mathbf{x}_I) = \left[1\ x\ y\ x^2 \ldots y^n\right]^{\mathrm{T}}$ is a column vector containing the complete n^{th} order monomials, and $\mathbf{b}^{\mathrm{T}}(\mathbf{x})$ is a row vector of coefficients to be determined. The correction function $\mathbf{b}^{\mathrm{T}}(\mathbf{x})\,\mathbf{H}(\mathbf{x}-\mathbf{x}_I)$ allows the approximation to reproduce any linear combination of monomials contained in $\mathbf{H}(\mathbf{x}-\mathbf{x}_I)$. The kernel function $\Phi_\rho(\mathbf{x}-\mathbf{x}_I)$ has compact support measure ρ, and the order of continuity of $\Phi_\rho(\mathbf{x}-\mathbf{x}_I)$ determines the continuity of $\Psi_I(\mathbf{x})$.

The coefficient vector $\mathbf{b}^{\mathrm{T}}(\mathbf{x})$ is obtained by enforcing the following reproducing conditions:

$$\sum_{I=1}^{NP} \Psi_I(\mathbf{x})\, x_{1I}^i x_{2I}^j = x_1^i x_2^j \quad 0 \le i+j \le n. \tag{13}$$

Substituting (12) into (13), the RK shape functions are constructed as

$$\Psi_I(\mathbf{x}) = \mathbf{H}(\mathbf{0})^{\mathrm{T}}\mathbf{M}(\mathbf{x})^{-1}\mathbf{H}(\mathbf{x}-\mathbf{x}_I)\,\Phi_\rho(\mathbf{x}-\mathbf{x}_I) \tag{14}$$

where

$$\mathbf{M}(\mathbf{x}) = \sum_{I=1}^{NP} \mathbf{H}(\mathbf{x}-\mathbf{x}_I)\,\mathbf{H}^{\mathrm{T}}(\mathbf{x}-\mathbf{x}_I)\,\Phi_\rho(\mathbf{x}-\mathbf{x}_I) \tag{15}$$

is called the moment matrix.

3.2 *Implicit Gradient Reproducing Kernel Particle Method*

The stabilization method SU/PG requires constructing second order derivatives of the approximation functions in (14), and for GL/S and SGS, third order derivatives are required, and construction of higher order derivatives of meshfree shape functions can be particularly expensive due to the need to take derivatives of $\mathbf{M}(\mathbf{x})^{-1}$. To avoid this computational expense, the implicit gradient reproducing kernel particle method is introduced. The basic idea is to achieve the same form of stabilized test functions (7) without explicit differentiation. To accomplish this, the following modification to the reproducing condition (13) is introduced for construction of the test functions:

$$\sum_{I=1}^{NP} \tilde{\Psi}_I^\alpha(\mathbf{x})\, x_{1I}^i x_{2I}^j = x_1^i x_2^j + \tau\mathbb{L}\left(x_1^i x_2^j\right), \quad 0 \le i+j \le n \tag{16}$$

where $\tilde{\Psi}_I^{\alpha}$ is the IG-RKPM shape function and $\mathbb{L}$ is the operator in (8). The reproducing condition (16) can be expressed in a generalized fashion as

$$\sum_{I=1}^{NP} \tilde{\Psi}_I^{\alpha}(\mathbf{x})\, x_{1I}^i x_{2I}^j = x_1^i x_2^j + \tau \sum_{i+j=0}^{n} \alpha_{ij} D_{ij}\left(x_1^i x_2^j\right), \quad 0 \leq i+j \leq n \tag{17}$$

where $D_{ij} = \partial^{i+j}/\partial x_1^i \partial x_2^j$ and the coefficients α_{ij} are determined based on the operator $\mathbb{L}$. The above can then be recast as [17]:

$$\sum_{I=1}^{NP} \tilde{\Psi}_I^{\alpha}(\mathbf{x})\,(x_1 - x_{1I})^i (x_2 - x_{2I})^j = \delta_{i0}\delta_{j0} + \tau\alpha_{ij}(-1)^{i+j} i! j!, \quad 0 \leq i+j \leq n. \tag{18}$$

Consider now the implicit gradient shape function in the following form:

$$\tilde{\Psi}_I^{\alpha}(\mathbf{x}) = \mathbf{b}_{\alpha}^{\mathrm{T}}(\mathbf{x})\, \mathbf{H}(\mathbf{x} - \mathbf{x}_I)\, \Phi_{\rho}(\mathbf{x} - \mathbf{x}_I) \tag{19}$$

where $\mathbf{b}_{\alpha}^{\mathrm{T}}(\mathbf{x})$ is a row vector of coefficients which satisfy (17). Following the same procedure as the previous sub-section, the coefficients $\mathbf{b}_{\alpha}^{\mathrm{T}}(\mathbf{x})$ can be obtained, resulting in the following construction for the implicit gradient approximation:

$$\tilde{\Psi}_I^{\alpha}(\mathbf{x}) = \mathbf{H}_{\alpha}^{\mathrm{T}} \mathbf{M}(\mathbf{x})^{-1} \mathbf{H}(\mathbf{x} - \mathbf{x}_I)\, \Phi_{\rho}(\mathbf{x} - \mathbf{x}_I) \tag{20}$$

where $\mathbf{H}_{\alpha}$ is a matrix containing terms from the right-hand side of (18). The values in this matrix are presented in Table 1.

Remark Comparing (20) to (14), it can be seen that the first term on the right-hand side of (20) is the only modification of the standard RK approximation. Stabilization using IG-RKPM is thus very straightforward compared to the explicit version, can be added to existing codes easily, and introduces virtually no additional cost. This is in contrast to explicit differentiation, which requires considerable more computational expense and implementation effort.

An important component of meshfree formulations is the selection of domain integration. For Gaussian integration, high order rules are necessary in order to ensure solution accuracy, while nodal integration yields poor accuracy and is unstable without special treatment [20, 21]. Herein we employ higher order SCNI [22, 23] which provides accurate and stable solutions and avoids the need of expensive high order quadrature.

Table 1 Implicit gradient RKPM vector.

Method	$\mathbf{H}_{\alpha}^{\mathrm{T}}$
SU/PG [1]	$[1, \; -\tau a_1, -\tau a_2, 0, \; \ldots \;, \; 0]$
G/LS [2]	$[1, \; -\tau a_1, -\tau a_2, -2\tau k, -2\tau k, 0, \; \ldots \;, \; 0]$
SGS [3]	$[1, \; -\tau a_1, -\tau a_2, 2\tau k, 2\tau k, \; 0, \ldots \;, \; 0]$

3.3 Selection of the Stabilization Parameter

The selection of the stabilization parameter was traditionally based on obtaining the exact solution at nodes in the one-dimensional Dirichlet problem of (1) in the absence of a source term [1]. The exact solution for this particular case is

$$u^e(x) = C_1 + C_2 e^{ax/k} \tag{21}$$

where C_1 and C_2 depend on the prescribed boundary conditions. Consider the trial and test functions approximated by

$$\begin{aligned} u^h &= \sum_{I=1}^{NP} \Psi_I u_I, \\ \tilde{v}^h &= \sum_{I=1}^{NP} \tilde{\Psi}_I u_I, \end{aligned} \tag{22}$$

where $\{\Psi_I\}_{I=1}^{NP}$ is the set of standard trail shape functions, e.g., finite element or RK approximations, $\{\tilde{\Psi}_I\}_{I=1}^{NP}$ is the set of corresponding stabilized test shape functions, and NP is the number of approximation functions.

Substitution of (22) into (11) yields

$$\sum_{J=1}^{NP} B\left(\tilde{\Psi}_I, \Psi_J\right) u_J = 0, \quad \forall I. \tag{23}$$

Due to the lack of Kronecker delta property in the RKPM shape functions, the relationship between the generalized coordinates u_I^e and the exact nodal values $\widehat{u}_I^e \equiv u^e(x_I) = A + Be^{ax_I/k}$ must be considered:

$$u_I^e = \sum_{J=1}^{NP} {\Lambda_{IJ}}^{-1} \widehat{u}_I^e \tag{24}$$

where $\Lambda_{IJ} = \Psi_J(x_I)$. Note that for linear finite elements, $\Lambda_{IJ} = \delta_{IJ}$. Substitution of the above into (23), and considering the partition of unity in the trial functions, the constants associated with the boundary conditions disappear:

$$\sum_{J=1}^{NP} \left\{ \int_\Omega k\tilde{\Psi}_{I,x}\Psi_{J,x} d\Omega + \int_\Omega a\tilde{\Psi}_I \Psi_{J,x} d\Omega \right\} E_J = 0, \qquad \forall I \tag{25}$$

where $E_I = \sum_{J=1}^{NP} \Lambda_{IJ}{}^{-1} e^{ax_J/k}$. Using a stabilized shape function with nodal stabilization parameter τ_I:

$$\tilde{\Psi}_I = \Psi_I + \tau_I \mathbb{L}\Psi_I \tag{26}$$

we have, picking row I in (25):

$$\tau_I = -\frac{\int\limits_\Omega k\Psi_{I,x}\overline{E}\mathrm{d}\Omega + \int\limits_\Omega a\Psi_I\overline{E}\mathrm{d}\Omega}{\int\limits_\Omega k\{\mathbb{L}\Psi_I\}_{,x}\overline{E}\mathrm{d}\Omega + \int\limits_\Omega a\mathbb{L}\Psi_I\overline{E}\mathrm{d}\Omega} \tag{27}$$

where $\overline{E}_I(x) = \sum_{J=1}^{NP} \Psi_{J,x}E_J$. For linear finite element methods in uniform discretizations, the above equation is not a function of nodal index, and yields the classical stability parameter for SU/PG [1]:

$$\tau = \frac{h}{2a}\left(\coth\left(Pe_h\right) - \frac{1}{Pe_h}\right). \tag{28}$$

In [11] it has been discussed that for RKPM, parameters in the form of (27) depend on the approximation functions on the global level due to the transformation between generalized values u_I and nodal values $u^h(x_I)$ and cannot be simplified to a form such as (28). Thus, the generalization of the condition (27) for higher spatial dimensions is not straightforward. Herein, we employ the classical parameter (28) which is often generalized to higher spatial dimensions as [1]:

$$\tau = \frac{h}{2a}\left(\coth\left(Pe_{h,a}\right) - \frac{1}{Pe_{h,a}}\right) \tag{29}$$

where $Pe_{h,a} = \|\mathbf{a}\| h_a/2k$, and h_a is the grid (or element) dimension along the direction of advection. For RKPM, the spacing h_a may be taken as the length of a representative nodal domain in the direction of the advection. The parameter (29) has been shown to be suitable for RKPM with linear basis, so long as the support parameter is not larger than roughly three times the nodal spacing [11].

4 Numerical Examples

In all numerical examples, RK approximations with linear basis and quartic B-spline kernels with a normalized support of two are employed. In this case, all stabilization methods with implicit gradients are identical due to the order of the basis vector used. In the comparisons made, "RKPM" denotes RKPM with no stabilization, "SU/PG RKPM" denotes RKPM with standard (explicit) SU/PG stabilization, and "IG-RKPM" denotes implicit gradient RKPM. For domain integration, high order SCNI [20] is employed with 2nd order Gaussian quadrature in each nodal cell.

4.1 One-dimensional Model Problem

Consider the one-dimensional homogenous version of the model problem (1) on a unit domain with $a/k = 200.0$, and $s = 1.0$, giving a grid Péclet number of 5.0 with 21 nodes employed. RKPM, RKPM with SU/PG stabilization, and IG-RKPM are compared in Fig. 1. RKPM without stabilization yields large spurious oscillations in the solution. The two stabilized methods give solutions which agree with the exact solution, and very little difference between the two is observed, indicating that a large increase in efficiency can be obtained with little lost by employing implicit stabilization.

4.2 Advection Skew to the Discretization with Outflow Boundary

Here a two-dimensional Dirichlet version of (1) with advection skew to the discretization is considered, as shown in Fig. 2, where the jump in inflow boundary condition along $x = 0.0$ is located at $y = 0.2$. A characteristic of this problem is that both internal layers and boundary layers exist in the solution. The boundary layers cause difficulty for numerical methods, just as in the one-dimensional case. Constant advection and isotropic diffusion parameters are chosen as $\mathbf{a} = (\cos\ \theta,\ \sin\ \theta)$ and $k = 10^{-6}$, respectively, and the domain is taken to be $[0.0, 1.0] \times [0.0, 1.0]$ so that the flow is convection-dominated. The domain is discretized by 31×31 nodes, which gives a grid Péclet number much larger than unity.

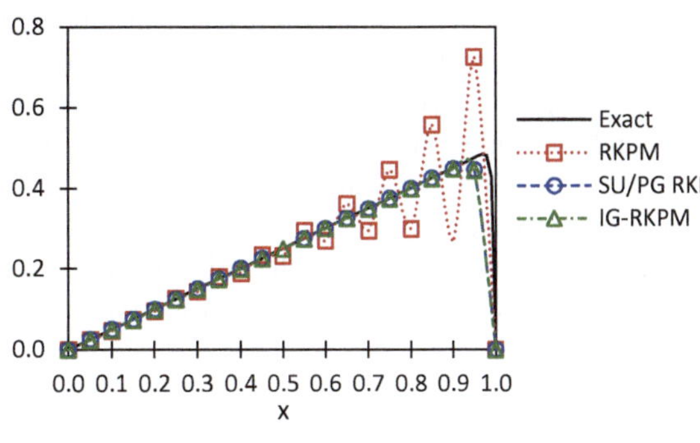

Fig. 1 Comparison between RKPM, explicit SU/PG, and IG-RKPM in the model problem. Markers indicate the solution at nodal locations.

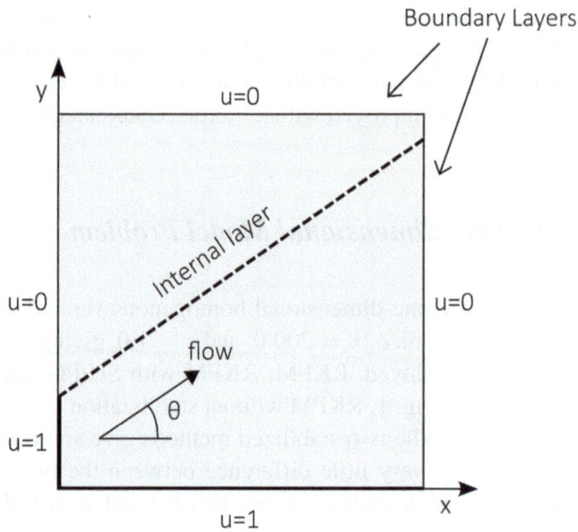

Fig. 2 Problem statement for advection skew to the discretization.

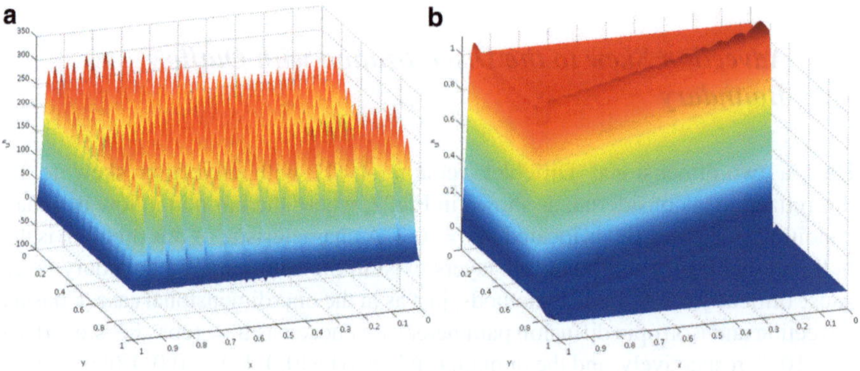

Fig. 3 Advection skew to the discretization with $\theta = \text{atan}(0.5)$ for (a) RKPM, and (b) IG-RKPM.

First, the case of $\theta = \text{atan}(0.5)$ is considered, with RKPM and IG-RKPM employed. As seen in Fig. 3, the magnitude of the RKPM solution is several orders of magnitude greater than the exact solution (essentially pure advection of the boundary condition), while IG-RKPM gives a stable solution. The values of $\theta = \text{atan}(1.0)$ and $\theta = \text{atan}(2.0)$ are then considered with IG-RKPM, where the method exhibits stability in the presence of the fine boundary layers as shown in Fig. 4. The slight overshoots in the solution are expected in linear methods which do not have the variation diminishing property.

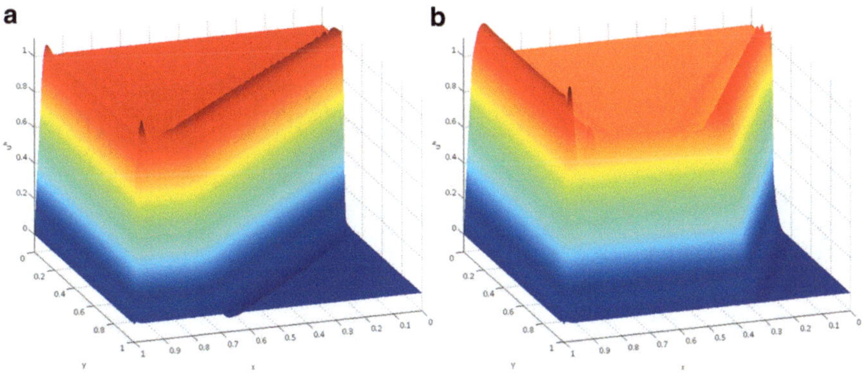

Fig. 4 Advection skew to the discretization with IG-RKPM for (a) $\theta = \text{atan}(1.0)$, and (b) $\theta = \text{atan}(2.0)$.

Fig. 5 Thermal boundary layer problem statement.

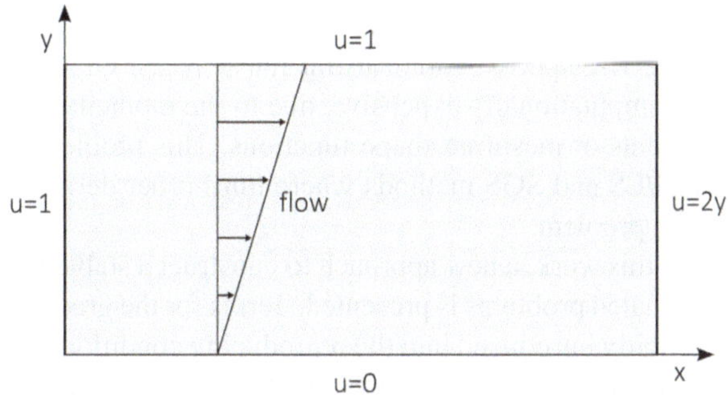

4.3 *Thermal Boundary Layer Problem*

Consider the problem statement shown in Fig. 5 with linearly distributed advection $\mathbf{a} = (2.0y, 0.0)$, and diffusivity $k = 7.0 \cdot 10^{-4}$. The domain is taken as $[0.0, 1.0] \times [0.0, 0.5]$ and is discretized by 31×16 nodes. This problem can be interpreted as one exhibiting a thermal boundary layer on a steady flow between two plates, where the top plate has unit velocity and the bottom plate is fixed. The grid Péclet number calculated from the advection speed at the top surface of the domain and the chosen nodal spacing is larger than unity. The RKPM and IG-RKPM methods are considered, with the results shown in Fig. 6. Again, RKPM gives an oscillatory solution while IG-RKPM gives a stable solution.

5 Conclusion

Meshfree methods offer smooth approximation spaces suitable for the gradient-type stabilization employed for convection-dominated problems. While for linear finite elements, all of the standard stabilization methods coincide due to vanishing higher

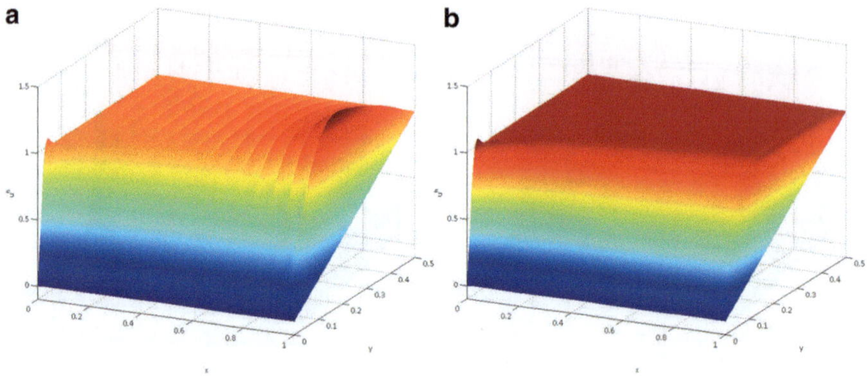

Fig. 6 Solutions for the thermal boundary layer problem for (a) RKPM and (b) IG-RKPM.

order derivatives on element interiors, these methods can be properly constructed by the RK approximation using linear basis. Gradient terms necessary, however, are computationally expensive, due to the moment matrix inversion involved in the gradients of meshfree shape functions. This problem is particularly exacerbated in the G/LS and SGS methods where third order derivatives appear in the weak form of the problem.

In this work, a new approach to construct a stable RKPM method for convection-dominated problems is presented. Terms for the gradient-type stabilized methods are implicitly introduced into the reproducing conditions under a unified framework that can include the SU/PG, G/LS, and SGS methods. The implicit gradients completely circumvent the costly derivatives otherwise necessary for stabilization. The only deviation from the standard RKPM method is the modification of the constant vector in the RK approximation for the test functions, and thus virtually no additional computational cost is introduced, and implementation is simple. The benchmark problems tested showed good performance of the proposed method and agreement with solutions by explicit stabilization.

Acknowledgements The support of this work by US Army Engineer Research and Development Center under contract W912HZ-07-C-0019 is greatly acknowledged.

References

1. Brooks, A.N., Hughes, T.J.R.: Streamline upwind/Petrov-Galerkin formulations for convection dominated flows with particular emphasis on the incompressible Navier-Stokes equations. Comput. Methods Appl. Mech. Eng. **32**, 199–259 (1982)
2. Hughes, T.J.R., Franca, L.P., Hulbert, G.M.: A new finite element formulation for computational fluid dynamics: VIII. The Galerkin/least-squares method for advective-diffusive equations. Comput. Methods Appl. Mech. Eng. **73**, 173–189 (1989)

3. Franca, L.P., Frey, S.L., Hughes, T.J.R.: Stabilized finite element methods: I. Application to the advective-diffusive model. Comput. Methods Appl. Mech. Eng. **95**, 253–276 (1992)
4. Johnson, C., Nävert, U., Pitkäranta, J.: Finite element methods for linear hyperbolic problems. Comput. Methods Appl. Mech. Eng. **45**, 285–312 (1984)
5. Hughes, T.J.R.: Multiscale phenomena: Green's functions, the Dirichlet-to-Neumann formulation, subgrid scale models, bubbles and the origins of stabilized methods. Comput. Methods Appl. Mech. Eng. **127**, 387–401 (1995)
6. Hughes, T.J.R., Feijóo, G.R., Mazzei, L., Quincy, J.B.: The variational multiscale method—a paradigm for computational mechanics. Comput. Methods Appl. Mech. Eng. **166**, 3–24 (1998)
7. Franca, L.P., Farhat, C.: Bubble functions prompt unusual stabilized finite element methods. Comput. Methods Appl. Mech. Eng. **123**, 299–308 (1995)
8. Brezzi, F., Bristeau, M.O., Franca, L.P., Mallet, M., Rogé, G.: A relationship between stabilized finite element methods and the Galerkin method with bubble functions. Comput. Methods Appl. Mech. Eng. **96**, 117–129 (1992)
9. Baiocchi, C., Brezzi, F., Franca, L.P.: Virtual bubbles and Galerkin-least-squares type methods (Ga. LS). Comput. Methods Appl. Mech. Eng. **105**, 125–141 (1993)
10. Günther, F., Liu, W.K., Diachin, D., Christon, M.A.: Multi-scale meshfree parallel computations for viscous, compressible flows. Comput. Methods Appl. Mech. Eng. **190**, 279–303 (2000)
11. Fries, T.P., Matthies, H.G.: A stabilized and coupled meshfree/meshbased method for the incompressible Navier–Stokes equations—part I: Stabilization. Comput. Methods Appl. Mech. Eng. **195**, 6205–6224 (2006)
12. Fries, T.P., Matthies, H.G.: A stabilized and coupled meshfree/meshbased method for the incompressible Navier–Stokes equations—Part II: Coupling. Comput. Methods Appl. Mech. Eng. **195**, 6191–6204 (2006)
13. Huerta, A., Fernández-Méndez, S.: Time accurate consistently stabilized mesh-free methods for convection dominated problems. Int. J. Numer. Methods Eng. **56**, 1225–1242 (2003)
14. Chi, S.W., Chen, J.S., Hu, H.Y., Yang, J.P.: A gradient reproducing kernel collocation method for boundary value problems. Int. J. Numer. Methods Eng. **93**, 1381–1402 (2013)
15. Li, S., Liu, W.K.: Reproducing kernel hierarchical partition of unity. Part I: Formulation and theory. Int. J. Numer. Methods Eng. **45**, 251–288 (1999)
16. Li, S., Liu, W.K.: Reproducing kernel hierarchical partition of unity. Part II: Applications. Int. J. Numer. Methods Eng. **45**, 289–317 (1999)
17. Chen, J.S., Zhang, X., Belytschko, T.: An implicit gradient model by a reproducing kernel strain regularization in strain localization problems. Comput. Methods Appl. Mech. Eng. **193**, 2827–2844 (2004)
18. Liu, W.K., Jun, S., Zhang, Y.F.: Reproducing kernel particle methods. Int. J. Numer. Methods Fluids **20**, 1081–1106 (1995)
19. Chen, J.S., Pan, C., Wu, C.T., Liu, W.K.: Reproducing kernel particle methods for large deformation analysis of nonlinear structures. Comput. Methods Appl. Mech. Eng. **139**, 195–227 (1996)
20. Dolbow, J., Belytschko, T.: Numerical integration of the Galerkin weak form in meshfree methods. Comput. Mech. **23**, 219–230 (1999)
21. Beissel, S., Belytschko, T.: Nodal integration of the element-free Galerkin method. Comput. Methods Appl. Mech. Eng. **139**, 49–74 (1996)
22. Chen, J.S., Wu, C.T., Yoon, S., You, Y.: A stabilized conforming nodal integration for Galerkin mesh-fee methods. Int. J. Numer. Methods Eng. **50**, 435–466 (2001)
23. Chen, J.S., Hu, W., Puso, M.: Orbital HP-clouds for solving Schrödinger equation in quantum mechanics. Comput. Methods Appl. Mech. Eng. **196**, 3693–3705 (2007)

Flow Analysis of a Wave-Energy Air Turbine with the SUPG/PSPG Method and DCDD

Lucio Cardillo, Alessandro Corsini, Giovanni Delibra, Franco Rispoli, and Tayfun E. Tezduyar

Abstract We present flow analysis of a wave-energy air turbine, specifically a Wells turbine. The analysis is based on the Streamline-Upwind/Petrov-Galerkin (SUPG) and Pressure-Stabilizing/Petrov-Galerkin (PSPG) methods and discontinuity-capturing directional dissipation (DCDD). The DCDD, first introduced to complement the SUPG/PSPG method in computation of incompressible flows in the presence of sharp solution gradients, was also shown to perform well in turbulent-flow test computations when compared to the Smagorinsky large eddy simulation model. Our computational analysis of the Wells turbine here, with results that compare favorably to the available experimental data, shows that the DCDD method performs well also in turbomachinery flows.

1 Introduction

Computational flow analysis of renewable-energy systems has reached a level of sophistication where it can now play a significant role in performance and design studies. Recent advances in aerodynamic and fluid–structure interaction analysis of wind turbines (see, for example, [1–11]) is a good illustration of that. Here we focus on flow analysis of a wave-energy air turbine, specifically a Wells turbine [12], which has been the subject of a number of computational studies [13–16]. The Wells turbine rotor consists of multiple symmetrical airfoil blades arranged around a hub. The baseline blade profile mostly comes from the NACA four-digit series [17].

Design of a Wells turbine is challenging because of the strong dependence of the bidirectional flow rate on the actual sea conditions. Rotating at fixed speed, the

L. Cardillo • A. Corsini (✉) • G. Delibra • F. Rispoli
Dipartimento di Ingegneria Meccanica e Aerospaziale, Sapienza University of Rome, Via Eudossiana, 18, I-00184 Rome, Italy
e-mail: alessandro.corsini@uniroma1.it

T.E. Tezduyar
Mechanical Engineering, Rice University – MS 321, 6100 Main Street, Houston, TX 77005, USA
e-mail: tezduyar@tafsm.org

Y. Bazilevs, K. Takizawa (eds.), *Advances in Computational Fluid-Structure Interaction and Flow Simulation*, Modeling and Simulation in Science, Engineering and Technology, DOI 10.1007/978-3-319-40827-9_4

turbine is capable of operating with acceptable efficiency only within a limited range of flow rate around the peak-efficiency point. This is due to aerodynamic instabilities linked to the onset of blade stall. For an accurate representation of the unsteady flow behavior, the computational method used in flow analysis of Wells turbines, and most turbomachinery for that matter, needs to be effective in addressing the turbulent-flow nature of the problem.

Many computational challenges involved in analysis of complex flow problems with turbulence have been addressed with a number of advanced numerical methods, including large eddy simulation (LES) models and stabilized and multiscale methods. In turbomachinery research, LES-based methods have been proven to be effective in increasing the solution accuracy compared to the Reynolds-averaged Navier–Stokes (RANS) models and being still affordable. Tucker [18] reviewed the LES and hybrid LES models in turbomachinery applications, focusing on the Reynolds number effects on coarse grids. Jakirlic et al. [19] studied the flow in a swirl combustor with a zonal LES–RANS approach. Borello et al. [20] used advanced LES models to compute the aerodynamics of an axial fan. Krappel et al. [21] investigated hydraulic turbines by means of a hybrid LES–RANS model to predict the complex unsteady separation at partial-load conditions, proving the approach to perform well where RANS models failed.

In finite element computation of flow problems, the Streamline-Upwind/Petrov-Galerkin (SUPG) formulation of incompressible flows [22, 23], the SUPG formulation of compressible flows [24–26], and the Pressure-Stabilizing/Petrov-Galerkin (PSPG) formulation of incompressible flows [23, 27] are some of the most prevalent stabilized methods. They are the predecessors of the variational multiscale (VMS) [28–31] and space–time VMS (ST-VMS) [32, 33] methods. The VMS and ST-VMS methods serve also as turbulence models and have been successful in computational analysis of many complex flow problems (see, for example, [7, 9–11, 34–42]).

The discontinuity-capturing directional dissipation (DCDD) stabilization was introduced in [23] to complement the SUPG/PSPG method in computation of flow fields with sharp gradients. The DCDD stabilization takes effect where there is a sharp gradient in the velocity field and introduces dissipation in the direction of that gradient. The way the DCDD stabilization is added to the finite element formulation precludes augmentation of the SUPG effect by the DCDD effect when the advection and discontinuity directions coincide. The DCDD stabilization, together with the SUPG/PSPG method, was also shown, first in [43–45], to perform well in turbulent-flow test computations when compared to the Smagorinsky LES model. The test computations for turbulent channel flow reported in [43–45] were preliminary and at friction Reynolds number $Re_\tau = 395$. A more detailed computational analysis of turbulent channel flow, at $Re_\tau = 180$, was reported in [46], with the DCDD again performing well compared to the Smagorinsky LES model. Here we perform the flow analysis of the Wells turbine based on the SUPG/PSPG method and DCDD stabilization.

We provide the governing equations in Section 2, and describe the SUPG/PSPG method in Section 3. The DCDD stabilization, together with validation results from

[46], is presented in Section 4. Computational analysis of the Wells turbine is reported in Section 5, and the concluding remarks are given in Section 6.

2 Governing Equations

Let $\Omega \subset \mathbf{R}^{n_{sd}}$ be the spatial domain with boundary Γ, and $(0, T)$ be the time domain. The Navier–Stokes equations of incompressible flows can be written on Ω and $\forall t \in (0, T)$ as

$$\rho\left(\frac{\partial \mathbf{u}}{\partial t} + \mathbf{u} \cdot \nabla \mathbf{u} - \mathbf{f}\right) - \nabla \cdot \boldsymbol{\sigma} = 0, \tag{1}$$

$$\nabla \cdot \mathbf{u} = 0, \tag{2}$$

where ρ, $\mathbf{u}$, and $\mathbf{f}$ are the density, velocity, and the external force, respectively. The stress tensor $\boldsymbol{\sigma}$ is defined as

$$\boldsymbol{\sigma}(p, \mathbf{u}) = -p\mathbf{I} + 2\mu\boldsymbol{\varepsilon}(\mathbf{u}). \tag{3}$$

Here p is the pressure, $\mathbf{I}$ the identity tensor, $\mu = \rho\nu$ the viscosity, ν the kinematic viscosity, and $\boldsymbol{\varepsilon}(\mathbf{u})$ is the strain-rate tensor:

$$\boldsymbol{\varepsilon}(\mathbf{u}) = \frac{1}{2}\left((\nabla \mathbf{u}) + (\nabla \mathbf{u})^T\right). \tag{4}$$

The essential and natural boundary conditions for Eq. (1) are represented as

$$\mathbf{u} = \mathbf{g} \text{ on } \Gamma_g, \qquad \mathbf{n} \cdot \boldsymbol{\sigma} = \mathbf{h} \text{ on } \Gamma_h, \tag{5}$$

where Γ_g and Γ_h are complementary subsets of the boundary Γ, $\mathbf{n}$ is the unit normal vector, and $\mathbf{g}$ and $\mathbf{h}$ are given functions. A divergence-free velocity field $\mathbf{u}_0(\mathbf{x})$ is specified as the initial condition.

3 SUPG/PSPG Method

3.1 Stabilized Formulation

Given Eqs. (1)–(2), we form some suitably defined finite-dimensional trial solution and test function spaces for velocity and pressure: $\mathcal{S}_{\mathbf{u}}^h$, $\mathcal{V}_{\mathbf{u}}^h$, $\mathcal{S}_p^h$, and $\mathcal{V}_p^h = \mathcal{S}_p^h$. The stabilized finite element formulation of Eqs. (1)–(2) can be written as follows: find $\mathbf{u}^h \in \mathcal{S}_{\mathbf{u}}^h$ and $p^h \in \mathcal{S}_p^h$ such that $\forall \mathbf{w}^h \in \mathcal{V}_{\mathbf{u}}^h$ and $\forall q^h \in \mathcal{V}_p^h$:

$$\int_{\Omega} \mathbf{w}^h \cdot \rho \left(\frac{\partial \mathbf{u}^h}{\partial t} + \mathbf{u}^h \cdot \nabla \mathbf{u}^h - \mathbf{f}^h \right) d\Omega + \int_{\Omega} \boldsymbol{\varepsilon}(\mathbf{w}^h) : \boldsymbol{\sigma}(p^h, \mathbf{u}^h) d\Omega - \int_{\Gamma_\mathrm{h}} \mathbf{w}^h \cdot \mathbf{h}^h d\Gamma$$

$$+ \int_{\Omega} q^h \nabla \cdot \mathbf{u}^h d\Omega + \sum_{e=1}^{n_{el}} \int_{\Omega^e} \frac{1}{\rho} \left[\tau_{\mathrm{SUPG}} \rho \mathbf{u}^h \cdot \nabla \mathbf{w}^h + \tau_{\mathrm{PSPG}} \nabla q^h \right] \cdot \left[\text{Ł}(p^h, \mathbf{u}^h) - \rho \mathbf{f}^h \right] d\Omega$$

$$+ \sum_{e=1}^{n_{el}} \int_{\Omega^e} \nabla \mathbf{w}^h : \boldsymbol{\sigma}_{\mathrm{SGS}} d\Omega = 0, \tag{6}$$

where

$$\text{Ł}(q^h, \mathbf{w}^h) = \rho \left(\frac{\partial \mathbf{w}^h}{\partial t} + \mathbf{u}^h \cdot \nabla \mathbf{w}^h \right) - \nabla \cdot \boldsymbol{\sigma}(q^h, \mathbf{w}^h), \tag{7}$$

$$\boldsymbol{\sigma}_{\mathrm{SGS}} = \rho \nu_{\mathrm{SGS}} \mathscr{E}. \tag{8}$$

Here n_{el} is the number of elements and τ_{SUPG} and τ_{PSPG} are the SUPG and PSPG stabilization parameters. The last term in Eq. (6) represents the subgrid scale (SGS) turbulence model, which will be either the DCDD stabilization or the Smagorinsky LES model in this paper. The symbols $\boldsymbol{\sigma}_{\mathrm{SGS}}$, ν_{SGS}, and $\mathscr{E}$ denote the SGS stress tensor, effective SGS kinematic viscosity, and the SGS strain-rate tensor. We will define ν_{SGS} and $\mathscr{E}$ in Section 4.

3.2 Stabilization Parameters

Various ways of calculating the stabilization parameters for incompressible flows can be found in detail in [23, 47]. In this section we focus on the versions of the stabilization parameters denoted by the subscript "UGN," namely the "UGN/RGN"-based stabilization parameters. For that, we first define the unit vectors **s** and **r**:

$$\mathbf{s} = \frac{\mathbf{u}^h}{\|\mathbf{u}^h\|}, \tag{9}$$

$$\mathbf{r} = \frac{\nabla \|\mathbf{u}^h\|}{\|\nabla \|\mathbf{u}^h\|\|}. \tag{10}$$

The components of $(\tau_{\mathrm{SUPG}})_{\mathrm{UGN}}$ corresponding to the advection-, transient-, and diffusion-dominated limits were defined in [23] as follows:

$$\tau_{\mathrm{SUGN1}} = \left(\sum_{a=1}^{n_{en}} |\mathbf{u}^h \cdot \nabla N_a| \right)^{-1}, \tag{11}$$

$$\tau_{\mathrm{SUGN2}} = \frac{\Delta t}{2}, \tag{12}$$

$$\tau_{\mathrm{SUGN3}} = \frac{h_{\mathrm{RGN}}^2}{4\nu}, \tag{13}$$

where n_{en} is the number of element nodes, N_a is the interpolation function associated with node a, Δt is the time-step size, and the "element length" h_{RGN} is defined as

$$h_{\mathrm{RGN}} = 2\left(\sum_{a=1}^{n_{en}} |\mathbf{r}\cdot\nabla N_a|\right)^{-1}. \tag{14}$$

Based on Eq. (11), the "element length" h_{UGN} is defined as

$$h_{\mathrm{UGN}} = 2\|\mathbf{u}^h\|\tau_{\mathrm{SUGN1}}. \tag{15}$$

Although writing a direct expression for τ_{SUGN1} as given by Eq. (11) was pointed out in [23], the element length definition one obtains by combining Eq. (11) and Eq. (15) was first introduced (as a direct expression for h_{UGN}) in [48]. The expression for h_{RGN} as given by Eq. (14) was first introduced in [23]. It was noted in [23] that h_{UGN} and h_{RGN} can be viewed as the local length scales corresponding to the advection- and diffusion-dominated limits, respectively. We now define $(\tau_{\mathrm{SUPG}})_{\mathrm{UGN}}$ and $(\tau_{\mathrm{PSPG}})_{\mathrm{UGN}}$ as follows:

$$(\tau_{\mathrm{SUPG}})_{\mathrm{UGN}} = \left(\frac{1}{\tau_{\mathrm{SUGN1}}^r} + \frac{1}{\tau_{\mathrm{SUGN2}}^r} + \frac{1}{\tau_{\mathrm{SUGN3}}^r}\right)^{-\frac{1}{r}}, \tag{16}$$

$$(\tau_{\mathrm{PSPG}})_{\mathrm{UGN}} = (\tau_{\mathrm{SUPG}})_{\mathrm{UGN}}. \tag{17}$$

Equation (16) is based on the inverse of $(\tau_{\mathrm{SUPG}})_{\mathrm{UGN}}$ being defined as the r-norm of the vector with components $\frac{1}{\tau_{\mathrm{SUGN1}}}$, $\frac{1}{\tau_{\mathrm{SUGN2}}}$ and $\frac{1}{\tau_{\mathrm{SUGN3}}}$. We note that the higher the integer r is, the sharper the switching between τ_{SUGN1}, τ_{SUGN2}, and τ_{SUGN3} becomes. This "*r-switch*" was introduced in [17]. Typically, $r = 2$. The expression for τ_{SUGN3} given by Eq. (13) was proposed in [23]. The "SUPG viscosity" ν_{SUPG} is defined as

$$\nu_{\mathrm{SUPG}} = \tau_{\mathrm{SUPG}}\|\mathbf{u}^h\|^2. \tag{18}$$

The ST versions of τ_{SUGN1}, τ_{SUGN2}, τ_{SUGN3}, $(\tau_{\mathrm{SUPG}})_{\mathrm{UGN}}$, and $(\tau_{\mathrm{PSPG}})_{\mathrm{UGN}}$, given respectively by Eqs. (11), (12), (13), (16), and (17), were defined in [23].

4 DCDD Stabilization

The DCDD stabilization was proposed in [23]. In describing it, we first define the "DCDD viscosity" ν_{DCDD} and the DCDD stabilization parameter τ_{DCDD}:

$$\nu_{\mathrm{DCDD}} = \tau_{\mathrm{DCDD}} \|\mathbf{u}^h\|^2, \tag{19}$$

$$\tau_{\mathrm{DCDD}} = \frac{h_{\mathrm{DCDD}}}{2u_{\mathrm{ref}}} \frac{\|\boldsymbol{\nabla}\|\mathbf{u}^h\|\| h_{\mathrm{DCDD}}}{u_{\mathrm{ref}}}, \tag{20}$$

where

$$h_{\mathrm{DCDD}} = h_{\mathrm{RGN}}. \tag{21}$$

Here u_{ref} is a reference velocity (such as $\|\mathbf{u}^h\|$ at the inflow, or the difference between the estimated maximum and minimum values of $\|\mathbf{u}^h\|$). Combining Eqs. (19) and (20), we obtain

$$\nu_{\mathrm{DCDD}} = \frac{1}{2}\left(\frac{\|\mathbf{u}^h\|}{u_{\mathrm{ref}}}\right)^2 (h_{\mathrm{DCDD}})^2 \|\boldsymbol{\nabla}\|\mathbf{u}^h\|\|. \tag{22}$$

Then the DCDD stabilization is defined as

$$S_{\mathrm{DCDD}} = \sum_{e=1}^{n_{el}} \int_{\Omega^e} \rho \boldsymbol{\nabla}\mathbf{w}^h : \left([\nu_{\mathrm{DCDD}}\mathbf{rr} - \boldsymbol{\kappa}_{\mathrm{CORR}}] \cdot \boldsymbol{\nabla}\mathbf{u}^h \right) d\Omega, \tag{23}$$

where $\boldsymbol{\kappa}_{\mathrm{CORR}}$ is defined as

$$\boldsymbol{\kappa}_{\mathrm{CORR}} = \nu_{\mathrm{DCDD}} (\mathbf{r} \cdot \mathbf{s})^2 \mathbf{ss}. \tag{24}$$

4.1 Comparison of the DCDD and LES Models

Table 1 is a summary of how the DCDD stabilization and Smagorinsky LES model compare. In the DCDD stabilization, the DCDD viscosity depends on the local

Table 1 DCDD stabilization and Smagorinsky LES model. Here l_0, t_0^{-1}, and f_{d} are the length scale, time scale, and damping function of the model, and C_{SMAG} is the Smagorinsky coefficient

ν_{SGS}	$\frac{1}{2}(f_{\mathrm{DCDD}} h_{\mathrm{DCDD}})^2 \|\boldsymbol{\nabla}\|\mathbf{u}^h\|\|$	$(f_{\mathrm{SMAG}} C_{\mathrm{SMAG}} \Delta_{\mathrm{SMAG}})^2 \left(2\boldsymbol{\varepsilon}(\mathbf{u}^h) : \boldsymbol{\varepsilon}(\mathbf{u}^h)\right)^{\frac{1}{2}}$
l_0	$h_{\mathrm{DCDD}} = 2\left(\sum_{a=1}^{n_{en}} \lvert\mathbf{r} \cdot \boldsymbol{\nabla} N_a\rvert\right)^{-1}$	$C_{\mathrm{SMAG}} \Delta_{\mathrm{SMAG}} = C_{\mathrm{SMAG}} (\Delta x \Delta y \Delta z)^{\frac{1}{3}}$
t_0^{-1}	$\|\boldsymbol{\nabla}\|\mathbf{u}^h\|\|$	$\left(2\boldsymbol{\varepsilon}(\mathbf{u}^h) : \boldsymbol{\varepsilon}(\mathbf{u}^h)\right)^{\frac{1}{2}}$
f_{d}	$f_{\mathrm{DCDD}} = \frac{\|\mathbf{u}^h\|}{u_{\mathrm{ref}}}$	$f_{\mathrm{SMAG}} = \left[1 - e^{\left(\frac{y^+}{26}\right)}\right]$
$\mathscr{E}$	$\left[\mathbf{rr} - (\mathbf{r} \cdot \mathbf{s})^2\, \mathbf{ss}\right] \cdot \boldsymbol{\nabla}\mathbf{u}^h$	$2\boldsymbol{\varepsilon}(\mathbf{u}^h)$

Table 2 DCDD validation. Reynolds numbers, dimensions of the computational domain (scaled by δ), number of grid points in x, y, and z directions, and other features of the spatial discretization. The mesh is uniform in the x and z directions for the LES and DCDD computations and is made of hexahedral elements

	DNS [49]	LES	DCDD
Re_τ	180	174.6	176.2
Re_b	$\approx$ 5,600	5,600	5,600
$L_x \times L_y \times L_z$	$4\pi \times 2 \times 2\pi$	$2\pi \times 2 \times 4/3\,\pi$	$2\pi \times 2 \times 4/3\,\pi$
$n_x \times n_y \times n_z$	$192 \times 129 \times 160$	$33 \times 119 \times 33$	$33 \times 119 \times 33$
$(\Delta y^+)_{min}$	0.05	0.96	0.97
$(\Delta y^+)_{max}$	4.4	3.67	3.70
Δx^+	$\approx$ 12	34.3	34.6
Δz^+	$\approx$ 7	22.9	23.1

solution gradient, the velocity profile, and the element length in the direction of the solution gradient. The way the DCDD stabilization is added precludes augmentation of the SUPG effect by the DCDD effect when the advection and discontinuity directions coincide.

4.2 *DCDD Validation*

The validation we present here is a summary of the computation reported in [46]. The test problem is a fully developed unidirectional 3D plane channel flow characterized by a friction Reynolds number of $Re_\tau = \langle u_\tau \rangle_{xzt} \delta / \nu = 180$. The channel walls are at $y = \pm\delta$, the flow is in x direction, and $\langle \cdots \rangle_{xzt}$ denotes averaging over x, z, and t. The Reynolds number based on the bulk velocity (u_b), defined as $Re_b = u_b 2\delta/\nu$, is 5,600. We compare the results obtained with the LES model and DCDD stabilization to direct numerical simulation (DNS) results from [49]. Table 2 shows the Reynolds numbers, dimensions of the computational domain, and the features of the spatial discretization. The boundary conditions are no-slip at the channel walls and spatial periodicity in the x and z directions. In the LES computation, $C_{\mathrm{SMAG}} = 0.15$.

Table 3 shows various flow parameters averaged over x, z, and t, including $\langle u_c \rangle_{xzt}$, which is the averaged value of the velocity at the center plane ($y = 0$), and the skin friction coefficient, $C_f = 2\left(\langle u_\tau \rangle_{xzt}\right)^2 / u_b^2$. Table 3 also shows some experimental values from [50]. For all parameters, the DCDD values are the ones very slightly closer to the DNS and experimental values.

Table 3 DCDD validation. Flow parameters averaged over x, z, and t

	Dean's correlation [50]	DNS [49]	LES	DCDD
$u_b/\langle u_\tau \rangle_{xzt}$	–	15.63	15.91	15.78
$\langle u_c \rangle_{xzt}/\langle u_\tau \rangle_{xzt}$	–	18.20	18.36	18.32
$\langle u_c \rangle_{xzt}/u_b$	1.158	1.164	1.154	1.161
C_f	8.44×10^{-3}	8.18×10^{-3}	7.90×10^{-3}	8.03×10^{-3}

Table 4 Wells turbine. Design parameters and geometry details

Design parameters	
Blade profile	NACA0015
Flow rate (m^3/s)	0.85
Rotor speed (rpm)	3,000
Wave specific power (kW/m)	5
Wave height (m)	0.8
Wave period (s)	4
Geometry details	
Casing diameter (mm)	500
Hub-to-casing diameter ratio	0.75
Blade count	9
Blade chord (mm)	109
Blade height (mm)	60
Tip clearance (mm)	2.5

5 Computation

5.1 Problem Description

The equations are solved in the rotating reference frame of the turbine rotor. Consequently, the computations are based on the version of Eq. (6) that includes the non-inertial terms coming from using such a reference frame. In the implementation of the stabilized formulations, the non-inertial terms are just added to the other source terms. Alternatively, the rotation can be handled with moving-mesh methods such as the arbitrary Lagrangian–Eulerian (ALE) methods (see, for example, [1, 4, 5, 7]) or the ST methods (see, for example, [1–3, 6–11]). We compute the flow in a Wells turbine originally designed for low flow rates, typical of Mediterranean conditions [51]. The design power is 1.5 kW at 3,000 rpm. Design parameters and geometry details are given in Table 4. A full-scale prototype was tested in laboratory in steady-state operation, using a centrifugal fan to drive the air to the turbine. The characteristic curves from measurements at 8 different flow rates are shown in Figure 1.

The computational domain encompasses the annular passage between the casing and the hub, bounded laterally by the two periodicity boundaries. These two boundaries are located at 38% of the blade chord from the leading and trailing

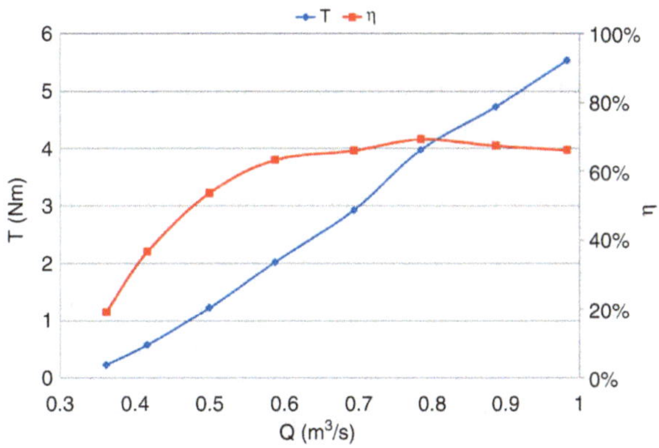

Fig. 1 Wells turbine. Characteristic curves of torque and efficiency from measurements [51]

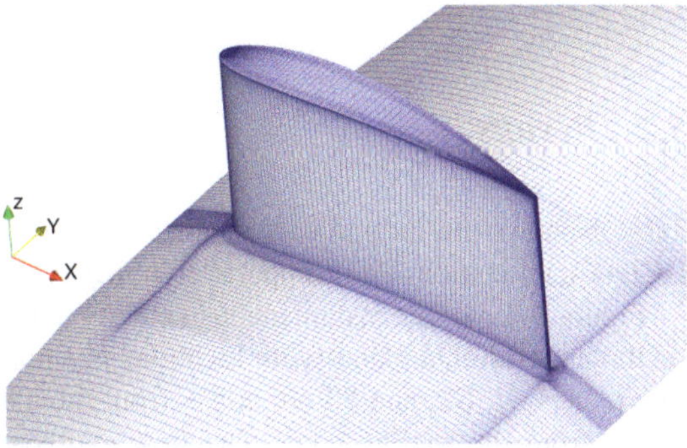

Fig. 2 Wells turbine. Surface mesh near the blade

edges at mid-span radius. In the flow direction, the domain stretches one chord length upstream and three chord lengths downstream from the blade section. We use a structured mesh consisting of about 3 million hexahedral elements. The distribution of the nodes in the axial direction is 26% upstream of the blade, and 74% downstream. The mesh has increased refinement near the solid surfaces, including the blade tip, casing wall, and the blade surfaces. Near those surfaces, the minimum grid spacing (scaled by the blade chord) is 2×10^{-3}. The maximum aspect ratio for the elements is 64, and the average aspect ratio is 5. From a posteriori calculations, the maximum $y+$ value for the nodes closest to the solid surfaces is less than 5. Figure 2 shows the surface mesh near the blade. Mesh independence was checked by computing with 1.2M, 1.8M, and 2.7M nodes, using the rotor torque as the convergence parameter. The torque difference between the computations with 1.8M and 2.7M nodes was less than 1%. The hexahedral mesh we use for the computation reported here has 2.7M nodes.

As the boundary conditions, we specify the relative velocity at the inflow section, no-slip conditions on all solid surfaces, flow periodicity at the matching periodicity boundaries, and zero stress at the outflow section. The velocity we specify at the inflow boundary was obtained from the measured inlet velocity profile reported in [52] by a modification making it consistent with the sea conditions observed at a

Mediterranean location [53]. The inlet distributions of the turbulence quantities were extracted from flow simulation [54] of an industrial ducted turbomachinery with identical hub-to-casing diameter ratio and an upstream spinner cone. The simulation is carried out during an interval of the expiration phase of the breathing cycle, when the water column rise forces airflow through the turbine. The computation is carried out at a flow rate of 0.93 m^3/s, which is 9.4% above the design condition. The air density and kinematic viscosity are 1.2 kg/m^3 and 1.51×10^{-5} m^2/s. The Reynolds number based on the tip speed is 2.594×10^6.

The time integration is a second-order implicit Runge–Kutta method, with a time-step size of 1.38×10^{-5} s, which translates to 4.34×10^{-3} radians per time step. The number of nonlinear iterations per time is 6. In solving the linear equation systems, we use 5 GMRES iterations with SOR preconditioning. The computation is carried out for 3 rotor revolutions.

5.2 *Results*

Table 5 shows the turbine performance obtained experimentally [51] and with the DCDD stabilization. The comparison indicates a very reasonable accuracy.

Figure 3 shows the instantaneous velocity magnitude behind the blade, displayed on three meridional planes with locations given as a percentage of the blade chord. The planes S1, S2, and S3 are at 5% from the leading edge, at the mid-chord, and at 5% from the trailing edge. The flow evolves on the blade suction side under the influence of a number of aerodynamic factors. On S1, we see a significant acceleration at the leading edge due to the flow turning around the thick leading edge. Notably, the flow does not accelerate uniformly along the radius but undergoes a more intense pressure drop at the blade tip. At about mid-chord, two regions of flow separation appear, one at the hub and the other at the tip. As shown on S2, the blade boundary layer flow is already fully separated and the swirling flow area extends to mid-span, as expected in near-stall operation. On S3, the instantaneous velocity field shows clearly two co-rotating vortex systems, one at each end wall.

Figure 4 shows the isosurfaces of $Q = \frac{1}{2}\left(\|\boldsymbol{\Omega}(\mathbf{u}^h)\|^2 - \|\boldsymbol{\varepsilon}(\mathbf{u}^h)\|^2\right)$ [20, 55], where $\boldsymbol{\Omega}(\mathbf{u}^h) = \frac{1}{2}\left((\nabla\mathbf{u}^h) - (\nabla\mathbf{u}^h)^T\right)$. The isosurfaces shown are for a scaled value of 3,000. The scaling is based on the tip velocity and casing diameter. The vortex patterns we see in Figure 4 are consistent with the observations we made for the velocity magnitude, including the flow acceleration at the leading edge and the vortex streams at the hub end wall and blade tip. The vortex visualization provides

Table 5 Wells turbine. Performance

	Experiment [51]	DCDD
Torque (Nm)	5.2	5.1
Efficiency	0.59	0.62

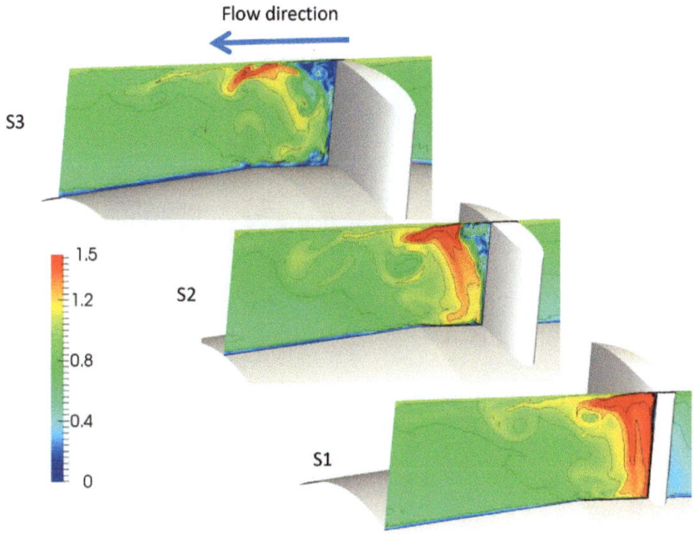

Fig. 3 Wells turbine. Instantaneous velocity magnitude behind the blade, displayed on three meridional planes

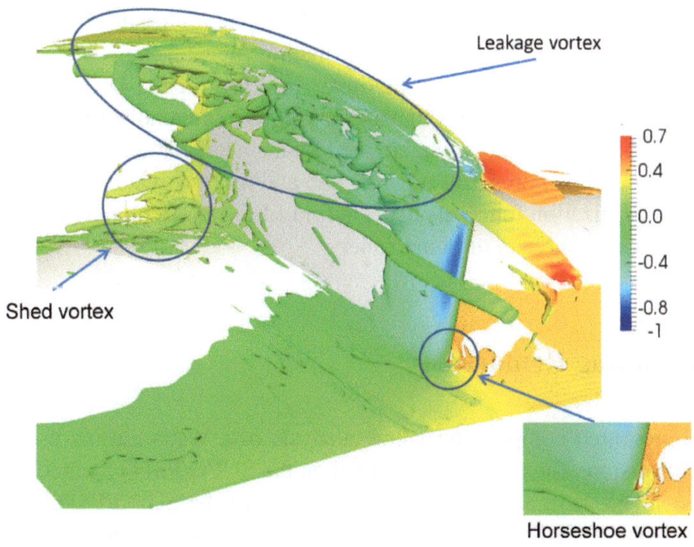

Fig. 4 Wells turbine. Isosurfaces for a scaled Q value of 3,000. The isosurfaces are colored with the scaled pressure, where the scaling is based on the tip velocity

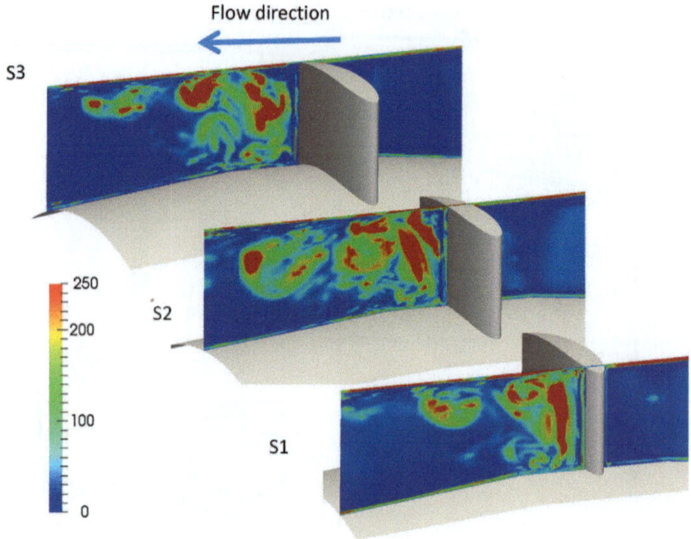

Fig. 5 Wells turbine. Instantaneous distribution of ν_{DCDD}/ν on S1, S2, and S3

a more vivid picture of the complexity of the vortical structures, such as those at the blade tip, which are the precursors of the blade stall at higher flow rates.

Figure 5 shows the instantaneous distribution of ν_{DCDD}/ν on S1, S2, and S3. The distribution provides hints on the inner workings of the DCDD stabilization. The model adapts the ν_{DCDD}/ν values from 5 to 10 in the undisturbed flow regions, to 100 in the blade boundary layer strained by the acceleration at the leading edge, to 200 in the swirling cores driven by the hub and tip flow separation. In that sense, the DCDD stabilization functions like a hybrid LES–RANS model [56] suitable for unsteady flow structures in turbomachinery.

6 Concluding Remarks

We have presented computational flow analysis of a Wells turbine used in wave-energy conversion. The analysis was based on the SUPG/PSPG method and DCDD stabilization. We have also provided an overview of the SUPG/PSPG method and the DCDD. The DCDD was originally introduced to complement the SUPG/PSPG method in computation of incompressible flows with sharp solution gradients. It was also shown earlier to perform well in turbulent-flow test computations when compared to the Smagorinsky LES model. Those test computations were for 3D plane channel flow at a friction Reynolds number of $Re_\tau = 180$. The computational analysis we presented here for a Wells turbine near-stall conditions, with an outcome that is quite consistent with the experimental data, showed that the DCDD, in combination with the SUPG/PSPG method, performs well also in turbomachinery flows. We believe that in this class of turbulent-flow problems, the DCDD acts like a hybrid LES–RANS model, giving us a computational analysis method that has reasonable accuracy and is still affordable.

Acknowledgements The authors acknowledge MIUR support under the project Ateneo and the Visiting Professor Program at University of Rome "La Sapienza."

References

1. Bazilevs, Y., Hsu, M.-C., Akkerman, I., Wright, S., Takizawa, K., Henicke, B., Spielman, T., Tezduyar, T.E.: 3D simulation of wind turbine rotors at full scale. Part I: geometry modeling and aerodynamics. Int. J. Numer. Methods Fluids **65**, 207–235 (2011)
2. Takizawa, K., Henicke, B., Tezduyar, T.E., Hsu, M.-C., Bazilevs, Y.: Stabilized space–time computation of wind-turbine rotor aerodynamics. Comput. Mech. **48**, 333–344 (2011)
3. Takizawa, K., Henicke, B., Montes, D., Tezduyar, T.E., Hsu, M.-C., Bazilevs, Y.: Numerical-performance studies for the stabilized space–time computation of wind-turbine rotor aerodynamics. Comput. Mech. **48**, 647–657 (2011)
4. Hsu, M.-C., Akkerman, I., Bazilevs, Y.: Wind turbine aerodynamics using ALE-VMS: validation and role of weakly enforced boundary conditions. Comput. Mech. **50**, 499–511 (2012)
5. Hsu, M.-C., Bazilevs, Y.: Fluid–structure interaction modeling of wind turbines: simulating the full machine. Comput. Mech. **50**, 821–833 (2012)
6. Bazilevs, Y., Hsu, M.-C., Takizawa, K., Tezduyar, T.E.: ALE-VMS and ST-VMS methods for computer modeling of wind-turbine rotor aerodynamics and fluid–structure interaction. Math. Models Methods Appl. Sci. **22**(supp02), 1230002 (2012)
7. Bazilevs, Y., Takizawa, K., Tezduyar, T.E.: Computational Fluid–Structure Interaction: Methods and Applications. Wiley, Chichester (2013). ISBN: 978-0470978771
8. Takizawa, K., Tezduyar, T.E., McIntyre, S., Kostov, N., Kolesar, R., Habluetzel, C.: Space–time VMS computation of wind-turbine rotor and tower aerodynamics. Comput. Mech. **53**, 1–15 (2014)
9. Takizawa, K., Bazilevs, Y., Tezduyar, T.E., Hsu, M.-C., Øiseth, O., Mathisen, K.M., Kostov, N., McIntyre, S.: Engineering analysis and design with ALE-VMS and space–time methods. Arch. Comput. Methods Eng. **21**, 481–508 (2014)
10. Bazilevs, Y., Takizawa, K., Tezduyar, T.E., Hsu, M.-C., Kostov, N., McIntyre, S.: Aerodynamic and FSI analysis of wind turbines with the ALE-VMS and ST-VMS methods. Arch. Comput. Methods Eng. **21**, 359–398 (2014)
11. Takizawa, K., Tezduyar, T.E., Mochizuki, H., Hattori, H., Mei, S., Pan, L., Montel, K.: Space–time VMS method for flow computations with slip interfaces (ST-SI). Math. Models Methods Appl. Sci. **25**, 2377–2406 (2015)
12. Raghunathan, S.: The Wells air turbine for wave energy conversion. Prog. Aerosp. Sci. **31**, 335–386 (1995)
13. Kim, T.H., Setoguchi, T., Kaneko, K., Raghunathan, S.: Numerical investigation on the effect of blade sweep on the performance of Wells turbine. Renew. Energy **25**, 235–248 (2002)
14. Thakker, A., Abdulhadi, R.: The performance of Wells turbine under bi-directional airflow. Renew. Energy **33**, 2467–2474 (2008)

15. Corsini, A., Marchegiani, A., Minotti, S., Rispoli, F.: On the use of blade sweep in Wells turbines for small power generation. In: Third International Conference on Applied Energy, Perugia, Italy (2011)
16. Corsini, A., Rispoli, F., Tezduyar, T.E.: Computer modeling of wave-energy air turbines with the SUPG/PSPG formulation and discontinuity-capturing technique. J. Appl. Mech. **79**, 010910 (2012)
17. Jacobs, E.N., Ward, K.E., Pinkerton, R.M.: The characteristics of 78 related airfoil sections from tests in the variable-density wind tunnel. Report 460, NACA (1933)
18. Tucker, P.G.: Computation of unsteady turbomachinery flows: part 2, LES and hybrids. Prog. Aerosp. Sci. **47**, 546–569 (2011)
19. Jakirlic, S., Zürker, R.J., John-Puthenveettil, G., Kniesner, B., Tropea, C.: Computational modelling of flow and scalar transport accounting for near-wall turbulence with relevance to gas turbine combustors. Flow Combust. Adv. Gas Turbine Combust. **1581**, 263–294 (2012)
20. Borello, D., Corsini, A., Delibra, G., Fiorito, M., Sheard, A.G.: Large eddy simulation of a tunnel ventilation fan. J. Fluids Eng. **135**, 071102 (2013)
21. Krappel, T., Ruprecht, A., Riedelbauch, S.: Flow simulation of Francis turbines using hybrid RANS-LES turbulence models. High Perform. Comput. Sci. Eng. **14**, 417–431 (2014)
22. Brooks, A.N., Hughes, T.J.R.: Streamline upwind/Petrov-Galerkin formulations for convection dominated flows with particular emphasis on the incompressible Navier-Stokes equations. Comput. Methods Appl. Mech. Eng. **32**, 199–259 (1982)
23. Tezduyar, T.E.: Computation of moving boundaries and interfaces and stabilization parameters. Int. J. Numer. Methods Fluids **43**, 555–575 (2003)
24. Hughes, T.J.R., Tezduyar, T.E.: Finite element methods for first-order hyperbolic systems with particular emphasis on the compressible Euler equations. Comput. Methods Appl. Mech. Eng. **45**, 217–284 (1984)
25. Tezduyar, T.E., Senga, M.: Stabilization and shock-capturing parameters in SUPG formulation of compressible flows. Comput. Methods Appl. Mech. Eng. **195**, 1621–1632 (2006)
26. Tezduyar, T.E., Senga, M., Vicker, D.: Computation of inviscid supersonic flows around cylinders and spheres with the SUPG formulation and YZβ shock-capturing. Comput. Mech. **38**, 469–481 (2006)
27. Tezduyar, T.E.: Stabilized finite element formulations for incompressible flow computations. Adv. Appl. Mech. **28**, 1–44 (1992)
28. Hughes, T.J.R.: Multiscale phenomena: green's functions, the Dirichlet-to-Neumann formulation, subgrid scale models, bubbles, and the origins of stabilized methods. Comput. Methods Appl. Mech. Eng. **127**, 387–401 (1995)
29. Hughes, T.J.R., Oberai, A.A., Mazzei, L.: Large eddy simulation of turbulent channel flows by the variational multiscale method. Phys. Fluids **13**, 1784–1799 (2001)
30. Bazilevs, Y., Calo, V.M., Cottrell, J.A., Hughes, T.J.R., Reali, A., Scovazzi, G.: Variational multiscale residual-based turbulence modeling for large eddy simulation of incompressible flows. Comput. Methods Appl. Mech. Eng. **197**, 173–201 (2007)
31. Bazilevs, Y., Akkerman, I.: Large eddy simulation of turbulent Taylor–Couette flow using isogeometric analysis and the residual–based variational multiscale method. J. Comput. Phys. **229**, 3402–3414 (2010)
32. Takizawa, K., Tezduyar, T.E.: Multiscale space–time fluid–structure interaction techniques. Comput. Mech. **48**, 247–267 (2011)
33. Takizawa, K., Tezduyar, T.E.: Space–time fluid–structure interaction methods. Math. Models Methods Appl. Sci. **22**(supp02), 1230001 (2012)
34. Takizawa, K., Bazilevs, Y., Tezduyar, T.E., Long, C.C., Marsden, A.L., Schjodt, K.: ST and ALE-VMS methods for patient-specific cardiovascular fluid mechanics modeling. Math. Models Methods Appl. Sci. **24**, 2437–2486 (2014)
35. Takizawa, K.: Computational engineering analysis with the new-generation space–time methods. Comput. Mech. **54**, 193–211 (2014)
36. Takizawa, K., Tezduyar, T.E., Buscher, A., Asada, S.: Space–time fluid mechanics computation of heart valve models. Comput. Mech. **54**, 973–986 (2014)

37. Takizawa, K., Tezduyar, T.E., Buscher, A.: Space–time computational analysis of MAV flapping-wing aerodynamics with wing clapping. Comput. Mech. **55**, 1131–1141 (2015)
38. Bazilevs, Y., Korobenko, A., Deng, X., Yan, J., Kinzel, M., Dabiri, J.O.: FSI modeling of vertical-axis wind turbines. J. Appl. Mech. **81**, 081006 (2014)
39. Bazilevs, Y., Korobenko, A., Deng, X., Yan, J.: Novel structural modeling and mesh moving techniques for advanced FSI simulation of wind turbines. Int. J. Numer. Methods Eng. **102**, 766–783 (2015)
40. Takizawa, K., Tezduyar, T.E., Kuraishi, T.: Multiscale ST methods for thermo-fluid analysis of a ground vehicle and its tires. Math. Models Methods Appl. Sci. **25**, 2227–2255 (2015)
41. Bazilevs, Y., Korobenko, A., Yan, J., Pal, A., Gohari, S.M.I., Sarkar, S.: ALE–VMS formulation for stratified turbulent incompressible flows with applications. Math. Models Methods Appl. Sci. **25**, 2349–2375 (2015)
42. Bazilevs, Y., Takizawa, K., Tezduyar, T.E.: New directions and challenging computations in fluid dynamics modeling with stabilized and multiscale methods. Math. Models Methods Appl. Sci. **25**, 2217–2226 (2015)
43. Rispoli, F., Borrelli, P., Tezduyar, T.E.: Discontinuity-capturing directional dissipation (DCDD) in computation of turbulent flows. In: Proceedings of the European Congress on Computational Methods in Applied Sciences and Engineering, ECCOMAS 2004 (CD-ROM), Jyvaskyla (2004)
44. Rispoli, F., Borrelli, P., Tezduyar, T.E.: Computation of turbulent flows with the discontinuity-capturing directional dissipation DCDD. In: Proceedings of the 6th World Congress on Computational Mechanics (CD-ROM), Beijing (2004)
45. Rispoli, F., Borrelli, P., Tezduyar, T.E.: DCDD in finite element computation of turbulent flows. In: Proceedings of the 6th World Congress on Computational Mechanics. Tsinghua University Press/Springer, Beijing (2004)
46. Rispoli, F., Corsini, A., Tezduyar, T.E.: Finite element computation of turbulent flows with the discontinuity-capturing directional dissipation (DCDD). Comput. Fluids **36**, 121–126 (2007)
47. Tezduyar, T.E., Osawa, Y.: Finite element stabilization parameters computed from element matrices and vectors. Comput. Methods Appl. Mech. Eng. **190**, 411–430 (2000)
48. Tezduyar, T.E., Park, Y.J.: Discontinuity capturing finite element formulations for nonlinear convection-diffusion-reaction equations. Comput. Methods Appl. Mech. Eng. **59**, 307–325 (1986)
49. Kim, J., Moin, P., Moser, R.: Turbulence statistics in fully developed channel flow at low Reynolds number. J. Fluid Mech. **177**, 133–166 (1987)
50. Dean, R.B.: Reynolds number dependence of skin friction and other bulk flow variables in two-dimensional rectangular duct flow. J. Fluids Eng. **10**, 215–223 (1978)
51. Bassetti, M., Corsini, A., Delibra, G., Rispoli, F., Ruggeri, M., Venturini, P.: Design and verification of a micro Wells turbine for Mediterranean operations. In: 11th European Turbomachinery Conference, Madrid (2015)
52. Gato, L.M.C., Warfield, V., Thakker, A.: Performance of a high-solidity Wells turbine for an OWC wave power plant. J. Energy Resour. Technol. **118**, 263–268 (1996)
53. Corsini, A., Marro, E., Rispoli, F., Tortora, E.: Space-time mapping of wave energy conversion potential in Mediterranean Sea states. In: ASME-ATI-UIT Conference on Thermal and Environmental Issues in Energy Systems, Sorrento (2010)
54. Corsini, A., Rispoli, F.: Using sweep to extend stall-free operational range in axial fan rotors. J. Power Energy **218**, 129–139 (2004)
55. Haller, G.: An objective definition of a vortex. J. Fluid Mech. **525**, 1–26 (2005)
56. Fröhlich, J., von Terzi, D.: Hybrid LES/RANS methods for the simulation of turbulent flows. Prog. Aerosp. Sci. **44**, 349–377 (2008)

The Advection–Diffusion Analysis of Smoke Flows Around a Body

Takashi Nomura, Hiroshi Hasebe, and Takehiro Kobayashi

Abstract The corrosion due to airborne sea salt is among significant problems for bridges located in coastal area. In order to evaluate its advection diffusion behavior, a special wind tunnel experiment is conducted in which the flow around a square cylinder is visualized by smoke. The brightness data of each frame recorded by a mono-chromatic high-speed camera is digitized and transformed to concentration of the smoke. The SUPG finite element analysis of the advection diffusion equation is conducted and compared with the experimental data. Consequently, the results of the isotropic diffusion analysis is not satisfactory, but the turbulent diffusion analysis gives the result closest to the experiment.

1 Introduction

Maintenance of infrastructures has become important in our modern society in order to keep our level of life satisfactorily within a limited public budget. In case of bridge maintenance, one of the key issues is to prevent its corrosion. The corrosion due to airborne sea salt is among significant problems for the bridges located in coastal area. Especially, it is a serious problem in Japan because many land traffics concentrate in narrow coastal area due to the mountainous topography. In order to make reasonable maintenance plan against the corrosion by airborne sea salt, it is necessary to evaluate the amount of the airborne sea salt which is flying to the bridges as well as adhered on their surfaces.

On-site measurement of the amount of airborne sea salt is conducted in Japan. There are two methods which are widely used in order to measure the amount of airborne sea salt. One is the Doken (Public Works Research Institute) tank method,

T. Nomura (✉) • H. Hasebe
Department of Civil Engineering, College of Science and Technology, Nihon University, Tokyo, Japan
e-mail: nomura@civil.cst.nihon-u.ac.jp

T. Kobayashi
CyberAgent, Inc., Tokyo, Japan

Y. Bazilevs, K. Takizawa (eds.), *Advances in Computational Fluid-Structure Interaction and Flow Simulation*, Modeling and Simulation in Science, Engineering and Technology, DOI 10.1007/978-3-319-40827-9_5

the other is the dry gauze method (JIS-Z2382) [1]. The Doken tank method uses the special sampler tank called "Doken tank" to capture the airborne sea salt. The dry gauze method exposes a dry gauze-patch near a bridge surface. However, because we can measure the amount of airborne sea salt only at the position in which these apparatuses located, it is difficult to reveal the advection diffusion behavior of airborne sea salt around a bridge girder. Therefore, various methods to predict the detailed behavior of airborne sea salt have been developed.

For example, Iwasaki et al. [2] proposed a method to evaluate partial corrosion level from the observation data of a bridge exposed in airborne sea salts. Moreover, Iwasaki et al. [3] conducted a numerical simulation to compare the simulation data to his observation data. Chen et al. [4] proposed a prediction model for the distribution of airborne sea salts in the coastal region of northern Taiwan based on the result of several on-site measurements. Noguchi et al. [5] evaluated the amount of airborne sea salts adhered to surfaces of bridges by using the CFD method considering deposition and collision mechanisms of airborne sea salt particles.

We are developing an experimental method to evaluate the advection diffusion behavior of airborne sea salt around bridge girders [6]. In our method, a wind tunnel experiment is conducted using a special flow visualization technique. From each frame of the motion pictures of the smoke flow around cylindrical specimens, the distribution of smoke concentration is evaluated through the image processing technique described in the following section.

The present article describes another approach of ours based on the numerical flow analysis. We have tried to simulate the advection diffusion phenomena of the visualized smoke flow around a square cylinder.

2 The Objective Visualized Flow

2.1 The Experimental Apparatus

Figure 1 shows schematically our experimental method for the flow visualization around a cylindrical specimen. A fog generator is deployed at the upwind end of an open circuit wind tunnel. The smoke for the flow visualization is once stored in a chamber and poured to the wind tunnel. The smoke flow around a test body is illuminated by a laser sheet and the visualized flow images are recorded by a monochromatic high-speed camera. The free stream velocity is 0.5 m/s. The frame rate is 250 frames per second. The injection time of the smoke is 5 s.

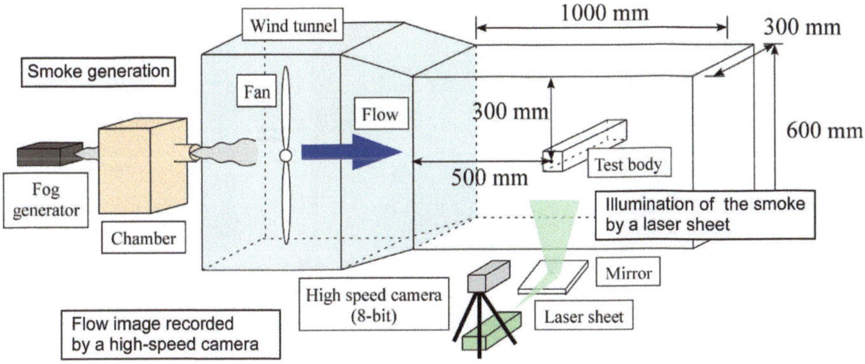

Fig. 1 Experimental setup of the smoke flow visualization.

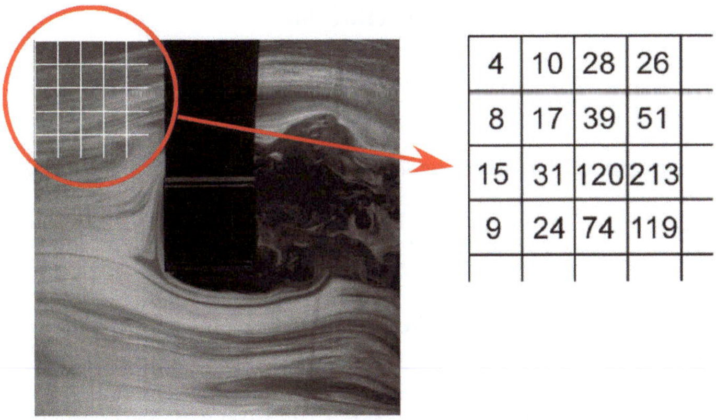

Fig. 2 An instantaneous image of smoke flow and its brightness digits.

2.2 Image Processing Technique

Each recorded frame has a resolution of 640 pixels × 640 pixels of 8 bit dots to represent the brightness of the dot as shown in Figure 2. We assume that the brightness represents the concentration of smoke. By converting these digits to smoke concentration, we can obtain a series of digital data to represent unsteady advection–diffusion process of smoke flow around the test body.

Since the laser sheet spreads like a fan from the probe tip of laser sheet generator, the brightness of the laser sheet attenuates as the distance from the tip increases. Therefore, each frame image of the visualized smoke flow is corrected by multiplying a correction factor to each pixel digit as shown in Figure 3. This distribution of correction factor is obtained from the time-averaged image of the visualized smoke flow without test body.

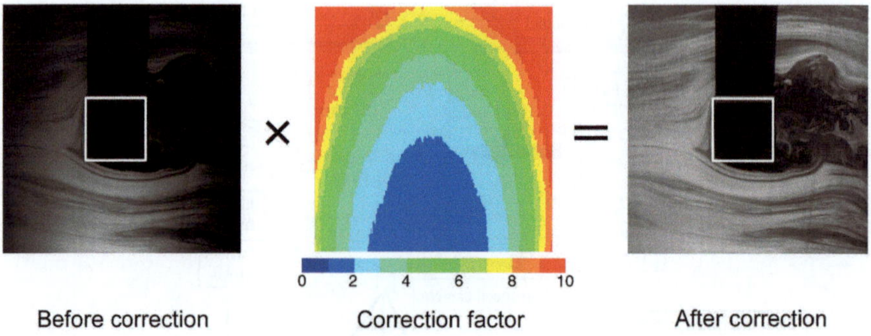

Fig. 3 Correction of the laser light attenuation in the brightness image.

3 The Advection–Diffusion Analysis

3.1 The Basic Equation and the Numerical Method

The basic equation of the advection–diffusion analysis is given as follows:

$$\frac{\partial \phi}{\partial t} + \nabla\left(u_i\phi - K\nabla\phi\right) = 0 \tag{1}$$

where ϕ is the concentration, u_i is the advection velocity, and K is the coefficient of diffusivity.

The basic equation was discretized by the Streamline Upwind/Petrov–Galerkin method (SUPG method) [7] using the hexahedral finite element. The time integration method is the forward Euler method.

3.2 The Advection Velocity

For the advection velocity of the advection–diffusion analysis, we have conducted a turbulent flow analysis of the flow around a square cylinder at the Reynolds number 2000 which is identical to our flow visualization experiment.

The modified k-ε model using the Kato–Launder model is employed as the basic equations of the turbulent flow analysis. The transport equations of the turbulent kinetic energy k and the dissipation rate ε are converted to the logarithmic form [8]. The SUPG method is employed for discretization.

The finite element mesh and a snap shot of the computed streamlines are shown in Figure 4. The computed vortex shedding period was 0.85 s which is close to the experimental value 0.9 s.

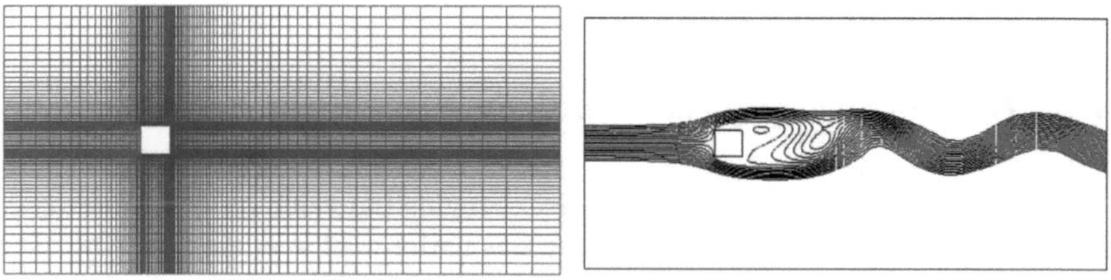

Fig. 4 The finite element mesh of the turbulent flow analysis around a square cylinder and a snapshot of the computed streamlines.

3.3 *The Boundary Condition of the Advection–Diffusion Analysis*

The same finite element mesh as Figure 4 is used for the advection–diffusion analysis of the smoke flow. Figure 5 shows the boundary conditions. The time history of concentration along the upstream boundary $\overline{\phi}(t)$ is given as the parabolic function of time t for $0 \leq t \leq 6.7$s and $\overline{\phi}(t) = 0$ for $t > 6.7s$ as shown in Figure 6. This parabolic function is a simplified curve of the measured time history at the point A of Figure 6. The point A is located at the distance of $D/2$ in the upstream of the square cylinder where D (= 6 cm) is the side length of the square section. This time history $\overline{\phi}(t)$ is specified to the all nodes along the line at the distance of $D/2$ in the upstream in the analysis as shown in Figure 5. Therefore, the upstream region beyond this line is out of consideration in the present advection–diffusion analysis.

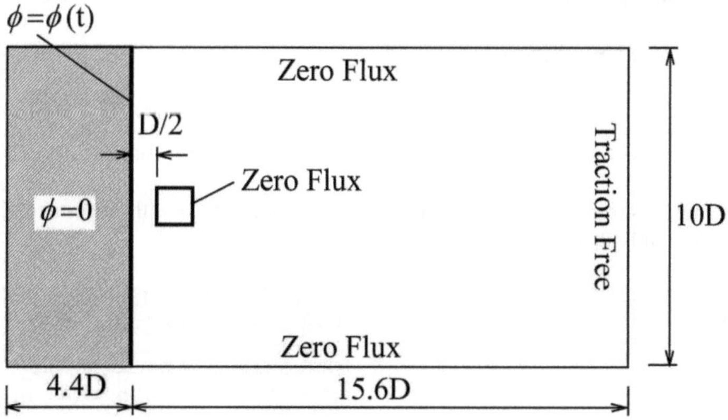

Fig. 5 The boundary conditions of the advection diffusion problem (D = 6 cm is the side length of the square cylinder).

4 The Computed Results

4.1 The Isotropic Diffusion Analysis

We have conducted a series of analysis by changing the value of the diffusivity K in Equation (1) under the assumption that K is uniform all over the flow field. The cases of four different values of $K = 0.1,\ 1.0,\ 5.0,$ and $10.0\ \mathrm{cm}^2/\mathrm{s}$ are computed.

Figure 7 shows two instantaneous spatial distributions of the computed smoke concentration for the case of $K = 1.0\ \mathrm{cm}^2/\mathrm{s}$. At $t = 4.0$s, the smoke concentration at the upstream boundary is high and most of the supplied smoke flows with main streams towards the downstream open boundary. At $t = 9.0$s, the supply of the smoke at the upstream boundary has ended at $t = 6.7$s as shown in Figure 6 and smoke in the main stream region has already flowed out through the downstream open boundary. Some smoke is left around the square cylinder since the flow speed in this area is relatively slow.

Figure 8 shows the four cases of computed time histories of the nodal concentration at the two points indicated in Figure 8 in comparison with the experimental time histories. At point C, the time histories of the four cases of diffusivity K are

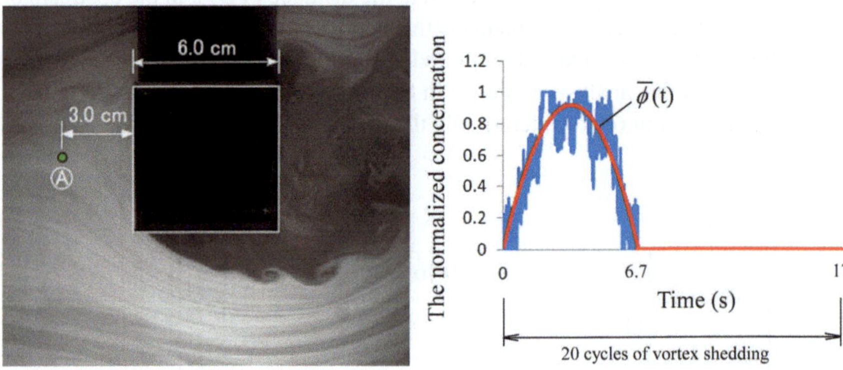

Fig. 6 The time history of concentration along the upstream boundary; the measured time history at the point A is simplified to $\overline{\phi}(t)$.

Fig. 7 The computed distribution of smoke concentration of the case of isotropic diffusion $K = 1.0\ \mathrm{cm}^2/\mathrm{s}$ at $t = 4.0$s and 9.0s.

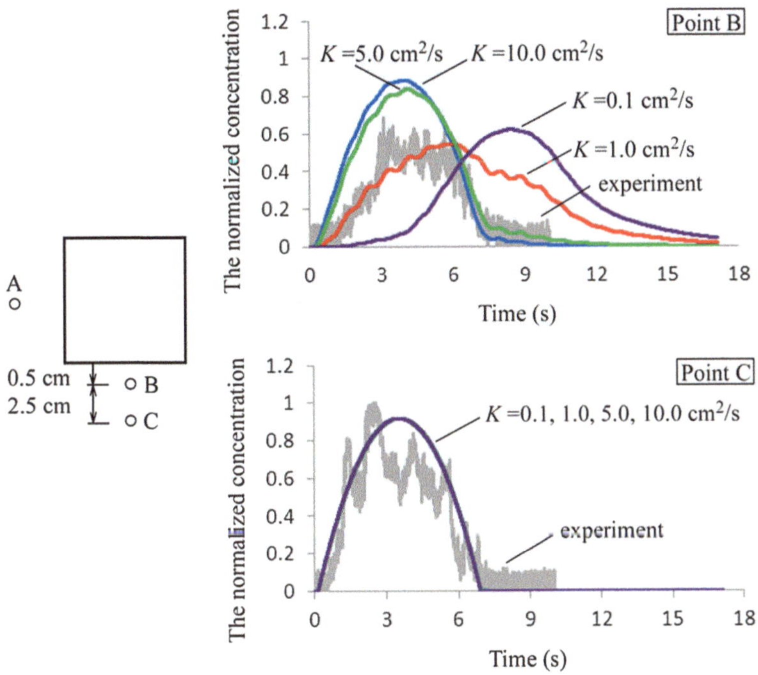

Fig. 8 The computed time histories of smoke concentration at the point B and C of the isotropic diffusion cases.

almost identical since this location is outside of the separation shear layer from the leading corner of the square cylinder. The four time histories reasonably follow the experimental time history. The fine fluctuation in the experiment is not shown in the computation because the flow analysis is the Reynolds averaged turbulence model.

On the contrary, at point B which is inside the separation shear layer, the computed time histories are different for different values of diffusivity K. The time history of $K = 1.0\ \mathrm{cm}^2/\mathrm{s}$ follows the experimental history well from the initiation to the highest value thought the duration is much longer than the experiment.

4.2 *The Turbulent Diffusion Analysis*

As the computations of the isotropic diffusion are not satisfactory, we have introduced the turbulent diffusivity to the advection–diffusion analysis. A typical turbulent diffusivity K_t is given as follows:

$$K_t = \frac{\nu_t}{S_{ct}} \tag{2}$$

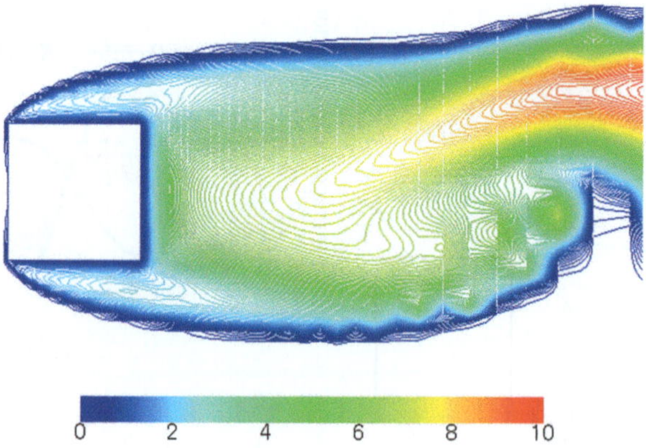

Fig. 9 The distribution of turbulent diffusivity K_t at $t = 0.9$ s.

where ν_t is the eddy viscosity and S_{ct} is the turbulent Schmidt number. The standard value of S_{ct} is 0.9.

The eddy viscosity ν_t can be defined as follows:

$$\nu_t = C_\mu \frac{k^2}{\varepsilon} \tag{3}$$

where k is the turbulent kinetic energy, ε is the dissipation rate, and $C_\mu = 0.09$ is the $k - \varepsilon$ model constant.

The eddy viscosity ν_t can be computed from the $k - \varepsilon$ model analysis. Figure 9 shows an example distribution of turbulent diffusivity K_t in our computation. The turbulent diffusivity K_t is high at the region where two separation shear layers encounter in the wake of the square cylinder.

Figure 10 shows the computed time histories of smoke concentration at the same two points as Figure 8. The time history at point B of the case of turbulent diffusivity K_t shows improved duration though the maximum value is a little higher than that of the experiment.

Figure 11 shows the computed instantaneous distribution of smoke concentration around the square cylinder in the cases of turbulent diffusivity K_t and isotropic diffusivity $K = 1.0\ \text{cm}^2/\text{s}$ at $t = 9.0$s. In case of the isotropic diffusion, the smoke still remains along the upper and lower side of the cylinder. On the contrary, in case of the turbulent diffusivity, the smoke becomes faint because of harder mixing process than the isotropic diffusivity.

5 Concluding Remarks

A finite element analysis of the advection–diffusion equation is conducted to simulate smoke concentration around a square cylinder. The analysis is compared with a digitized data of smoke flow visualization of a wind tunnel experiment. The modified $k - \varepsilon$ model is used to obtain the advection velocity data.

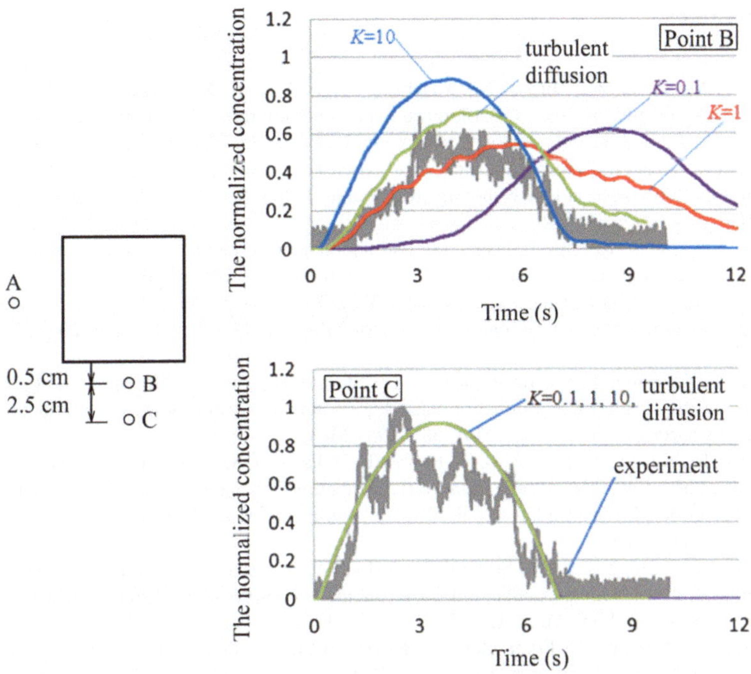

Fig. 10 The computed time histories of smoke concentration at the point B and C of the isotropic diffusion case.

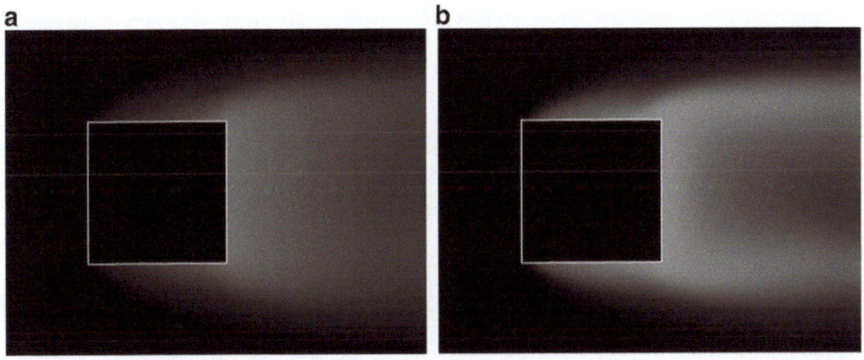

Fig. 11 The computed smoke concentration around the square cylinder at $t = 9.0$ s. (a) Turbulent diffusivity K_t. (b) Isotropic diffusivity $K = 1.0\ \text{cm}^2/\text{s}$.

For the diffusivity of the advection–diffusion equation, the isotropic diffusivity and the turbulent diffusivity are employed and the computed results are compared. It revealed that the analysis using the turbulent diffusivity gives the concentration behavior closest to the experiment.

References

1. Takebe, M., Ohya, M., Hirose, N., Adachi, R., Ago, Y., Doi, K., Iwatani, Y., Kitagawa, N., Kimura, Y., Ochibe, K., Ota, J.: Difference in precipitation rates of air-borne salts collected by the dry gauze method and the Doken tank method. Corros. Sci. **52**, 2928–2935 (2010)
2. Iwasaki, E., Kage, I., Kato, M., Nakanishi, K., Niwa, H.: A feature of corrosion and its evaluation around the cross-section of weathering steel bridge. J. Jpn. Soc. Civil Eng. **A-66**(2), 297–311 (2010)
3. Iwasaki, E., Ito, S., Kojima, Y., Nagai, M.: Estimation of airborne sea salt around a bridge section by numerical simulation. J. Jpn. Soc. Civil Eng. **A-66**(4), 752–766 (2010)
4. Chen, Y.S., Chiu, H.J., Chan, Y.W., Chang, Y.C., Yang, C.C.: Prediction model of air-borne salt distribution in the coastal region of northern Taiwan. J. Mar. Sci. Technol. **20**(3), 259–268 (2012)
5. Noguchi, K., Kang, Y., Okuda, S., Kurata, N., Shirato, H., Yagi, T., Morishita, T., Tanaka, Y.: Study on improvement of estimation accuracy of amount of air-borne sea salt adhered to a structural surface. J. Struct. Eng. **59A**, 585–595 (2013)
6. Hasebe, H., Kosugi, T., Hatamoto, S., Kawai, T., Nakayama, S., Nomura, T.: Image processing technique of smoke flow visualization for the prediction of behavior of airborne sea salt around a bridge girder. In: Proceedings of the Eighth Asia-Pacific Conference on Wind Engineering, Chennai, India (2013)
7. Brooks, A.N., Hughes, T.J.R.: Streamline upwind/Petrov-Galerkin formulations for convection dominated flows with particular emphasis on the incompressible Navier-Stokes equations. Comput. Methods Appl. Mech. Eng. **32**, 199–259 (1982)
8. Hasebe, H., Nomura, T.: Finite element analysis of 2D turbulent flows using the logarithmic form of the model. Wind Struct. **12**, 21–48 (2009)

Finite Element Computation of Buzz Instability in Supersonic Air Intakes

V.M. Krushnarao Kotteda and Sanjay Mittal

Abstract An intake is a duct designed to provide sufficient amount of air at low speed to the engine face, at high efficiency and with little distortion. The physical shape of the intake is fundamentally the same as that of a convergent-divergent nozzle. Inlets may have different shape and size depending on the speed of the aircraft. An inlet must be able to sustain high pressure that exists at the engine face. Supersonic intakes may be associated with flow instability at relatively high back pressure at the engine face. The buzz instability involves periodic filling and discharge of the plenum chamber, complex shock-boundary layer interaction, shear layer/slip stream-boundary layer interaction, transient shock movement and flow separation. It adversely affects the mass flow entering the engine and may lead to combustion instability, engine surge and flame out. It can also lead to deterioration of the performance of propulsion system, thus causing catastrophic loss in thrust. In this article we describe our computational efforts in understanding flows in intakes of supersonic vehicles. The flow instabilities in a Y-intake and mixed compression intake are studied numerically via a stabilized finite element method.

1 Introduction

Supersonic flow entering an aircraft engine needs to be slowed down to low subsonic speed for efficient functioning of the compressor, turbine and combustor. A supersonic intake slows down the flow by efficiently converting its kinetic energy to higher temperature and pressure. The intake can be one of the various types: external, internal and mixed compression intake [1, 2]. The simplest external compression intake is of the pitot type. In this intake, the compression of the flow takes place via a normal shock. As a result, the flow downstream of the shock is subsonic. However, the pressure loss across the normal shock is rather high. This leads to relatively large loss of thrust by the engine for high flight Mach number. Therefore, despite its simplicity, a pitot intake is generally used only if the flight

---

V.M.K. Kotteda • S. Mittal (✉)
Department of Aerospace Engineering, IIT Kanpur, Kanpur 208016, India
e-mail: muralik@iitk.ac.in; smittal@iitk.ac.in

Y. Bazilevs, K. Takizawa (eds.), *Advances in Computational Fluid-Structure Interaction and Flow Simulation*, Modeling and Simulation in Science, Engineering and Technology, DOI 10.1007/978-3-319-40827-9_6

Mach number (M_∞) is below 1.6. For $M_\infty > 1.6$, a multi-shock system inlet is used where the compression takes place through a series of oblique shocks followed by a normal shock outside the inlet duct. It is also possible to design an intake where the compression takes place within the duct. The internal compression intakes are associated with starting-up problems and require a complex shock control system for starting the inlet. For $M_\infty > 2.5$, a mixed compression intake is preferred where the compression takes place inside as well as outside the inlet. The mixed compression intake is characterized by multiple reflected oblique shocks in the convergent portion and a weak terminal normal shock immediately downstream of the throat. Mixed compression intakes are less susceptible to unstarting problem compared to internal compression intakes.

Owing to the curved shape of the fuselage and to meet the requirement of compactness the shape of an external intake, operating at $M_\infty < 1.6$, is generally curved with increasing area from inlet to exit. Two types of commonly used intakes are: Y-intake and S-intake. Y-intakes consist of a pair of ducts in the wing root or on the two sides of a fuselage, feeding a single engine via a common section of duct. They are also referred to as twin intakes or bifurcated intakes. As the sideslip angle (β) increases, the mass flow through the windward duct increases while it decreases in the leeward duct. Buzz instability may appear in a Y-duct for relatively large values of back pressure at the engine face [3–8]. The instability is accompanied by an asymmetry of the flow in which one limb operates at a higher mass flow than the other. Asymmetric flow in the two ducts of the Y-intake produces an imbalance of pressure forces on the airplane which may lead to severe problems with the stability and control of the aircraft [9].

A typical mixed compression inlet [1, 10] consists of two parts: supersonic diffuser just upstream of the throat and a subsonic diffuser that lies downstream of the throat section. A mixed compression intake may also be associated with buzz instability for relatively large pressure at engine face. The instability was first reported by Oswatitsch [11]. The tests were conducted for supersonic missile flights for Mach number in the range of 2.5–3.0. He observed buzz during his experiments, but disregarded it. Later (1945–1955), buzz was observed in many of the extensive experiments that were conducted at NASA, Langley on external compression inlets. Ferri and Nucci [12] conducted detailed experiments on an axisymmetrical external compression air intake. The occurrence of buzz instability was attributed to the velocity discontinuity across vortex sheet originating at the intersection of conical shock and strong shock ahead of the intake entrance. The fluctuations began when this vortex sheet approached the inner surface of cowling. This type of oscillations is referred to as Ferri type buzz. Dailey [13] conducted experiments on an axisymmetrical external compression intake with an engine incorporated in the design. He attributed the origin of the buzz oscillations to a random pressure pulse from subsonic diffuser. This type of oscillations is referred to as Dailey type buzz.

Flow in an intake involves complex flow features such as shock emanating from the engine cowl, multiple shock reflections, terminal/bow shock and shock-boundary layer interactions and associated flow separations. In this work, we investigate buzz instability in Y-duct and mixed compression intakes. The effect of sideslip angle on the performance of the Y-intake is studied at various back pressure ratios. The effect of bleed [14, 15] and its locations on the performance of the mixed compression intake is also investigated. The equations governing the flow are the compressible Navier-Stokes equations in the conservation law form. A stabilized finite element formulation based on conservation variables is used to solve the flow equations. The SUPG (Streamline-Upwind/Petrov-Galerkin) stabilization method [16–20] is employed to stabilize the computations against spurious numerical oscillations due to advection dominated flows. A shock capturing term is added to the formulation to provide stability to the computations in the presence of discontinuities and large gradients in the flow [20–24].

2 The Governing Equations

The Navier-Stokes equations governing the fluid flow, in the conservation form, are given as

$$\frac{\partial \rho}{\partial t} + \frac{\partial}{\partial x_i}(\rho u_i) = 0, \tag{1}$$

$$\frac{\partial}{\partial t}(\rho u_i) + \frac{\partial}{\partial x_j}(\rho u_j u_i) = -\frac{\partial p}{\partial x_i} + \frac{\partial \tau_{ji}}{\partial x_j}, \tag{2}$$

$$\frac{\partial}{\partial t}\left[\rho\left(e + \frac{1}{2}u_i u_i\right)\right] + \frac{\partial}{\partial x_j}\left[\rho u_j\left(h + \frac{1}{2}u_i u_i\right)\right] = -\frac{\partial q_j}{\partial x_j} + \frac{\partial}{\partial x_j}(u_i \tau_{ij}), \tag{3}$$

where ρ, $\mathbf{u}$, p, τ, e, h and $\mathbf{q}$ are the density, velocity, pressure, viscous stress tensor, total energy per unit mass, enthalpy per unit mass and heat flux vector, respectively. The viscous stress tensor is defined as

$$\tau_{ij} = \mu\left(\frac{\partial u_i}{\partial x_j} + \frac{\partial u_j}{\partial x_i}\right) + \lambda \frac{\partial u_k}{\partial x_k}\delta_{ij}, \tag{4}$$

where μ is the coefficient of viscosity and δ_{ij} is the Kronecker delta. It is assumed that λ is related to μ by the Stokes hypothesis as $\lambda = -\frac{2}{3}\mu$. The total energy per unit mass is defined as $e = c_v\theta$ where θ is the temperature and c_v is the specific heat at constant volume. The enthalpy per unit mass is defined as $h = c_p\theta$ where c_p is the specific heat at constant pressure. The fluid is assumed to be calorifically perfect. The equation of state, for the ideal gas, is $p = \rho R\theta$ where R is the ideal gas constant. Assuming the medium to be isotropic, the heat flux vector is related to the temperature gradient as $q_j = -\kappa\frac{\partial \theta}{\partial x_j}$, where κ is the heat conductivity. The Prandtl number, $P_r = \frac{c_p\mu}{\kappa}$, relates the heat conductivity to the fluid viscosity, and is assumed to be a constant ($= 0.72$). The coefficient of dynamic viscosity depends on the temperature and is assumed to follow the Sutherland's law.

2.1 The Quasi-Linear Form of Flow Equations

The flow equations (1)–(3) can be rewritten in the following form:

$$\frac{\partial \mathbf{U}}{\partial t} + \frac{\partial \mathbf{F}_i}{\partial x_i} - \frac{\partial \mathbf{E}_i}{\partial x_i} = \mathbf{0} \qquad \text{on } \Omega \times (0, T), \tag{5}$$

where $\mathbf{U} = (\rho, \rho u_1, \rho u_2, \rho e)^T$ is the vector of conservation variables in two dimensions. $\mathbf{F}_i$ and $\mathbf{E}_i$ are the Euler and viscous flux vectors, respectively, defined as

$$\mathbf{F}_i = (\rho u_i, \ \ \rho u_i u_1 + \delta_{i1} p, \ \ \rho u_i u_2 + \delta_{i2} p, \ \ u_i(\rho e + p))^T, \tag{6}$$

$$\mathbf{E}_i = (0, \ \ \tau_{i1}, \ \ \tau_{i2}, \ \ -q_i + \tau_{ik} u_k)^T. \tag{7}$$

Equation (5) is rewritten in a quasi-linear form:

$$\frac{\partial \mathbf{U}}{\partial t} + \mathbf{A}_i \frac{\partial \mathbf{U}}{\partial x_i} - \frac{\partial}{\partial x_i}\left(\mathbf{K}_{ij} \frac{\partial \mathbf{U}}{\partial x_j}\right) = \mathbf{0} \qquad \text{on } \Omega \times (0, T), \tag{8}$$

where $\mathbf{A}_i = \frac{\partial \mathbf{F}_i}{\partial \mathbf{U}}$ is the Euler Jacobian matrix, and $\mathbf{K}_{ij}$ is the diffusivity matrix such that $\mathbf{K}_{ij} \frac{\partial \mathbf{U}}{\partial x_j} = \mathbf{E}_i$. The boundary conditions are represented as $\mathbf{U} = \mathbf{g}$ on Γ_g and $\mathbf{n} \cdot \mathbf{E} = \mathbf{h}$ on Γ_h where $\mathbf{n}$ is the unit normal to the boundary of the domain. The initial condition on the $\mathbf{U}$ is specified as $\mathbf{U}(\mathbf{x}, 0) = \mathbf{U}_0$ on Ω.

3 Finite Element Formulation

Consider a finite element discretization of the domain Ω into sub-domains $\Omega^e, e = 1, 2, 3, \ldots, n_{el}$, where n_{el} is the number of elements. Based on this discretization, we define the finite element trial function space $\mathcal{S}^h$ and weighting function space $\mathcal{V}^h$ for conservation variables. These function spaces are selected by taking the Dirichlet boundary condition into account as subsets of $[\mathbf{H}^{1h}(\Omega)]^{n_{dof}}$, where $\mathbf{H}^{1h}(\Omega)$ is the finite dimensional function space over Ω and n_{dof} is the number of degrees of freedom.

The stabilized finite element formulation of equation (8) is given as follows: find $\mathbf{U}^h \in \mathcal{S}^h : \forall\, \mathbf{W}^h \in \mathcal{V}^h$ the following is satisfied:

$$\begin{aligned}
&\int_\Omega \mathbf{W}^h \cdot \left(\frac{\partial \mathbf{U}^h}{\partial t} + \mathbf{A}_i^h \frac{\partial \mathbf{U}^h}{\partial x_i}\right) d\Omega + \int_\Omega \left(\frac{\partial \mathbf{W}^h}{\partial x_i}\right) \cdot \left(\mathbf{K}_{ij}^h \frac{\partial \mathbf{U}^h}{\partial x_j}\right) d\Omega \\
&+ \sum_{e=1}^{n_{el}} \int_{\Omega^e} \tau (\mathbf{A}_k^h)^T \left(\frac{\partial \mathbf{W}^h}{\partial x_k}\right) \cdot \left[\frac{\partial \mathbf{U}^h}{\partial t} + \mathbf{A}_i^h \frac{\partial \mathbf{U}^h}{\partial x_i} - \frac{\partial}{\partial x_i}\left(\mathbf{K}_{ij}^h \frac{\partial \mathbf{U}^h}{\partial x_j}\right)\right] d\Omega \\
&+ \sum_{e=1}^{n_{el}} \int_{\Omega^e} \delta \left(\frac{\partial \mathbf{W}^h}{\partial x_i}\right) \cdot \left(\frac{\partial \mathbf{U}^h}{\partial x_i}\right) d\Omega = \int_{\Gamma_h} \mathbf{W}^h \cdot \boldsymbol{h}^h d\Gamma.
\end{aligned} \tag{9}$$

In the variational formulation given by Eq. (9), the first two terms together with the right-hand side term constitute the Galerkin formulation of the problem. The first series of element level integrals in Eq. (9), and involving τ, are the SUPG stabilization terms. They stabilize the computations against node-to-node spatial oscillations in the advection dominated flows. The SUPG formulation for the convection dominated flows was introduced by Hughes and Brooks [16] and Brooks and Hughes [17]. The SUPG method for the compressible Euler equations was developed by Tezduyar and Hughes [18] and Hughes and Tezduyar [19].

The SUPG stabilizations may not be able to effectively control the localized oscillations that arise due to sharp layers and discontinuities. Hughes et al. [21] introduced a discontinuity-capturing term that senses the sharp layers in the solution and acts only in the local region. In that sense, unlike the SUPG, it is a non-linear method even when it is applied to a linear advection-diffusion equation. The SUPG formulation along with the discontinuity-capturing term was recast in the entropy variables formulation and can be found in the work by Hughes et al. [22]. It was shown in Le Beau and Tezduyar [23] and Le Beau et al. [24] that the accuracy of the SUPG formulation in conservation variables when supplemented with the discontinuity-capturing term, introduced originally for entropy variables formulation, gives comparable accuracy. The second series of element level integrals in the formulation given by Eq. (9) are the shock capturing terms that stabilize the computations in the presence of sharp gradients [23]. The stabilization parameters τ and δ are the ones that were used in our earlier work [15, 20, 25–27] and similar to those employed by Aliabadi and Tezduyar [28]. The application of this formulation to compute various internal and external flows can be found in our earlier articles [14, 15, 20, 25–27, 29–32]. The time discretization of the variational formulation given by Eq. (9) is done via the generalized trapezoidal rule. For unsteady computations, we employ a second-order accurate-in-time procedure.

4 Flow in a Y-Intake

The geometry of the intake used for the present computations is adapted from the one used by Jolly et al. [8] for their laboratory experiments. We have considered the Y-duct geometry obtained by the intersection of the duct with a plane passing through its half height. The details of the geometry can be found in our other work [32]. The length of the intake duct up to the merger section is used as the characteristic length to non-dimensionalize all the length scales. At larger values of back pressure (p_b), flow reversal might occur at the exit of the intake. This poses numerical difficulty in terms of mass entering the computational domain through a boundary on which outflow conditions are imposed. To circumvent this situation a duct of length 0.6 units is attached at the exit of the intake. p_b is specified at the end of this duct except when the flow is supersonic. The width of the duct at the inlet is 0.1156 units. The ratio of the outlet to inlet area is 1.15. The spacing between the left and right limb at the inflow is 0.4 units. The length of the splitter plate at the lip of the inlet is 0.0572 units. An important quantity that governs the flow in the intake is p_b/p_i. This represents the ratio of the back pressure to the free stream static pressure (p_i). The free stream Mach number is 1.5. The Reynolds number (*Re*), based on the width of the duct at the inlet, the free stream speed and viscosity, is 1×10^5.

The performance of the intake is studied for $2.1 \leq p_b/p_i \leq 3.44$ and $0° \leq \beta \leq 5°$. Various flow regimes are observed. The onset of instability occurs at $p_b/p_i = 3.122$ for $\beta = 5°$. It is associated with flow separation inside the leeward limb of the Y-intake. The nature of this flow separation is periodic in time and causes alternate filling and discharge of the plenum. It is also associated with asymmetry in the movement of the bow shock upstream of the windward and leeward limbs. The onset of instability occurs at relatively high back pressure ratio for low β. Beyond a certain p_b/p_i, the bow shock moves far upstream and leads to unstarting of the intake. The intake unstarts at $p_b/p_i = 3.36$ for $\beta = 5°$. The critical p_b/p_i for unstart of the intake decreases with increase in β.

The onset of buzz instability in the intake is observed at $p_b/p_i = 3.122$ for $\beta = 5°$. The flow and the related parameters for this case are shown in Fig. 1. The first three frames show a buzz cycle. In the first and last flow frames, the bow shocks are at their most downstream locations while they are at their most upstream locations in the second frame. The net flow in the windward limb is always positive. However, the leeward limb experiences a reverse flow during part of the buzz cycle as also seen in the time history of the mass flow rate. Streamlines as well as vorticity fields are shown in the right column of Fig. 1. The upstream movement of the bow shock can be seen in the second frame. It is accompanied with an upstream movement of the vortex, located in the duct beyond the merger section, and the reverse flow in the leeward limb. The negative mass flow rate in the leeward limb corresponding to the second frame can be seen in the time history. The bow shocks upstream of the windward and leeward limbs move back and forth leading to very large amplitude oscillations in the mass flow rate delivered to the engine face. The buzz cycle is fairly periodic and repeats itself. Interestingly, the net mass flow rate through the intake is still positive and the engine continues to receive air flow, but at a reduced level. More details and a large range of flow parameters can be seen in other work [32].

5 Flow in a Mixed Compression Intake

The geometry used in the computation is the one utilized by Anderson and Wong [33] for their experimental investigations. The details of the geometry can be found in earlier papers [14, 15, 27, 30, 31]. The length of the intake is used as the characteristic length to non-dimensionalize all the length scales. The throat area of the original geometry used by Anderson and Wong [33] is not sufficient to start the intake without boundary layer bleed. The separation of the boundary layer is fuelled by the shocks associated with the ramp and cowl surfaces. The thick boundary layer reduces the effective throat area that is not enough to allow the start-up shock to

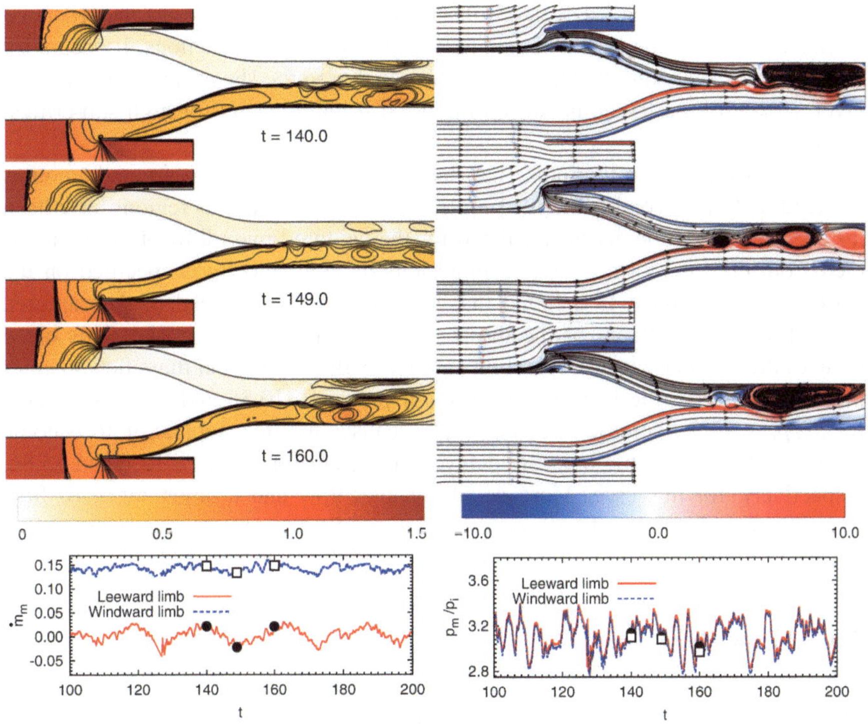

Fig. 1 $M_\infty = 1.5$, $Re = 1 \times 10^5$, $\beta = 5°$, $p_b/p_i = 3.122$ flow in the Y-intake: Mach number field (left) and vorticity along with streamlines (right) at various time instants during onset of buzz instability. Time histories of the mass flow rate in the windward and leeward limb (left) and pressure at the middle of the windward and leeward limb (right) at the merger section are shown in the last row. The time instants at which the flow is shown are marked on these plots.

pass through. Therefore, the intake unstarts. Anderson and Wong [33] had to utilize bleed in the original geometry for starting the flow. Bleed is utilized to control the boundary layer separation and, therefore, to overcome the start-up problem of the intake. A certain percentage of the intake capture mass flow is bled out through two sections on each of the ramp and inner surface of the cowl wall. The free stream Mach number of the flow entering the inlet is 3.0. Re, based on the length of the inlet, the free stream speed and viscosity, is 1×10^6. A duct is attached at the exit of the intake and the back pressure is specified at the end of this duct. The length of the duct is twice that of the intake. Back pressure is specified at the exit of the duct except when the outflow is supersonic. As back pressure ratio increases, the normal shock moves upstream toward the throat. At the critical condition, it is located at the throat of the intake, at least in a time-averaged sense. Further increase in the p_b/p_i pushes the normal shock upstream of the throat leading to sub-critical state. In certain cases the sub-critical operation of the intake can lead to unsteady flow, also known as buzz. Two kinds of unsteady oscillations are observed: little buzz [12] and big buzz [13]. The little buzz arises out of shear layer instabilities of the separated boundary layer or due to slip stream as a result of shock interactions. It occurs for low sub-critical condition. The big buzz arises due to the shock-induced boundary layer separation on the compression ramps. It usually occurs for high sub-critical conditions.

The performance of the intake has been studied for various p_b/p_i and amounts of bleed. Different combinations of mass flow bled from various regions have been tested. The bleed velocity is calculated for the captured mass flow of air, which is bled from the intake walls, over a finite length. This velocity is specified on the entire bleed boundary. It is found that a minimum of 6% bleed of the captured mass flow rate, upstream of the throat, is needed to start the intake. This bleed reduces the boundary layer separation enough to allow sufficient mass flow to enter the engine face. Little buzz is observed for $p_b/p_i = 10.36$ and big buzz occurs for $p_b/p_i = 11.2$. Increase in bleed to 15% causes the flow separation to become milder. The little and big buzz are delayed and occur for $p_b/p_i = 12.6$ and 13.86, respectively, for this case. More details on this flow and a larger range of flow parameters can be found in our earlier work [14, 15, 30, 31].

5.1 Little Buzz

The Mach number field at various time instants during one oscillation of little buzz oscillation is shown in Fig. 2. The flow at $t = 33.45$ corresponds to, approximately, the maximum mass flow rate through the throat. It is also associated with the lowest pressure at the inlet of the intake in the buzz cycle. The separated shear layer becomes unstable downstream of the throat. A jet-like flow structure is formed that constricts the flow in the subsonic portion of the diffuser. As a result, the shock is pushed upstream of the throat leading to a sub-critical situation. It is well known that the location of shock upstream of the throat in the convergent part of the intake is unstable. The separated boundary layer on the cowl wall plays an important role in determining the position of the normal shock. This can be observed clearly in the figure. The growth of the separated boundary layer causes the normal shock to move further upstream. Consequently, the mass flow entering the intake reduces. The increase in the angle of the oblique shock on the first ramp can be clearly observed at $t = 33.45$. The flow at $t = 34.15$ corresponds to low mass flow rate through the throat. It is also associated with close to the highest pressure at the inlet of the intake in the buzz cycle. Once sufficient pressure builds up at the inlet of the intake, the flow once again begins and the normal shock moves toward the throat and inside the intake. The cycle is again repeated. The oscillation cycle includes expelling of shock from throat region, increasing the shock angle of first ramp and returning of shock back to the throat region. The cyclic nature of the flow separation at the cowl lip can also be seen in the figure.

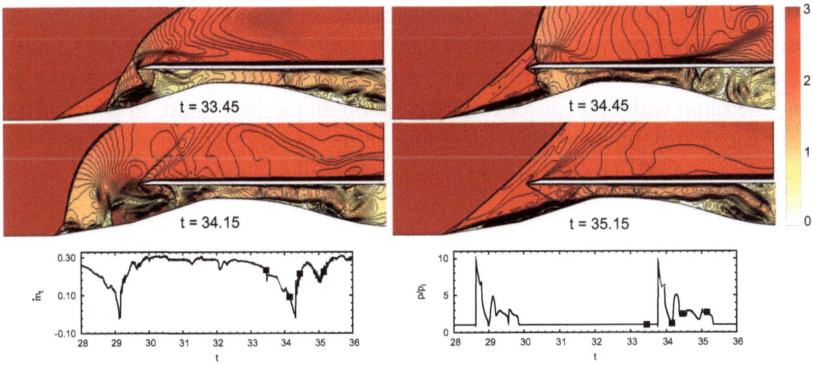

Fig. 2 $M_\infty = 3.0$, $Re = 1 \times 10^6$, $p_b/p_i = 12.6$ flow in the air intake with 15% bleed: the Mach number field for the unsteady flow at various time instants during one little buzz cycle. Also shown in the bottommost row are the time histories of the pressure at a point located at (0.2, 0.1) and the mass flow rate at the throat. The time instants at which the flow is shown are marked on these plots.

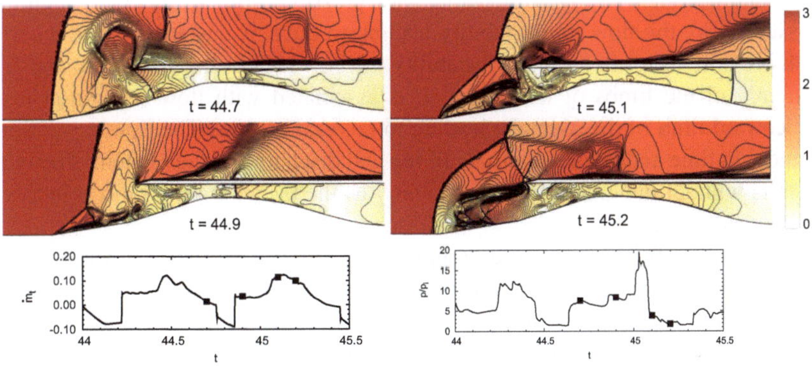

Fig. 3 $M_\infty = 3.0$, $Re = 1 \times 10^6$, $p_b/p_i = 11.2$ flow in the air intake with 6% bleed: the Mach number field for the unsteady flow at various time instants during one big buzz cycle. Also shown in the bottommost row are the time histories of the pressure at a point located at (0.2, 0.1) and the mass flow rate at the throat. The time instants at which the flow is shown are marked on these plots.

5.2 *Big Buzz*

The Mach number field at various time instants during one cycle of big buzz is shown in Fig. 3. The shock in the convergent part of the intake is unstable and moves upstream causing the boundary layer on the cowl and ramp surfaces to separate. The separated regions block the flow in the intake. A bow shock forms upstream of the first ramp can be seen at $t = 44.7$. At $t = 44.9$, vortices due to flow separation in the intake spill over outside the cowl and carried by the flow. As the bow shock moves downstream, the oblique shock at the first ramp reappears and its intersection with the bow shock leads to the formation of a lambda shock, which also moves downstream toward the throat. In the meantime, a pressure pulse originating in the subsonic diffuser moves upstream. It can be seen at $t = 45.1$. It hits the moving lambda shock and moves further upstream through the separated boundary layer along the ramp wall and causes flow separation on the first ramp. In addition to the earlier shock structure, the flow separation at the ramp results in a new bow shock as seen at $t = 45.2$. The buzz cycle is self-sustained and repeats itself.

6 Conclusions

A stabilized finite element method is used to investigate the buzz instability in two types of supersonic intakes. The viscous flow in a Y-intake in two dimensions is studied for $2.1 \leq p_b/p_i \leq 3.44$ and $0° \leq \beta \leq 5°$. The free stream Mach number of the flow entering the intake is 1.5 and the Reynolds number, based on the width of the duct at the inlet, is 1×10^5. The flow is unsteady at low back pressure ratios and becomes steady at moderate back pressure ratio. At relatively high back pressure ratio, the flow loses stability leading to buzz. It unstarts beyond a certain back pressure ratio. In the buzz instability the bow shocks oscillate back and forth upstream of the limbs of the intake. It is associated with flow separation inside the leeward limb of the Y-intake leading to reversal of flow. The flow separation is periodic in time and causes alternate filling and discharge of the plenum. The intake unstarts without undergoing buzz instability as the back pressure ratio is increased for zero sideslip angle.

The viscous flow in a mixed compression intake is studied for various bleed configurations. The free stream Mach number is 3.0 and the Reynolds number, based on the captured area of the intake, is 1×10^6. The back pressure is applied at the downstream end of the duct and progressively increased to study the flow for various values of back pressures. As the back pressure is increased, the normal shock moves upstream toward the throat. At the critical condition it is located at the throat of the intake, at least in a time-averaged sense. Further increase in the back pressure pushes the intake to sub-critical state and leads to one of the two situations. In the first case, the normal shock is completely expelled out of the intake leading to spillage of flow over the cowl and unstarting of intake. In the second situation, buzz is observed. This corresponds to oscillation of the normal shock as well as the flow in the convergent part of the intake. It is also associated with cyclic filling and discharging of the intake.

References

1. Seddon, J., Goldsmith, E.L.: Intake Aerodynamics, vol. 2. Blackwell Science, London (1999)
2. El-Sayed, A.F.: Aircraft Propulsion and Gas Turbine Engines. CRC Press, Boca Raton (2008)
3. Martin, N.J., Holzhauser, C.A.: An experimental investigation at large scale of single and twin NACA submerged side intakes at several angles of sideslip. NACA RM–A9F20 (1949)

4. Martin, N.J., Holzhauser, C.A.: Analysis of factors influencing the stability characteristics of symmetrical twin-intake air-induction systems. NACA TN–2049 (1950)
5. Seddon, J., Trebble, W.: Model test on the asymmetry of airflow occurring in twin-intake systems at subsonic speeds. Aeronautical Research Council Reports and Memoranda–2910 (1955)
6. Mossman, E.A., Pfyl, F.A., Lazzeroni, F.A.: Experimental investigation at mach numbers from 0 to 1.9 of trapezoidal and circular side inlets for a fighter-type airplane. NACA RM A55D27 (1955)
7. Anderson, W.E., Perkins, E.W.: Effects of unsymmetrical air-flow characteristics of twin-intake air-induction systems on airplane static stability at supersonic speeds. NACA TM X–94 (1959)
8. Jolly, R., Pai, T.G., Jayasimha, P.: Design and development of a bifurcated Y-duct intake for a modern combat aircraft. J. Aeronaut. Soc. India **54**(2), 125–131 (2002)
9. Nugent, J.: Nonsteady flow of a twin-duct propulsion system and airplane coupling slightly below a Mach number of 2.0. NASA TM X-54 (1959)
10. Hill, P.G., Peterson, C.R.: Mechanics and Thermodynamics of Propulsion. Addison-Wesley Publishing Company, Reading, MA (1992)
11. Oswatitsch, K.: Pressure recovery in missiles with reaction propulsion at high supersonic speeds. NACA TM–1140 (1947)
12. Ferri, A., Nucci, R.: The origin of aerodynamic instability of supersonic inlets at subcritical conditions. NACA RM–L50K30 (1951)
13. Dailey, C.L.: Supersonic diffuser instability. J. Aeronaut. Sci. **22**, 733–749 (1955)
14. Vivek, P., Mittal, S.: Buzz instability in a mixed-compression air intake. J. Propul. Power **25**, 819–822 (2009)
15. Kotteda, V.M.K., Mittal, S.: Viscous flow in a mixed compression intake. Int. J. Numer. Meth. Fluids **67**, 1393–1417 (2011)
16. Hughes, T.J., Brooks, A.: A multidimensional upwind scheme with no crosswind diffusion. Finite element methods for convection dominated flows, AMD, vol. 34, pp. 19–35 (1979)
17. Brooks, A.N., Hughes, T.J.R.: Streamline upwind/Petrov-Galerkin formulations for convection dominated flows with particular emphasis on the incompressible Navier-Stokes equations. Comput. Methods Appl. Mech. Eng. **32**, 199–259 (1982)
18. Tezduyar, T.E., Hughes, T.J.R.: Finite element formulations for convection dominated flows with particular emphasis on the compressible Euler equations. In: Proceedings of AIAA 21st Aerospace Sciences Meeting, Reno, NV, AIAA Paper 83-0125 (1983)
19. Hughes, T.J.R., Tezduyar, T.E.: Finite element methods for first-order hyperbolic systems with particular emphasis on the compressible Euler equations. Comput. Methods Appl. Mech. Eng. **45**, 217–284 (1984)
20. Mittal, S.: Finite element computation of unsteady viscous compressible flows. Comput. Methods Appl. Mech. Eng. **157**, 151–175 (1998)
21. Hughes, T.J.R., Mallet, M., Mizukami, A.: A new finite element formulation for computational fluid dynamics: II. Beyond SUPG. Comput. Methods Appl. Mech. Eng. **54**, 341–355 (1986)
22. Hughes, T.J.R., Mallet, M.: A new finite element formulation for computational fluid dynamics: IV. A discontinuity-capturing operator for multidimensional advective-diffusive systems. Comput. Methods Appl. Mech. Eng. **58**, 329–336 (1986)
23. Le Beau, G.J., Tezduyar, T.E.: Finite element computation of compressible flows with the SUPG formulation. In: Dhaubhadel, M., Engelman, M.S., Reddy, J.N. (eds.) Advances in Finite Element Analysis in Fluid Dynamics, vol. 123, pp. 21–27. American Society of Mechanical Engineers, Fluids Engineering Division (Publication) FED, New York (1991)
24. Le Beau, G.J., Ray, S., Aliabadi, S., Tezduyar, T.: SUPG finite element computation of compressible flows with the entropy and conservation variables formulations. Comput. Methods Appl. Mech. Eng. **104**(3), 397–422 (1993)
25. Mittal, S.: Finite element computation of unsteady viscous transonic flows past stationary airfoils. Comput. Mech. **21**, 172–188 (1998)
26. Mittal, S., Yadav, S.: Computation of flows in supersonic wind-tunnels. Comput. Methods Appl. Mech. Eng. **191**(6), 611–634 (2001)

27. Kotteda, V.M.K., Mittal, S.: Stabilized finite-element computation of compressible flow with linear and quadratic interpolation functions. Int. J. Numer. Meth. Fluids **75**(4), 273–294 (2014)
28. Aliabadi, S.K., Tezduyar, T.E.: Space-time finite element computation of compressible flows involving moving boundaries and interfaces. Comput. Methods Appl. Mech. Eng. **107**, 209–223 (1993)
29. Mittal, S.: Computation of internal and external compressible flows using EDICT. Int. J. Comput. Fluid Dyn. **14**, 225–241 (2001)
30. Jain, M.K., Mittal, S.: Euler flow in a supersonic mixed-compression inlet. Int. J. Numer. Meth. Fluids **50**(12), 1405–1423 (2006)
31. Kotteda, V.M.K., Mittal, S.: Computation of turbulent flow in a mixed compression intake. Int. J. Adv. Eng. Sci. Appl. Math. **6**(3–4), 126–141 (2014)
32. Kotteda, V.M.K., Mittal, S.: Flow in a Y-intake at supersonic speeds. J. Propuls. Power **32**, 171–187 (2015)
33. Anderson, W.E., Wong, N.D.: Experimental investigation of a large scale, two-dimensional, mixed-compression inlet system: Performance at design conditions, $M_\infty = 3.0$. NASA TM X-2016 (1970)

SUPG/PSPG Computational Analysis of Rain Erosion in Wind-Turbine Blades

Alessio Castorrini, Alessandro Corsini, Franco Rispoli, Paolo Venturini, Kenji Takizawa, and Tayfun E. Tezduyar

Abstract Wind-turbine blades exposed to rain can be damaged by erosion if not protected. Although this damage does not typically influence the structural response of the blades, it could heavily degrade the aerodynamic performance, and therefore the power production. We present a method for computational analysis of rain erosion in wind-turbine blades. The method is based on a stabilized finite element fluid mechanics formulation and a finite element particle-cloud tracking method. Accurate representation of the flow would be essential in reliable computational turbomachinery analysis and design. The turbulent-flow nature of the problem is dealt with a RANS model and SUPG/PSPG stabilization, the particle-cloud trajectories are calculated based on the flow field and closure models for the turbulence–particle interaction, and one-way dependence is assumed between the flow field and particle dynamics. The erosion patterns are then computed based on the particle-cloud data.

1 Introduction

Wind turbines operating at high latitudes or off-shore are exposed to heavy rain. In large horizontal-axis wind turbines (HAWT), at normal operating conditions, the blade-tip velocity range is 90–110 m/s, causing rain-driven erosion that influences the overall turbine performance. Various field studies [1, 2] show that the maximum power in a turbine with deeply eroded blades could be down as much as 20% of

A. Castorrini • A. Corsini (✉) • F. Rispoli • P. Venturini
Dipartimento di Ingegneria Meccanica e Aerospaziale, Sapienza University of Rome, Via Eudossiana, 18, I-00184 Rome, Italy
e-mail: alessandro.corsini@uniroma1.it

K. Takizawa
Department of Modern Mechanical Engineering, Waseda University, 1-6-1 Nishi-Waseda, Shinjuku-ku, Tokyo 169-8050, Japan
e-mail: Kenji.Takizawa@tafsm.org

T.E. Tezduyar
Mechanical Engineering, Rice University – MS 321, 6100 Main Street, Houston, TX 77005, USA
e-mail: tezduyar@tafsm.org

Y. Bazilevs, K. Takizawa (eds.), *Advances in Computational Fluid-Structure Interaction and Flow Simulation*, Modeling and Simulation in Science, Engineering and Technology, DOI 10.1007/978-3-319-40827-9_7

the rated nominal power. Wood [3] and 3M [2] reported that, in severe climates, serious damage to the blade leading edge can be seen after two years of operation. These findings indicate the critical role erosion could play in the scheduling of the wind-turbine maintenance and even in the early stages of the wind-turbine design.

As in any turbomachinery erosion problem, the dynamics of the dispersed phase carried by the turbulent flow is affected by the inertia and drift velocity of the particles, and the physical and chemical properties of both the particles and the target surface play a role. Therefore accurate representation of the particle-laden flow is essential in reliable computational turbomachinery analysis and design, together with case-specific erosion models [4, 5].

Motivated by in-service erosion in aero-engines, in 1970s several researchers started developing methods for computing the particle trajectories and related erosion in gas turbine components. For example, Tabakoff and coworkers [6–8] carried out particle-trajectory computations in axial and centrifugal turbomachinery, substantiating the interaction between the particle dynamics and the inertial forces in rotating cascades. Later numerical studies [9, 10] focused on various modeling aspects of sand erosion. In simulation of particle-laden flows, Corsini et al. [11] pointed out turbulence–particle closure modeling as one of the main computational challenges. In wind-turbine erosion, on the other hand, most work reported in the open literature focused on maintenance and technology issues related to protective coatings [1, 2, 12–17]. While rain erosion in wings has been studied systematically in dedicated experimental test rigs [13, 18, 19], for wind turbines the experiments have been performed on blade specimens rather than the whole blade [1, 12].

In this paper, we present a computational method for simulation of particle-laden flows in turbomachinery. The emphasis here is on the turbulence–particle interaction and rain-driven erosion in a HAWT designed with a rated power of 6 MW [20]. The method is based on a stabilized finite element fluid mechanics formulation and a finite element particle-cloud tracking (PCT) method. The PCT method was originally formulated by Baxter and Smith [21], and then further developed [22, 23] and improved to obtain statistically independent results [24]. One-way dependence is assumed between the flow field and particle dynamics, that is, particle (and cloud) motion is driven by the flow but the flow does not see the particles. The trajectory of the particle-cloud center is calculated with a finite element method. The discrete representation of the cloud is based on the elements of the particle mesh inside the cloud with trajectory-dependent radius. The tracking method accounts for the drifting-velocity gradient in the near-wall regions [22, 25].

The turbulent-flow nature of the problem is dealt with a 3D Reynolds-Averaged Navier–Stokes (RANS) model and the Streamline-Upwind/Petrov-Galerkin (SUPG) [26] and Pressure-Stabilizing/Petrov-Galerkin (PSPG) [27, 28] stabilizations. These are complemented with the "DRDJ" stabilization [29–33]. The stabilization and discontinuity-capturing parameters to be used with the SUPG and PSPG formulations received much attention (see, for example, [29, 34–54]). Here we use the ones given in [39]. The particle-cloud trajectories are calculated based on the flow field and a closure model for the turbulence–particle interaction. The closure model utilizes the scale-separation feature of the variational multiscale (VMS) method [55]. We use an in-house parallel finite element solver [56].

In Section 2, we provide an overview of the mathematical model, including the RANS and PCT models. The SUPG/PSPG stabilized formulations for the Navier–Stokes and RANS closure equations are described in Section 3. In Section 4, we describe the discretized particle equations, including the turbulence–particle interaction. The rain-drop erosion model is described in Section 5. The computations are presented in Section 6, and the concluding remarks are given in Section 7.

2 Mathematical Model

2.1 *Fluid-Phase RANS Model for Incompressible Turbulent Flows*

Let $\Omega \subset R^{nsd}$ be the spatial domain with boundary Γ, and $(0, T)$ be the time domain. The unsteady RANS equations of incompressible turbulent flows can be written on Ω and $\forall t \in (0, T)$ as

$$\rho\left(\frac{\partial \boldsymbol{u}}{\partial t} + \boldsymbol{u}\cdot\nabla\boldsymbol{u} - \mathscr{F}\right) - \nabla\cdot\boldsymbol{\sigma} = \boldsymbol{0}, \tag{1}$$

$$\nabla\cdot\boldsymbol{u} = 0, \tag{2}$$

$$\rho\left(\frac{\partial \boldsymbol{\phi}}{\partial t} + \boldsymbol{u}\cdot\nabla\boldsymbol{\phi} - \boldsymbol{B}_{k\varepsilon}\boldsymbol{\phi} - \mathscr{F}_{k\varepsilon}\right) - \nabla\cdot\left(\rho(\nabla\boldsymbol{\phi})\boldsymbol{\nu}_{k\varepsilon}\right) = \boldsymbol{0}, \tag{3}$$

where ρ, $\boldsymbol{u}$, and $\boldsymbol{\phi} = (k, \tilde{\varepsilon})^T$ are the density, velocity, and turbulence closure variables, and k and $\tilde{\varepsilon}$ are the turbulent kinetic energy and homogeneous dissipation. The symbols $\mathscr{F}$ and $\mathscr{F}_{k\varepsilon}$ represent the vector of external forces and the source vector of turbulence closure equations.

As given in Corsini and Rispoli [57], $\mathscr{F}$ accounts for the volume sources related to the second- and third-order terms in the non-isotropic stress–strain relation [58]. The force vector reads as

$$\begin{aligned}
\mathscr{F} = \nabla\cdot\Bigg(&-0.1\nu_t\tau\left(\boldsymbol{\varepsilon}(\boldsymbol{u})\cdot\boldsymbol{\varepsilon}(\boldsymbol{u}) - \boldsymbol{\varepsilon}(\boldsymbol{u}):\boldsymbol{\varepsilon}(\boldsymbol{u})\frac{1}{3}\boldsymbol{I}\right)\\
&+0.1\nu_t\tau\left(\overline{\boldsymbol{\omega}}(\boldsymbol{u})\cdot\boldsymbol{\varepsilon}(\boldsymbol{u}) + (\overline{\boldsymbol{\omega}}(\boldsymbol{u})\cdot\boldsymbol{\varepsilon}(\boldsymbol{u}))^T\right)\\
&+0.26\nu_t\tau\left(\overline{\boldsymbol{\omega}}(\boldsymbol{u})\cdot\overline{\boldsymbol{\omega}}(\boldsymbol{u}) - \overline{\boldsymbol{\omega}}(\boldsymbol{u}):\overline{\boldsymbol{\omega}}(\boldsymbol{u})\frac{1}{3}\boldsymbol{I}\right)\\
&-10c_\mu^2\nu_t\tau^2\left(\boldsymbol{\varepsilon}(\boldsymbol{u})\cdot\boldsymbol{\varepsilon}(\boldsymbol{u})\cdot\overline{\boldsymbol{\omega}}(\boldsymbol{u}) + (\boldsymbol{\varepsilon}(\boldsymbol{u})\cdot\boldsymbol{\varepsilon}(\boldsymbol{u})\cdot\overline{\boldsymbol{\omega}}(\boldsymbol{u}))^T\right)\\
&-5c_\mu^2\nu_t\tau^2\boldsymbol{\varepsilon}(\boldsymbol{u}):\boldsymbol{\varepsilon}(\boldsymbol{u})\boldsymbol{\varepsilon}(\boldsymbol{u})\\
&+5c_\mu^2\nu_t\tau^2\overline{\boldsymbol{\omega}}(\boldsymbol{u}):\overline{\boldsymbol{\omega}}(\boldsymbol{u})\boldsymbol{\varepsilon}(\boldsymbol{u})\Bigg).
\end{aligned} \tag{4}$$

Table 1 Turbulence closure coefficients

C_μ	$\frac{0.3(1-\exp(-0.36/\exp(-0.75max(\hat{e},\hat{\overline{\omega}}))))}{1+0.35(max(\hat{e},\hat{\overline{\omega}}))^{1.5}}$
f_μ	$1-\exp\left(-(Re_t/90)^{0.5}-(Re_t/400)^2\right)$
$C_{\varepsilon 1}$	1.44
$C_{\varepsilon 2}$	1.92
$f_{\varepsilon 2}$	$1-0.3\exp\left(-Re_t^2\right)$
σ_ε	1.3
σ_k	1.0

Here $\boldsymbol{\varepsilon}(\boldsymbol{u}) = (\nabla \boldsymbol{u}) + (\nabla \boldsymbol{u})^T$ is twice the strain-rate tensor, $\overline{\boldsymbol{\omega}} = (\nabla \boldsymbol{u}) - (\nabla \boldsymbol{u})^T$ is twice the vorticity tensor, ν_t is the turbulent kinematic viscosity defined as $\nu_t = c_\mu f_\mu \tau k$, and $\tau = k/\tilde{\varepsilon}$ is the turbulence time scale, with c_μ and f_μ and other closure coefficients for the turbulence model [58] used given in Table 1. In Table 1, $Re_t = k^2/(\nu\tilde{\varepsilon})$ is the turbulence Reynolds number, with ν being the molecular viscosity, and $\hat{e}$ and $\hat{\overline{\omega}}$ are, respectively, the strain-rate and vorticity invariants defined as $\hat{e} = \tau\sqrt{0.5\boldsymbol{\varepsilon}(\boldsymbol{u}) : \boldsymbol{\varepsilon}(\boldsymbol{u})}$ and $\hat{\overline{\omega}} = \tau\sqrt{0.5\overline{\boldsymbol{\omega}}(\boldsymbol{u}) : \overline{\boldsymbol{\omega}}(\boldsymbol{u})}$.

The source vector $\mathscr{F}_{k\varepsilon}$ is defined as

$$\mathscr{F}_{k\varepsilon} = \begin{bmatrix} P_k - D \\ c_{\varepsilon l} P_k \frac{\tilde{\varepsilon}}{k} + E \end{bmatrix}, \tag{5}$$

where $P_k = \boldsymbol{R} : \nabla \boldsymbol{u}$ is the production of turbulent kinetic energy, with $\boldsymbol{R}$ being the Reynolds stress tensor, $D = 2\nu\|\nabla\sqrt{k}\|^2$, and $E = 0.0022\hat{e}k\tau\nu_t||\nabla \cdot (\nabla \boldsymbol{u})||^2$.

The stress tensor is defined as

$$\boldsymbol{\sigma}(p, \boldsymbol{u}) = -\left(p + \frac{2}{3}\rho k\right)\boldsymbol{I} + \rho\nu_u \boldsymbol{\varepsilon}(\boldsymbol{u}), \tag{6}$$

with $\nu_u = \nu + \nu_t$.

The diffusion terms in the turbulence closure equations are represented with the diffusivity matrix defined as

$$\boldsymbol{\nu_{k\varepsilon}} = \begin{bmatrix} \nu + \frac{\nu_t}{\sigma_k} & 0 \\ 0 & \nu + \frac{\nu_t}{\sigma_\varepsilon} \end{bmatrix}, \tag{7}$$

with the values of the coefficients σ_k and σ_ε given in Table 1.

The reaction terms, absorption like in Eq. (3), account for the dissipation-destruction matrices and are defined as

$$\boldsymbol{B_{k\varepsilon}} = \begin{bmatrix} B_k & 0 \\ 0 & B_\varepsilon \end{bmatrix}, \tag{8}$$

with

$$B_k = \frac{\tilde{\varepsilon}}{k}, \quad B_\varepsilon = c_{\varepsilon 2} f_{\varepsilon 2} \frac{\tilde{\varepsilon}}{k}, \tag{9}$$

and the values of the coefficients $c_{\varepsilon 2}$ and $f_{\varepsilon 2}$ are given in Table 1.

The essential and natural boundary conditions for Eqs. (1) and (2) are

$$\boldsymbol{u} = \boldsymbol{g} \text{ on } \Gamma_g \quad \text{and} \quad \boldsymbol{\phi} = \boldsymbol{g}_{k\varepsilon} \text{ on } \Gamma_{gk\varepsilon}, \tag{10}$$

$$\boldsymbol{n} \cdot \boldsymbol{\sigma} = \boldsymbol{h} \text{ on } \Gamma_k \quad \text{and} \quad \boldsymbol{n} \cdot (\rho(\boldsymbol{\nabla}\boldsymbol{\phi})\boldsymbol{\nu}_{k\varepsilon} = \boldsymbol{0} \text{ on } \Gamma_{hk\varepsilon}, \tag{11}$$

where Γ_g, $\Gamma_{gk\varepsilon}$, Γ_h, and $\Gamma_{hk\varepsilon}$ are the subsets of the boundary Γ, $\boldsymbol{n}$ is the direction normal to the boundary, and $\boldsymbol{g}$, $\boldsymbol{g}_{k\varepsilon}$, and $\boldsymbol{h}$ are given functions representing the essential and natural boundary conditions.

2.2 *Dispersed-Phase Model*

Particle trajectories are simulated in a Lagrangian reference frame. Since particle concentration in this kind of applications is very small (i.e., less than 10^{-6} in the particle volume fraction), a one-way dependence approach can be used [59]. That is, the flow field affects particle motion but particles do not affect the flow field. The concept of one-way dependence has been used in other computational engineering analyses. For example, in [60], the concept was used for computing the aerodynamic forces acting on the suspension lines of spacecraft parachutes, where the suspension lines are assumed to have no influence on the flow field. We use the PCT model [61] to simulate a large number of particles without tracking them individually. The PCT approach was used in turbulent particle dispersion [21, 24, 62–64] and validated in turbomachinery and biomass furnaces [65, 66]. In the PCT model, each trajectory is not related to a particle, but to a group of particles (a "cloud"), thus representing the evolution of the cloud position at time t:

$$\boldsymbol{x}_c = \int_0^t \boldsymbol{v}_c dt' + (\boldsymbol{x}_c)_0. \tag{12}$$

Here, subscript c refers to the cloud, $\boldsymbol{v}_c$ is the velocity of the cloud, and $(\boldsymbol{x}_c)_0$ is the initial position of the cloud, which is at the inflow boundary in our computations.

The equation of motion for the cloud is given by the Basset–Boussinesq–Oseen formulation, which, with one-way dependence hypothesis according to Armenio and Fiorotto [67], reads as

$$\frac{d\boldsymbol{v}_c}{dt} = \tau_R^{-1} \left(\langle \boldsymbol{u} \rangle - \boldsymbol{v}_c\right) + \langle \boldsymbol{f} \rangle + \left(1 - \frac{\rho}{\rho_p}\right) \boldsymbol{g}, \tag{13}$$

where $\langle\ \rangle$ denote ensemble average of the enclosed quantity (defined later), $\boldsymbol{f}$ is the centrifugal and Coriolis forces, ρ_p is the particle material density, and τ_R is the particle relaxation time, which, for spherical particles, reads as

$$\tau_R^{-1} = \frac{3}{4d_p} C_D \frac{\rho}{\rho_p} \|\langle \boldsymbol{u} \rangle - \boldsymbol{v}_c\|. \tag{14}$$

Here, d_p is the particle diameter and C_D is the drag coefficient based on the particle Reynolds number $Re_p = \frac{\|\langle \boldsymbol{u}\rangle - \boldsymbol{v}_c\| d_p}{\nu}$, first introduced in [68]. The initial condition for $\boldsymbol{v}_c$ is given as $\boldsymbol{v}_c(0) = \langle \boldsymbol{u}\rangle|_{t=0}$.

The ensemble average for the dispersed phase within the cloud is defined according to the hypothesis of independent statistical events, and for any quantity θ it reads as

$$\langle\theta\rangle = \frac{\int_{\Omega_c} \theta PDF(\boldsymbol{x}, t) d\Omega}{\int_{\Omega_c} PDF(\boldsymbol{x}, t) d\Omega}. \tag{15}$$

Here, θ is ensemble-averaged quantity, Ω_c is the cloud domain, and $PDF(\boldsymbol{x}, t)$ is the multi-variate probability density function of the dispersed phase. This definition of the ensemble average is appropriate for stationary and non-stationary quantities, and also for both continuous and discontinuous quantities.

The PCT approach assumes that particle position distribution within a cloud is Gaussian, and the cloud size varies in time according to the properties of the flow. To this end, the *PDF* describing the particle distribution within the cloud reads as

$$PDF(\boldsymbol{x}, t) = \frac{1}{(2\pi)^{1/2}\sigma} exp\left(-\frac{1}{2}\left(\frac{\|\boldsymbol{x} - \boldsymbol{x}_c\|}{\sigma}\right)^2\right). \tag{16}$$

Here, σ is the square root of the variance of particle position, which accounts for the turbulent dispersion of particles. We will define it in Section 4. The cloud size (i.e., cloud radius) is taken as 3σ, and that gives us Ω_c. Each cloud is assumed to consist of perfectly spherical particles with the same chemical and physical characteristics.

Combining Eqs. (13) and (14), we obtain

$$\frac{d\boldsymbol{v}_c}{dt} = C'_D \|\langle \boldsymbol{u}\rangle - \boldsymbol{v}_c\| \left(\langle \boldsymbol{u}\rangle - \boldsymbol{v}_c\right) + \langle \boldsymbol{f}\rangle + \left(1 - \frac{\rho}{\rho_p}\right)\boldsymbol{g}, \tag{17}$$

where

$$C'_D = \frac{3}{4d_p} C_D \frac{\rho}{\rho_p}. \tag{18}$$

3 SUPG/PSPG Formulation of Fluid Mechanics Equations of Turbulent Flows

3.1 Stabilized Formulations

In describing the SUPG/PSPG formulation of Eqs. (1), (2), and (3), we assume that we have constructed some suitably defined finite-dimensional trial solution and test function spaces S_u^h, S_p^h, S_ϕ^h and V_u^h, V_p^h, V_ϕ^h. The SUPG/PSPG formulation reads as follows: find $\boldsymbol{u}^h \in S_u^h$, $p^h \in S_p^h$, $\boldsymbol{\phi}^h \in S_\phi^h$, such that $\forall \boldsymbol{w}^h \in V_u^h$, $\forall q^h \in V_p^h$, and $\forall \boldsymbol{\psi}^h \in V_\phi^h$:

$$\int_\Omega \boldsymbol{w}^h \cdot \rho \left(\frac{\partial \boldsymbol{u}^h}{\partial t} + \boldsymbol{u}^h \cdot \nabla \boldsymbol{u}^h - \mathscr{F}^h \right) d\Omega$$
$$+ \int_\Omega \boldsymbol{\epsilon}(\boldsymbol{w}^h) : \boldsymbol{\sigma}(p^h, \boldsymbol{u}^h) d\Omega - \int_{\Gamma_h} \boldsymbol{w}^h \cdot \boldsymbol{h}^h d\Gamma + \int_\Omega q^h \nabla \cdot \boldsymbol{u}^h d\Omega$$
$$+ \sum_{e=1}^{n_{el}} \int_{\Omega^e} \boldsymbol{P}^{stab}(\boldsymbol{w}^h, q^h) \cdot \left(Ł(p^h, \boldsymbol{u}^h) - \rho \mathscr{F}^h \right) d\Omega = 0, \quad (19)$$

where

$$Ł(q^h, \boldsymbol{w}^h) = \rho \left(\frac{\partial \boldsymbol{w}^h}{\partial t} + \boldsymbol{u}^h \cdot \nabla \boldsymbol{w}^h \right) - \nabla \cdot \boldsymbol{\sigma}(q^h, \boldsymbol{w}^h), \quad (20)$$

and

$$\int_\Omega \boldsymbol{\psi}^h \cdot \rho \left(\frac{\partial \boldsymbol{\phi}^h}{\partial t} + \boldsymbol{u}^h \cdot \nabla \boldsymbol{\phi}^h + \boldsymbol{B}_{k\varepsilon}^h \boldsymbol{\phi}^h - \mathscr{F}_{k\varepsilon}^h \right) d\Omega$$
$$+ \int_\Omega \nabla \boldsymbol{\psi}^h : \left(\rho (\nabla \boldsymbol{\phi}^h) \boldsymbol{\nu}_{k\varepsilon}^h \right) d\Omega$$
$$- \sum_{e=1}^{n_{el}} \int_{\Omega^e} \boldsymbol{P}_{k\varepsilon}^{stab}(\boldsymbol{\psi}^h) \cdot \left(Ł_{k\varepsilon}(\boldsymbol{\phi}^h) - \rho \mathscr{F}_{k\varepsilon}^h \right) d\Omega$$
$$+ \sum_{e=1}^{n_{el}} \int_{\Omega^e} \boldsymbol{K}_{k\varepsilon}^{DC} \rho \nabla \boldsymbol{\psi}^h : \nabla \boldsymbol{\phi}^h d\Omega = 0, \quad (21)$$

where

$$Ł_{k\epsilon}(\boldsymbol{\phi}^h) = \rho \left(\frac{\partial \boldsymbol{\phi}^h}{\partial t} + \boldsymbol{u}^h \cdot \nabla \boldsymbol{\phi}^h + \boldsymbol{B}_{k\epsilon}^h \boldsymbol{\phi}^h \right) - \nabla \cdot \left(\rho (\nabla \boldsymbol{\phi}^h) \boldsymbol{\nu}_{k\epsilon}^h \right). \quad (22)$$

In calculation of $\mathscr{F}_{k\epsilon}^h$, we calculate $\nabla \cdot (\nabla \boldsymbol{u}^h)$ in the E term by first calculating the nodal values of $\nabla \boldsymbol{u}^h$ by least-squares projection and then taking the divergence of the interpolated value of $\nabla \boldsymbol{u}^h$.

In Eqs. (19), (20), (21), and (22), $\boldsymbol{P}^{stab}$, $\boldsymbol{P_{k\epsilon}}^{stab}$, and $\boldsymbol{K_{k\epsilon}}^{DC}$ are the SUPG/PSPG stabilization operators and the discontinuity-capturing (DC) matrix of the DRDJ stabilization. The vectors $\boldsymbol{P}^{stab}$ and $\boldsymbol{P_{k\epsilon}}^{stab}$ take the forms

$$\boldsymbol{P}^{stab}(\boldsymbol{w}^h, q^h) = \tau_{SUPG} (\boldsymbol{u}^h \cdot \nabla) \boldsymbol{w}^h + \frac{\tau_{PSPG}}{\rho} \nabla q^h, \quad (23)$$

$$\boldsymbol{P}_{k\epsilon}^{stab}(\boldsymbol{\psi}^h) = \begin{bmatrix} \tau_{SUPG-k} & 0 \\ 0 & \tau_{SUPG-\epsilon} \end{bmatrix} (\boldsymbol{u}^h \cdot \nabla)\boldsymbol{\psi}^h. \tag{24}$$

Here τ_{SUPG} and τ_{PSPG} are the SUPG and PSPG stabilization parameters. These are defined in Section 3.2.

The DC matrix terms are defined as

$$\begin{bmatrix} \kappa_{DRDJ-k} & 0 \\ 0 & \kappa_{DRDJ-\epsilon} \end{bmatrix}. \tag{25}$$

Here κ_{DRDJ-k} and $\kappa_{DRDJ-\epsilon}$ are the DRDJ diffusivities (see [29–33]).

3.2 *Stabilization Parameters*

We first define the element length [29] in the advection-dominated limit:

$$h_{UGN} = 2\left(\sum_{a}^{nen} |\boldsymbol{s} \cdot \nabla N_a|\right)^{-1}, \tag{26}$$

where $\boldsymbol{s}$ is the unit vector in direction of the velocity, *nen* is the number of element nodes, and N_a is the interpolation function associated with node a.

In the diffusion-dominated limit, the element lengths [39] are defined as follows:

$$h_{RGN} = 2\left(\sum_{a}^{nen} |\boldsymbol{r} \cdot \nabla N_a|\right)^{-1}, \tag{27}$$

$$h_{RGN-k} = 2\left(\sum_{a}^{nen} |\boldsymbol{r}_k \cdot \nabla N_a|\right)^{-1}, \tag{28}$$

$$h_{RGN-\varepsilon} = 2\left(\sum_{a}^{nen} |\boldsymbol{r}_\epsilon \cdot \nabla N_a|\right)^{-1}, \tag{29}$$

where $\boldsymbol{r}$, $\boldsymbol{r}_k$, and $\boldsymbol{r}_\epsilon$ are the unit vectors in the direction of the solution gradient:

$$\boldsymbol{r} = \frac{\nabla \|\boldsymbol{u}\|}{\|\nabla \|\boldsymbol{u}\|\|}, \ \boldsymbol{r}_k = \frac{\nabla |k|}{\|\nabla |k|\|}, \ \boldsymbol{r}_\epsilon = \frac{\nabla |\tilde{\epsilon}|}{\|\nabla |\tilde{\epsilon}|\|}. \tag{30}$$

The components of τ_{SUPG} corresponding to the advection-, transient-, and diffusion-dominated limits were defined in [39, 45] as follows:

$$\tau_{SUGN1} = \left(\sum_a^{nen} |\boldsymbol{u} \cdot \nabla N_a| \right)^{-1} = \frac{h_{UGN}}{2\|\boldsymbol{u}\|}, \tag{31}$$

$$\tau_{SUGN2} = \frac{\Delta t}{2}, \tag{32}$$

$$\tau_{SUGN3} = \frac{h_{RGN}^2}{4\nu}, \ \tau_{SUGN3-k} = \frac{h_{RGN-k}^2}{4\nu_k}, \ \tau_{SUGN3-\epsilon} = \frac{h_{RGN-\epsilon}^2}{4\nu_\epsilon}. \tag{33}$$

From these, the stabilization parameters are defined by using the r-switch [37]:

$$\tau_{SUPG} = \left(\frac{1}{\tau_{SUGN1}^{r_s}} + \frac{1}{\tau_{SUGN2}^{r_s}} + \frac{1}{\tau_{SUGN3}^{r_s}} \right)^{-\frac{1}{r_s}}, \tag{34}$$

$$\tau_{PSPG} = \tau_{SUPG}, \tag{35}$$

$$\tau_{SUPG-\phi} = \left(\frac{1}{\tau_{SUGN1}^{r_s}} + \frac{1}{\tau_{SUGN2}^{r_s}} + \frac{1}{\tau_{SUGN3-\phi}^{r_s}} \right)^{-\frac{1}{r_s}}. \tag{36}$$

Subscript ϕ generates the expressions corresponding to k and $\tilde{\epsilon}$. Typically, $r_s = 2$.

4 Discretized Particle Equations

In the discretized particle equations, ensemble averaging is carried out over the discretized cloud domain $\Omega_c = \bigcup_{e=1}^{n_{elc}} (\Omega_c)_e$, where $(\Omega_c)_e$ is the cloud element, and n_{elc} is the number of elements. The cloud elements come from a fixed mesh, which we call "particle mesh," and consist of the elements of that fixed mesh within the radius $\sigma = 3$. With that, the discretized version of ensemble averaging is written as

$$\langle \theta \rangle^h = \frac{\sum_{e=1}^{n_{elc}} \int_{(\Omega_c)_e} \theta PDF(\boldsymbol{x}, t) d\Omega}{\sum_{e=1}^{n_{elc}} \int_{(\Omega_c)_e} PDF(\boldsymbol{x}, t) d\Omega}, \tag{37}$$

where the element-level integration is performed by Gaussian quadrature.

4.1 Trajectory Calculation

Spatially discretized version of Eq. (17) is written as

$$\frac{d\boldsymbol{v_c}^h}{dt} = \boldsymbol{a}_c^h, \tag{38}$$

where

$$\boldsymbol{a}_c^h = C_D' \| \langle \boldsymbol{u} \rangle^h - \boldsymbol{v}_c^h \| \left(\langle \boldsymbol{u} \rangle^h - \boldsymbol{v}_c^h \right) + \langle \boldsymbol{f} \rangle^h + \left(1 - \frac{\rho}{\rho_p} \right) \boldsymbol{g}. \tag{39}$$

Time discretization of Eq. (38) is done with a predictor–multicorrector algorithm.
Predictor stage:

$$(\boldsymbol{v}_c^h)_{n+1}^0 = (\boldsymbol{v}_c^h)_n + (\boldsymbol{a}_c^h)_n \Delta t. \tag{40}$$

Multicorrector stage:

$$(\boldsymbol{v}_c^h)_{n+1}^{i+1} = (\boldsymbol{v}_c^h)_n + \left((\boldsymbol{a}_c^h)_n + (\boldsymbol{a}_c^h)_{n+1}^i \right) \frac{\Delta t}{2}. \tag{41}$$

Here the superscript n is the time level, and the superscript i is the counter for the multiple corrections. We stop the corrections when

$$\frac{(\boldsymbol{v}_c^h)_{n+1}^{i+1} - (\boldsymbol{v}_c^h)_{n+1}^i}{(\boldsymbol{v}_c^h)_{n+1}^{i+1}} \leq 2 \times 10^{-2}. \tag{42}$$

At each time step, the PCT model requires the computation of the cloud mean position and radius, and the identification of the elements contained within the cloud volume. This is done with the search algorithm described in [25].

4.2 Turbulence–Particle Interaction Parameters

The variance is taken to be dependent upon the Lagrangian time scale of the particle-laden flow, τ_L, and, according to Baxter [61], its Markovian approximation reads as

$$\sigma^2 = 2(v')_c^2 \tau_L^2 \left(\frac{t}{\tau_L} - (1 - e^{-t/\tau_L}) \right) + \sigma_0^2, \tag{43}$$

where τ_L is defined as

$$\tau_L = max(\tau, \tau_p) = max(\tau, \tau_R), \tag{44}$$

with τ_p given as $\tau_p = \tau_R$ and τ is defined below. The fluctuating component of the particle velocity for the cloud, driven by the turbulence–particle interaction [69], reads as

$$(v')_c^2 = (u')_c^2 \left(1 - e^{-\tau/\tau_p} \right). \tag{45}$$

We adopt the definitions for $(u')^2_c$ and τ based on the VMS approach first proposed in [55] and further developed for RANS computations [31, 45, 70]. The definition for $(u')^2_c$ is based on the VMS scale separation $\mathbf{u} = \mathbf{u}^h + \mathbf{u}'$, where $\mathbf{u}^h$ is the resolved flow velocity and $\mathbf{u}'$ is the fine-scale flow velocity modeled as

$$\mathbf{u}' = -\frac{1}{\rho}\tau_{SUPG}\left(\text{Ł}(p^h, \mathbf{u}^h) - \rho\mathscr{F}^h\right). \tag{46}$$

Then the definitions of the VMS turbulence–particle interaction parameters become

$$(u')^2_{c-VMS} = \left\langle \|\mathbf{u}'\|^2 \right\rangle^h, \tag{47}$$

$$\tau = \tau_{SUPG}. \tag{48}$$

5 Rain-Drop Erosion Model

Based on the computed data from the flow and particle computations, we can compute the erosion rate on the blade surface. Keegan et al. [71] provide a review of the available models for rain-erosion prediction. According to Evans et al. [72], a threshold damage velocity can be computed as

$$v_D \approx 1.41\left(\frac{K_m^2 c_m}{\rho_w^2 c_w^2 d_w}\right)^{1/3}. \tag{49}$$

Here ρ_w and c_w are the density of the water and the compressional wave speed in water, and d_w is the droplet diameter; K_m is the fracture toughness of the target material, and c_m is the Rayleigh wave velocity of the target material, defined as

$$c_m = \left(\frac{0.862 + 1.14\nu_m}{1 + \nu_m}\right)\left(\frac{E_m}{2(1 + \nu_m)\rho_m}\right), \tag{50}$$

with ρ_m, ν_m, and E_m being the density, Poisson's ratio, and Young's modulus for the target material.

The threshold damage velocity is the minimum impact velocity of a rain drop that causes erosion damage to the blade. In quantifying the damage, we will use v_D in combination with other computed data. As a first approximation, one can write the impact force of a droplet as

$$F_i = \frac{m_w v_i^2}{d_w}, \tag{51}$$

with m_w representing the mass of a water droplet, and v_i its impact velocity. Combining Eqs. (49) and (51), the minimum impact force that causes damage to the blade, F_D, can be computed. Assuming that the damage is proportional to the impact force and the number of droplets impacting, the damage during the time step Δt can be predicted as

$$\Delta D = \Delta n_w \frac{F_i}{F_D} H\left(\frac{F_i}{F_D} - 1\right), \tag{52}$$

where Δn_w is the number of droplets impacting during Δt, per unit surface area, and $H(\cdot)$ is the Heaviside step function. We use Eq. (52) to predict the erosion patterns. The impact count, n_w, and damage, D, are calculated by summing Δn_w and ΔD over the number of time steps the PCT computation is carried out.

6 Computations

The equations are solved in the rotating reference frame of the turbine rotor. Consequently, the computations are based on the version of Eq. (1) that includes the non-inertial terms, and in the implementation of the stabilized formulations these terms are just added to the other source terms. Alternatively, the arbitrary Lagrangian–Eulerian (ALE) techniques as in [73–76], or the space–time technique as in [52, 53, 73, 76–81], can be employed to handle the rotation. We solve the Navier–Stokes and turbulence closure equations in a fully coupled fashion. The linear solver uses 5 outer and 5 inner GMRES iterations, with SOR preconditioning.

6.1 Description of the Wind Turbine

In this study, we design and use a blade for a 6MW, 3-blade HAWT rotor. The main characteristics of the rotor are given in Table 2. Table 3 provides the airfoil cross-sections used in the chord and twist design, which is based on the Blade Element

Table 2 Rotor characteristics, where "TSR" is the tip-speed ratio

Rated wind speed (m/s)	12
Design rated power (kW)	6,000
Rotor radius (m)	61
Number of blades	3
Rated rotor speed (rpm)	15
Nominal TSR	8
Hub radius (m)	3

Table 3 Airfoil cross-sections of the blade

Radial position (m)	Airfoil
10.0	DU 99-W-405
17.5	DU 99-W-350
22.5	DU 97-W-300
29.0	DU 91-W2-250
37.0	DU 93-W-210
46.0–60.0	NACA 64-618

Momentum (BEM) theory with tip and root loss correction factors [82]. Specifically, we use the same airfoils as those in the 5MW NREL [83] wind turbine.

In the flow computation the mesh is unstructured, with 4.76 million nodes, 4.24 million hexahedral elements, 2.28 million tetrahedral elements, and 0.09 million pyramidal elements. The mesh is structured in the PCT computation, with 1.24 million nodes and 1.20 million hexahedral elements. The PCT domain surrounding the blade extends 3 mean chords from the leading-edge side, and 2 mean chords from the other sides. This domain size for the PCT computation is sufficient in using a large-enough portion of the already computed flow field around the blade. Globally, 50 million droplets, each having a 2 mm diameter, enter the domain in 10 identical clouds (see Fig. 1). Each cloud has an initial radius of 7.5 m, with the initial positions arranged in such a way that we have a uniform rain-drop distribution along the blade. The initial velocity of each cloud is equal to the flow velocity at the cloud center. Droplet and target-material properties are given in Table 4.

6.2 *Results*

6.2.1 Comparison to BEM Computation

We first compare, at the nominal operating point, the result from the SUPG/PSPG computation to data from a standard BEM computation with NREL FAST[84]. Table 5 shows the out-of-plane force and torque acting on the blade, obtained from the SUPG/PSPG and BEM computations.

The difference between the torques predicted is less than 3%, and the difference between the out-of-plane forces is even less. Table 6 shows, for the SUPG/PSPG computation, the components of the forces and moments acting on a single blade, together with the rotor thrust and power output.

6.2.2 Erosion Patterns

Fig. 2 shows the normalized damage rate, and thus the erosion patterns, on the blade surface, while Fig. 3 provides the same data at three different blade sections. The normalization is with respect to the maximum value computed on the blade surface. Erosion is mainly concentrated on the leading edge, in particular at the blade tip,

Fig. 1 Mesh for the PCT computation. The arrows represent the cloud initial velocities

Table 4 Droplet and target-material properties

Water density (kg/m^3)	1,000
Droplet diameter (mm)	2
Compressional wave speed in water (m/s)	1,490
Density of target material (kg/m^3)	1,150
Fracture toughness of target material (MPa m$^{1/2}$)	1.0
Young's modulus of target material (GPa)	3.32
Poisson ratio of target material	0.38

where flow (and droplet) speed is maximum. Flow separation is seen mainly in the root region, however, because the relative flow velocity is very low there, the effect of these flow instabilities on the erosion patterns is minimal. These findings match those observed in actual applications, where the zone most affected is usually the leading edge, in the mid-tip span region.

Table 5 Out-of-plane force and torque acting on a single blade, obtained from the SUPG/PSPG and BEM computations

	SUPG/PSPG	BEM
Out-of-plane force (kN)	280	278
Torque (kN m)	1, 282	1, 316

Table 6 Components of the forces and moments acting on a single blade, together with the rotor thrust and power output, obtained from the SUPG/PSPG computation

	X	Y	Z
Pressure forces (kN)	4.3	−39.7	279.9
Pressure moments (kN m)	−38.8	−11,069.4	−1,355.4
Viscous forces (kN)	0.1	1.9	0.4
Viscous moments (kN m)	−0.4	−11.9	73.7
Rotor thrust (kN)			840.9
Rotor power output (kW)			6,039.9

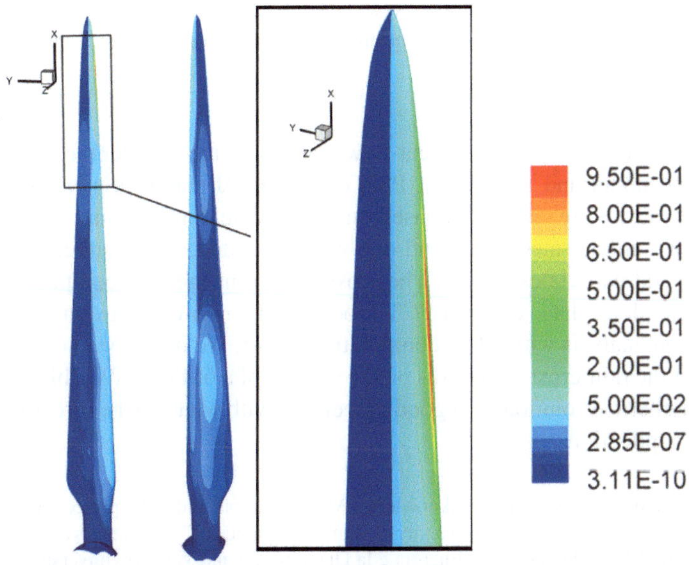

Fig. 2 Normalized erosion rate on the blade surface. Suction side, pressure side, and zoom on the maximum-damage region

7 Concluding Remarks

We have extended the method we developed for computational analysis of particle-laden flows in turbomachinery to computational analysis of rain erosion in wind-turbine blades. The two main components of the method are a stabilized finite element fluid mechanics formulation and a finite element PCT method. Accurate simulation of the flow would be essential in reliable computational wind-turbine

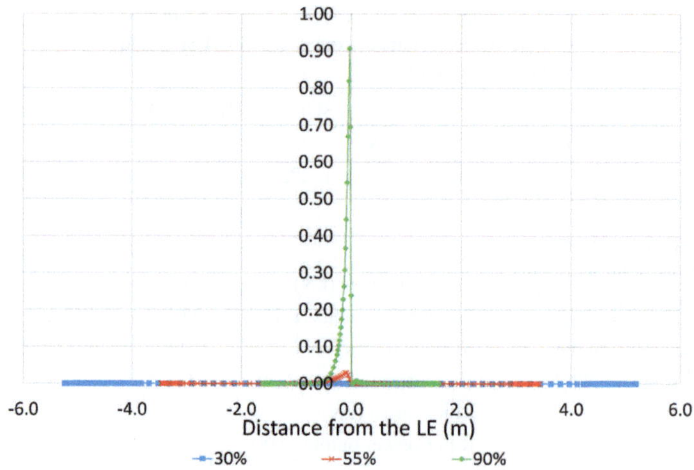

Fig. 3 Normalized erosion rate at 30%, 55%, and 90% of the blade span, as a function of the distance from the leading edge (LE). The positive and negative distance values are for the suction and pressure sides, respectively

analysis and design. The turbulent-flow nature of the problem is dealt with a 3D RANS model and the SUPG, PSPG, and DRDJ stabilizations. The particle-cloud trajectories are calculated based on the flow field and closure models for the turbulence–particle interaction, and one-way dependence is assumed between the flow field and particle dynamics. The erosion patterns are then computed based on the particle-cloud data. Using a PCT model of the rain-droplet dynamics enables rain-erosion analysis with a low computational cost. We have tested our method on predicting the rain-erosion patterns on the surface of a 6MW HAWT blade, showing that the numerical solution is in good agreement with what can be directly observed in actual applications.

Acknowledgements The authors acknowledge MIUR support under the project Ateneo and the Visiting Professor Programme at University of Rome "La Sapienza." The authors also acknowledge support from SED (Soluzioni per l'Energia e la Diagnostica) and Waseda University, Department of Modern Mechanical Engineering.

References

1. Powell, S.: 3M wind blade protection coating. Industrial Marketing Presentation W4600, 3M (2011)
2. 3M: A 3M study is the first to show the effects of erosion on wind turbine efficiency. [Online] www.pressebox.com/pressrelease/3m-deutschland-gmbh/A-3M-Study-Is-the-First-to-Show-the-Effects-of-Erosion-on-Wind-Turbine-Efficiency/boxid/445007 (2011)
3. Wood, K.: Blade repair: closing the maintenance gap. Technical Report, Composites Technology (2011)

4. Tabakoff, W., Kotwal, R., Hamed, A.: Erosion study of different materials affected by coal ash particles. Wear **52**, 161–173 (1979)
5. Keegan, M.H., Nash, D.H., Stack, M.M.: On erosion issues associated with the leading edge of wind turbine blades. J. Phys. D Appl. Phys. **46**, 383001 (2013)
6. Grant, G., Tabakoff, W.: Erosion prediction in turbomachinery resulting from environmental solid particles. J. Aircr. **12**, 471–478 (1975)
7. Hussein, M.F., Tabakoff, W.: Computation and plotting of solid particle flow in rotating cascades. Comput. Fluids **2**, 1–15 (1974)
8. Elfeki, S., Tabakoff, W.: Erosion study of radial flow compressor with splitters. J. Turbomach. **109**, 62–69 (1987)
9. Ghenaiet, A., Tan, S.C., Elder, R.L.: Experimental investigation of axial fan erosion and performance degradation. Proc. Inst. Mech. Eng. Part A J. Power Energy **218**, 437–446 (1987)
10. Ghenaiet, A.: Numerical simulations of flow and particle dynamics within a centrifugal turbomachine. Compressors and Their Systems, vol. 218 (2005). IMechE Paper No. C639-52
11. Corsini, A., Rispoli, F., Sheard, A.G., Takizawa, K., Tezduyar, T.E., Venturini, P.: A variational multiscale method for particle-cloud tracking in turbomachinery flows. Comput. Mech., **54**, 1191–1202 (2014). doi: 10.1007/s00466-014-1050-0
12. Haag, M.: Advances in leading edge protection of wind turbine blades. EWEA Annual Wind Energy Event, Vienna (2013)
13. Weigel, W.: Advanced rotor blade erosion protection system. Technical Report, Kaman Aerospace Corporation (1996)
14. TGM-Services: Blade erosion, 2011 [Online]. Available: http://tgmwind.com/bladeerosion.html# bladeerosion
15. Henkel: Blade maintenance, 2013 [Online]. Available: http://www.henkelna.com/industrial/blade-maintenance-19836.htm
16. Broadwind-Energy: Blade services, 2012 [Online]. Available: http://www.bwen.com/WindTurbineBladeServices_7777.aspx
17. Ropeworks: Blade repair and maintenance services, 2011 [Online]. Available: http://www.ropeworks.com/service_wind_blade.htm
18. Gohardani, O.: Impact of erosion testing aspects on current and future flight conditions. Prog. Aerosp. Sci. **47**, 280–303 (2011)
19. University of Dayton Research Institute: Rain erosion test facility, 2013 [Online] Available: http://www.udri.udayton.edu/NONSTRUCTURALMATERIALS/COATINGS/Pages/RainErosionTestFacility.aspx
20. Corsini, A., Castorrini, A., Morei, E., Rispoli, F., Sciulli, F., Venturini, P.: Modeling of rain drop erosion in a multi-MW wind turbine. ASME Turbo Expo, Montreal (2015)
21. Baxter, L.L., Smith, P.J.: Turbulent dispersion of particles: the STP model. Energy Fuels **7**, 852–859 (1993)
22. Venturini, P.: Modelling of particle-wall deposition in two-phase gas-solid flows. Ph.D. thesis, Sapienza University of Rome (2010)
23. Cardillo, L., Corsini, A., Delibra, G., Rispoli, F., Sheard, A.G., Venturini, P.: Simulation of particle-laden flows in a large centrifugal fan for erosion prediction. In: 58th American Society of Mechanical Engineers Turbine and Aeroengine Congress, Düsseldorf (2015)
24. Kaer, S.K.: Numerical investigation of ash deposition in straw-fired furnaces. Ph.D. thesis, Aalborg University (2001)
25. Corsini, A., Marchegiani, A., Rispoli, F., Venturini, P.: Predicting blade leading edge erosion in an axial induced draft fan. ASME J. Eng. Gas Turbines Power, **134**, 042601 (1993)
26. Brooks, A.N., Hughes, T.J.R.: Streamline upwind/Petrov-Galerkin formulations for convection dominated flows with particular emphasis on the incompressible Navier-Stokes equations. Comput. Methods Appl. Mech. Eng. **32**, 199–259 (1982)
27. Tezduyar, T.E.: Stabilized finite element formulations for incompressible flow computations. Adv. Appl. Mech. **28**, 1–44 (1992). doi: 10.1016/S0065-2156(08)70153-4

28. Tezduyar, T.E., Mittal, S., Ray, S.E., Shih, R.: Incompressible flow computations with stabilized bilinear and linear equal-order-interpolation velocity-pressure elements. Comput. Methods Appl. Mech. Eng. **95**, 221–242 (1992). doi: 10.1016/0045-7825(92)90141-6
29. Tezduyar, T.E., Park, Y.J.: Discontinuity capturing finite element formulations for nonlinear convection-diffusion-reaction equations. Comput. Methods Appl. Mech. Eng. **59**, 307–325 (1986). doi: 10.1016/0045-7825(86)90003-4
30. Corsini, A., Rispoli, F., Santoriello, A., Tezduyar, T.E.: Improved discontinuity-capturing finite element techniques for reaction effects in turbulence computation. Comput. Mech. **38**, 356–364 (2006). doi:10.1007/s00466-006-0045-x
31. Corsini, A., Menichini, C., Rispoli, F., Santoriello, A., Tezduyar, T.E.: A multiscale finite element formulation with discontinuity capturing for turbulence models with dominant reaction like terms. J. Appl. Mech. **76**, 021211 (2009). doi:10.1115/1.3062967
32. Corsini, A., Iossa, C., Rispoli, F., Tezduyar, T.E.: A DRD finite element formulation for computing turbulent reacting flows in gas turbine combustors. Comput. Mech. **46**, 159–167 (2010). doi:10.1007/s00466-009-0441-0
33. Corsini, A., Rispoli, F., Tezduyar, T.E.: Stabilized finite element computation of NOx emission in aero-engine combustors. Int. J. Numer. Methods Fluids **65**, 254–270 (2011). doi:10.1002/fld.2451
34. Tezduyar, T.E., Park, Y.J., Deans, H.A.: Finite element procedures for time-dependent convection-diffusion-reaction systems. Int. J. Numer. Methods Fluids **7**, 1013–1033 (1987). doi:10.1002/fld.1650071003
35. Hughes, T.J.R., Tezduyar, T.E.: Finite element methods for first-order hyperbolic systems with particular emphasis on the compressible Euler equations. Comput. Methods Appl. Mech. Eng. **45**, 217–284 (1984). doi:10.1016/0045-7825(84)90157-9
36. Le Beau, G.J., Ray, S.E., Aliabadi, S.K., Tezduyar, T.E.: SUPG finite element computation of compressible flows with the entropy and conservation variables formulations. Comput. Methods Appl. Mech. Eng. **104**, 397–422 (1993). doi:10.1016/0045-7825(93)90033-T
37. Tezduyar, T.E., Osawa, Y.: Finite element stabilization parameters computed from element matrices and vectors. Comput. Methods Appl. Mech. Eng. **190**, 411–430 (2000) doi:10.1016/S0045-7825(00)00211-5
38. Akin, J.E., Tezduyar, T., Ungor, M., Mittal, S.: Stabilization parameters and Smagorinsky turbulence model. J. Appl. Mech. **70**, 2–9 (2003). doi:10.1115/1.1526569
39. Tezduyar, T.E.: Computation of moving boundaries and interfaces and stabilization parameters. Int. J. Numer. Methods Fluids **43**, 555–575 (2003). doi:10.1002/fld.505
40. Akin, J.E., Tezduyar, T.E.: Calculation of the advective limit of the SUPG stabilization parameter for linear and higher-order elements. Comput. Methods Appl. Mech. Eng. **193**, 1909–1922 (2004). doi:10.1016/j.cma.2003.12.050
41. Tezduyar, T.E., Senga, M.: Stabilization and shock-capturing parameters in SUPG formulation of compressible flows. Comput. Methods Appl. Mech. Eng. **195**, 1621–1632 (2006). doi:10.1016/j.cma.2005.05.032
42. Tezduyar, T.E.: Finite elements in fluids: stabilized formulations and moving boundaries and interfaces. Comput. Fluids **36**, 191–206 (2007). doi:10.1016/j.compfluid.2005.02.011
43. Tezduyar, T.E., Senga, M.: SUPG finite element computation of inviscid supersonic flows with YZβ shock-capturing. Comput. Fluids **36**, 147–159 (2007). doi:10.1016/j.compfluid.2005.07.009
44. Tezduyar, T.E., Senga, M., Vicker, D.: Computation of inviscid supersonic flows around cylinders and spheres with the SUPG formulation and YZβ shock-capturing. Comput. Mech. **38**, 469–481 (2006). doi:10.1007/s00466-005-0025-6
45. Rispoli, F., Corsini, A., Tezduyar, T.E.: Finite element computation of turbulent flows with the discontinuity-capturing directional dissipation (DCDD). Comput. Fluids **36**, 121–126 (2007). doi:10.1016/j.compfluid.2005.07.004
46. Bazilevs, Y., Calo, V.M., Tezduyar, T.E., Hughes, T.J.R.: YZβ discontinuity-capturing for advection-dominated processes with application to arterial drug delivery. Int. J. Numer. Methods Fluids **54**, 593–608 (2007). doi:10.1002/fld.1484

47. Bazilevs, Y., Calo, V.M., Cottrell, J.A., Hughes, T.J.R., Reali, A., Scovazzi, G.: Variational multiscale residual-based turbulence modeling for large eddy simulation of incompressible flows. Comput. Methods Appl. Mech. Eng. **197**, 173–201 (2007)
48. Hughes, T.J.R., Scovazzi, G., Tezduyar, T.E.: Stabilized methods for compressible flows. J. Sci. Comput. **43**, 343–368 (2010). doi:10.1007/s10915-008-9233-5. doi:10.1007/s10915-008-9233-5
49. Hsu, M.-C., Bazilevs, Y., Calo, V.M., Tezduyar, T.E., Hughes, T.J.R.: Improving stability of stabilized and multiscale formulations in flow simulations at small time steps. Comput. Methods Appl. Mech. Eng. **199**, 828–840 (2010). doi:10.1016/j.cma.2009.06.019
50. Tezduyar, T.E.: Comments on adiabatic shock capturing in perfect gas hypersonic flows. Int. J. Numer. Methods Fluids **66**, 935–938 (2011). doi:10.1002/fld.2293
51. Takizawa, K., Tezduyar, T.E.: Multiscale space-time fluid-structure interaction techniques. Comput. Mech. **48**, 247–267 (2011). doi:10.1007/s00466-011-0571-z
52. Takizawa, K., Henicke, B., Tezduyar, T.E., Hsu, M.-C., Bazilevs, Y.: Stabilized space-time computation of wind-turbine rotor aerodynamics. Comput. Mech. **48**, 333–344 (2011). doi:10.1007/s00466-011-0589-2
53. Takizawa, K., Henicke, B., Montes, D., Tezduyar, T.E., Hsu, M.-C., Bazilevs, Y.: Numerical-performance studies for the stabilized space-time computation of wind-turbine rotor aerodynamics. Comput. Mech. **48**, 647–657 (2011). doi:10.1007/s00466-011-0614-5
54. Kler, P.A., Dalcin, L.D., Paz, R.R., Tezduyar, T.E.: SUPG and discontinuity-capturing methods for coupled fluid mechanics and electrochemical transport problems. Comput. Mech. **51**, 171–185 (2013). doi:10.1007/s00466-012-0712-z
55. Hughes, T.J.R.: Multiscale phenomena: Green's functions, the Dirichlet-to-Neumann formulation, subgrid scale models, bubbles, and the origins of stabilized methods. Comput. Methods Appl. Mech. Eng. **127**, 387–401 (1995)
56. Borello, D., Corsini, A., Rispoli, F.: A finite element overlapping scheme for turbomachinery flows on parallel platforms. Comput. Fluids **32**, 1017–1047 (2003)
57. Corsini, A., Rispoli, F.: Flow analyses in a high-pressure axial ventilation fan with a non-linear eddy viscosity closure. Int. J. Heat Fluid Flow **17**, 108–155 (2005)
58. Craft, T.J., Launder, B.E., Suga, K.: Development and application of a cubic eddy-viscosity model of turbulence. Int. J. Heat Fluid Flow **17**, 108–155 (1996)
59. Lain, S., Sommerfeld, M.: Turbulence modulation in dispersed two-phase flow laden with solids from a Lagrangian perspective. Int. J. Heat Fluid Flow **24**, 616–625 (2003)
60. Tezduyar, T.E., Takizawa, K., Moorman, C., Wright, S., Christopher, J.: Space-time finite element computation of complex fluid–structure interactions. Int. J. Numer. Methods Fluids **64**, 1201–1218 (2010). doi:10.1002/fld.2221
61. Baxter, L.L.: Turbulent transport of particles. Ph.D. thesis, Brigham Young University (1989)
62. Wang, L.P.: On the dispersion of heavy particles by turbulent motion. Ph.D. thesis, Washington State University (1990)
63. Litchford, L.J., Jeng, S.M.: Efficient statistical transport model for turbulent particle dispersion in sprays. AIAA J. **29**, 1443–1451 (1991)
64. Jain, S.: Three-dimensional simulation of turbulent particle dispersion. Ph.D. thesis, University of Utah (1995)
65. Borello, D., Venturini, P., Rispoli, F., Saavedra, G.Z.R.: Prediction of multiphase combustion and ash deposition within a biomass furnace. Appl. Energy **101**, 413–422 (2013)
66. Venturini, P., Borello, D., Iossa, C.V., Lentini, D., Rispoli, F.: Modelling of multiphase combustion and deposit formation and deposit formation in a biomass-fed boiler. Energy **35**, 3008–3021 (2010)
67. Armenio, V., Fiorotto, V.: The importance of the forces acting on particles in turbulent flows. Phys. Fluids **13**, 2437–2440 (2001)
68. Schiller, L., Naumann, A.: Uber die grundlegenden berechnungen bei der schwekraftaubereitung. Z. Ver. Dtsch. Ing. **77**, 318–320 (1933)
69. Smith, P.J.: 3-D turbulent particle dispersion submodel development. Quarterly progress report, Department of Energy, Pittsburgh Energy Technology Center (1991)

70. Corsini, A., Rispoli, F., Santoriello, A.: A variational multiscale high-order finite element formulation for turbomachinery flow computations. Comput. Methods Appl. Mech. Eng. **194**, 4797–4823 (2005)
71. Keegan, M.H., Nash, D.H., Stack, M.M.: Numerical modelling of hailstone impact on the leading edge of a wind turbine blade. EWEA Annual Wind Energy Event, Vienna (2013)
72. Evans, A.G., Gulden, M.E., Eggum, G.E., Rosenblatt, M.: Impact damage in brittle materials in the elastic response regime. Technical Report SC5023, Rockwell International Science Centre (1976)
73. Bazilevs, Y., Hsu, M.-C., Akkerman, I., Wright, S., Takizawa, K., Henicke, B., Spielman, T., Tezduyar, T.E.: 3D simulation of wind turbine rotors at full scale. Part I: geometry modeling and aerodynamics. Int. J. Numer. Methods Fluids **65**, 207–235 (2011). doi:10.1002/fld.2400
74. Hsu, M.-C., Akkerman, I., Bazilevs, Y.: Wind turbine aerodynamics using ALE-VMS: validation and role of weakly enforced boundary conditions. Comput. Mech. **50**, 499–511 (2012)
75. Hsu, M.-C., Bazilevs, Y.: Fluid-structure interaction modeling of wind turbines: simulating the full machine. Comput. Mech. **50**, 821–833 (2012)
76. Bazilevs, Y., Takizawa, K., Tezduyar, T.E.: Computational Fluid-Structure Interaction: Methods and Applications. Wiley, New York (2013). ISBN 978-0470978771
77. Bazilevs, Y., Hsu, M.-C., Takizawa, K., Tezduyar, T.E.: ALE-VMS and ST-VMS methods for computer modeling of wind-turbine rotor aerodynamics and fluid-structure interaction. Math. Models Methods Appl. Sci. **22**, 1230002 (2012). doi:10.1142/S0218202512300025
78. Takizawa, K., Tezduyar, T.E., McIntyre, S., Kostov, N., Kolesar, R., Habluetzel, C.: Space-time VMS computation of wind-turbine rotor and tower aerodynamics. Comput. Mech. **53**, 1–15 (2014). doi:10.1007/s00466-013-0888-x
79. Takizawa, K., Bazilevs, Y., Tezduyar, T.E., Hsu, M.-C., Øiseth, O., Mathisen, K.M., Kostov, N., McIntyre, S.: Engineering analysis and design with ALE-VMS and space-time methods. Arch. Comput. Meth. Eng. **21**, 481–508 (2014). doi:10.1007/s11831-014-9113-0
80. Bazilevs, Y., Takizawa, K., Tezduyar, T.E., Hsu, M.-C., Kostov, N., McIntyre, S.: Aerodynamic and FSI analysis of wind turbines with the ALE-VMS and ST-VMS methods. Arch. Comput. Meth. Eng. **21**, 359–398 (2014). doi:10.1007/s11831-014-9119-7
81. Takizawa, K., Tezduyar, T.E., Mochizuki, H., Hattori, H., Mei, S., Pan, L., Montel, K.: Space–time VMS method for flow computations with slip interfaces (ST-SI). Math. Models Methods Appl. Sci. **25**, 2377–2406 (2015). doi: 10.1142/S0218202515400126
82. Glauert, H.: Windmills and Fans. Springer, Berlin (1935)
83. Jonkman, J., Butterfield, S., Musial, W., Scott, G.: Definition of a 5-MW reference wind turbine for offshore system development. Technical Report NREL/TP-500-38060, National Renewable Energy Laboratory (2009)
84. NWTC Information Portal (FAST v7). https://nwtc.nrel.gov/FAST7 (2016)

The Multi-Moment Finite Volume Solver for Incompressible Navier-Stokes Equations on Unstructured Grids

Bin Xie and Feng Xiao

Abstract This chapter presents a robust and accurate finite volume method (FVM) for incompressible viscous fluid dynamics on unstructured grids. Different from conventional FVM where the volume integrated average (VIA) value is the only computational variable, the present formulation treats both VIA and the point value (PV) as the computational variables which are updated separately at each time step. The VIA is computed from a finite volume scheme of flux form, and is thus numerically conservative. The PV is updated from the differential form of the governing equation that does not have to be conservative but can be solved in a very efficient way. Including PV as the additional variable enables us to make higher-order reconstructions over compact mesh stencil to improve the accuracy, and moreover, the resulting numerical model is more robust for unstructured grids. The presented numerical framework provides an easy and straightforward approach to design schemes that well balance the numerical accuracy and computational cost.

Numerical results of several benchmark tests are also presented to verify the proposed numerical method as an accurate and robust solver for incompressible flows on unstructured grids.

1 Introduction

The finite volume method (FVM) has become the main stream approach in computational fluid dynamics (CFD). Due to its rigorous numerical conservativeness and flexibility to adapt both structured and unstructured grids, FVM has been applied to a wide spectrum of engineering applications in the presence of complex geometric configurations. However, developing high-order finite volume model on unstructured grids is not a trivial task. Conventional FVM requires wide stencil to generate high-order reconstructions as in the k-exact finite volume method [1] and the weighted essentially non-oscillatory method [7, 9]. As mentioned in [7], particular attention must be paid to choose the admissible stencils, or the

B. Xie • F. Xiao (✉)
Tokyo Institute of Technology, 4259 Nagatsuta Midori-ku, Yokohama 226-8502, Japan
e-mail: xie.b.aa@m.titech.ac.jp; xiao@es.titech.ac.jp

Y. Bazilevs, K. Takizawa (eds.), *Advances in Computational Fluid-Structure Interaction and Flow Simulation*, Modeling and Simulation in Science, Engineering and Technology, DOI 10.1007/978-3-319-40827-9_8

reconstruction on unstructured grids might be of illness otherwise. On the other hand, numerical schemes with compact stencil and more locally defined degrees of freedom (DOFs), such as the discontinuous Galerkin (DG) method [3, 4] and the spectral finite volume (SV) method [17], have gained great success in solving convection-dominated flows. However, not as many as implementations of these methods are found for the incompressible Navier-Stokes equations due to algorithmic complexity.

In this chapter, we present a novel numerical formulation, volume integrated average and point value based multi-moment (VPM) method, to solve incompressible unsteady Navier-Stokes equations. Adding the point value (PV) of the velocity field at the cell vertices as a new DOF in addition to the volume integrated average (VIA), we can construct higher-order polynomials over compact stencils for unstructured grids of arbitrary element shape. In comparison with the conventional FVM, with a modest increase in algorithmic complexity and computational cost, the VPM method achieves significant improvements in numerical accuracy and robustness.

2 Multi-Moment Reconstructions on Unstructured Grids

The computational domain is divided into non-overlapping elements Ω_i ($i = 1, 2, \cdots, I$) of different types, shown in Fig. 1 in 2 and 3 dimensions, for example. We denote the vertices at (x_{ik}, y_{ik}, z_{ik}) by θ_{ik}, $k = 1, 2, \cdots, K$, and K being the total number of vertices. The vertices are connected by boundary segments Γ_{ij} ($j = 1, 2, \cdots, J$) which are either a straight line (2D) or a plane (3D) separating elements Ω_i and Ω_{ij}, where the cell index "ij" is re-numbered in respect to the target cell Ω_i. The topological relation with its global index is realized through a connection table. We also denote the outward normal unit vector of segment Γ_{ij} by $\mathbf{n}_{ij} = (n_{xij}, n_{yij}, n_{zij})$.

Being the computational variables in the present formulation, the volume integrated average (VIA) and point value (PV) at the cell vertices, which are considered as different moments of physical field $\phi(x, y, z, t)$, are defined as

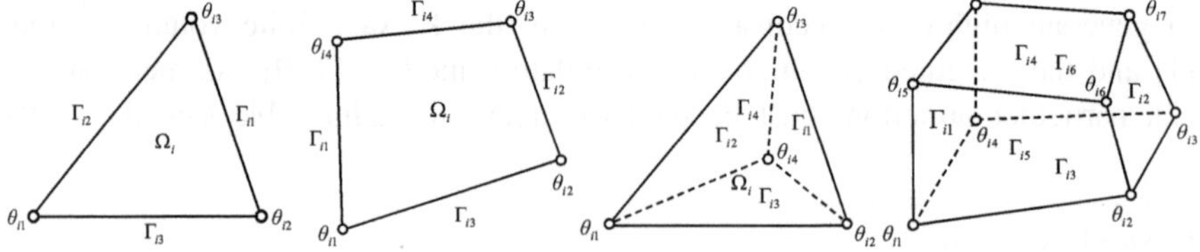

Fig. 1 Typical unstructured elements (from left to right): triangular, quadrilateral, tetrahedral, and hexahedral elements

$$\overline{\phi}_i(t) \equiv \frac{1}{|\Omega_i|}\int_{\Omega_i}\phi(x,y,z,t)d\Omega, \tag{1}$$

$$\phi_{ij}(t) \equiv \phi(x_{ik}, y_{ik}, z_{ik}, t),\ k = 1, 2, \cdots, K.$$

The piecewise reconstruction polynomial $\Phi(x, y.z)$ for physical field $\phi(x, y, z)$ on cell Ω_i is written in a form of basis function with respect to different moments as

$$\Phi_i(x,y,z) = \sum_{k=1}^{K}\phi_{ik}\Psi_{ik} + \overline{\phi}_i\overline{\Psi}_i + \sum_{\alpha=x,y,z}\phi_{\alpha i}\Psi_{\alpha i} + \sum_{\alpha,\beta=x,y,z}\phi_{\alpha\beta i}\Psi_{\alpha\beta i}, \tag{2}$$

where the PVs at the vertices, ϕ_{ij}, and the VIA, $\overline{\phi}_i$, as defined in (1), are the computational variables which are updated at every step. The derivatives at the cell center, $\phi_{\alpha i}$ and $\phi_{\alpha\beta i}$, are computed from the PVs and VIAs of the neighboring cells (see [20, 21] for details).

It is noted that the number of cell-wisely available DOFs might be different for different cell shapes. So, the reconstructed function (2) varies depending on the type of grid element. The reconstructed functions for the elements shown in Fig. 1 are given in [20, 21] with the basis function in local coordinates.

Once the multi-moment reconstruction function is made, we can use it to approximate the operators of the governing equations.

3 Numerical Formulation for Incompressible Navier-Stokes Equations

We consider the incompressible unsteady Navier-Stokes equations,

$$\nabla \cdot \mathbf{u} = 0, \tag{3}$$

$$\frac{\partial \mathbf{u}}{\partial t} + \nabla \cdot (\mathbf{u} \otimes \mathbf{u}) = -\frac{1}{\rho}\nabla p + \nu\nabla^2\mathbf{u}, \tag{4}$$

where $\mathbf{u} = (u, v, w)$ is the velocity vector with components u, v, and w in x, y, and z directions, respectively. p is the pressure, ρ the density, and ν the kinematic viscosity.

The projection method of Chorin [2] provides a simple and robust solution procedure for incompressible flow and is adopted in this paper. We briefly summarize the numerical steps to update the velocity from time level m ($t = t^m$) to $m + 1$ ($t = t^m + \Delta t$) as follows:

1. Given the velocity field $\mathbf{u}^m$ at step m, compute the intermediate velocity $\mathbf{u}^*$ from the momentum equation (4) without the pressure gradient term,

$$\frac{\partial \mathbf{u}}{\partial t} = -\nabla \cdot (\mathbf{u}^m \otimes \mathbf{u}^m) + \nu\nabla^2\mathbf{u}^m. \tag{5}$$

2. The intermediate velocity $\mathbf{u}^*$ usually does not satisfy the mass conservation equation (3), and must be corrected by the projection shown in the next step. For that purpose, we solve the pressure field at step $(m+1)$ from the following Poisson equation,

$$\nabla \cdot \left(\frac{1}{\rho} \nabla p^{m+1} \right) = \frac{1}{\Delta t} \nabla \cdot \mathbf{u}^*. \tag{6}$$

3. Correct the velocity by the projection step,

$$\frac{\mathbf{u}^{m+1} - \mathbf{u}^*}{\Delta t} = -\frac{1}{\rho} \nabla p^{m+1}. \tag{7}$$

It is straightforward to show that the updated velocity satisfies $\nabla \cdot \mathbf{u}^{m+1} = 0$.

Our major interest is to develop an approach for the spatial discretization for the solution procedure by using multi-moment reconstruction formulation shown above.

3.1 The Scheme for Advection-Diffusion Equation

We write (5) as an advection-diffusion equation,

$$\frac{\partial \phi}{\partial t} = -\nabla \cdot (\phi \mathbf{u}) + \nu \nabla^2 \phi, \tag{8}$$

where ϕ is the physical quantity, which stands for either u, v, or w in Navier-Stokes equations.

Making use of the underlying idea of the CIP (Constrained Interpolation Profile) method [22, 23] where different moments are updated simultaneously in time, two types of moments, i.e., the volume integrated average (VIA) and the point value (PV) of the physical field ϕ, are treated as the prognostic variables and updated separately. Their governing equations can be immediately obtained from definition (1) and (8), i.e.,

$$\frac{d\overline{\phi}_i}{dt} = -\frac{1}{|\Omega_i|} \oint_{\Gamma_i} (\phi \mathbf{u} \cdot \mathbf{n}) \, d\Gamma + \frac{\nu}{|\Omega_i|} \oint_{\Gamma_i} (\mathbf{n} \cdot \nabla \phi) \, d\Gamma \tag{9}$$

for VIA moment and

$$\frac{d\phi_{ij}}{dt} = -\mathbf{u}_{ij} \cdot (\nabla \phi)_{ij} + \nu (\nabla^2 \phi)_{ij}, \tag{10}$$

for PV moment. Note that we have used Gauss divergence theorem and assumed a constant kinematic viscosity to yield (9).

Next, we describe the spatial discretization for the terms on the right-hand sides of (9) and (10).

The total advection flux in (10) is approximated by summing up the fluxes across each boundary segments,

$$\oint_{\Gamma_i} (\phi \mathbf{u} \cdot \mathbf{n})\, d\Gamma = \sum_{j=1}^{J} \left(|\Gamma_{ij}| \overline{\phi \mathbf{u}}_{\Gamma_{ij}} \cdot \mathbf{n}_{ij} \right) \approx \sum_{j=1}^{J} \left(|\Gamma_{ij}| \overline{\phi}_{\Gamma_{ij}} \overline{\mathbf{u}}_{\Gamma_{ij}} \cdot \mathbf{n}_{ij} \right), \tag{11}$$

where $|\Gamma_{ij}|$ represents the area of cell surface Γ_{ij}, $\overline{\phi}_{\Gamma_{ij}}$ the average value of ϕ on Γ_{ij}, and $\overline{\mathbf{u}}_{\Gamma_{ij}} = (\overline{u}_{\Gamma_{ij}}, \overline{v}_{\Gamma_{ij}}, \overline{w}_{\Gamma_{ij}})$ the corrected divergence-free velocity.

$$\overline{\phi}_{\Gamma_{ij}} = \frac{1}{|\Gamma_{ij}|} \int_{\Gamma_{ij}} \Phi_{\mathrm{iup}}(x, y, z) d\Gamma, \tag{12}$$

where Φ_{iup} denotes the reconstruction function in (2) over the upwinding cell, i.e.,

$$\mathrm{iup} = \begin{cases} = i, \ \text{for } \mathbf{n}_{ij} \cdot \overline{\mathbf{u}}_{\Gamma_{ij}} > 0; \\ = ij, \ \text{otherwise.} \end{cases} \tag{13}$$

The total diffusion flux in (10) for cell Ω_i is approximated by

$$\oint_{\Gamma_i} (\mathbf{n} \cdot \nabla \phi)\, d\Gamma = \sum_{j=1}^{J} \left(|\Gamma_{ij}| \left(\overline{\phi}_{x\Gamma_{ij}} n_{xij} + \overline{\phi}_{y\Gamma_{ij}} n_{yij} + \overline{\phi}_{z\Gamma_{ij}} n_{zij} \right) \right), \tag{14}$$

where $\left(\overline{\phi}_{x\Gamma_{ij}}, \overline{\phi}_{y\Gamma_{ij}}, \overline{\phi}_{z\Gamma_{ij}} \right)$ represent the averaged values of (ϕ_x, ϕ_y, ϕ_z) on surface Γ_{ij}, which are computed from the reconstruction functions Φ_i and Φ_{ij} over two neighboring cells sharing surface segment Γ_{ij}, i.e.,

$$\overline{\phi}_{\alpha\Gamma_{ij}} = \frac{1}{2} \left(\frac{1}{|\Gamma_{ij}|} \int_{\Gamma_{ij}} \Phi_{\alpha i}(x, y, z) d\Gamma + \frac{1}{|\Gamma_{ij}|} \int_{\Gamma_{ij}} \Phi_{\alpha ij}(x, y, z) d\Gamma \right) \tag{15}$$

where α denotes either x, y, or z.

The PV moment at the vertices (ϕ_{ik}) is predicted from the differential form equation (10), where the gradients of the advection terms are computed from a weighted upwinding scheme,

$$\left(\frac{\partial \phi}{\partial \alpha} \right)_{ik} = \sum_{l=1}^{L} \omega_{ikl} \Phi_{\alpha ikl}(\theta_{ik}), \tag{16}$$

for derivative with respect to $\alpha = x, y$, or z. Ω_{ikl} $(l = 1, 2, \cdots, L)$ represent the union of cells that share vertex θ_{ik}, and Φ_{ikl} the reconstruction functions (2) on cell Ω_{ikl}. The weight ω_{ikl} is computed by

$$\omega_{ikl} = \frac{\max\left(0, \mathbf{u}_{ik} \cdot \overrightarrow{\theta_{cikl}\theta_{ik}}\right)}{\sum_{l=1}^{L} \max\left(0, \mathbf{u}_{ik} \cdot \overrightarrow{\theta_{cikl}\theta_{ik}}\right)}, \tag{17}$$

where $\overrightarrow{\theta_{cikl}\theta_{ik}}$ denotes the vector from the center of cell Ω_{ikl} to vertex θ_{ik}. It is clear from (17) that the gradient computed in the upwinding cell is used.

Using (16), the PVs can be computed very efficiently once the piecewise reconstructions (2) are made.

Given the net diffusion fluxes calculated by (14) for all cells, Ω_{ikl}, which share vertex θ_{ik}, the point-wise value of the diffusion term at θ_{ik} can be simply obtained from the time-evolution converting (TEC) formula described in [18, 19].

To be more precise, the time tendency of VIA in this case is computed by

$$\delta_t \overline{\phi}_i = \frac{\nu}{|\Omega_i|} \oint_{\Gamma_{ij}} (\mathbf{n} \cdot \nabla \phi) \, d\Gamma, \tag{18}$$

which is already obtained by (14). The time tendency of the PV due to the Laplace operator is then computed from the TEC formula described in [18, 19], which reads

$$\nu(\nabla^2 \phi)_{ik} = \delta_t \phi_{ik} = TEC(\delta_t \overline{\phi}_{ikl}). \tag{19}$$

3.2 *The Scheme for Pressure Poisson Equation*

In the present formulation, we only use the VIA moment for pressure. The Poisson equation (6) is recast in an integral form,

$$\oint_{\Gamma_i} \left(\frac{1}{\rho} \mathbf{n} \cdot \nabla p \right) d\Gamma = \frac{1}{\Delta t} \oint_{\Gamma_i} (\mathbf{n} \cdot \mathbf{u}^*) \, d\Gamma. \tag{20}$$

Discretizing (20) over cell Ω_i yields

$$\sum_{j=1}^{J} \left(\frac{|\Gamma_{ij}|}{\rho_{ij}} \left(\frac{\partial \overline{p}}{\partial r} \right)_{ij} \mathbf{e}_{ij} \cdot \mathbf{n}_{ij} \right) = \frac{1}{\Delta t} \sum_{j=1}^{J} \left(|\Gamma_{ij}| (\overline{u}^*_{\Gamma_{ij}} n_{xij} + \overline{v}^*_{\Gamma_{ij}} n_{yij} + \overline{w}^*_{\Gamma_{ij}} n_{zij}) \right), \tag{21}$$

where $(\overline{u}^*_{\Gamma_{ij}}, \overline{v}^*_{\Gamma_{ij}}, \overline{w}^*_{\Gamma_{ij}})$ on the right-hand side is calculated as

$$\overline{\phi}_{\Gamma_{ij}} = \frac{1}{2} \left(\frac{1}{|\Gamma_{ij}|} \int_{\Gamma_{ij}} \phi_i(x, y, z) d\Gamma + \frac{1}{|\Gamma_{ij}|} \int_{\Gamma_{ij}} \phi_{ij}(x, y, z) d\Gamma \right) \tag{22}$$

where ϕ denotes u^*, v^*, or w^* component of immediate velocity $\mathbf{u}^*$. The pressure gradient is approximated by a linear interpolation spanned over two neighboring cells sharing edge Γ_{ij}. We denote $(\frac{\partial \overline{p}}{\partial r})_{ij}\mathbf{e}_{ij}$ as the pressure gradient along line segment $\mathbf{r}_{ij} = \mathbf{r}(\theta_{cij}) - \mathbf{r}(\theta_{ci})$ with $\mathbf{r}(\theta_{cij}) = (x_{cij}, y_{cij}, z_{cij})$ and $\mathbf{r}(\theta_{ci}) = (x_{ci}, y_{ci}, z_{cij})$ being the position vectors of the centroids of cells Ω_{ij} and Ω_i, respectively. The unit vector of $\mathbf{r}_{ij}$ is denoted by $\mathbf{e}_{ij} = (e_{xij}, e_{yij}, e_{zij}) = \mathbf{r}_{ij}/|\mathbf{r}_{ij}|$.

As derived in [21], we calculate the orthogonal component implicitly using the values at level $m+1$ and the non-orthogonal correction explicitly with the values at level m,

$$\begin{aligned}\left(\frac{\partial \overline{p}}{\partial r}\right)_{ij}\mathbf{e}_{ij}\cdot\mathbf{n}_{ij} &\approx \frac{1}{\mathbf{e}_{ij}\cdot\mathbf{n}_{ij}}\left(\frac{\partial \overline{p}}{\partial r}\right)_{ij}^{m+1} + \left(\frac{\partial \overline{p}}{\partial r}\right)_{ij}^{m}\mathbf{e}_{ij}\cdot\mathbf{n}_{\parallel ij} \\ &= \frac{1}{\mathbf{e}_{ij}\cdot\mathbf{n}_{ij}}\frac{(\overline{p}_{ij}^{m+1} - \overline{p}_i^{m+1})}{e_{xij}(x_{cij}-x_{ci}) + e_{yij}(y_{cij}-y_{ci}) + e_{zij}(z_{cij}-z_{ci})} \\ &+ \frac{(\overline{p}_{ij}^{m} - \overline{p}_i^{m})}{e_{xij}(x_{cij}-x_{ci}) + e_{yij}(y_{cij}-y_{ci}) + e_{zij}(z_{cij}-z_{ci})}\mathbf{n}_{\parallel ij}\cdot\mathbf{e}_{ij}.\end{aligned} \tag{23}$$

Thus, Poisson equation (21) for pressure is finally written as

$$\sum_{j=1}^{J}\frac{\alpha_{ij}|\Gamma_{ij}|}{\mathbf{e}_{ij}\cdot\mathbf{n}_{ij}}(\overline{p}_{ij}^{m+1} - \overline{p}_i^{m+1}) = \beta_i \tag{24}$$

where

$$\alpha_{ij} = \frac{1}{\rho_{ij}\left(e_{xij}(x_{cij}-x_{ci}) + e_{yij}(y_{cij}-y_{ci}) + e_{zij}(z_{cij}-z_{ci})\right)}$$

and

$$\beta_i = \sum_{j=1}^{J}\left(|\Gamma_{ij}|\left(\frac{1}{\Delta t}(\overline{u}^*_{\Gamma_{ij}}n_{xij} + \overline{v}^*_{\Gamma_{ij}}n_{yij} + \overline{w}^*_{\Gamma_{ij}}n_{zij}) - \frac{1}{\rho_{ij}}\alpha_{ij}(\overline{p}_{ij}^{m} - \overline{p}_i^{m})\mathbf{n}_{\parallel ij}\cdot\mathbf{e}_{ij}\right)\right).$$

Equation (24) is a simultaneously linked linear algebraic equation for pressure $\overline{p}_i^{m+1}$ and can be solved by iterative solvers.

3.3 Projection of the Velocity Field

As the final step in the solution procedure, the velocity field must be projected onto a divergence-free subset. From the pressure field computed above, we firstly evaluate the pressure gradients on cell surfaces $\left(\overline{\nabla p}\right)_{\Gamma_{ij}}$, and then find the volume-averaged pressure gradient $\left(\overline{\nabla p}\right)_i = \left(\frac{\partial p}{\partial x}, \frac{\partial p}{\partial y}, \frac{\partial p}{\partial z}\right)_i$ to update the VIA of velocity field.

According to [6, 8], the volume-average pressure gradient $\left(\overline{\nabla p}\right)_i$ should be calculated from surrounding surface pressure gradient $\left(\overline{\nabla p}\right)_{\Gamma_{ij}}$ by surface-volume weighted least square method that minimizes

$$\sum_{j=1}^{J}\left(\left(\overline{\nabla p}\right)_i\cdot\mathbf{n}_{ij}-\left(\overline{\nabla p}\right)_{\Gamma_{ij}}\cdot\mathbf{n}_{ij}\right)^2\left|\Gamma_{ij}\right|,$$

which yields

$$\sum_{j=1}^{J}\left|\Gamma_{ij}\right|\begin{pmatrix} n_{xij}n_{xij} & n_{xij}n_{yij} & n_{xij}n_{zij}\\ n_{xij}n_{yij} & n_{yij}n_{yij} & n_{yij}n_{zij}\\ n_{xij}n_{zij} & n_{yij}n_{zij} & n_{zij}n_{zij}\end{pmatrix}\cdot\begin{pmatrix}\overline{\frac{\partial p}{\partial x}}\\ \overline{\frac{\partial p}{\partial y}}\\ \overline{\frac{\partial p}{\partial z}}\end{pmatrix}_i=\sum_{j=1}^{J}\begin{pmatrix}\left|\Gamma_{ij}\right|n_{xij}\left(\overline{\frac{\partial p}{\partial x}}\right)_{\Gamma_{ij}}^{m+1}\\ \left|\Gamma_{ij}\right|n_{yij}\left(\overline{\frac{\partial p}{\partial y}}\right)_{\Gamma_{ij}}^{m+1}\\ \left|\Gamma_{ij}\right|n_{zij}\left(\overline{\frac{\partial p}{\partial z}}\right)_{\Gamma_{ij}}^{m+1}\end{pmatrix}.$$

The volume-average pressure gradients $\left(\overline{\frac{\partial p}{\partial x}},\overline{\frac{\partial p}{\partial y}},\overline{\frac{\partial p}{\partial z}}\right)_i$ are then obtained. Given the pressure gradients, we updated the volume average velocity as

$$\overline{u}_i^{m+1}=\overline{u}_i^*-\frac{\Delta t}{\rho}\left(\overline{\frac{\partial p}{\partial x}}\right)_i,\tag{25}$$

$$\overline{v}_i^{m+1}=\overline{v}_i^*-\frac{\Delta t}{\rho}\left(\overline{\frac{\partial p}{\partial y}}\right)_i,\tag{26}$$

$$\overline{w}_i^{m+1}=\overline{w}_i^*-\frac{\Delta t}{\rho}\left(\overline{\frac{\partial p}{\partial z}}\right)_i.\tag{27}$$

Once the time increments of VIA for velocity field are known, we update the PVs of velocity at the cell vertices by the time-evolution converting (TEC) formula [21],

$$u_{ik}^{m+1}=u_{ik}^*+TEC\left(\overline{u}_{ikl}^{m+1}-\overline{u}_{ikl}^*\right),\tag{28}$$

$$v_{ik}^{m+1}=v_{ik}^*+TEC\left(\overline{v}_{ikl}^{m+1}-\overline{v}_{ikl}^*\right).\tag{29}$$

$$w_{ik}^{m+1}=w_{ik}^*+TEC\left(\overline{w}_{ikl}^{m+1}-\overline{w}_{ikl}^*\right).\tag{30}$$

4 Numerical Results

Numerical experiments were carried out to evaluate the numerical method presented in this chapter. We quantify the numerical errors by following norms:

$$E(L_1) = \frac{\sum_{i=1}^{N_e}(|\phi_{ni} - \phi_{ei}||\Omega_i|)}{\sum_{i=1}^{N_e}(|\phi_{ei}||\Omega_i|)} \text{ and } E(L_2) = \sqrt{\frac{\sum_{i=1}^{N_e}((\phi_{ni} - \phi_{ei})^2|\Omega_i|)}{\sum_{i=1}^{N_e}(\phi_{ei}^2|\Omega_i|)}}, \quad (31)$$

where ϕ_{ni} and ϕ_{ei} denote numerical and exact solutions, respectively.

To evaluate the convergence rate of the advection scheme presented in section 3.1, we computed the advection transport of a sine function as in [9] with gradually refined grids. A 3rd-order Runge-Kutta time integration scheme [14] was used. The L_1 and L_2 errors and the convergence rates are given in Table 1 for grids of different resolutions. We see a uniform 3rd-order convergence rate on different unstructured grids.

To evaluate the accuracy of the whole incompressible fluid solver, we computed the Taylor vortex problem [6, 13]. Numerical errors and convergence rates for both velocity and pressure fields are shown in Fig. 2 in comparison with the results with conventional finite volume method with QUICK[12], superbee-TVD [15], and MUSCL [16] schemes. We observe that VPM method is much more accurate than the conventional FVM models. For example, VPM results on a coarse mesh with 3604 cells have less numerical errors than those of FVM models computed on a much finer grid (57670 cells). We also show the comparison of computational cost in Table 2, which reveals that VPM model gains significant improvement in numerical accuracy at a modest increase of computational cost in terms of both elapse time and required memory.

As a three-dimensional test, we simulated viscous flows passing a sphere with different Reynolds numbers, which is widely used to evaluate numerical codes for incompressible viscous flows. We plot the streamlines on the cross section cutting through the sphere center for different Reynolds numbers in Fig. 3 which visually agree well with those in [10]. The quantitative verification is given in Table 3 for the Reynolds numbers of 250 and 300 over which the flow transits from a steady non-axisymmetric regime to an unsteady periodical regime. It is observed that the drag and lift coefficients for both cases are in excellent agreement with the existing researches, and so is Strouhal number for 300 Reynolds number case.

Table 1 Numerical errors and convergence orders of the advection scheme

Triangle element			Quadrilateral element		
Cell number	L_1 error	L_1 order	Cell number	L1 error	L_1 order
226	1.288×10^{-1}	–	100	1.429×10^{-1}	—
894	1.767×10^{-2}	2.89	400	1.815×10^{-2}	2.98
3588	2.183×10^{-3}	3.01	1600	2.261×10^{-3}	2.99
14412	2.732×10^{-4}	2.99	6400	2.821×10^{-4}	3.00
57518	3.513×10^{-5}	2.96	25600	3.536×10^{-5}	3.00

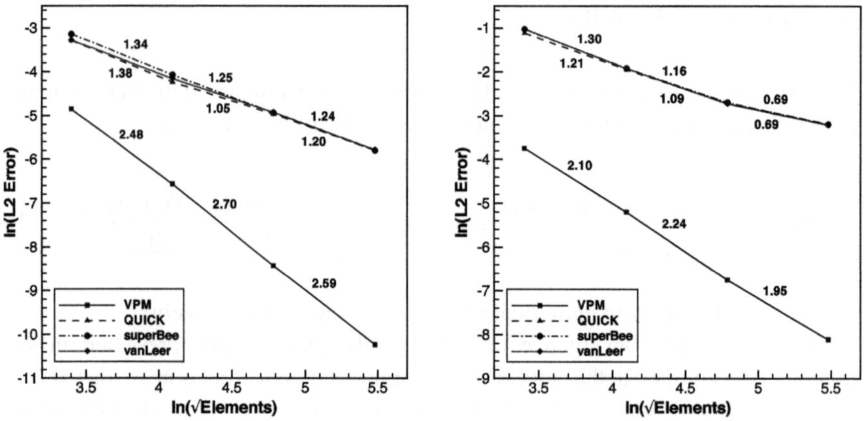

Fig. 2 The numerical error and convergence rate of velocity (left) and pressure (right) for Taylor vortex problem

5 Summary

We have developed and tested a novel numerical solver for incompressible Navier-Stokes equations on unstructured grids of different shapes of elements. The numerical formulation, the so-called volume integrated average and point value based multi-moment (VPM) method, adds the point values at the vertices of each grid element as new computational variables besides the volume average values in the conventional finite volume method. The present formulation gains a great deal in numerical accuracy and robustness at a modest increase in computational complexity and cost in comparison with the conventional finite volume method. Being well balanced among solution quality and computational cost, the present formulation can be expected as a promising framework for practical applications.

Table 2 Comparisons between VPM and conventional FVM models in terms of computational cost and numerical accuracy

	QUICK				VPM			
Elements	DOFs	Time(s)	Error of U	Error of p	DOFs	Time(s)	Error of U	Error of p
902	902	4.24	5.379×10^{-2}	5.781×10^{-1}	1394	5.03	7.851×10^{-3}	2.819×10^{-2}
3604	3604	12.43	2.219×10^{-2}	2.436×10^{-1}	5487	16.27	1.410×10^{-3}	6.603×10^{-3}
14362	14362	47.78	6.920×10^{-3}	6.743×10^{-2}	21704	68.25	2.183×10^{-4}	1.194×10^{-3}
57670	57670	226.32	1.642×10^{-3}	2.582×10^{-2}	86826	317.6	4.121×10^{-5}	3.779×10^{-4}

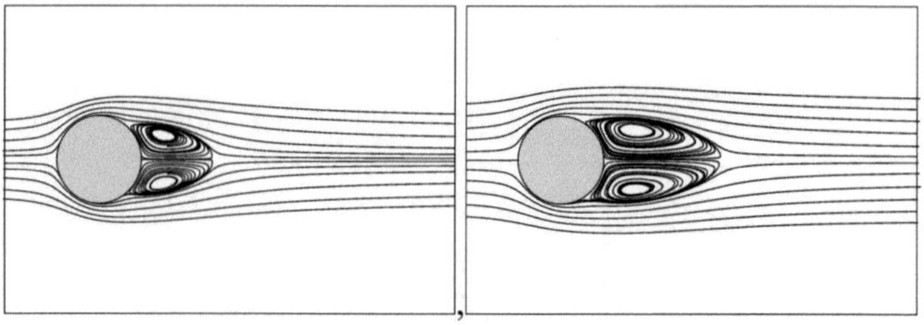

Fig. 3 Streamlines of 3D viscous flows passing a sphere for Re=100 (left) and 200 (right)

Table 3 Comparison of the numerical results for viscous flows of different Reynolds numbers passing a sphere

	Re	C_D	C_L	St
Present scheme	250	0.698	0.060	
	300	0.654	0.065	0.135
Ref. [11]	250	0.701	0.059	
	300	0.657	0.067	0.134
Ref. [10]	250	0.70	0.062	
	300	0.656	0.069	0.137
Ref. [5]	250	0.70	0.062	
	300	0.655	0.065	0.136

Acknowledgements This work was supported in part by JSPS KAKENHI (Grant Numbers 24560187, 15H03916, and 15J09915).

References

1. Barth, T.J., Frederickson, P.O.: High-order solution of the Euler equations on unstructured grids using quadratic reconstruction. AIAA Paper No. 90-0013 (1990)
2. Chorin, A.J.: Numerical solution of the Navier–Stokes equations. Math. Comput. **22**, 745–762 (1968)
3. Cockburn, B., Shu, C.W.: TVB Runge–Kutta local projection discontinuous Galerkin finite element method for conservation laws II: general framework. Math. Comput. **52**, 411–435 (1989)
4. Cockburn, B., Shu, C.W.: The Runge–Kutta discontinuous Galerkin method for conservation laws V: the multidimensional systems. J. Comput. Phys. **141**, 199–224 (1998)
5. Constantinescuand, G.S., Squires, K.D.: LES and DES investigations of turbulent flow over a sphere. AIAA Paper No. 2000–0540 (2000)
6. Ferrer, E., Willden, R.: A high order discontinuous Galerkin finite element solver for the incompressible Navier-Stokes equations. Comput. Fluids **46**, 224–230 (2011)
7. Friedrich, O.: Weighted essentially non-oscillatory schemes for the interpolation of mean values on unstructured grids. J. Comput. Phys. **144**, 194–212 (1998)
8. Herrmann, M.: A balanced force refined level set grid method for two-phase flows on unstructured flow solver grids. J. Comput. Phys. **227**, 2674–2706 (2008)
9. Hu, C., Shu, C.-W.: Weighted essentially non-oscillatory schemes on triangular meshes, J. Comput. Phys. **150**, 97–127 (1999)

10. Johnson, T.A., Patel, V.C.: Flow past a sphere up to a Reynolds number of 300. J. Fluid Mech. **378**, 19–70 (1999)
11. Kim, J., Kim, D., Choi, H.: An immersed-boundary finite-volume method for simulations of flow in complex geometries. J. Comput. Phys. **171**, 132–150 (2001)
12. Leonard, B.P.: A stable and accurate convective modelling procedure based on quadratic upstream interpolation. Comput. Methods Appl. Mech. Eng. **19**, 59–98 (1979)
13. Shahbazi, K., Fischer, P., Ethier, C.: A high-order discontinuous Galerkin method for the unsteady incompressible Navier-Stokes equations. J. Comput. Phys. **222**, 391–407 (2007)
14. Shu, C.W.: Total-variation-diminishing time discretizations. SIAM J. Sci. Stat. Comput. **9**, 1073–1084 (1988)
15. Sweby, P.K.: High resolution schemes using flux limiters for hyperbolic conservation laws. SIAM J. Numer. Anal. **21** 995–1011 (1984)
16. Van Leer, B.: Towards the ultimate conservative difference scheme. IV. A new approach to numerical convection. J. Comput. Phys. **23**, 276–299 (1977)
17. Wang, Z.J., Liu, Y.: Spectral (finite) volume method for conservation laws on unstructured grids II: extension to two-dimensional scalar equation. J. Comput. Phys. **179**, 665–697 (2002)
18. Xiao, F.: Unified formulation for compressible and incompressible flows by using multi integrated moment method: one-dimensional inviscid compressible flow. J. Comput. Phys. **195**, 629–654 (2004)
19. Xiao, F., Akoh, R., Ii, S.: Unified formulation for compressible and incompressible flows by using multi-integrated moments II: multi-dimensional version for compressible and incompressible flows. J. Comput. Phys. **213**, 31–56 (2006)
20. Xie, B., Xiao, F.: Two and three dimensional multi-moment finite volume solver for incompressible Navier–Stokes equations on unstructured grids with arbitrary quadrilateral and hexahedral elements. Comput. Fluids **104**, 40–54 (2014)
21. Xie, B., Ii, S., Ikebata, A., Xiao, F.: A multi-moment finite volume method for incompressible Navier–Stokes equations on unstructured grids: volume-average/point-value formulation. J. Comput. Phys. **227**, 138–162 (2014)
22. Yabe, T., Aoki, T.: A universal solver for hyperbolic-equations by cubic-polynomial interpolation. 1. One-dimensional solver. Comput. Phys. Commun. **66**, 219–232 (1991)
23. Yabe, T., Xiao, F., Utsumi, T.: The constrained interpolation profile method for multiphase analysis. J. Comput. Phys. **169**, 556–593 (2001)

An Immersogeometric Method for the Simulation of Turbulent Flow Around Complex Geometries

Fei Xu, David Kamensky, Vasco Varduhn, Chenglong Wang, Sean A. Wasion, Bryann Sotomayor-Rinaldi, Carolyn N. Darling, Dominik Schillinger, and Ming-Chen Hsu

Abstract In this chapter we summarize a recently proposed immersogeometric method for the simulation of incompressible flow around geometrically complex objects. The method immerses the objects into unfitted tetrahedral finite elements meshes and weakly enforces Dirichlet boundary conditions on the surfaces of the objects. Adaptively refined quadrature rules are used to faithfully capture the flow domain geometry in the discrete problem without modifying the unfitted finite element mesh. A variational multiscale formulation which provides accuracy and robustness in both laminar and turbulent flow conditions is employed. We assess the accuracy of the method by analyzing the flow around an immersed sphere for a wide range of Reynolds numbers. We show that flow quantities of interest are in very good agreement with reference values obtained from standard boundary-fitted approaches. Our results also show that the faithful representation of the geometry in intersected elements is critical for accurate flow analysis. We demonstrate the potential of our proposed method for high-fidelity industrial scale simulations by performing an aerodynamic analysis of a full-scale agricultural tractor.

F. Xu • C. Wang • S.A. Wasion • B. Sotomayor-Rinaldi • C.N. Darling • M.-C. Hsu (✉)
Department of Mechanical Engineering, Iowa State University, 2025 Black Engineering, Ames, IA 50011, USA
e-mail: jmchsu@iastate.edu

D. Kamensky
Institute for Computational Engineering and Sciences, The University of Texas at Austin, 201 East 24th St, Stop C0200, Austin, TX 78712, USA

V. Varduhn • D. Schillinger
Department of Civil, Environmental, and Geo-Engineering, University of Minnesota, 500 Pillsbury Drive S.E., Minneapolis, MN 55455, USA

Y. Bazilevs, K. Takizawa (eds.), *Advances in Computational Fluid-Structure Interaction and Flow Simulation*, Modeling and Simulation in Science, Engineering and Technology, DOI 10.1007/978-3-319-40827-9_9

1 Introduction

Immersed methods approximate the solution of boundary value problems on analysis meshes that do not necessarily conform to the boundary of the domain. Such methods have greater geometric flexibility than their boundary-fitted counterparts and circumvent the meshing obstacles that frequently impede analysis of problems posed on geometrically complex domains. In the context of finite elements [1], several variants of immersed methods for fluids have been explored over the last decade. Löhner et al. [2] adapted kinetic and kinematic enforcement of boundary conditions used in immersed boundary methods [3–5] for use in adaptive nodal finite element grids. Glowinski et al. [6] simulated viscous flow interacting with rigid particles by forcing the rigid body motion in each particle subdomain onto the overlapping fluid field via a distributed Lagrange multiplier field. Zhang et al. [7] proposed the immersed finite element method (IFEM) to use a flexible Lagrangian solid mesh that moves on top of a background Eulerian fluid mesh.

In addition, several researchers designed immersed methods that resolve immersed boundaries and introduce weak coupling schemes for velocity and stress fields directly at the interface. Baaijens [8] and Parussini [9] combined the fictitious domain approach with Lagrange multiplier fields at the interface for immersed thin and volumetric structures. Gerstenberger and Wall [10] combined Lagrange multiplier fields with interface enrichments of the velocity and pressure fields in the sense of the extended finite element method [11] to ensure the separation of physical and fictitious domains. Rüberg and Cirak [12, 13] combined weak Nitsche-type coupling methods at the interface with Cartesian B-spline finite elements for moving boundary and FSI problems.

This work presents an immersed method for solving incompressible flow on tetrahedral finite element meshes. The proposed method combines a variational multiscale (VMS) formulation of incompressible flow [14, 15], consistent weak enforcement of boundary conditions [16, 17], and a geometrically accurate representation of the fluid domain in the integration of the variational problem on elements that straddle the domain boundary. We emphasize the implications of the latter, highlighting the importance of accurately describing the geometry in intersected elements for obtaining accurate flow solutions. We follow our previous work [18] in denoting immersed methods that accurately represent the geometry of the domain as *immersogeometric* methods.

A pioneering instantiation of the immersogeometric concept is the finite cell method (FCM) [19, 20]. The FCM represents the geometry of the domain in intersected elements by adaptive quadrature points, such that the geometric accuracy can be increased by adding additional levels of quadrature points, if needed. Since its inception, significant efforts have been invested to further develop the FCM. Technical improvements include the weak imposition of boundary and coupling conditions [21], local refinement schemes [22], large deformation analysis [23], and improved quadrature rules for intersected elements [24]. A concise summary of the FCM and related developments and applications can be found in [25].

This chapter is organized as follows. In Sect. 2, we describe the variational problem under consideration and its discrete formulation. We also detail the implementation of a key technical component—the tetrahedral FCM. Section 3 focuses on the canonical benchmark of the flow around a sphere. In Sect. 4, we apply our method to the flow analysis of the full-scale tractor, illustrating the potential of immersogeometric methods for high-fidelity aerodynamic analysis of complex geometry problems. Section 5 draws conclusions and motivates future work.

2 Variational Problem and Implementations

2.1 Variational Multiscale Formulation with Weakly Enforced Boundary Conditions

Consider a collection of disjoint elements $\{\Omega^e\}$, $\cup_e \Omega^e \subset \mathbb{R}^d$, with closures covering the fluid domain: $\Omega \subset \cup_e \overline{\Omega^e}$. Note that Ω^e is not necessarily a subset of Ω. Let $\mathcal{V}_u^h$ and $\mathcal{V}_p^h$ be the discrete velocity and pressure spaces of functions supported on these elements. The VMS discretization of incompressible Navier–Stokes equations is stated as: Find $\mathbf{u}^h \in \mathcal{V}_u^h$ and $p^h \in \mathcal{V}_p^h$ such that for all $\mathbf{w}^h \in \mathcal{V}_u^h$ and $q^h \in \mathcal{V}_p^h$:

$$B^{\text{VMS}}\left(\{\mathbf{w}^h, q^h\}, \{\mathbf{u}^h, p^h\}\right) - F^{\text{VMS}}\left(\{\mathbf{w}^h, q^h\}\right) = 0, \tag{1}$$

where

$$\begin{aligned} B^{\text{VMS}}\left(\{\mathbf{w}^h, q^h\}, \{\mathbf{u}^h, p^h\}\right) &= \int_\Omega \mathbf{w}^h \cdot \rho \left(\frac{\partial \mathbf{u}^h}{\partial t} + \mathbf{u}^h \cdot \nabla \mathbf{u}^h\right) d\Omega \\ &+ \int_\Omega \boldsymbol{\varepsilon}(\mathbf{w}^h) : \boldsymbol{\sigma}\left(\mathbf{u}^h, p^h\right) d\Omega + \int_\Omega q^h \, \nabla \cdot \mathbf{u}^h d\Omega \\ &- \sum_e \int_{\Omega^e \cap \Omega} \left(\mathbf{u}^h \cdot \nabla \mathbf{w}^h + \frac{\nabla q^h}{\rho}\right) \cdot \mathbf{u}' d\Omega - \sum_e \int_{\Omega^e \cap \Omega} p' \, \nabla \cdot \mathbf{w}^h d\Omega \\ &+ \sum_e \int_{\Omega^e \cap \Omega} \mathbf{w}^h \cdot (\mathbf{u}' \cdot \nabla \mathbf{u}^h) d\Omega - \sum_e \int_{\Omega^e \cap \Omega} \frac{\nabla \mathbf{w}^h}{\rho} : \left(\mathbf{u}' \otimes \mathbf{u}'\right) d\Omega \\ &+ \sum_e \int_{\Omega^e \cap \Omega} \left(\mathbf{u}' \cdot \nabla \mathbf{w}^h\right) \overline{\tau} \cdot \left(\mathbf{u}' \cdot \nabla \mathbf{u}^h\right) d\Omega, \end{aligned} \tag{2}$$

and

$$F^{\text{VMS}}\left(\{\mathbf{w}^h, q^h\}\right) = \int_\Omega \mathbf{w}^h \cdot \rho \mathbf{f} d\Omega + \int_{\Gamma^N} \mathbf{w}^h \cdot \mathbf{h} d\Gamma. \tag{3}$$

In (2), $\mathbf{u}'$ is defined as

$$\mathbf{u}' = -\tau_{\mathrm{M}} \left(\rho \left(\frac{\partial \mathbf{u}^h}{\partial t} + \mathbf{u}^h \cdot \nabla \mathbf{u}^h - \mathbf{f} \right) - \nabla \cdot \boldsymbol{\sigma} \left(\mathbf{u}^h, p^h \right) \right) \tag{4}$$

and p' is given by

$$p' = -\rho\, \tau_{\mathrm{C}}\, \nabla \cdot \mathbf{u}^h. \tag{5}$$

In the above equations, the terms integrated over element interiors may be interpreted both as stabilization and as a turbulence model [15, 26–30]. τ_{M}, τ_{C}, and $\overline{\tau}$ are the stabilization parameters. Their detailed expression used in this work can be found in [31].

The standard way of imposing Dirichlet boundary conditions in (1) is to enforce them strongly by ensuring that they are satisfied by all trial solution functions. This is not feasible in immersed methods. We replace the strong enforcement by weakly enforced Dirichlet boundary conditions in the sense of Nitsche's method proposed by Bazilevs et al. [16, 32]. The semi-discrete problem becomes

$$\begin{aligned}
& B^{\mathrm{VMS}} \left(\{\mathbf{w}^h, q^h\}, \{\mathbf{u}^h, p^h\} \right) - F^{\mathrm{VMS}} \left(\{\mathbf{w}^h, q^h\} \right) \\
& - \int_{\Gamma^D} \mathbf{w}^h \cdot \left(-p^h \mathbf{n} + 2\mu\, \boldsymbol{\varepsilon}(\mathbf{u}^h)\, \mathbf{n} \right) d\Gamma - \int_{\Gamma^D} \left(2\mu\, \boldsymbol{\varepsilon}(\mathbf{w}^h)\, \mathbf{n} + q^h \mathbf{n} \right) \cdot \left(\mathbf{u}^h - \mathbf{g} \right) d\Gamma \\
& - \int_{\Gamma^{D,-}} \mathbf{w}^h \cdot \rho \left(\mathbf{u}^h \cdot \mathbf{n} \right) \left(\mathbf{u}^h - \mathbf{g} \right) d\Gamma \\
& + \int_{\Gamma^D} \tau_{\mathrm{TAN}}^B \left(\mathbf{w}^h - \left(\mathbf{w}^h \cdot \mathbf{n} \right) \mathbf{n} \right) \cdot \left(\left(\mathbf{u}^h - \mathbf{g} \right) - \left(\left(\mathbf{u}^h - \mathbf{g} \right) \cdot \mathbf{n} \right) \mathbf{n} \right) d\Gamma \\
& + \int_{\Gamma^D} \tau_{\mathrm{NOR}}^B \left(\mathbf{w}^h \cdot \mathbf{n} \right) \left(\left(\mathbf{u}^h - \mathbf{g} \right) \cdot \mathbf{n} \right) d\Gamma \; = \; 0.
\end{aligned} \tag{6}$$

In the above equation, $\Gamma^{D,-}$ is the *inflow* part of Γ^D, on which $\mathbf{u}^h \cdot \mathbf{n} < 0$. τ_{TAN}^B and τ_{NOR}^B are stabilization parameters that need to be chosen element-wise as a compromise between the following two requirements. If they are too large, they assume a penalty-type character, affecting the conditioning of the stiffness matrix and overshadowing the variational consistency. If they are too small, the solution of (6) is unstable.

An advantage of weakly enforced Dirichlet boundary conditions is the release of the point-wise no-slip condition at the boundary of the fluid domain. Although maybe counter-intuitive at first sight, this violation of the no-slip boundary condition is in fact desirable, as it imitates the presence of a boundary layer [17, 32]. For immersogeometric methods, weakly enforced boundary conditions are particularly attractive as the additional Nitsche terms in (6) are formulated independently of the mesh. In contrast to strong enforcement, which relies on boundary-fitted meshes to impose Dirichlet boundary conditions on the discrete solution space, the Nitsche terms in (6) also hold for intersected elements, where the domain boundary does not coincide with element boundaries.

2.2 Implementation of the Tetrahedral Finite Cell Method

Our immersogeometric method largely draws on the FCM [25], which is based on the fictitious domain concept illustrated in Fig. 1. Its main idea is to extend the original fluid domain to a more tractable shape then discretize the embedding domain into elements irrespective of the geometric boundary of potentially complex embedded objects. This introduces elements that are intersected by the geometric boundary, which creates complex integration domains in intersected elements. The present contribution applied the FCM with unstructured meshes of tetrahedral elements to simulations of incompressible flow, where the flexibility of unstructured tetrahedral meshes is useful for boundary layer refinement.

To ensure the geometrically accurate evaluation of volume integrals in intersected elements, we adapt the sub-cell based adaptive quadrature scheme of the Cartesian FCM [20] to the tetrahedral case. The basic concept is to increase quadrature points around immersed geometric boundaries, so that arbitrary integration domains resulting from the intersecting boundary can be taken into account accurately. This is achieved by recursively splitting intersected cells into sub-tetrahedra. At each level, only those sub-tetrahedra intersected by the boundary are further split. For clarity, we illustrate the quadrature scheme based on adaptive sub-cells for triangles in 2D in Fig. 1. We emphasize that splitting is performed on the quadrature level only and does not affect the basis functions, which are still defined on the original tetrahedral element.

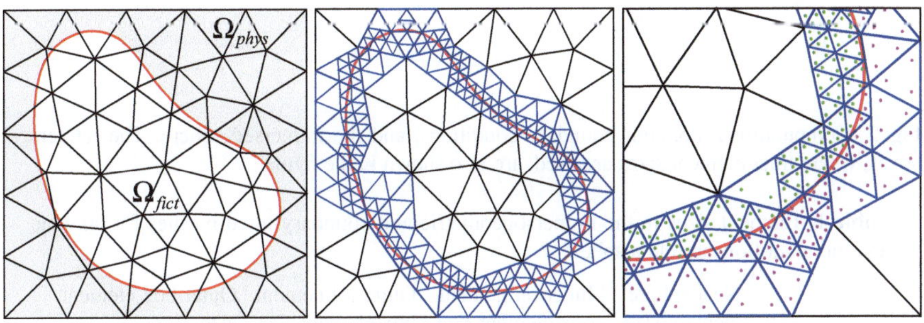

Fig. 1 The concept of physical and fictitious domains and quadrature scheme based on adaptive sub-cells (blue lines). Quadrature points within the fluid domain (marked in pink) are used in the numerical integration. Quadrature points outside (marked in green) are discarded

3 Benchmark Example: Flow Around a Sphere

We use the flow around a sphere at Reynolds numbers $Re = 100$ and 3700 as a benchmark to assess the accuracy of our immersogeometric method in both laminar and turbulent flow regimes. We compute reference results using the same variational framework with standard boundary-fitted tetrahedral meshes that are comparable in terms of overall mesh resolution and boundary layer mesh grading. Figure 2 illustrates the dimensions of the computational domain, the location and size of the immersed sphere, and the boundary conditions. The radius of the sphere, the inflow velocity, and the fluid density are all one, and the Reynolds number is the inverse of the viscosity. The inlet boundary condition and the slip boundary condition on the lateral faces are strongly enforced, while the no-slip/no-penetration condition $\mathbf{u} = \mathbf{0}$ on the surface of the sphere is enforced weakly. The element sizes used in the immersogeometric and boundary-fitted mesh generations for laminar and turbulent flow cases are shown in Table 1.

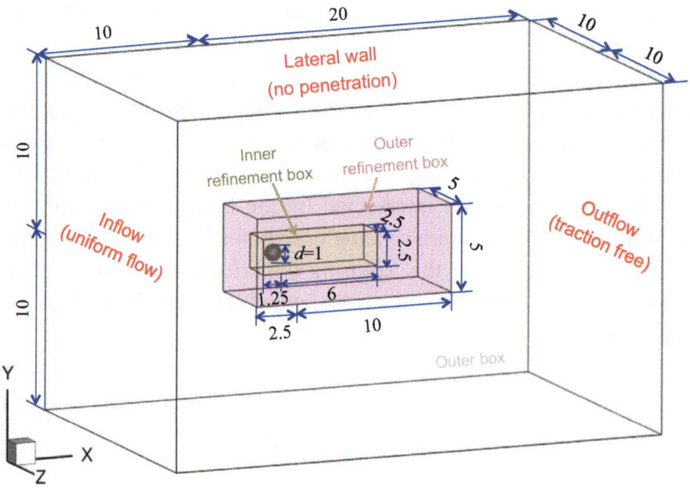

Fig. 2 Computational domain, boundary conditions, and the immersed sphere. The refinement boxes, where finer element sizes are used, are also shown in the figure

Table 1 Element sizes in the immersogeometric and boundary-fitted meshes for laminar and turbulent flow around a sphere

	Near sphere element size	Inner refinement box element size	Outer refinement box element size	Outer box element size
Laminar	0.005	0.05	$0.2/\sqrt{2}$	0.8
Turbulent	0.004	0.04	$0.16/\sqrt{2}$	0.8

3.1 Immersogeometric Results for Laminar Flow

Figure 3 shows the drag coefficient C_D, computed with our immersogeometric method and different levels of adaptive quadrature sub-cells, for flow around a sphere at $Re = 100$. The results are compared with the boundary-fitted reference value. Taking more sub-cell levels into account increases the accuracy of the domain integration, which is directly linked to geometric accuracy. From the results in Fig. 3, we clearly see that an increased geometric resolution is crucial to achieve accurate flow solutions.

Increasing the number of adaptive sub-cell levels becomes expensive for larger l, because the number of quadrature points increases exponentially [23]. We observe that from $l = 0$ to $l = 1$ there is a significant improvement. We still obtain an improvement from $l = 1$ to $l = 2$, but the difference between $l = 2$ and $l = 4$ is very small. Taking into account the increase in computing time (see Fig. 3), we conclude that $l = 2$ levels of adaptive sub-cells represent a good balance between computational cost and geometric accuracy for the present immersogeometric method.

3.2 Immersogeometric Results for Turbulent Flow

For assessing the accuracy of our immersogeometric method for turbulent flows, we carry out the computation of flow around a sphere at $Re = 3700$. Figure 4 shows a visualization of the immersogeometric result of instantaneous vortical structures, which illustrates the three-dimensional chaotic nature of turbulent flow in the wake of the sphere. Time-averaged quantities of interest are reported in Table 2, computed

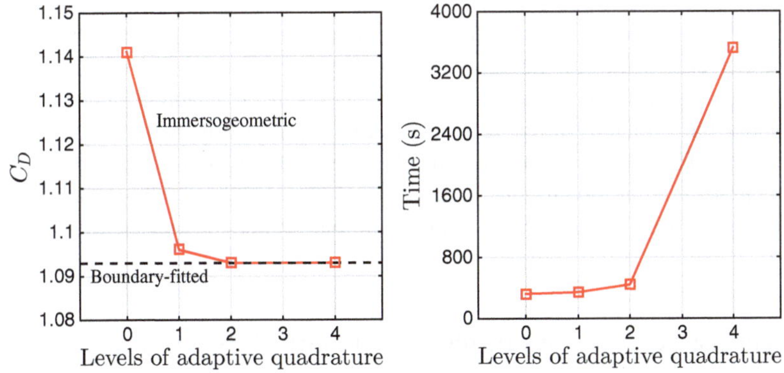

Fig. 3 Drag coefficient C_D and computing time required to run 50 steps computed with different levels of adaptive quadrature sub-cells for flow around a sphere at $Re = 100$. Dashed line is the reference C_D computed with boundary-fitted mesh

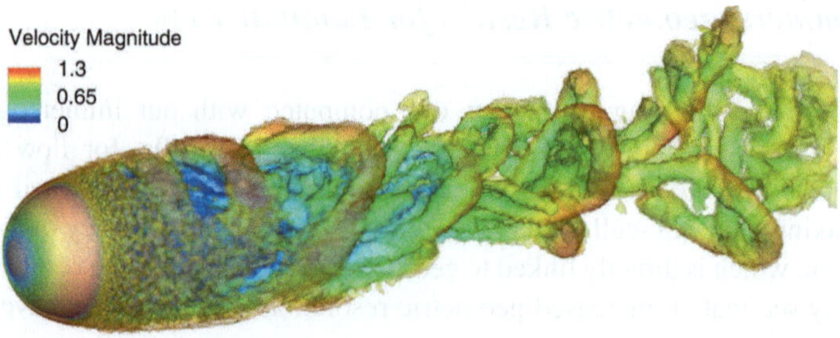

Fig. 4 Visualization of the instantaneous vortical structures for flow at $Re = 3700$, colored by velocity magnitude

Table 2 Comparison of the time-averaged drag coefficient $\overline{C}_D$, Strouhal number $\overline{St}$, non-dimensional length $\overline{L}/d$ of the recirculation bubble, and pressure coefficient $\overline{C}_{pb}$ at an azimuthal angle of 180°for turbulent flow around a sphere at $Re = 3700$

	$\overline{C}_D$	$\overline{L}/d$	$\overline{St}$	$\overline{C}_{pb}$
Immersogeometric ($l = 0$)	0.399	2.26	0.205	−0.254
Immersogeometric ($l = 1$)	0.397	2.26	0.208	−0.258
Immersogeometric ($l = 2$)	0.393	2.27	0.218	−0.217
Boundary-fitted	0.393	2.27	0.217	−0.215
DNS (Rodriguez et al. [33])	0.394	2.28	0.215	−0.207
VMS (Bazilevs et al. [34])	0.392	2.28	0.221	−0.207

with our immersogeometric and boundary-fitted methods, and compare them with reference values in [33, 34]. We observe that the immersogeometric results converge to the boundary-fitted reference values when l is increased from 0 to 2, i.e., under the refinement of adaptive quadrature sub-cells. We also find that all quantities obtained with $l = 2$ are in good agreement with the values reported in the literature.

Figures 5 shows the distribution of the time-averaged pressure coefficient over the upper crown line of the sphere along the main flow direction. The pressure coefficient distribution obtained with $l = 2$ levels of adaptive quadrature sub-cells fits the boundary-fitted reference curve significantly better than the other two cases. This confirms that a faithful representation of the geometry in terms of accurate volume quadrature in intersected elements is a key requirement for obtaining accurate turbulent flow results with our immersogeometric method.

4 Industrial Scale Example: Turbulent Flow Around a Tractor

Typical vehicle designs lead to very complex fluid domain boundaries. This constitutes a major obstacle for the transfer of fluid domains into boundary-fitted computational meshes. An example is the agricultural tractor shown in Fig. 6,

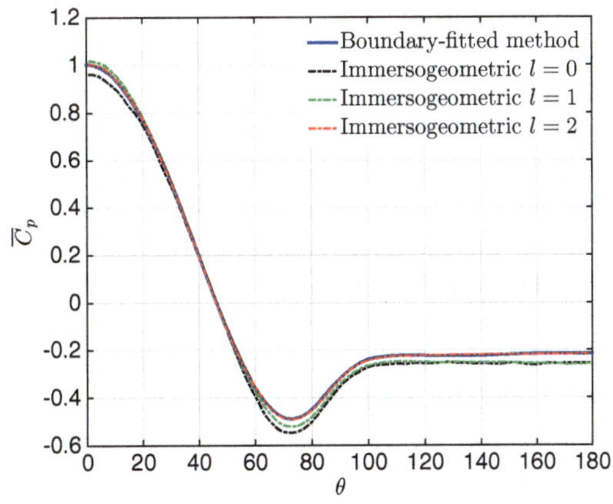

Fig. 5 Distribution of pressure coefficient over the upper crown line of a sphere at $Re = 3700$

Fig. 6 Time-averaged pressure distribution over the tractor surface, computed with the immersogeometric mesh

which incorporates many geometrically complex details. In this section, we use the tractor to demonstrate how immersion of complex geometries into an unfitted mesh can alleviate many challenges of standard boundary-fitted mesh generation in the context of large-scale industrial applications.

4.1 Generating Immersogeometric and Boundary-Fitted Meshes

We generate an adaptive immersogeometric mesh of the tractor using an open-source mesh generator Gmsh [35], leveraging its refinement capability. A zoom of the final immersogeometric mesh is shown in Fig. 7. We locally refine the tetrahedral mesh close to the tractor surface for capturing boundary layers. In all intersected elements we add two levels of adaptive quadrature sub-cells to accurately integrate the volume integrals. To obtain simulation results based on standard boundary-fitted meshes for comparison, we use a commercial mesh generator ANSA [36] due to its robustness in generating boundary-fitted meshes for complex geometries. We ensure that the local refinement pattern close to the tractor and ground surfaces and in the wake of the tractor is comparable to the immersogeometric mesh. A zoom of our boundary-fitted mesh can be seen in Fig. 7.

We note that a significant advantage of the immersogeometric workflow is its geometric flexibility. For example, it enables us to impose a uniform mesh size along the immersed tractor surface regardless of fine-scale geometric features. We can therefore easily control the mesh size resolution independently of the geometry, for example, to obtain a coarser mesh for fast preliminary design studies. This is not possible in boundary-fitted analysis, where a coarser mesh requires geometry operations first to remove all geometric features that are of finer scale than the targeted minimum element length.

4.2 Comparison of Immersogeometric and Boundary-Fitted Results

We specify a uniform inflow with a streamwise velocity of 11.176 m/s ahead of the tractor. The air density and dynamic viscosity are 1.177 kg/m^3 and 1.846×10^{-5} kg/(m·s), respectively. The Reynolds number, based on the streamwise length of the tractor, is approximately 3×10^6. Figure 8 shows the instantaneous vortical structures of the highly turbulent flow around the tractor. Snapshots of the velocity magnitude on planar cuts at different heights above the ground are shown in Figure 9. Detailed flow features such as flow around the pipe and mirrors can be clearly seen. Figure 6 shows the static pressure over the tractor surface. We observe that the pressure distribution over geometric details of the tractor surface is captured well.

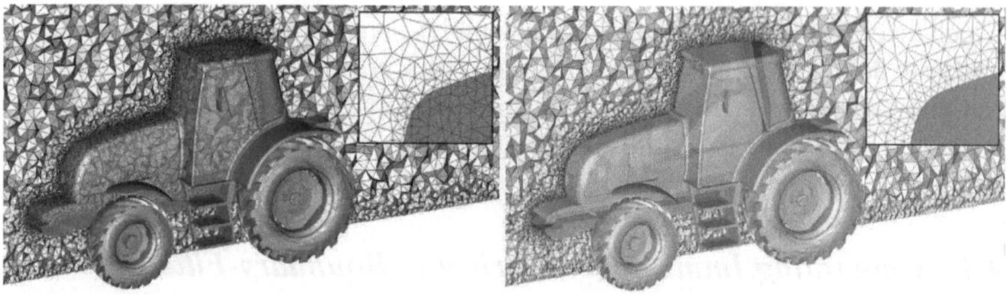

Fig. 7 Locally refined tetrahedral meshes of the fluid domain for aerodynamic analysis of the tractor. Left: immersogeometric mesh. Right: standard boundary-fitted mesh

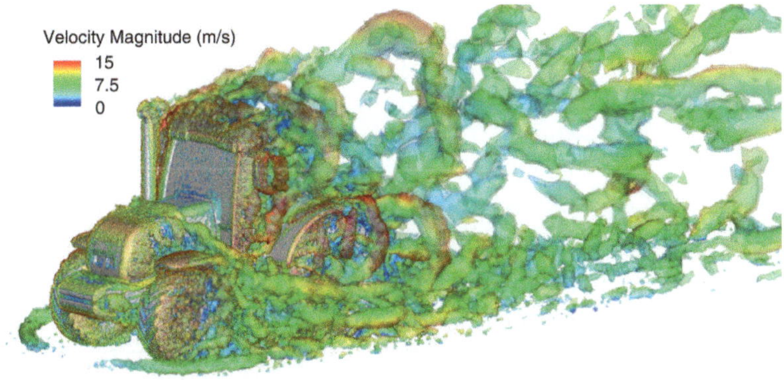

Fig. 8 Visualization of the immersogeometric result of instantaneous vortical structures for turbulent flow around the tractor, colored by velocity magnitude

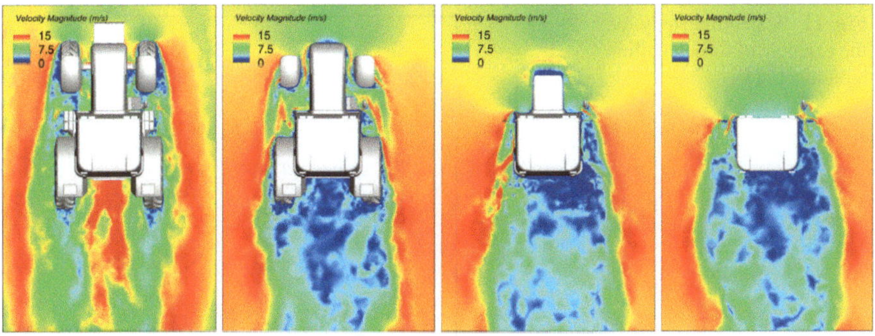

Fig. 9 Snapshots of the instantaneous velocity magnitude at different heights above the ground

To assess the accuracy of the immersogeometric results, we plot the distribution of the time-averaged pressure coefficient $\overline{C}_P = 2(\overline{P} - P_\infty)/\rho U^2$ along curves over the tractor top and bottom surfaces in Fig. 10. Overall good agreement between the immersogeometric and boundary-fitted methods is observed. This shows that our immersogeometric method is able to achieve accurate flow solutions near the boundary of an immersed object, where all elements are intersected, for high Reynolds number turbulent flow problems.

The use of integration sub-cells improves the immersogeometric solution quality, but comes at the cost of additional computation during the finite element assembly procedures. The additional computations are in the form of quadrature over elements, and do not require any communication between subdomains. The impact of this additional cost on wall clock time may therefore be mitigated by simply partitioning the mesh into more subdomains (each assigned to a processor core). To demonstrate this, we perform a scalability study of our immersogeometric method using the tractor example presented in this section. The computations are carried out on the Lonestar Linux cluster [37]. The results are shown in Figure 11

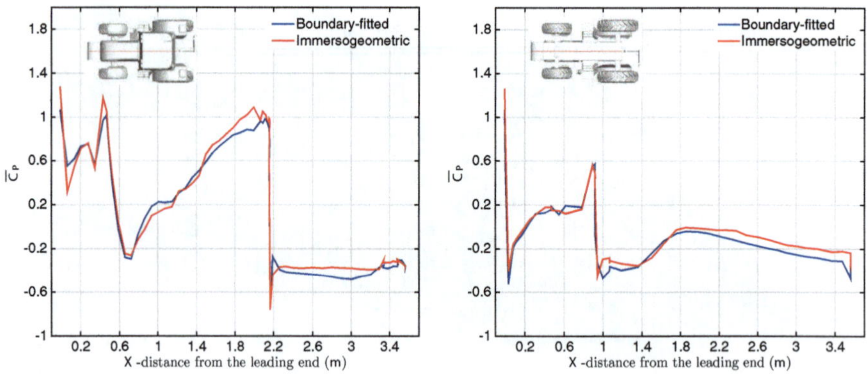

Fig. 10 Time-averaged pressure coefficient $\overline{C}_P$ plotted along two curves over the tractor top (left) and bottom (right) surfaces

Fig. 11 Scalability study for the example of turbulent flow around a tractor

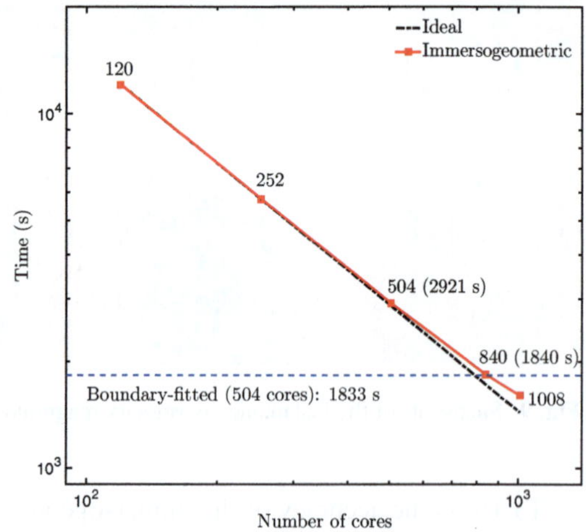

and demonstrate nearly ideal scaling. Although the immersogeometric case takes more wall clock time than the boundary-fitted case when 504 partitions are used for both, equivalent times may be achieved by simply partitioning the immersogeometric mesh into more subdomains. As the cost of access to supercomputing resources drops, it is not computer time but rather the time of human analysts—which is often spent on mesh design—that dominates overall analysis cost. We therefore believe that it is useful to investigate numerical methods that require moderate increases in computation but drastically simplify mesh generation.

5 Conclusions and Future Work

In this chapter, we presented an immersogeometric framework for analyzing incompressible flows around geometrically complex objects immersed in unfitted tetrahedral finite element meshes. The main components of this framework are the variational multiscale method, the weak enforcement of boundary conditions, an adaptive quadrature scheme for the integration of intersected elements, and the local refinement of areas with boundary layers. We examined in detail two representative example problems: flow around a sphere and aerodynamic analysis of a tractor. We showed that the immersogeometric solutions were in good agreement with reference solutions. We also demonstrated that such agreement is *not* achieved without the faithful representation of surface geometry provided by our approach. The tractor analysis indicates that our immersogeometric method can greatly simplify the mesh generation process for industrial turbulent flow problems without sacrificing accuracy.

Some future research directions in immersogeometric CFD include development of advanced quadrature schemes for intersected elements that are geometrically faithful and computationally efficient, improved weak boundary and coupling conditions that limit the dependence on stabilization parameters and maintain a good conditioning of the system matrix, boundary layer refinement strategies with anisotropic meshes, and efficient treatment of moving immersed boundaries. The last of these points is of particular importance to immersogeometric FSI analysis [18, 38–40].

References

1. Hughes, T.J.R.: The Finite Element Method. Linear Static and Dynamic Finite Element Analysis. Prentice-Hall, Englewood Cliffs, NJ (1987)
2. Löhner, R., Baum, J.D., Mestreau, E., Sharov, D., Charman, C., Pelessone, D.: Adaptive embedded unstructured grid methods. Int. J. Numer. Methods Eng. **60**(3), 641–660 (2004)
3. Peskin, C.S.: The immersed boundary method. Acta Numer. **11**, 479–517 (2002)
4. Mittal, R., Iaccarino, G.: Immersed boundary methods. Annu. Rev. Fluid Mech. **37**, 239–261 (2005)
5. Sotiropoulos, F., Yang, X.: Immersed boundary methods for simulating fluid–structure interaction. Prog. Aerosp. Sci. **65**, 1–21 (2014)
6. Glowinski, R., Pan, T.-W., Hesla, T.I., Joseph, D.D.: A distributed Lagrange multiplier/fictitious domain method for particulate flows. Int. J. Multiphase Flow **25**(5), 755–794 (1999)
7. Zhang, L., Gerstenberger, A., Wang, X., Liu, W.K.: Immersed finite element method. Comput. Methods Appl. Mech. Eng. **193**, 2051–2067 (2004)
8. Baaijens, F.P.T.: A fictitious domain/mortar element method for fluid–structure interaction. Int. J. Numer. Methods Fluids **35**(7), 743–761 (2001)
9. Parussini, L.: Fictitious domain approach via Lagrange multipliers with least squares spectral element method. J. Sci. Comput. **37**(3), 316–335 (2008)
10. Gerstenberger, A., Wall, W.A.: Enhancement of fixed-grid methods towards complex fluid–structure interaction applications. Int. J. Numer. Methods Fluids **57**, 1227–1248 (2008)

11. Moës, N., Dolbow, J., Belytschko, T.: A finite element method for crack growth without remeshing. Int. J. Numer. Methods Eng. **46**, 131–150 (1999)
12. Rüberg, T., Cirak, F.: Subdivision-stabilised immersed B-spline finite elements for moving boundary flows. Comput. Methods Appl. Mech. Eng. **209–212**, 266–283 (2012)
13. Rüberg, T., Cirak, F.: A fixed-grid b-spline finite element technique for fluid–structure interaction. Int. J. Numer. Methods Fluids **74**(9), 623–660 (2014)
14. Hughes, T.J.R., Mazzei, L., Jansen, K.E.: Large eddy simulation and the variational multiscale method. Comput. Vis. Sci. **3**, 47–59 (2000)
15. Bazilevs, Y., Calo, V.M., Cottrel, J.A., Hughes, T.J.R., Reali, A., Scovazzi, G.: Variational multiscale residual-based turbulence modeling for large eddy simulation of incompressible flows. Comput. Methods Appl. Mech. Eng. **197**, 173–201 (2007)
16. Bazilevs, Y., Hughes, T.J.R.: Weak imposition of Dirichlet boundary conditions in fluid mechanics. Comput. Fluids **36**, 12–26 (2007)
17. Hsu, M.-C., Akkerman, I., Bazilevs, Y.: Wind turbine aerodynamics using ALE–VMS: validation and the role of weakly enforced boundary conditions. Comput. Mech. **50**, 499–511 (2012)
18. Kamensky, D., Hsu, M.-C., Schillinger, D., Evans, J.A., Aggarwal, A., Bazilevs, Y., Sacks, M.S., Hughes, T.J.R.: An immersogeometric variational framework for fluid–structure interaction: application to bioprosthetic heart valves. Comput. Methods Appl. Mech. Eng. **54**(4), 1055–1071 (2014)
19. Parvizian, J., Düster, A., Rank, E.: Finite cell method: *h*- and *p*- extension for embedded domain methods in solid mechanics. Comput. Mech. **41**, 122–133 (2007)
20. Düster, A., Parvizian, J., Yang, Z., Rank, E.: The finite cell method for three-dimensional problems of solid mechanics. Comput. Methods Appl. Mech. Eng. **197**, 3768–3782 (2008)
21. Ruess, M., Schillinger, D., Bazilevs, Y., Varduhn, V., Rank, E.: Weakly enforced essential boundary conditions for NURBS-embedded and trimmed NURBS geometries on the basis of the finite cell method. Int. J. Numer. Methods Eng. **95**(10), 811–846 (2013)
22. Schillinger, D., Rank, E.: An unfitted *hp* adaptive finite element method based on hierarchical B-splines for interface problems of complex geometry. Comput. Methods Appl. Mech. Eng. **200**(47–48), 3358–3380 (2011)
23. Schillinger, D., Ruess, M., Zander, N., Bazilevs, Y., Düster, A., Rank, E.: Small and large deformation analysis with the *p*- and B-spline versions of the finite cell method. Comput. Mech. **50**(4), 445–478 (2012)
24. Yang, Z., Ruess, M., Kollmannsberger, S., Düster, A., Rank, E.: An efficient integration technique for the voxel-based finite cell method. Int. J. Numer. Methods Eng. **91**, 457–471 (2012)
25. Schillinger, D., Ruess, M.: The finite cell method: a review in the context of higher-order structural analysis of CAD and image-based geometric models. Arch. Comput. Meth. Eng. **22**(3), 391–455 (2015)
26. Brooks, A.N., Hughes, T.J.R.: Streamline upwind/Petrov-Galerkin formulations for convection dominated flows with particular emphasis on the incompressible Navier–Stokes equations. Comput. Methods Appl. Mech. Eng. **32**, 199–259 (1982)
27. Tezduyar, T.E.: Stabilized finite element formulations for incompressible flow computations. Adv. Appl. Mech. **28**, 1–44 (1992)
28. Tezduyar, T.E., Osawa, Y.: Finite element stabilization parameters computed from element matrices and vectors. Comput. Methods Appl. Mech. Eng. **190**, 411–430 (2000)
29. Hughes, T.J.R., Mazzei, L., Oberai, A.A., Wray, A.: The multiscale formulation of large eddy simulation: decay of homogeneous isotropic turbulence. Phys. Fluids **13**, 505–512 (2001)
30. Hsu, M.-C., Bazilevs, Y., Calo, V.M., Tezduyar, T.E., Hughes, T.J.R.: Improving stability of stabilized and multiscale formulations in flow simulations at small time steps. Comput. Methods Appl. Mech. Eng. **199**, 828–840 (2010)
31. Xu, F., Schillinger, D., Kamensky, D., Varduhn, V., Wang, C., Hsu, M.-C.: The tetrahedral finite cell method for fluids: Immersogeometric analysis of turbulent flow around complex geometries. Comput. Fluids (2015). http://dx.doi.org/10.1016/j.compfluid.2015.08.027

32. Bazilevs, Y., Michler, C., Calo, V.M., Hughes, T.J.R.: Isogeometric variational multiscale modeling of wall-bounded turbulent flows with weakly enforced boundary conditions on unstretched meshes. Comput. Methods Appl. Mech. Eng. **199**, 780–790 (2010)
33. Rodriguez, I., Borell, R., Lehmkuhl, O., Perez Segarra, C.D., Oliva, A.: Direct numerical simulation of the flow over a sphere at $Re = 3700$. J. Fluid Mech. **679**, 263–287 (2011)
34. Bazilevs, Y., Yan, J., de Stadler, M., Sarkar, S.: Computation of the flow over a sphere at $Re = 3700$: A comparison of uniform and turbulent inflow conditions. J. Appl. Mech. **81**(12), 121003 (2014)
35. Geuzaine, C., Remacle, J.-F.: Gmsh: A 3-D finite element mesh generator with built-in pre-and post-processing facilities. Int. J. Numer. Methods Eng. **79**(11), 1309–1331 (2009)
36. BETA CAE Systems S.A. – ANSA pre-processor: http://www.beta-cae.gr/ansa.htm (2015)
37. Texas Advanced Computing Center (TACC): http://www.tacc.utexas.edu (2015)
38. Hsu, M.-C., Kamensky, D., Bazilevs, Y., Sacks, M.S., Hughes, T.J.R.: Fluid–structure interaction analysis of bioprosthetic heart valves: significance of arterial wall deformation. Comput. Mech. **54**(4), 1055–1071 (2014)
39. Hsu, M.-C., Kamensky, D., Xu, F., Kiendl, J., Wang, C., Wu, M.C.H., Mineroff, J., Reali, A., Bazilevs, Y., Sacks, M.S.: Dynamic and fluid–structure interaction simulations of bioprosthetic heart valves using parametric design with T-splines and Fung-type material models. Comput. Mech. **55**, 1211–1225 (2015)
40. Kamensky, D., Evans, J.A., Hsu, M.-C.: Stability and conservation properties of collocated constraints in immersogeometric fluid–thin structure interaction analysis. Commun. Comput. Phys. (2015). http://dx.doi.org/10.4208/cicp.150115.170415s

Part II
CFD: Moving Boundaries and Interfaces

Numerical Simulation of the Behavior of a Rising Bubble by an Energy-Stable Lagrange-Galerkin Scheme

Masahisa Tabata

Abstract We consider two-fluid flow problems, where each fluid is governed by the Navier-Stokes equations and the surface tension proportional to the curvature acts on the interface. The domain which each fluid occupies is unknown, and the interface moves with the velocity of the particle on it. We have developed an energy-stable Lagrange-Galerkin finite element scheme for the two-fluid flow problems. It maintains not only the advantages of Lagrange-Galerkin method of the robustness to high-Reynolds numbers and of the symmetry of the resultant matrix but also the property of energy-stability under the condition of the smoothness of the interface. Here we perform numerical simulation of the behavior of a rising bubble by the scheme.

1 Introduction

We consider two-fluid flow problems, where each fluid is governed by the Navier-Stokes equations and the surface tension proportional to the curvature acts on the interface. The domain which each fluid occupies is unknown, and the interface moves with the velocity of the particle on it. While numerical solution of one-fluid flow problems governed by the Navier-Stokes equations has been successfully established from the point of stability and convergence, it is still not an easy task to construct numerical schemes solving the two-fluid flow problems. To the best of our knowledge there are no numerical schemes whose solutions are proved to converge to the exact one and there is very little discussion even for the stability of schemes [1]. As for the study from the engineering point of view we refer to [6, 12] and the bibliography therein. Recently we have developed an energy-stable Lagrange-Galerkin finite element scheme for the two-fluid flow problems [11]. The scheme is an extension of the energy-stable finite element scheme proposed by us [9, 10] to the Lagrange-Galerkin method. It maintains not only the advantages of

M. Tabata (✉)
Department of Mathematics, Waseda University, 3-4-1 Ohkubo,
Shinjuku, Tokyo 169-8555, Japan
e-mail: tabata@waseda.ac.jp

Y. Bazilevs, K. Takizawa (eds.), *Advances in Computational Fluid-Structure Interaction and Flow Simulation*, Modeling and Simulation in Science, Engineering and Technology, DOI 10.1007/978-3-319-40827-9_10

Lagrange-Galerkin method of the robustness to high-Reynolds numbers and of the symmetry of the resultant matrix but also the property of energy-stability under the condition of the smoothness of the interface. Here we perform numerical simulation of the behavior of a rising bubble by the scheme.

2 An Energy-Stable Lagrange-Galerkin Scheme

Let Ω be a bounded domain in $\mathbf{R}^2$ with piecewise smooth boundary Γ, and $(0, T)$ a time interval. The domain Ω is occupied by $m + 1$ immiscible incompressible viscous fluids. Each fluid k, whose density and viscosity are ρ_k and μ_k, occupies an unknown domain $\Omega_k(t)$ at time t. Fluid $k(= 1, \cdots, m)$ is surrounded by fluid 0, and the surface tension acts on the interface $\Gamma_k(t)$. Let the coefficient of the surface tension be σ_k. $\Gamma_k(t)$ is expressed as a closed curve,

$$\Gamma_k(t) = \{\chi_k(s, t);\ s \in [0, 1)\},$$

where

$$\chi_k : [0, 1) \times (0, T) \to \mathbb{R}^2,\ \chi(1, t) = \chi(0, t) \quad (t \in (0, T))$$

is a function to be determined. $\Omega_k(t)$, $k = 1, \cdots, m$, is the interior of $\Gamma_k(t)$, and fluid 0 occupies

$$\Omega_0(t) = \Omega \setminus \bigcup \{\bar{\Omega}_k(t);\ k = 1, \cdots, m\}.$$

Unknown functions (u, p), velocity and pressure,

$$u : \Omega \times (0, T) \to \mathbb{R}^2, \quad p : \Omega \times (0, T) \to \mathbb{R}$$

and χ_k satisfy the system of equations,

$$\rho_k \left\{ \frac{\partial u}{\partial t} + (u \cdot \nabla) u \right\} - \nabla \Big[2\mu_k D(u) \Big] + \nabla p = \rho_k f, \quad x \in \Omega_k(t),\ t \in (0, T) \tag{1a}$$

$$\nabla \cdot u = 0, \quad x \in \Omega_k(t),\ t \in (0, T) \tag{1b}$$

$$[u] = 0,\ [-pn + 2\mu D(u) n] = \sigma_k \kappa n, \quad x \in \Gamma_k(t),\ t \in (0, T) \tag{1c}$$

$$\frac{\partial \chi_k}{\partial t} = u(\chi_k, t), \quad s \in [0, 1),\ t \in (0, T) \tag{1d}$$

$$u \cdot n = 0,\ D(u) n \parallel n,\ x \in \Gamma,\ t \in (0, T) \tag{1e}$$

$$u = u^0, \quad x \in \Omega,\ t = 0 \tag{1f}$$

$$\chi_k = \chi_k^0, \quad s \in [0,1), \ t = 0, \tag{1g}$$

where $k = 0, \cdots, m$ in (1a) and (1b), $k = 1, \cdots, m$ in (1c), (1d), and (1g), and

$$f : \Omega \times (0,T) \to \mathbb{R}^2, \quad u^0 : \Omega \to \mathbb{R}^2, \quad \chi_k^0 : [0,1) \to \mathbb{R}^2$$

are given functions; f is an acceleration, u^0 is an initial velocity, χ_k^0 is a function showing the initial interface position. $[\cdot]$ means the difference of the values approached from both sides to the interface, κ is the curvature of the interface, and n is the unit normal. On the boundary of Ω the slip boundary condition (1e) is imposed.

Let X, V, Q, and Φ be function spaces defined by

$$X = \{\chi \in H^1(0,1)^2; \chi(1) = \chi(0)\}, \quad \Phi = L^\infty(\Omega),$$
$$V = \{v \in H^1(\Omega)^2; v \cdot n = 0 \ (x \in \Gamma)\}, \quad Q = L_0^2(\Omega).$$

Let X_h, Φ_h, V_h, and Q_h be finite-dimensional approximation spaces of X, Φ, V and Q. Let Δt be a time increment and $N_T = \lfloor T/\Delta t \rfloor$. We seek approximate solutions χ_h^n, ρ_h^n, μ_h^n, u_h^n, and p_h^n at $t = n\Delta t$ in X_h, Φ_h, Φ_h, V_h, and Q_h, respectively. More precisely, these approximate function spaces are constructed as follows. Dividing the domain Ω into a union of triangles, we use $P0$, $P2$, and $P1$ finite element spaces for Φ_h, V_h, and Q_h, respectively. They are fixed for all time steps n. On the other hand, X_h is composed of functions obtained by the parameterization of polygons. We denote by $\{s_i^n \in [0,1]; i = 0, \cdots, N_x^n\}$ the set of parameter values such that $s_0^n = 0$ and $s_{N_x^n}^n = 1$ and that $\{\chi_h^n(s_i^n); i = 0, \cdots, N_x^n - 1\}$ are vertices of a polygon. The number N_x^n may change depending on n. Our scheme is to find

$$\{(\chi_h^n, \rho_h^n, \mu_h^n, u_h^n, p_h^n) \in X_h \times \Phi_h \times \Phi_h \times V_h \times Q_h; n = 1, \cdots, N_T\}$$

satisfying

$$\frac{\tilde{\chi}_h^n - \chi_h^{n-1}}{\Delta t} = \begin{cases} u_h^{n-1}(\chi_h^{n-1}), & \forall s_i^{n-1}, \ n = 1 \\ \frac{3}{2}u_h^{n-1}(\chi_h^{n-1}) - \frac{1}{2}u_h^{n-2}(\chi_h^{n-1} - \Delta t u_h^{n-1}(\chi_h^{n-1})), & \forall s_i^{n-1}, \ n \geq 2, \end{cases} \tag{2a}$$

$$\chi_h^n = \mathscr{X}_h(\tilde{\chi}_h^n), \quad \rho_h^n = \mathscr{R}_h(\chi_h^n), \quad \mu_h^n = \mathscr{M}_h(\chi_h^n), \tag{2b}$$

$$\left(\frac{\rho_h^n u_h^n - \sqrt{\rho_h^n \rho_h^{n-1}} u_h^n \circ X_h^{n-1} \sqrt{\gamma_h^{n-1}}}{\Delta t}, v_h\right) + a_1(\rho_h^n, u_h^{n-1}, u_h^n, v_h) + a_0(\rho_h^n, u_h^n, v_h)$$
$$+ b(v_h, p_h^n) + \Delta t \, d_h(u_h^n, v_h; \mathscr{C}_h^n) = \left(\rho_h^n \Pi_h f^n, v_h\right) - d_h(\chi_h^n, v_h; \mathscr{C}_h^n),$$
$$\forall v_h \in V_h, \tag{2c}$$

$$b(u_h^n, q_h) = 0, \quad \forall q_h \in Q_h \tag{2d}$$

subject to the initial conditions

$$\chi_h^0 = \Pi_h \chi^0, \quad u_h^0 = \Pi_h u^0, \tag{3}$$

where the symbol $\circ$ stands for the composition of two functions, X_h^{n-1} is a mapping defined by

$$X_h^{n-1}(x) = x - \Delta t u_h^{n-1}(x),$$

γ_h^{n-1} is the Jacobian of the mapping, and Π_h is the Lagrange interpolation operator to the corresponding finite-dimensional space. The linear forms a_1, a_0, b, and d_h are defined by

$$a_1(\rho, w, u, v) = \int_\Omega \frac{1}{2}\rho\Big\{[(w\cdot\nabla)u]\cdot v - [(w\cdot\nabla)v]\cdot u\Big\}dx, \tag{4}$$

$$a_0(\rho, u, v) = \int_\Omega 2\mu(\rho)D(u) : D(v)\, dx, \quad b(v, q) = -\int_\Omega (\nabla\cdot v)q\, dx,$$

$$d_h(\chi, v; \mathscr{C}_h) = \sum_{i=1}^{N_x} \sigma_0 \bar{D}_{\Delta s}\chi_i \cdot \bar{D}_{\Delta s} v_i \frac{(s_i - s_{i-1})^2}{|\chi_i - \chi_{i-1}|},$$

where $\mathscr{C}_h^n$ is a polygon associated with χ_h^n, $\bar{D}_{\Delta s}$ is the backward difference with respect to the parameter s, $\mathscr{X}_h$ is an operator which adjusts the position of the interface, $\mathscr{R}_h$ and $\mathscr{M}_h$ are operators which define average constants on each element. For the details of the linear forms and the operators refer to [10].

The Lagrange-Galerkin method has nice features for the approximation of material derivative terms [2–5, 7, 8]. Since the basic idea is to approximate the particle path along characteristic curves, the method is robust for high Reynolds number problems. The scheme (2) has the following advantages:

- It is stable in the sense of energy if an integral of the square of approximate curvature of the interface remains bounded.
- Since the resultant matrix is symmetric, we can use efficient solvers for symmetric system of linear equations, e.g., MINRES.
- Since we use the interface-tracking method, we can distribute much more nodes on the interface than the level-set method.
- When it is applied to incompressible viscous one-fluid flow problems, the stability and convergence are assured.
- Since the main computation part is similar to that of the Stokes problem, the computation is light.

We apply this scheme to simulate the behavior of a rising bubble in fluids of different viscosities.

3 Numerical Results

Let $m = 1$ and set

$$\Omega = (0,1) \times (0,4),$$

$$\Omega_1 = \{(x_1, x_2);\ (x_1 - a)^2 + (x_2 - 2a)^2 < a^2\},\ a = \frac{1}{5},$$

$$\rho_0 = 100,\ \mu_0 = 5.0,\ 0.5,\ 0.05,\ 0.005,\ \rho_1 = 0.1,\ \mu_1 = 1.0,$$

$$f = (0, -1)^T,\ \sigma_1 = 2.0.$$

When the viscosity μ_0 of fluid 0 varies, we observe the change of the bubble movement depending on μ_0. We set the time increment $\Delta t = 1/16$. Figs. 1, 2, 3, and 4 show the time histories of the interfaces and streamlines when $\mu_0 = 5.0, 0.5, 0.05$, and 0.005. When $\mu_0 = 5.0$, that is, the viscosity is large, the rising speed of the bubble is slow and any wakes are hardly visible after the bubble. When $\mu_0 = 0.5$, that is, the viscosity decreases, the rising speed of the bubble increases and there appear wakes after the bubble. When $\mu_0 = 0.05$, that is, the viscosity is small, the rising speed of the bubble becomes high and there appears oscillation when the bubble rises up. When $\mu_0 = 0.005$, that is, the viscosity becomes smaller, the rising speed increases and the wake becomes larger. The wake has a pattern similar to the Kármán vortex street in the flow past a circular cylinder. Fig. 5 shows the trace of the centroid of the bubble in each case. Although the change of the position of the centroid of the bubble is not so remarkable compared to the shape, we can observe the increase of the distance from the center-line $x_1 = 1/2$ when μ_0 decreases.

Acknowledgements The author was supported by JSPS (Japan Society for the Promotion of Science) under Grants-in-Aid for Scientific Research (C), No. 25400212 and (S), No. 24224004 and under the Japanese-German Graduate Externship (Mathematical Fluid Dynamics) and by Waseda University under Project research, Spectral analysis and its application to the stability theory of the Navier-Stokes equations of Research Institute for Science and Engineering.

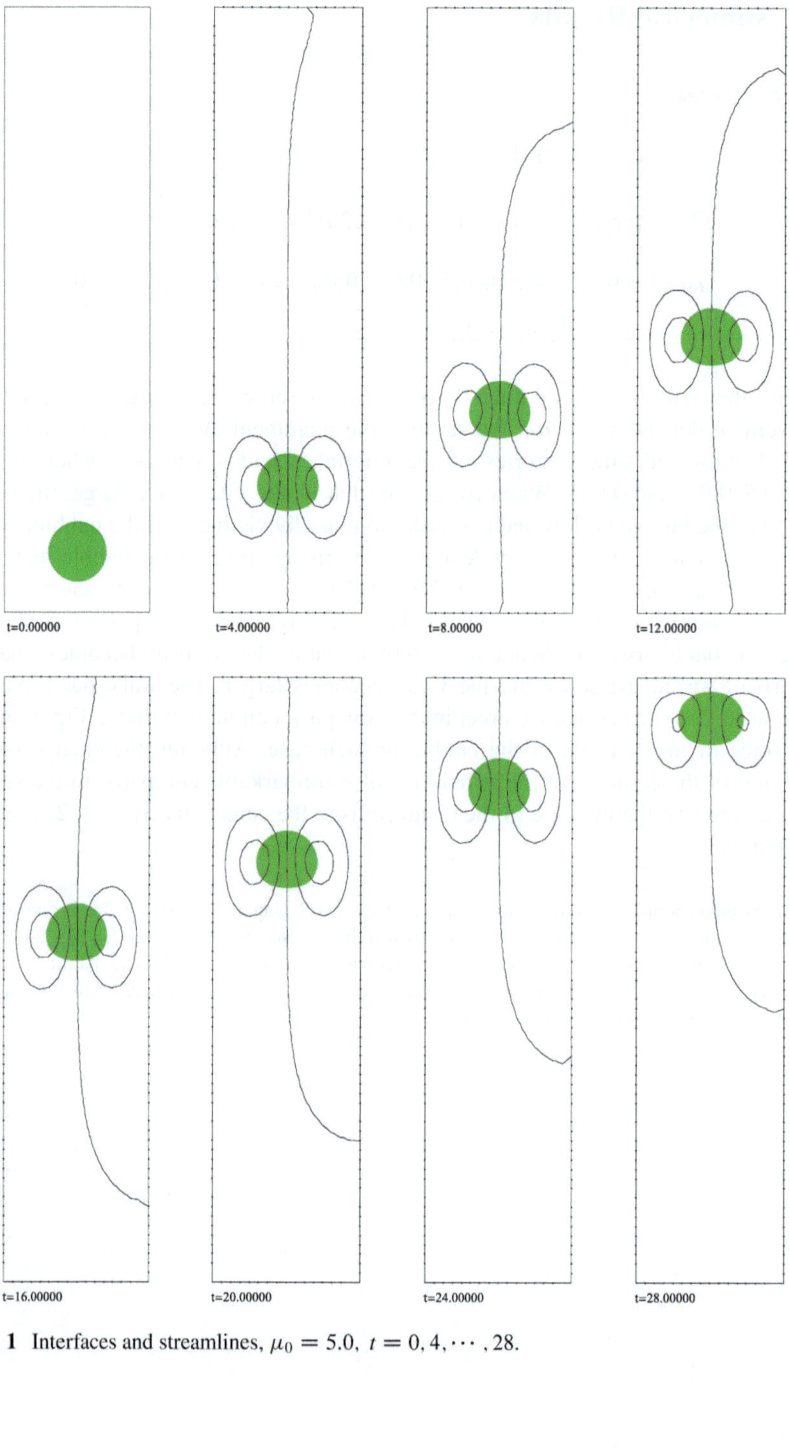

Fig. 1 Interfaces and streamlines, $\mu_0 = 5.0,\ t = 0, 4, \cdots, 28.$

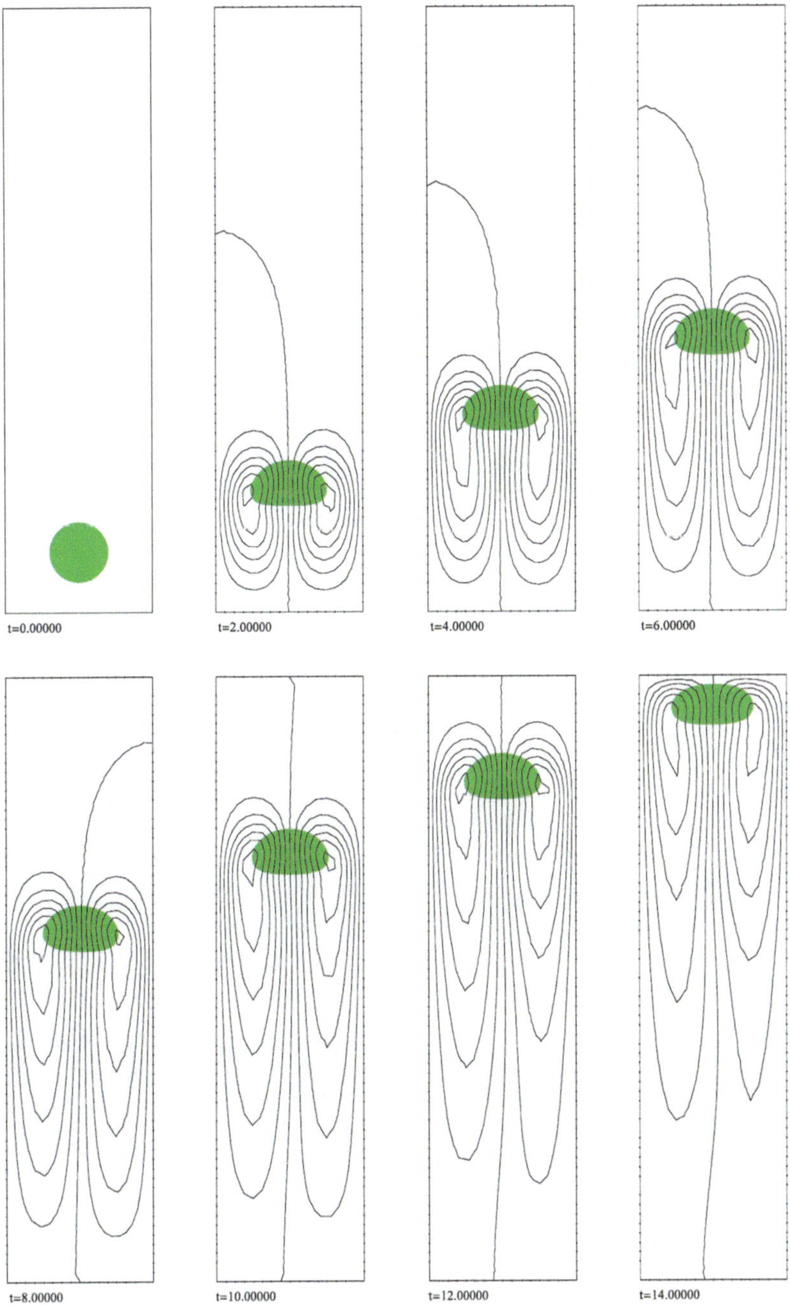

Fig. 2 Interfaces and streamlines, $\mu_0 = 0.5$, $t = 0, 2, \cdots, 14$.

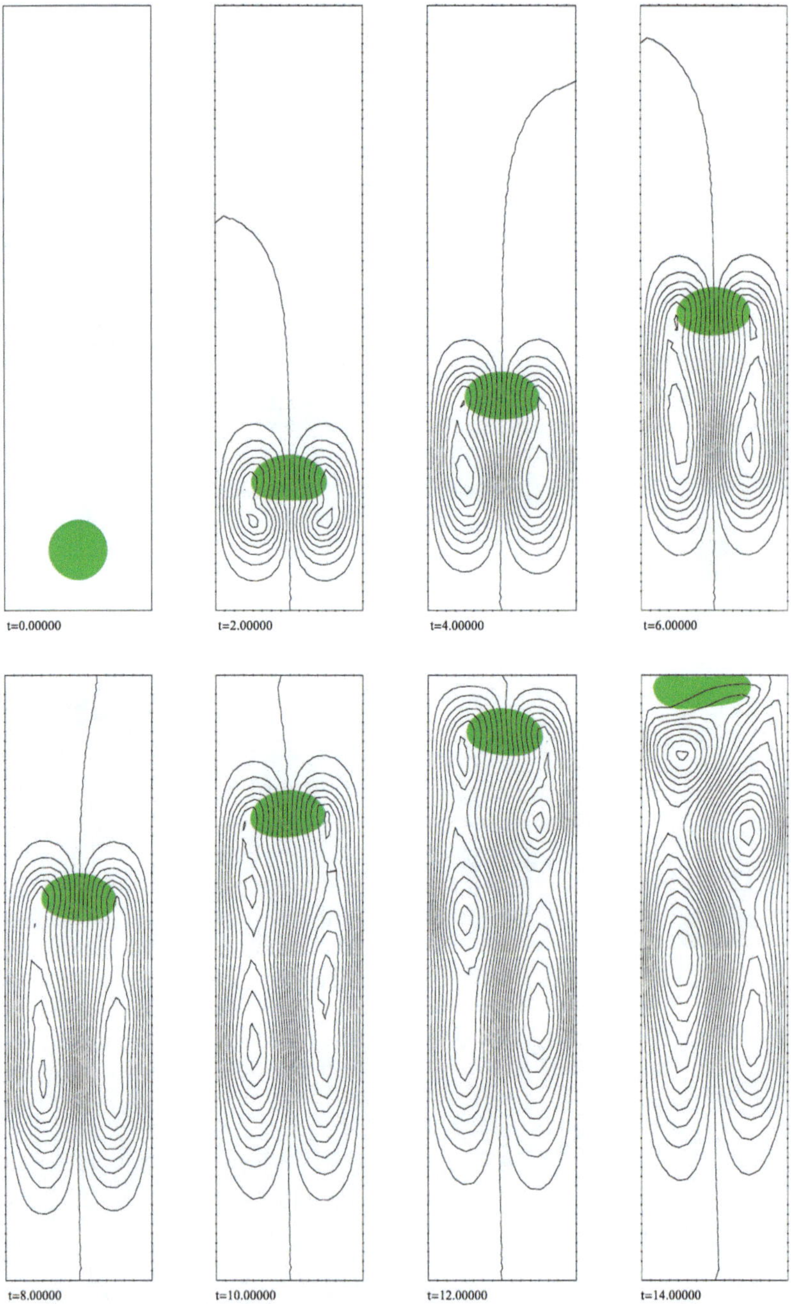

Fig. 3 Interfaces and streamlines, $\mu_0 = 0.05$, $t = 0, 2, \cdots, 14$.

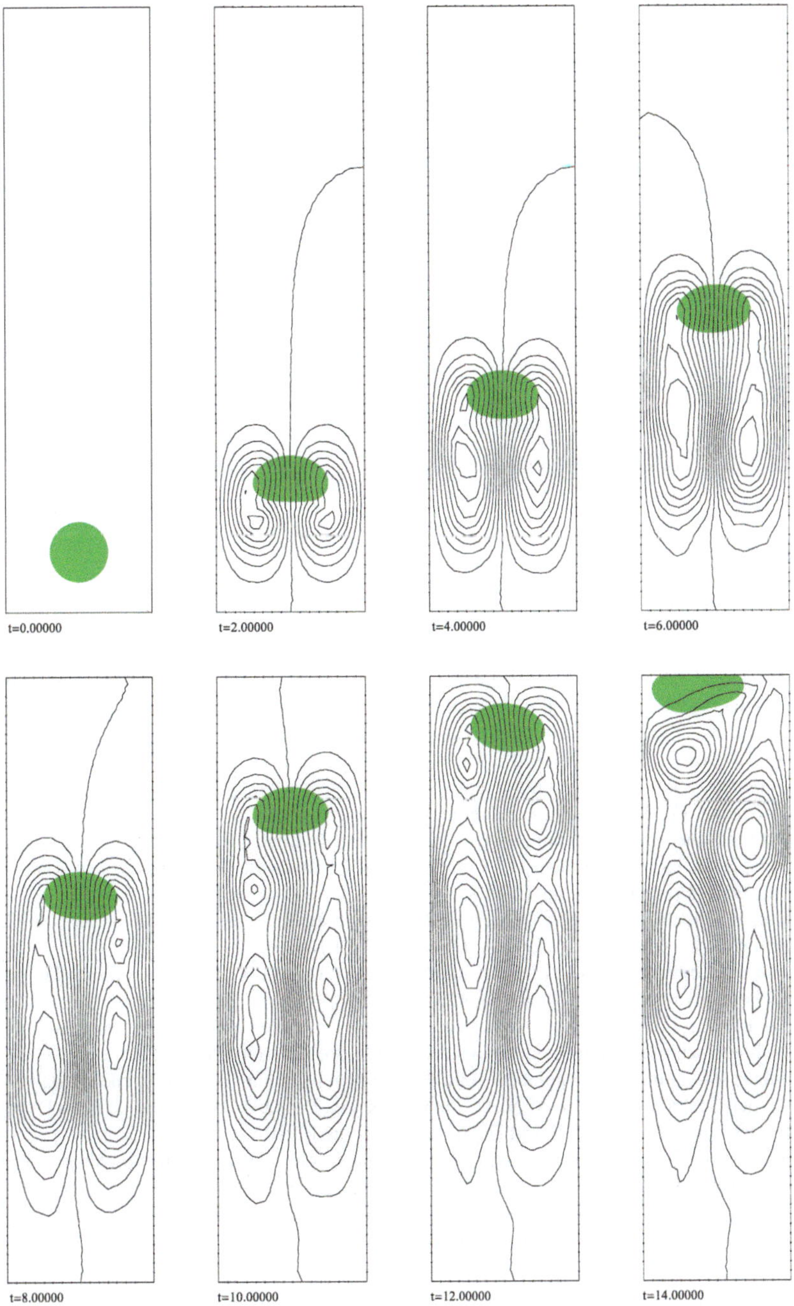

Fig. 4 Interface and streamlines, $\mu_0 = 0.005$, $t = 0, 2, \cdots, 14$.

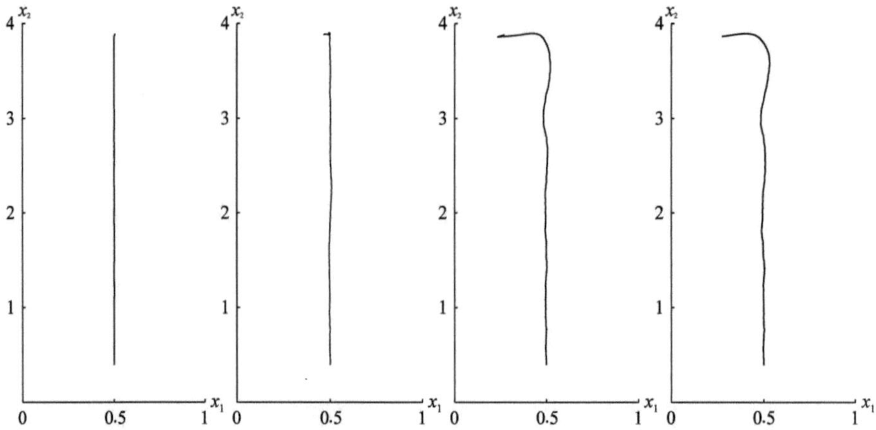

Fig. 5 Traces of the centroid of the bubble, $\mu_0 = 5,\ 0.5,\ 0.05,\ 0.005$ (from left to right).

References

1. Bänsch, C.: Finite element discretization of the Navier–Stokes equations with a free capillary surface. Numer. Math. **88**, 203–235 (2001)
2. Notsu, H., Tabata, M.: Error estimates of a pressure-stabilized characteristics finite element scheme for the Oseen equations. J. Sci. Comput. **65**, 940–955 (2015)
3. Notsu, H., Tabata, M.: Error estimates of a stabilized Lagrange–Galerkin scheme for the Navier–Stokes equations. ESAIM Math. Model. Numer. Anal. **50**, 361–380 (2016)
4. Pironneau, O.: Finite Element Methods for Fluids. Wiley, Chichester (1989)
5. Pironneau, O., M. Tabata, M.: Stability and convergence of a Galerkin-characteristics finite element scheme of lumped mass type. Int. J. Numer. Methods Fluids **64**, 1240–1253 (2010)
6. Prosperetti, A., Tryggvason, G.: Computational Methods for Multiphase Flow. Cambridge University Press, Cambridge (2009)
7. Rui, H., Tabata, M.: A mass-conservative characteristic finite element scheme for convection-diffusion problems. J. Sci. Comput. **43**, 416–432 (2010)
8. Süli, E.: Convergence and nonlinear stability of the Lagrange–Galerkin method for the Navier–Stokes equations. Numer. Math. **53**, 459–483 (1988)
9. Tabata, M.: Finite element schemes based on energy-stable approximation for two-fluid flow problems with surface tension. Hokkaido Math. J. **36**, 875–890 (2007)
10. Tabata, M.: Numerical simulation of fluid movement in an hourglass by an energy-stable finite element scheme. In: Hafez, M.N., Oshima, K., Kwak, D. (eds.) Computational Fluid Dynamics Review 2010, pp. 29–50. World Scientific, Singapore (2010)
11. Tabata, M.: Energy-stable Lagrange-Galerkin schemes for two-fluid flow problems (to appear)
12. Tezduyar, T.E., Behr, M., Liou, J.: A new strategy for finite element computations involving boundaries and interfaces - the deforming-spatial-domain /space-time procedure: I. Comput. Methods Appl. Mech. Eng. **94**, 339–351 (1992)

A Numerical Review of Multi-Fluid SPH Algorithms for High Density Ratios

Jan-Philipp Fürstenau, Bircan Avci, and Peter Wriggers

Abstract The smoothed particle hydrodynamics (SPH) method is a recent numerical technique and particularly well suited for the simulation of free surface flows. The SPH method is also a good technique for the simulation of multi-fluid problems, as in that approach the fluid interface is resolved by freely moving particles. Due to high density gradients at the interface, the SPH multi-fluid models generally show instabilities for high density ratios. In the current work, the multi-fluid models of Monaghan and Rafiee [7], Colagrossi and Landrini [2], and Hu and Adams [6] are reviewed and benchmarked. The focus of this work is to survey the implementations with regard to stability, accuracy, and computational effort.

1 Introduction

Many technical fluid problems involve multi-fluid flows, examples can be found, for instance, in naval or process engineering, like the oil loss of a tanker or the gas–liquid flows in pipes. For the prediction or simulation of these flows, a multi-fluid flow solver is required that is usually implemented by extending a single-fluid solver. Especially in this context, the well known but relatively new smoothed particle hydrodynamics (SPH) method demonstrates its flexibility. Since through its particle nature the main routines of the SPH method can be extended easily to result in a free surface multi-fluid flow model, that can directly simulate multi-fluid interactions within the particle scale. However, numerical experiments on the resulting multi-fluid flows show stability problems for density ratios greater than 2 [2]. The instabilities in these cases result from the density gradients at the interfaces. In the current work multi-fluid models were implemented in the GPU-based open-source SPH framework DualSPHysics [3] that is basically designed for

J.-P. Fürstenau • B. Avci • P. Wriggers (✉)
Institute of Continuum Mechanics, Leibniz Universität Hannover, Appelstrasse 11, 30167 Hannover, Germany
e-mail: fuerstenau@ikm.uni-hannover.de; avci@ikm.uni-hannover.de; wriggers@ikm.uni-hannover.de

Y. Bazilevs, K. Takizawa (eds.), *Advances in Computational Fluid-Structure Interaction and Flow Simulation*, Modeling and Simulation in Science, Engineering and Technology, DOI 10.1007/978-3-319-40827-9_11

oceanographic problems. The focus of this work is to review and evaluate three multi-fluid SPH approaches with regard to stability and computational effort in an implementation on GPU.

2 SPH Basics

For isothermal problems the flow of a weakly compressible viscous fluid can be obtained by the following form of the Navier–Stokes equations:

$$\frac{d\mathbf{v}}{dt} = -\frac{1}{\rho}\nabla p + \nu\nabla^2\mathbf{v} + \mathbf{f} \tag{1}$$

with the continuity equation

$$\frac{d\rho}{dt} = -\rho\nabla\cdot\mathbf{v}, \tag{2}$$

where ρ, p, ν, $\mathbf{v}$, and $\mathbf{f}$ are fluid density, pressure, kinematic viscosity, velocity, and body force of an arbitrary point in the described fluid. For weakly compressible fluids commonly the following equation of state is evaluated to calculate the pressure with respect to the density [7]:

$$p(\rho) = \frac{\rho_0 c^2}{\gamma}\left[\left(\frac{\rho}{\rho_0}\right)^{\gamma} - 1\right], \tag{3}$$

where ρ_0 is the reference density of the fluid. The constant γ is a parameter depending on the compressibility of the fluid, where for a fluid like water or air a value $\gamma = 7$ and $\gamma = 1.4$ is chosen, respectively, according to the isentropic exponent of the fluid [2]. Within the SPH framework the speed of sound of the fluid c is used more as a numerical parameter, that has to be at least 10 times bigger than the maximum fluid speed to keep the density fluctuations below 1 % [7].

2.1 *Discretization*

The central idea of the SPH method is to introduce a kernel interpolant W to evaluate a field quantity A at a certain point in space $\mathbf{r}$ [4]

$$A(\mathbf{r}) = \int W(\mathbf{r} - \mathbf{r}')A(\mathbf{r}')d\mathbf{r}'. \tag{4}$$

As the field around the point is commonly unknown the continuum is discretized with movable particles that carry each field quantity of the fluid. This way, on the one hand, each particle represents an amount of fluid, which is moving in space, and on the other hand, an interpolation support for its neighbors. Thus, the particles appear like the scattered points in the Monte Carlo method [4]. In the SPH method, the integration for each particle i can be interpolated by a sum over its neighboring particles j. Finally, as discrete interpolation of the quantity A in (4) one obtains

$$\langle A(\mathbf{r})\rangle_i = \sum_j A(\mathbf{r}_j)W(\mathbf{r}_i - \mathbf{r}_j)V_j = \sum_j A(\mathbf{r}_j)W(\mathbf{r}_i - \mathbf{r}_j)\frac{m_j}{\rho_j}, \tag{5}$$

where $\mathbf{r}_j$, V_j, and m_j are positions, assigned volumes, and masses of the neighboring particles (see [4] for details). The kernel is also used for the calculation of the spatial derivatives of the field quantity A at point i

$$\langle \nabla A(\mathbf{r})\rangle_i = \sum_j A(\mathbf{r}_j)\nabla_i W(\mathbf{r}_i - \mathbf{r}_j)V_j = \sum_j A(\mathbf{r}_j)\nabla_i W(\mathbf{r}_i - \mathbf{r}_j)\frac{m_j}{\rho_j}. \tag{6}$$

In the following, for a better overview the field quantities are rewritten as $A(\mathbf{r}_i) = A_i$ and the kernel function between i and j is rewritten as $W(\mathbf{r}_i - \mathbf{r}_j) = W_{ij}$. With this notation equations (5) and (6) take the form

$$\langle A\rangle_i = \sum_j \frac{m_j}{\rho_j} A_j W_{ij} \quad \text{and} \quad \langle \nabla A\rangle_i = \sum_j \frac{m_j}{\rho_j} A_j \nabla_i W_{ij}. \tag{7}$$

For computational reasons the kernel must have a compact form and its global integral must yield "1". Many kernel functions are designed in a way that they fulfill these criteria, and in this paper the Wendland kernel is used as presented in [11]

$$W_{ij} = \alpha \begin{cases} \left(1 - \frac{r_{ij}}{2h}\right)^4 \left(\frac{2r_{ij}}{h} + 1\right) & \text{for } r_{ij} \leq 2h \\ 0 & \text{for } r_{ij} > 2h \end{cases} \tag{8}$$

where r_{ij} is the scalar particle distance and h is the constant delimiter of the kernel function for a compact kernel and is called the smoothing length. The parameter α depends on the dimensions of the problem, where $\alpha = 7/(4\pi h^2)$ for 2D and $\alpha = 21/(16\pi h^3)$ for 3D problems.

2.2 *Equation of Motion*

The pressure term of the Navier–Stokes momentum equation (1) for particle i can directly be interpolated with the kernel derivative (7) as

$$\left\langle \frac{1}{\rho} \nabla p \right\rangle_i = \frac{1}{\rho_i} \sum_j \frac{m_j}{\rho_j} p_j \nabla_i W_{ij}. \tag{9}$$

This equation is certainly not symmetrical, what results in an unphysical behavior [9]. For the symmetrization of the equation commonly the identity is used in the following form:

$$\frac{1}{\rho} \nabla p = \nabla \left(\frac{p}{\rho} \right) + \frac{p}{\rho^2} \nabla \rho \tag{10}$$

that can be interpolated as

$$\left\langle \frac{1}{\rho} \nabla p \right\rangle_i = \sum_j m_j \left(\frac{p_i}{\rho_i^2} + \frac{p_j}{\rho_j^2} \right) \nabla_i W_{ij}. \tag{11}$$

This formulation is symmetrical and represents the common form for the calculation of the pressure interaction between particles.

Also for the calculation of the viscous forces several formulations have been presented in the literature. In this work the viscosity approach of Monaghan [7] is consistently used, where the viscosity reads

$$\Pi_{ij} = -\frac{16 \nu_i \nu_j}{\nu_i \rho_i + \nu_j \rho_j} \frac{\mathbf{v}_{ij} \cdot \mathbf{r}_{ij}}{\left| \mathbf{r}_{ij} \right| h}. \tag{12}$$

Herein, $\mathbf{v}_{ij}$ and $\mathbf{r}_{ij}$ are defined as $\mathbf{v}_{ij} = \mathbf{v}_i - \mathbf{v}_j$ and $\mathbf{r}_{ij} = \mathbf{r}_i - \mathbf{r}_j$, ν_i and ν_j are the kinematic viscosities of the interacting fluids. After calculating Π_{ij} the viscous forces can be obtained by evaluating

$$\left\langle \nu \nabla^2 \mathbf{v} \right\rangle_i = -\sum_j m_j \Pi_{ij} \nabla_i W_{ij}, \tag{13}$$

which represents the discretized form of the viscosity part in equation (1) (for details see [1, 7]).

2.3 Continuity Equation

In the framework of the SPH method, the continuity equation (2) can be interpolated by using the kernel derivative

$$\frac{d\rho_i}{dt} = \rho_i \sum_j \frac{m_j}{\rho_j} \mathbf{v}_{ij} \nabla_i W_{ij}. \tag{14}$$

Another possibility for the calculation of the density is the direct evaluation of the kernel function (7) around particle i

$$\rho_i = \sum_j \frac{m_j}{\rho_j} \rho_j W_{ij} = \sum_j m_j W_{ij}. \tag{15}$$

However, equation (14) is used more often, as the temporal density evolution can be evaluated together with the particle acceleration whereas equation (15) has to be evaluated before the rest of the computations.

3 Multi-Fluid Implementations

The possibility of simulating multiple fluids is already included in the basic SPH formulations (13) and (14) or (15). Certainly, these equations expect a continuous field of variables without jumps. However, for multi-fluid implementations the continuity of the density field $\langle \rho \rangle$ is disturbed at the fluid interfaces. Also the density gradient $\langle \nabla \rho \rangle$ is affected by these discontinuities. The main tasks in simulating multi-fluid problems with high density ratios are

- to keep the density jump from being smoothed out,
- to stabilize the interface.

In this work, the multi-fluid approaches of Monaghan and Rafiee [7], Colagrossi and Landrini [2], and Hu and Adams [6] are implemented and compared. The approach of Grenier et al. [5] has also been taken into account, but due to its high computational effort, as already noticed in [7] and [9], it is not further considered here.

3.1 Monaghan and Rafiee Formulation

The simplest implementation in this paper concerning additional formulations and computational effort per particle and time step is based on the approach of Monaghan and Rafiee [7]. They propose an extended momentum equation with a modified pressure part which reads

$$\left\langle \frac{1}{\rho} \nabla p \right\rangle_i = - \sum_j m_j \left(\frac{p_i + p_j}{\rho_i \rho_j} + R_{ij} \right) \nabla_i W_{ij}. \tag{16}$$

In the modified pressure part a repulsive force R_{ij}, which is driven by the density ratio, is introduced to stabilize the interface

$$R_{ij} = \alpha \left| \frac{\rho_{0i} - \rho_{0j}}{\rho_{0i} + \rho_{0j}} \right| \left| \frac{p_i + p_j}{\rho_i \rho_j} \right|. \tag{17}$$

Here, the parameter α is of empirical type. In Monaghan and Rafiee [7] the continuity equation (14) is chosen as it can directly be used in the analysis without any alteration.

3.2 Colagrossi and Landrini Formulation

In the approach of Colagrossi and Landrini [2] a very similar pressure formulation as given in (16) is proposed which reads

$$\left\langle \frac{1}{\rho} \nabla p \right\rangle_i = -\frac{1}{\rho_i} \sum_j \frac{m_j}{\rho_j} \left(p_i + p_j\right) \nabla_i W_{ij}. \tag{18}$$

As continuity equation the authors employ equation (14). For the handling of high density ratios they propose a surface tension force. This surface tension is affecting the equation of state (3)

$$p(\rho) = \frac{\rho_0 c^2}{\gamma} \left[\left(\frac{\rho}{\rho_0} \right)^\gamma - 1 \right] - \bar{a} \rho^2 \tag{19}$$

and the pressure term (18)

$$\left\langle \frac{1}{\rho} \nabla p \right\rangle_i = -\frac{1}{\rho_i} \sum_j \frac{m_j}{\rho_j} \left(p_i + p_j\right) \nabla_i W_{ij} - \frac{\bar{a}}{\rho_i} \sum_j \frac{m_j}{\rho_j} \left(\rho_i^2 + \rho_j^2\right) \nabla_i W_{ij}. \tag{20}$$

Herein, $\bar{a}$ is an empirical constant chosen according to the problem type [2].

3.3 Hu and Adams Formulation

The approach of Hu and Adams [6] differs from the aforementioned approaches as in this case, the density is evaluated directly with the kernel based on equation (15). However, to handle the density jump at the interface, the masses of the neighboring particles are neglected and the sum only contains the kernel values. This is called the number density or specific volume σ, which is used to calculate the density by

$$\rho_i = m_i \sum_j W_{ij} = m_i \sigma_i. \tag{21}$$

The specific volume is also used to calculate the pressure term by evaluating

$$\left\langle \frac{1}{\rho} \nabla p \right\rangle_i = -\frac{1}{m_i} \sum_j \left(\frac{p_i}{\sigma_i^2} + \frac{p_j}{\sigma_j^2} \right) \nabla_i W_{ij}. \tag{22}$$

In this approach, as stabilization for high density ratio interactions a surface tension is used in terms of a continuous surface force model. At first, a color function c of each phase a is defined for each particle

$$c_i^a = \begin{cases} 1 & \text{if particle i belongs to phase a,} \\ 0 & \text{if particle i does not belong to phase a.} \end{cases} \tag{23}$$

This color field also has a gradient which exists when a particle i of phase a has a neighboring particle j of phase b. Then the gradient between phase a and b at particle i can be evaluated as follows:

$$\nabla c_i^{ab} = \sigma_i \sum_j \left(\frac{c_i^b}{\sigma_i^2} + \frac{c_j^b}{\sigma_j^2} \right) \nabla W, \qquad a \neq b. \tag{24}$$

This color gradient acts as an interface normal. A surface stress tensor Π_S can then be calculated by using the following expression:

$$\Pi_S^{ab} = \gamma_{ab} \frac{1}{|\nabla c^{ab}|} \left(\frac{1}{d} \mathbf{I} |\nabla c^{ab}|^2 - \nabla c^{ab} \nabla c^{ab} \right), \qquad a \neq b, \tag{25}$$

where d is the number of dimensions of the problem and γ_{ab} is the surface tension parameter of phase a with respect to phase b. This variable can differ for all phases, and γ_{ab} must not be equal to γ_{ba}. For this reason the summation of the color gradient has to be performed for all phases separately. Finally all stress tensors can be summed up to

$$\Pi_S = \sum_b \Pi_S^{ab}, \qquad a \neq b. \tag{26}$$

The force resulting from the surface tension on particle i is then obtained in a similar manner like for the pressure and one ends up with

$$\left\langle \frac{1}{\rho} \nabla p \right\rangle_i = -\frac{1}{m_i} \sum_j \left(\frac{p_i}{\sigma_i^2} + \frac{p_j}{\sigma_j^2} \right) \nabla_i W_{ij} - \frac{1}{m_i} \sum_j \left(\frac{\Pi_{Si}}{\sigma_i^2} + \frac{\Pi_{Sj}}{\sigma_j^2} \right) \cdot \nabla_i W_{ij}. \tag{27}$$

4 Numerical Simulations

To evaluate the multi-fluid approaches, two benchmark tests are performed. The first is the Rayleigh–Taylor instability test to examine the multi-fluid flow with a low density ratio ($\rho_2/\rho_1 = 1.8$). The second benchmark is the rising bubble example. In this case the density ratio of the fluid phases is high ($\rho_1/\rho_2 = 1000$), and the governing multi-fluid SPH equations require stabilization to ensure robustness. The used time integration scheme in the benchmarks is the Verlet algorithm [10], and the system boundaries are discretized by means of SPH boundary particles.

4.1 *Rayleigh–Taylor Instability*

The Rayleigh–Taylor instability test consists of two fluids that are placed on each other like shown in Fig. 1. The fluid domain is discretized by 120 x 240 particles. The parameters used in the numerical simulations are listed in the Appendix.

The obtained results for the three approaches are shown in Fig. 2. While performing the simulations it turned out that the approaches of Monaghan and Rafiee [7] and Colagrossi and Landrini [2] are rather stable regarding the time step size, whereas the approach of Hu and Adams [6] responds very sensitive in this regard. Fig. 3 shows the results when increasing the speed of sound of both fluids, c_1 and c_2, by a factor of 5. As one can see, with the approach of Hu and Adams [6] the fluids nearly did not move, because the fluid motion is totally damped out by oscillations of the particle velocities and densities.

In Table 1, the execution results of the computations are shown. The approaches of Monaghan and Rafiee [7] and Colagrossi and Landrini [2] result in the same numerical effort, as the basics of the methods are quite similar. The approach of Hu and Adams [6], on the other side, performs less time steps, but requires about 25 % more simulation time.

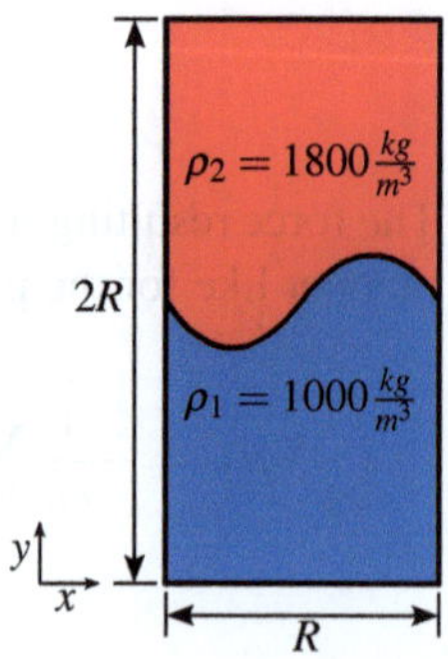

Fig. 1 Initial configuration of the fluids in the Rayleigh–Taylor instability benchmark. The initial interface is represented by $y = R\,[1 - 0.15\sin(2\pi x/R)]$.

Fig. 2 Simulation results at $t = 5\sqrt{R/g}$ [9] obtained with the approaches of Monaghan and Rafiee [7] (left), Colagrossi and Landrini [2] (middle), and Hu and Adams [6] (right) (Compared with level-set results of Grenier et al. [5] [black dots])

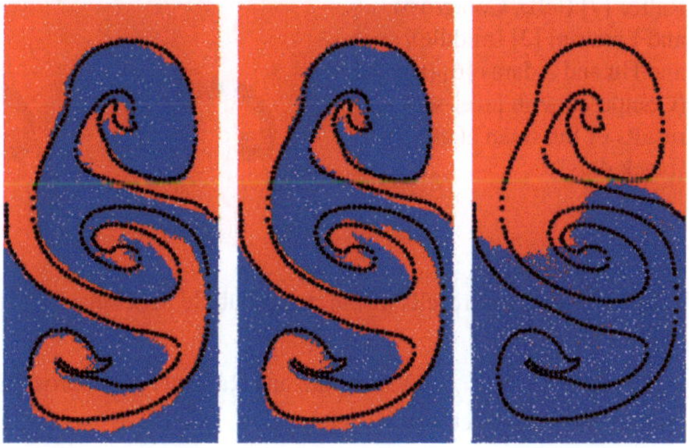

Fig. 3 Simulation results at $t = 5\sqrt{R/g}$ [9] with a 5 times higher speed of sound of both fluids obtained with the approaches of Monaghan and Rafiee [7] (left), Colagrossi and Landrini [2] (middle), and Hu and Adams [6] (right) (Compared with level-set results of Grenier et al. [5] [black dots])

Table 1 Execution results of the Rayleigh–Taylor instability carried out on a GeForce 840M

Approach	Monaghan and Rafiee [7]	Colagrossi and Landrini [2]	Hu and Adams [6]
Number of time steps	18,961	18,960	18,456
Simulation runtime [s]	115.72	117.95	145.93

4.2 *Rising Bubble*

The second test case is the rising bubble benchmark. In this case a bubble of air with radius R is placed in a box of water like shown in Fig. 4. The fluid domain is discretized by 121 x 201 particles. The used parameters for the simulations can also be found in the Appendix. As one can see in Fig. 5, all three approaches are capable of representing the rising bubble problem, but the effort for finding suitable stabilization parameters varies strongly from model to model. In the case of the approach of Hu and Adams [6], no parameter could be found in reasonable time which ensures both good stability and good accuracy.

All the stabilization parameters and the execution results related to this benchmark analysis are given in Table 2. As in the previous benchmark the approach of Hu and Adams [6] performs less time steps, but it requires about 75 % more simulation runtime compared to the other two. Mainly, this is due to the numerical extensive calculation of the surface tension force.

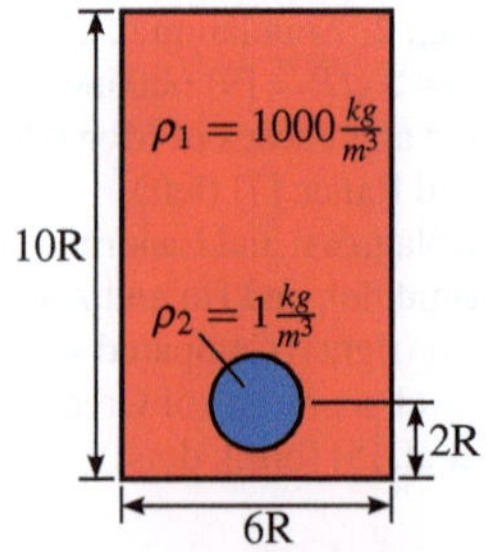

Fig. 4 Initial configuration of the rising bubble benchmark

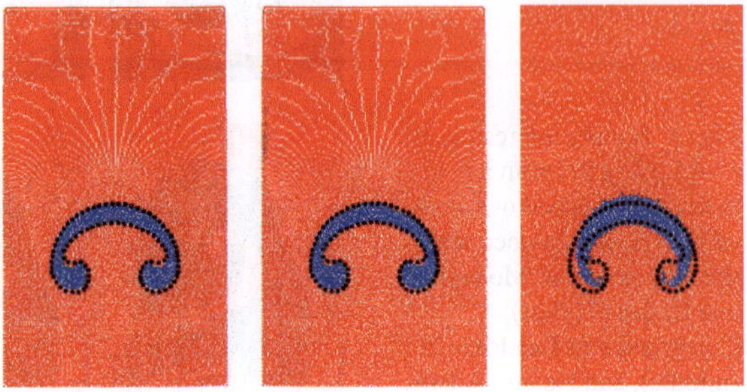

Fig. 5 Simulation results at $t = 4\sqrt{R/g}$ obtained with the approaches of Monaghan and Rafiee [7] (left), Colagrossi and Landrini [2] (middle), and Hu and Adams [6] (right) (Compared with level-set results of Sussman et al. [8] [black dots])

Table 2 Execution results of the rising bubble benchmark carried out on a GeForce GTX TITAN Black

Approach	Monaghan and Rafiee [7]	Colagrossi and Landrini [2]	Hu and Adams [6]
Stabilization parameters	$\alpha = 0.03$	$\bar{a}_1 = 0$	$\gamma_{12} = 0$
		$\bar{a}_2 = 3500$	$\gamma_{21} = 500$
Number of time steps	128,583	129,246	123,991
Simulation runtime [s]	217.73	242.64	376.17

5 Summary and Outlook

In the present work, three multi-fluid approaches for high density ratios were reviewed and realized in the GPU-based open-source SPH code DualSPHysics [3]. The resulting codes were investigated within two benchmarks, to expose their physical and numerical capabilities. The numerical simulations showed up some strengths and weaknesses of the basic equations and stabilization strategies of the approaches. In the next step the implementations will be applied to larger problems. As a first idea an oil leaking tanker was simulated to show the capabilities of the actually used GPUs. The tanker problem shown in Fig. 6 consists of about 44 million particles. The simulation was carried out on a GeForce GTX TITAN Black where the simulation runtime for 103,786 time steps was about 90.87 hours with the approach of Monaghan and Rafiee [7]. For the simulation of further and even more complex technical problems, it will be examined how the strengths of the three approaches can be combined to obtain a stable code for massively parallelized computing of multi-fluid flows with high density ratios.

Fig. 6 Simulation results of the oil leaking tanker problem at two different points in time

Appendix

The parameters used for the simulations shown in Section 4 are shown in Tables 3 and 4, where the parameters ν_1 and c_2 can be calculated using $Re = \sqrt{R^3 g}/\nu_1$ [9] and $c_2 = \sqrt{(\rho_1\gamma_2)/(\rho_2\gamma_1)}\, c_1$ [2]. In Table 4 ν_2 is calculated using $\nu_2 = 128\nu_1$ [9].

Table 3 Parameters used for the Rayleigh–Taylor instability benchmark

Re	g	R	$\gamma_1 = \gamma_2$	$\nu_1 = \nu_2$ [9]	c_1	c_2
420	$9.81\ \frac{kgm}{s^2}$	$1\ m$	7	$\approx 0.0075\ \frac{m^2}{s}$	$\approx 66.44\ \frac{m}{s}$	$\approx 49.5227\ \frac{m}{s}$

Table 4 Parameters used for the rising bubble benchmark

Re	g	R	γ_1	γ_2	ν_1	ν_2	c_1	c_2
1000	$9.81\ \frac{kgm}{s^2}$	$1\ m$	7	1.4	$\approx 0.0089\ \frac{m^2}{s}$	$\approx 1.1339\ \frac{m^2}{s}$	$\approx 148.94\ \frac{m}{s}$	$\approx 2106.32\ \frac{m}{s}$

References

1. Cleary, P.W., Monaghan, J.J.: Conduction modelling using smoothed particle hydrodynamics. J. Comput. Phys. **148**(1), 227–264 (1999)
2. Colagrossi, A., Landrini, M.: Numerical simulation of interfacial flows by smoothed particle hydrodynamics. J. Comput. Phys. **191**(2), 448–475 (2003)
3. Crespo, A.J.C., Domínguez, J.M., Rogers, B.D., Gómez-Gesteira, M., Longshaw, S., Canelas, R., Vacondio R., Barreiro, A., García-Feal, O.: DualSPHysics: open-source parallel CFD solver based on smoothed particle hydrodynamics (SPH). Comput. Phys. Commun. **187**, 204–216 (2015)
4. Gingold, R.A., Monaghan, J.J.: Smoothed particle hydrodynamics: theory and application to non-spherical stars. Mon. Not. R. Astron. Soc. **181**(3), 375–389 (1977)
5. Grenier, N., Antuono, M., Colagrossi, A., Le Touzé, Alessandrini, B.: An Hamiltonian interface SPH formulation for multi-fluid and free surface flows. J. Comput. Phys. **228**(22), 8380–8393 (2009)
6. Hu, X.Y., Adams, N.A.: A multi-phase SPH method for macroscopic and mesoscopic flows. J. Comput. Phys. **213**(2), 844–861 (2006)

7. Monaghan, J.J., Rafiee, A.: A simple SPH algorithm for multi-fluid flow with high density ratios. Int. J. Numer. Methods Fluids **71**(5), 537–561 (2013)
8. Sussman, M., Smereka, P., Osher, S.: A level set approach for computing solutions to incompressible two-phase flow. J. Comput. Phys. **114**(1), 146–159 (1994)
9. Szewc, K., Pozorski, J., Minier, J.-P.: A study on application of smoothed particle hydrodynamics to multi-phase flows. Int. J. Nonlinear Sci. Numer. Simul. **13**(6), 383–395 (2012)
10. Verlet, L.: Computer experiments on classical fluids. I. Thermodynamical properties of Lennard-Jones molecules. Phys. Rev. **159**(1), 98 (1967)
11. Wendland, H.: Piecewise polynomial, positive definite and compactly supported radial functions of minimal degree. Adv. Comput. Math. **4**(1), 389–396 (1995)

Self-Propulsion of a Killifish from Impulsive Starts

Yoichi Ogata and Takayuki Azama

Abstract The present study discusses an acceleration of a swimming killifish-like small fish by fluid force, using three-dimensional simulations with CIP method in combination with an immersed boundary method. Simulation results have indicated that time variation of a small fish speed in small Reynolds number can be estimated as a free fall model of a sphere, that is, the fluid force can be separated into positive part (thrust) and negative part (viscous drag). Thrust is almost independent of Reynolds number, but a coefficient of viscous drag is a decreasing function of Reynolds number when a swimming motion of a fish with a shape is determined.

1 Introduction

Applications of fish morphology and motion to robots have been attracting the attention as new underwater propelling systems which do not use screws in recent years. For example, robots which are simulative such as a dolphin and a turtle have been studied and developed [1, 2]. A lot of knowledge have been found from fish swimming so that we can design propelling systems and develop their swimming performances. For example, bending rules in natural propulsors as well as fish during steady motion have been found by a lot of observations; mean flexion ratio and angle are 0.65 and 26.5 deg, respectively [3], and the Strauhal number of most of the swimming animals is in the range of $0.25 < St < 0.35$ [4].

Lighthill [5] has indicated the fundamental theory of fish swimming and Newman has also proposed a generalized-slender body theory and fluid force for fish-like body [6]. These studies have suggested optimized Froude efficiency depending on motions of a slender body placed in a uniform potential flow.

Flow velocity fields around and behind a fish have been gradually made clear by measurements using such as particle image velocimetry [7], and robotic fish of which a tail dye tubes to visualize vortex structures in its wake can also obtain

Y. Ogata (✉) • T. Azama
Department of Mechanical System Engineering, Hiroshima University,
1-4-1 Kagamiyama, Higashi-Hiroshima, Hiroshima 152-8527, Japan
e-mail: yogata@hiroshima-u.ac.jp

Y. Bazilevs, K. Takizawa (eds.), *Advances in Computational Fluid-Structure Interaction and Flow Simulation*, Modeling and Simulation in Science, Engineering and Technology, DOI 10.1007/978-3-319-40827-9_12

hydrodynamic force by inner strain gauge [8]. On the contrary, numerical studies have also helped us in understanding the mechanism of fish swimming. For example, flow fields induced by a deforming fish-like body placed in a uniform flow have been simulated using moving grid system in two dimensions [9, 10]. Recent studies by three-dimensional simulations have shown correlation between fish acceleration and wake structure generated by a swimming fish [11, 12].

One of the authors has also investigated self-propulsion of a two-dimensional deforming NACA foil on the assumption of a small fish in not so large Reynolds number [13]. The present paper is extended to a three-dimensional small fish like a killifish accelerated by fluid forces in swimming. The constrained interpolation profile (CIP) method [14, 15] with an immersed boundary method [16, 17] is used to simulate interactions between a laminar flow and a deforming body accurately even in cartesian coordinate system. The CIP method has been applied to a lot of fluid–structure interaction (FSI) and successful results have been introduced [18, 19].

Basic equations and approaches for an accelerated fish are briefly introduced in Section 2. Section 3 displays flow fields arisen by a fish which impulsively starts deforming, and discusses fish acceleration. Concluding remarks are in Section 4.

2 Basic Equations and Approaches for an Accelerated Fish

Figure 1 shows a fish shape which we have used referring to a real killifish for three-dimensional simulation [20]. The fish is covered with triangle meshes for an immersed boundary method. Equation (1) is a function of the wavy lateral oscillation of the midline of the fish $h(x,t)$ with an amplitude $H(x)$ as follows:

$$h(x,t) = H(x)\cos\{\frac{2\pi}{\lambda}(X - f\lambda t)\}, \tag{1}$$

$$H(x) = 0.1\{\frac{13}{8}X^2 - \frac{33}{40}X + 0.2\} \tag{2}$$

$$= 0.1\{-(X-1)^2 + 1\}, \tag{3}$$

where λ, f, and $X = x - x_0$ are a wavelength, swimming frequency, and distance from the fish head located at x_0, respectively; the coordinate is normalized by fish length L, thus, the head and tail correspond to $X = 0$ and $X = 1$, respectively. Amplitude functions $H(x)$ of Eqs. (2) and (3) were suggested by Dong [21] and Akimoto [9], respectively.

Equation (2) is carangiform as the model (A) and Eq. (3) is anguilliform with larger deformation than Eq. (2) as the model (B) in Fig. 1. In the present paper, the fish is assumed to move in only x-direction by thrust F_x generated by cosine-wave deformation because time averages of lifts F_y and F_z during a period are negligible

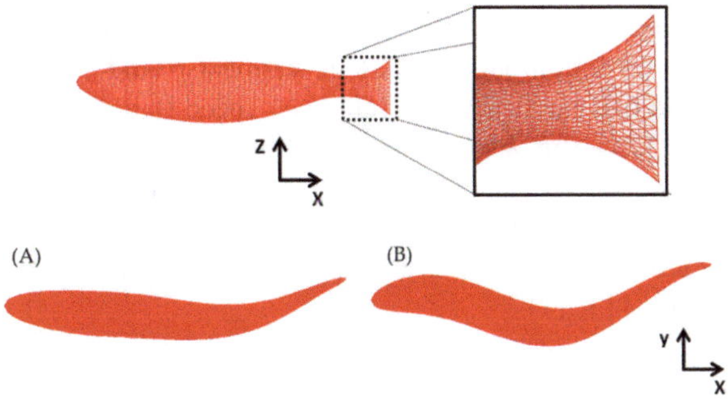

Fig. 1 Cross sections in lateral and top views of a killifish covered with triangle meshes for an IB method. Models (A) and (B) in the top view of *xy*-plane are snapshots of Eqs. (2) and (3), respectively

owing to the symmetry of the deformation. Therefore, the horizontal fish speed $U(t)$ is assumed to follow the Newton's law as follows:

$$M\frac{dU(t)}{dt} = F_x \equiv C_D\frac{1}{2}\rho_f U_0^2 L^2, \tag{4}$$

where $M = \rho_b V$ is mass of the fish with volume V; ρ_f and ρ_b are density of fluid and the fish body, respectively. Thus, Eq. (4) is normalized using a fish length L and a reference velocity U_0 as follows:

$$\gamma\hat{V}\frac{d\hat{U}(t)}{d\hat{t}} = \frac{1}{2}C_D, \tag{5}$$

where $\hat{t} = U_0 t/L$ is normalized time, and $\hat{U} = U/U_0$, $\hat{V} = V/L^3$ and $\gamma \equiv \rho_b/\rho_f$ are speed, volume, and the specific gravity of a fish, respectively. Here, $\gamma = 5.0$ is considered like a fish robot, and the volume $\hat{V}$ is 0.009 for the present model. From now on, the ˆ sign to represent dimensionless variables is dropped for simple notation. Drag coefficient C_D can be obtained by surface integral of pressure and viscous shear stress on the fish body, and then the fish speed U is also obtained by time integral of dU/dt in Eq. (5). Note that both C_D and dU/dt will be negative when a fish is accelerated in Fig. 1.

The virtual boundary (VB) method [16, 17] is employed to identify a deforming body in a fluid, and a non-inertial coordinate in which the swimming fish is fixed at a position. Therefore, the fluid velocity $\mathbf{u}$ is replaced with $\mathbf{u}' = \mathbf{u} - \mathbf{U}(t)$, where $\mathbf{U}(t) = (U, 0, 0)$, and the dimensionless incompressible Navier–Stokes equations in the coordinate system for the VB method can be described as follows:

$$\nabla \cdot \mathbf{u}' = 0, \tag{6}$$

$$\frac{\partial \mathbf{u}'}{\partial t} + \left(\mathbf{u}' \cdot \nabla\right)\mathbf{u}' = -\nabla p + \frac{1}{Re}\nabla^2\mathbf{u}' + \mathbf{G}' - \frac{d\mathbf{U}}{dt}, \tag{7}$$

thus, the inflow boundary is $u'(x = 0, t) = -U (\geq 0)$. Since there is no characteristic flow speed for impulsive start in stationary fluid, we take a reference phase speed of the fish $c_0 = f_0\lambda$ as a reference speed to define Reynolds number $Re \equiv c_0L/\nu = f_0L^2/\nu$ in $\lambda = L$. For example, $Re \cong 5000$ when a fish with the body length $L = 7$cm and frequency $f_0 = 1.0$Hz is considered in water $\nu = 1.0 \times 10^{-6} m^2/s$. The last term in the right-hand side of Eq. (7) can be obtained by Eq. (5). The additional force $\mathbf{G}'$ is added not only at points near the boundary but also inside the body [13, 17].

3 Acceleration Model of a Small Fish

Next, we consider an acceleration model of a three-dimensional killifish-like deforming model. The fish begins to deform at $t = 0$ in stationary fluid, following Eq. (1) with $H(x)$ of Eqs. (2) and (3). Each direction of the computational region is divided into nonuniform sections with minimum mesh sizes $(\Delta x_{min}, \Delta y_{min}, \Delta z_{min}) = \left(2.0 \times 10^{-2}, 8.7 \times 10^{-3}, 8.0 \times 10^{-3}\right)$ around the fish, and they become gradually large near boundaries of the region.

3.1 Vortical Structure on Wake of a Small Fish

Figure 2 displays snapshots of wake structure of both Eqs. (2) and (3) for $Re = 5000$ and $f = 3.0$. The structure is represented by isosurface of the second invariant $Q \equiv \left(\Omega_{ij}\Omega_{ij} - S_{ij}S_{ij}\right)/2 = 0.1$; $S_{ij} = \left(\partial u_i/\partial x_j + \partial u_j/\partial x_i\right)/2$ and $\Omega_{ij} = \left(\partial u_i/\partial x_j - \partial u_j/\partial x_i\right)/2$ are symmetric and asymmetric parts of the velocity gradient, respectively.

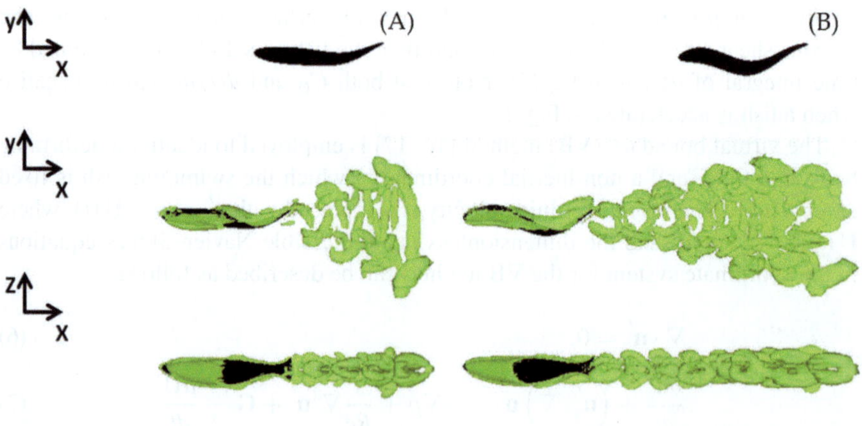

Fig. 2 Comparison of wake structure $Q = 0.1$ during swimming from different views. (A) : Eq. (2), (B) : Eq. (3)

The fish with Eq. (3) is accelerated faster than the one with Eq. (2) at the same number of kicks to swim. It is seen that strong vortex and clear vortex rings are generated on wake of the fish with Eq. (3) due to the deformation on the whole body. The strong vortex is considered to be the source of propulsion as reversed Karman vortex in two-dimensions.

3.2 Estimation of a Small Fish Speed; Re Dependence

It has been found in two-dimensional simulations of our previous study that the right-hand side of Eq. (5) can be estimated using the similar model to free fall of a sphere as follows [13]:

$$\gamma V \frac{dU}{dt} = \frac{1}{2} C_D \equiv a - bU - cU^2, \tag{8}$$

where (a, b, c) are undetermined positive coefficients. The study has shown that the third term of inertial drag is much smaller than the second term of viscous drag in Eq. (8) because the viscous drag dominates and the coefficient c is found to be very small in low Reynolds number. Therefore, it is also assumed in three-dimensions that the third term is negligible as $c = 0$ in Eq. (8).

The exact solution of Eq. (8) with $c = 0$ can be easily obtained as follows:

$$U(t)_{exa} = \frac{a}{b}\left[1 - exp\left(-\frac{b}{\gamma V} t\right)\right], \tag{9}$$

where coefficients (a, b) could be also estimated using least squares method so that U_{exa} by Eq. (9) is fit with numerical solutions U_{cal} in three-dimensions as two-dimensions.

Figure 3 shows dependence of time variation of fish speed U and coefficients (a, b) on Re from 300 to 12000, respectively, for the model (A) with $f = 1.0$. Dots and dash lines in Fig. 3(a) are U_{cal} and U_{exa}, respectively; each dot of U_{cal} is time average in each deformation period.

Figure 3(a) has indicated that Eq. (9) agrees with U_{cal} very well by finding appropriate coefficients (a, b) for each Re. The fish speed is large for large Re, and will converge a terminal speed $U_\infty = a/b$ as time goes on.

It is found in Fig. 3(b) that the coefficient a is almost independent of Re, while the coefficient b decreases and becomes an asymptotic value with increasing Re. The coefficient a which corresponds to the thrust part is considered to be determined by fish shape and deformation. On the other hand, the coefficient b can be expressed as a function of Re as follows:

$$b = \frac{C_1}{\sqrt{Re}}\left(1 + C_2\sqrt{Re}\right), \tag{10}$$

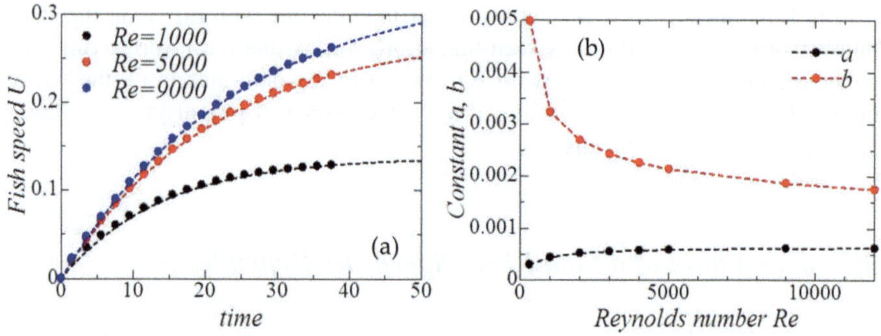

Fig. 3 (a) Time history of fish speed U for Eq. (2) with $f = 1.0$. Dots are simulation results and dash lines are Eq. (9) with appropriate (a, b). (b)dependence of (a, b) on *Re*

where (C_1, C_2) are almost $(0.65, 0.002)$ in the present condition. Figure 3 implies that, if we could find the pair (a, b) obtained by numerical solutions for two *Re* in which fish conditions such as swimming motion are determined, dependence of (a, b) on *Re* can be found and then, fish speed U for any small *Re* can be predicted using Eqs. (9) and (10).

4 Conclusion

An acceleration of a three-dimensional fish model of small size or low speed is discussed through FSI simulations, and a prediction model of fish speed is proposed.

Fluid force generated by fish deformation is decomposed to thrust part (positive term) and drag part (negative term). The model to predict fish speed is the same as our previous study of a deforming NACA foil in two-dimensions, and in good agreement with simulation results. However, the model might be applied to only small *Re* for laminar flow because of turbulent flows for larger *Re*, such as a fast fish like a dolphin. In addition, a fish actually has not only a caudal fin but also some fins, which also play very important roles for acceleration including a turn. Dependence of not only fish speed but also efficiency to save energy to swim on swimming frequency, amplitude function, and fish shape will be shown and discussed in detail in our future papers.

Acknowledgements I would like to thank Prof. Hosotani in Tsuyama National College of Technology, who has given our group a lot of information and discussion.

References

1. Hosotani, K., et al.: Development of simple underwater robots with oscillating propulsion fins to encourage interest in engineering. In: Proceedings of Techno-Ocean 2010 International Conference (2010)
2. Nakashima, M., et al.: Three-dimensional movement in water of the dolphin robot - control between two positions by roll and pitch combination. J. Rob. Mechatronics **18**, 347–355 (2006)
3. Kelsey, N. L., et al.: Bending rules for animal propulsion. Nat. Commun. (2014) doi:10.1038/ncomms4293
4. Eloy, C.: Optimal Strouhal number for swimming animcals. J. Fluids Struct. **30**, 205–218 (2012)
5. Lighthill, M.J.: Note on the swimming of slender fish. Fluid Mech. **9**, 305–317 (1960)
6. Newman, J.N.: The force on a slender fish-like body. J. Fluid Mech. **58**, 689–702 (1973)
7. Lauder, G.V., et al.: Fish biorobotics: kinematics and hydrodynamics of self-propulsion. J. Exp. Biol. **210**, 2767–2780 (2007)
8. Tan, G.-K., et al: Investigation of flow mechanism of a robotic fish swimming by using flow visualization synchronized with hydrodynamic force measurement. Exp. Fluids **43**, 811–821 (2007)
9. Akimoto, H., Miyata, H.: Finite-volume simulation of a flow about a moving body with deformation. In: Proceedings of the 5th International Symposium on Computational Fluid Dynamics, pp. 13–18 (1993)
10. Nakaoka, T., Toda, Y.: Laminar flow computation of fish-like motion wing. In: Proceedings of the Fourth International Offshore and Polar Engineering Conference, pp. 530–538 (1994)
11. Borazjani, I., Sotiropoulos, F.: On the role of form and kinematics on the hydrodynamics of self-propelled body/caudal fin swimming. J. Exp. Biol. **213**, 89–107 (2010)
12. Li, G., et al.: Body dynamics and hydrodynamics of swimming fish larvae: a computational study. J. Exp. Biol. **215**, 4015–4033 (2012)
13. Ogata, Y., Ogasawara, T.: Accelerated motions of a fish-like foil from impulsive starts to terminal states. J. Fluid Sci. Technol. **7**, 358–373 (2012)
14. Yabe, T., Aoki, T.: A universal solver for hyperbolic equations by cubic-polynomial interpolation I. One-dimensional solver. Comput. Phys. Commun. **66**, 219–232 (1991)
15. Xiao, F., et al.: Constructing oscillation preventing scheme for advection equation by rational function. Comput. Phys. Commun. **93**, 1–12 (1996)
16. Saiki, E.M., Biringen, S.: Numerical simulation of a cylinder in uniform flow: application of a virtual boundary method. J. Comput. Phys. **123**, 450–465 (1996)
17. Nishida, H., Tajiri, K.: Numerical simulation of incompressible flows around a fish model at low Reynolds number using seamless virtual boundary method. J. Fluid Sci. Technol. **4**, 500–511 (2009)
18. Yabe, T., et al.: The constrained interpolation profile method for multiphase analysis. J. Comput. Phys. **169**, 556–593 (2001)
19. Takizawa, K., et al.: Simulation and experiment on swimming fish and skimmer by CIP method. Comput. Struct. **83**, 397–408 (2005)
20. Azama, T., et al.: Correlation between vortical structure and propulsive efficiency of swimming fish. In: Proceedings of 28th Computational Fluid Dynamics 2014 (2014) [in Japanese]
21. Dong, G.-J., Lu, X.-Y.: Numerical analysis on the propulsive performance and vortex shedding of fish-like travelling wavy plate. Int. J. Numer. Methods Fluids **48**, 1351–1373 (2005)

New Directions in Space–Time Computational Methods

Kenji Takizawa and Tayfun E. Tezduyar

Abstract This is an overview of some of the new directions we have taken the space–time (ST) computational methods since 2010 in bringing solution and analysis to different classes of challenging engineering problems. The new directions include the variational multiscale (VMS) version of the Deforming-Spatial-Domain/Stabilized ST method, using NURBS basis functions in temporal representation of the unknown variables and motion of the solid surfaces and fluid mechanics meshes, ST techniques with continuous representation in time, ST interface-tracking with topology change, and the ST-VMS method for flow computations with slip interfaces. We describe these new directions and present a few examples.

1 Introduction

In computational engineering analysis, one frequently faces the challenges involved in solving flow problems with moving boundaries and interfaces (MBI). This wide class of problems include fluid–structure interaction (FSI), fluid–object interaction (FOI), fluid–particle interaction (FPI), free-surface and multi-fluid flows, and flows with solid surfaces in fast, linear, or rotational relative motion. The computational challenges still being addressed include accurately representing the small-scale flow patterns, which require a reliable multiscale method. They also include contact or near contact between moving solid surfaces and other cases of topology change (TC) or near TC, such as those in flapping-wing aerodynamics, wind-turbine aerodynamics, cardiovascular fluid mechanics, and thermo-fluid analysis of ground

K. Takizawa (✉)
Department of Modern Mechanical Engineering, Waseda University, 1-6-1 Nishi-Waseda, Shinjuku-ku, Tokyo 169-8050, Japan
e-mail: Kenji.Takizawa@tafsm.org

T.E. Tezduyar
Mechanical Engineering, Rice University – MS 321, 6100 Main Street, Houston, TX 77005, USA
e-mail: tezduyar@tafsm.org

Y. Bazilevs, K. Takizawa (eds.), *Advances in Computational Fluid-Structure Interaction and Flow Simulation*, Modeling and Simulation in Science, Engineering and Technology, DOI 10.1007/978-3-319-40827-9_13

vehicle tires. These four classes of problems played a key role in motivating the development of the space–time (ST) computational methods discussed in this article.

A method for flows with MBI can be viewed as an interface-tracking (moving-mesh) technique or an interface-capturing (nonmoving-mesh) technique, or possibly a combination of the two. In interface-tracking methods, as the interface moves, the mesh moves to follow (i.e., "track") the interface. The Arbitrary Lagrangian–Eulerian (ALE) finite element formulation [1] is the most widely used moving-mesh technique. It has been used for many flow problems with MBI, including FSI (see, for example, [2–6]). The Deforming-Spatial-Domain/Stabilized ST (DSD/SST) method [4, 7–11], introduced in 1992, is also a moving-mesh method. The costs and benefits of moving the fluid mechanics mesh to track a fluid–solid interface were articulated in many papers (see, for example, [8]).

Moving-mesh methods require mesh update methods. Mesh update consists of moving the mesh for as long as possible and remeshing as needed. With the key objectives being to maintain the element quality near solid surfaces and to minimize frequency of remeshing, a number of advanced mesh update methods [9, 12, 13] were developed in conjunction with the DSD/SST method, including those that minimize the deformation of the layers of small elements placed near solid surfaces.

Perceived challenges in mesh update are quite often given as reasons for avoiding interface-tracking methods in classes of problem that can be, and actually have been, solved with such methods. A robust moving-mesh method with effective mesh update can handle FSI or other MBI problems even when the solid surfaces undergo large displacements (see, for example, FPI [14, 15] with the number of particles reaching 1,000 [15], parachute FSI [9, 16], flapping-wing aerodynamics [17–22]), and wind-turbine rotor and tower aerodynamics [21, 23]. It can handle FSI or other MBI problems also even when the solid surfaces are in near contact or create near TC, if the "nearness" is sufficiently "near" for the purpose of solving the problem. Examples of such problems are FPI with collision between the particles [14, 15], parachute-cluster FSI with contact between the parachutes of the cluster [16], flapping-wing aerodynamics with the forewings and hindwings crossing each other very close [17–21], and wind-turbine rotor and tower aerodynamics with the blades passing the tower close [21, 23].

No method is a panacea for all classes of MBI problems. Some interfaces, such as those in splashing, might be too complex for an interface-tracking technique, requiring an interface-capturing technique. The Mixed Interface-Tracking/Interface-Capturing Technique (MITICT) [15] was introduced for computations that involve both fluid–solid interfaces that can be accurately tracked with a moving-mesh method and fluid–fluid interfaces that are too complex to be tracked. Those fluid–fluid interfaces are captured over the mesh tracking the fluid–solid interfaces. The MITICT was successfully tested in 2D computations with solid circles and free surfaces [24] and in 3D computation of ship hydrodynamics [25].

In some MBI problems with contact between the solid surfaces, the "nearness" that can be modeled with a moving-mesh method without actually bringing the surfaces into contact might not be "near" enough for the purpose of solving the problem. Cardiovascular FSI with heart valves is an example. The Fluid–Solid

Interface-Tracking/Interface-Capturing Technique (FSITICT) [26] was motivated by such FSI problems. In the FSITICT, we track the interface we can with a moving mesh, and capture over that moving mesh the interfaces we cannot track, specifically the interfaces where we need to have an actual contact between the solid surfaces. Two versions of the FSITICT were proposed, one with fixed-partitioning (FSITICT-FP), and one with adaptive-partitioning (FSITICT-AP). In the FSITICT-FP, the tracked/captured partitioning of the fluid–solid interface will be fixed during the computation. In the FSITICT-AP, the mesh will stop tracking the parts that have become too difficult to track, leaving them to be captured. This does not require remeshing. If any of the captured parts become suitable for tracking, a new mesh that is boundary-conforming for those parts will be generated and the tracking of those parts will start. An application of the FSITICT-FP was presented in [27], where an ALE method was used for interface-tracking, and a fully Eulerian approach for interface capturing, with some 2D test computations. This specific application was extended in [28] to 2D FSI models with flapping and contact. The FSITICT-FP was successfully extended in [29] to 3D FSI computation of a bioprosthetic heart valve. In that case the interface-tracking technique was an ALE method, and the interface-capturing technique was a variational immersed-boundary method.

Since its inception, the DSD/SST method has been applied to some of the most challenging flow problems with MBI. The classes of problems solved include the free-surface and multi-fluid flows [7, 24], FOI [7, 24], aerodynamics of flapping wings [17–22, 30], flows with solid surfaces in fast, linear, or rotational relative motion [14, 15, 23], compressible flows [14, 31], shallow-water flows [15, 32], FPI [14, 15], and FSI [16, 20, 30, 31, 33]. Much of the success with the DSD/SST method in recent years has been due to the new directions we have taken the ST methods in bringing solution and analysis to different classes of computationally challenging engineering problems.

The original DSD/SST method is based on the SUPG/PSPG stabilization, where "SUPG" and "PSPG" stand for the Streamline-Upwind/Petrov-Galerkin [34] and Pressure-Stabilizing/Petrov-Galerkin [7] methods. Starting in its early years, the DSD/SST method also included the "LSIC" (least-squares on incompressibility constraint) stabilization. The ST-VMS method [4, 10, 11] is the variational multiscale version of the DSD/SST method. It was called "DSD/SST-VMST" (i.e., the version with the VMS turbulence model) when it was first introduced in [10]. The VMS components are from the residual-based VMS method given in [35, 36]. We have been using the ST-VMS method since its inception in most of our computations. The original DSD/SST method was named "DSD/SST-SUPS" in [10] (i.e., the version with the SUPG/PSPG stabilization), which was also called "ST-SUPS" in [4].

The ST methods give us the option of using higher-order basis functions in time, including the NURBS, which have been used very effectively as spatial basis functions (see [37, 38]). We started using that option with the methods and concepts introduced in [10]. This of course increases the order of accuracy in the computations [39], and the desired accuracy can be attained with larger time steps, but there are positive consequences beyond that. The ST context provides us better accuracy and efficiency in temporal representation of the motion and deformation of the moving interfaces and volume meshes, and better efficiency in remeshing. This has been demonstrated in a number of 3D computations, specifically, flapping-wing aerodynamics [17–21], separation aerodynamics of spacecraft [40], and wind-turbine aerodynamics [21, 23]. The mesh update method based on using NURBS basis functions in mesh motion and remeshing was named "ST/NURBS Mesh Update Method (STNMUM)" in [23].

There are some advantages in using a discontinuous temporal representation in ST computations. For a given order of temporal representation, we can reach a higher-order accuracy than one would reach with a continuous representation of the same order. When we need to change the spatial discretization (i.e., remesh) between two ST slabs, the temporal discontinuity between the slabs provides a natural framework for that change. There are advantages also in continuous temporal representation. We obtain a smooth solution, NURBS-based when needed. We also can deal with the computed data in a more efficient way, because we can represent the data with fewer temporal control points, and that reduces the computer storage cost. These advantages motivated the development of the ST computation techniques with continuous temporal representation (ST-C) [41].

There are different types of nonmoving-mesh methods that can compute MBI problems involving an actual contact between solid surfaces or other cases of TC. Some of those methods give up on the accurate representation of the interface, and most give up on the consistent representation of the interface motion. The DSD/SST formulation does not need to give up on either, even where we have an actual contact or some other TC, provided that we can update the mesh even there. Using an ST mesh that is unstructured both in space and time, as proposed for contact problems in [15], would give us such a mesh update option. However, that would require a fully unstructured 4D mesh generation, and that is not easy in computing real-world problems. An ST interface-tracking method that can deal with TC was introduced in [42], and it is called ST-TC. It is a practical alternative to using unstructured ST meshes, but without giving up on the accurate representation of the interface or the consistent representation of the interface motion, even where there is an actual contact between solid surfaces or other TC. The ST-TC method is based on special mesh generation and update, and a master–slave system that maintains the connectivity of the "parent" mesh when there is a TC.

The ST Slip Interface (ST-SI) method [43] is the ST version of the ALE-VMS sliding-interface method [44]. It is for FSI and other MBI problems where one or more of the subdomains contain spinning structures, and the mesh covering a spinning structure spins with it, thus maintaining the high-resolution representation of the boundary layers near the structure. The subdomains are covered by meshes that do not match at the interface and have slip between them, and the ST-SI method carefully accounts for, in the ST context, the compatibility conditions for the velocity and stress between the two sides.

In Section 2 we briefly review the ST basis functions. The ST-VMS method is described in Section 3, and the STNMUM in Section 4. The ST-C, ST-TC, and ST-SI methods are described in Sections 5, 6, and 7. The examples are given in Section 8, and the concluding remarks in Section 9.

2 ST Basis Functions

The concept of using NURBS basis functions, in conjunction with the ST methods, in temporal representation of the unknown variables and motion of the solid surfaces and fluid meshes was introduced in [10]. An ST basis function can be written as a product of its spatial and temporal parts:

$$N_a^\alpha = T^\alpha(\theta) N_a(\boldsymbol{\xi}), \quad a = 1, 2, \ldots, n_{\text{en}}, \\ \alpha = 1, 2, \ldots, n_{\text{ent}}, \tag{1}$$

where $\theta \in [-1, \ 1]$ is the temporal element coordinate, and n_{en} and n_{ent} are the number of spatial and temporal element nodes. Figure 1 shows an example. Temporal NURBS basis functions can be used in an ST slab for the representation of the unknown variables and test functions as well as the spatial coordinates. As pointed out in [4, 10, 11, 23], different components (i.e., unknowns), and the corresponding test functions, can be discretized with different sets of temporal basis functions. This was shown in [4, 10, 11, 23] by introducing a secondary mapping $\Theta_\zeta(\theta) \in [-1, \ 1]$, where $\Theta_\zeta(\theta)$ is a strictly increasing function, and rewriting the generalized ST basis function for the element indices $(a, \ \alpha)$ as

$$\left(N_a^\alpha\right)_\zeta = T^\alpha\left(\Theta_\zeta(\theta)\right) N_a(\boldsymbol{\xi}). \tag{2}$$

For example, we can discretize time and position as

$$t = \sum_{\alpha=1}^{n_{\text{ent}}} T^\alpha(\Theta_t(\theta)) t^\alpha, \tag{3}$$

$$\mathbf{x} = \sum_{\alpha=1}^{n_{\text{ent}}} T^\alpha(\Theta_x(\theta)) \mathbf{x}^\alpha. \tag{4}$$

Here $\Theta_t(\theta)$ and $\Theta_x(\theta)$ are the secondary mappings for time and position, and t^α and $\mathbf{x}^\alpha$ are the time and position values corresponding to the basis function T^α.

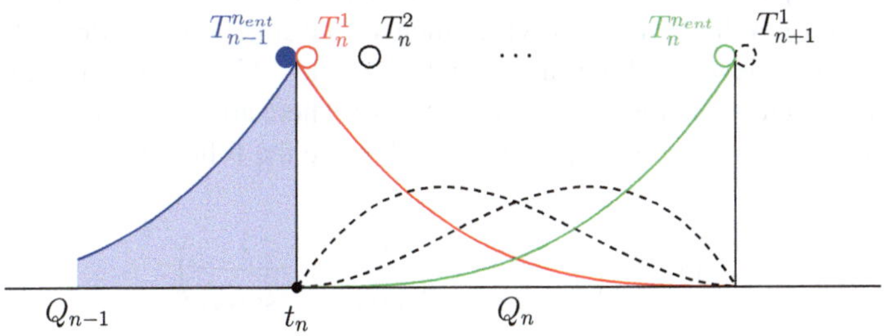

Fig. 1 Temporal NURBS basis functions.

3 ST-VMS Method

The ST-VMS method can be written as

$$\begin{aligned}
&\int_{Q_n} \mathbf{w}^h \cdot \rho \left(\frac{\partial \mathbf{u}^h}{\partial t} + \mathbf{u}^h \cdot \nabla \mathbf{u}^h - \mathbf{f}^h \right) \mathrm{d}Q + \int_{Q_n} \boldsymbol{\varepsilon}(\mathbf{w}^h) : \boldsymbol{\sigma}(\mathbf{u}^h, p^h) \mathrm{d}Q \\
&- \int_{(P_n)_\mathrm{h}} \mathbf{w}^h \cdot \mathbf{h}^h \mathrm{d}P + \int_{Q_n} q^h \nabla \cdot \mathbf{u}^h \mathrm{d}Q + \int_{\Omega_n} (\mathbf{w}^h)_n^+ \cdot \rho \left((\mathbf{u}^h)_n^+ - (\mathbf{u}^h)_n^- \right) \mathrm{d}\Omega \\
&+ \sum_{e=1}^{(n_\mathrm{el})_n} \int_{Q_n^e} \frac{\tau_\mathrm{SUPS}}{\rho} \left[\rho \left(\frac{\partial \mathbf{w}^h}{\partial t} + \mathbf{u}^h \cdot \nabla \mathbf{w}^h \right) + \nabla q^h \right] \cdot \mathbf{r}_\mathrm{M}(\mathbf{u}^h, p^h) \mathrm{d}Q \\
&+ \sum_{e=1}^{(n_\mathrm{el})_n} \int_{Q_n^e} \nu_\mathrm{LSIC} \nabla \cdot \mathbf{w}^h \rho r_\mathrm{C}(\mathbf{u}^h) \mathrm{d}Q - \sum_{e=1}^{(n_\mathrm{el})_n} \int_{Q_n^e} \tau_\mathrm{SUPS} \mathbf{w}^h \cdot \left(\mathbf{r}_\mathrm{M}(\mathbf{u}^h, p^h) \cdot \nabla \mathbf{u}^h \right) \mathrm{d}Q \\
&- \sum_{e=1}^{(n_\mathrm{el})_n} \int_{Q_n^e} \frac{\tau_\mathrm{SUPS}^2}{\rho} \mathbf{r}_\mathrm{M}(\mathbf{u}^h, p^h) \cdot \left(\nabla \mathbf{w}^h \right) \cdot \mathbf{r}_\mathrm{M}(\mathbf{u}^h, p^h) \mathrm{d}Q = 0, \qquad (5)
\end{aligned}$$

where

$$\mathbf{r}_\mathrm{M}(\mathbf{u}^h, p^h) = \rho \left(\frac{\partial \mathbf{u}^h}{\partial t} + \mathbf{u}^h \cdot \nabla \mathbf{u}^h - \mathbf{f}^h \right) - \nabla \cdot \boldsymbol{\sigma}(\mathbf{u}^h, p^h), \qquad (6)$$

$$r_\mathrm{C}(\mathbf{u}^h) = \nabla \cdot \mathbf{u}^h \qquad (7)$$

are the residuals of the momentum equation and incompressibility constraint. Here, ρ, $\mathbf{u}$, p, $\mathbf{f}$, $\boldsymbol{\sigma}$, $\boldsymbol{\varepsilon}$, and $\mathbf{h}$ are the density, velocity, pressure, external force, stress tensor, strain rate tensor, and the traction specified at the boundary. The test functions associated with the velocity and pressure are $\mathbf{w}$ and q. A superscript "h" indicates that the function is coming from a finite-dimensional space. The symbol Q_n represents the ST slice between time levels n and $n+1$, $(P_n)_\mathrm{h}$ is the part of the lateral boundary of that slice associated with the traction boundary condition $\mathbf{h}$, and Ω_n is the spatial domain at time level n. The superscript "e" is the ST element counter, and n_el is the number of ST elements. The functions are discontinuous in time at each time level, and the superscripts "−" and "+" indicate the values of the functions just below and just above the time level. There are various ways of defining the stabilization parameters τ_SUPS and ν_LSIC. We use the following expressions:

$$\tau_\mathrm{SUPS} = \left(\frac{1}{\tau_\mathrm{SUGN12}^2} + \frac{1}{\tau_\mathrm{SUGN3}^2} + \frac{1}{\tau_\mathrm{SUGN4}^2} \right)^{-\frac{1}{2}}, \qquad (8)$$

$$\tau_\mathrm{SUGN12} = \left(\sum_{\alpha=1}^{n_\mathrm{ent}} \sum_{a=1}^{n_\mathrm{ens}} \left| \frac{\partial N_a^\alpha}{\partial t} + \mathbf{u}^h \cdot \nabla N_a^\alpha \right| \right)^{-1}, \qquad (9)$$

$$\tau_{\text{SUGN3}} = \frac{h^2_{\text{RGN-MIN}}}{4\nu}, \tag{10}$$

$$\tau_{\text{SUGN4}} = \left\|\nabla \mathbf{u}^h\right\|_F^{-1}, \tag{11}$$

$$\nu_{\text{LSIC}} = \frac{h^2_{\text{RGN-MIN}}}{4\tau_{\text{SUPS}}}, \tag{12}$$

where n_{ens} and n_{ent} are the number of spatial and temporal element nodes, N_a^α is the basis function associated with spatial and temporal nodes a and α, $\|\cdot\|_F$ represents the Frobenius norm,

$$h_{\text{RGN-MIN}} = 2\left\|\left\{\nabla N_a^\alpha\right\}\right\|_2^{-1} \equiv 2\left(\lambda_{\max}\left(\sum_{\alpha=1}^{n_{\text{ent}}}\sum_{a=1}^{n_{\text{ens}}} \nabla N_a^\alpha \nabla N_a^\alpha\right)\right)^{-\frac{1}{2}}, \tag{13}$$

and $\lambda_{\max}(\cdot)$ represents the maximum eigenvalue of the symmetric tensor.

Remark 1. Most of these stabilization parameters originate from [8, 9].

Remark 2. The component τ_{SUGN4}, given by Eq. (11), was introduced in [45]. It was introduced based on the reasoning that the $\tau_{\text{SUPS}}\mathbf{w}^h \cdot \left(\mathbf{r}_{\text{M}} \cdot \nabla \mathbf{u}^h\right)$ term should not overwhelm the $\mathbf{w}^h \cdot \rho(\mathbf{u}^h \cdot \nabla \mathbf{u}^h)$ term.

Remark 3. A symmetric version of that τ_{SUGN4} was also introduced in [45]:

$$\tau_{\text{SUGN4}} = \|\boldsymbol{\varepsilon}(\mathbf{u}^h)\|_F^{-1}. \tag{14}$$

4 STNMUM

4.1 Mesh Computation and Representation

Given the fluid mechanics mesh on a moving solid surface, we compute the fluid mechanics volume mesh using the mesh moving methods [9, 12, 13] developed in conjunction with the DSD/SST method. As proposed in [46] and also described in [4], these mesh moving methods are used in computing the meshes that will serve as temporal control points. This allows us to do mesh computations with longer time in between, but get the mesh-related information, such as the coordinates and their time derivatives, from the temporal representation whenever we need. Of course this also reduces the storage amount and access associated with the meshes.

Remark 4. Getting the meshes used in the computations from the temporal representation can be done independent of which time direction was used in computing the control meshes. For example, in flapping-wing aerodynamics, the control meshes can be computed while the wings are flapping forward or backward in time.

4.2 Remeshing

In many computations remeshing becomes unavoidable. Two choices were proposed in [46] and also described in [4]. To explain those choices, let us assume that when we try to move from control mesh M_c^{β} to $M_c^{\beta+1}$, we find the quality of $M_c^{\beta+1}$ to be less than desirable. In the first choice, called "trimming," we remesh going back to $M_c^{\beta-p+1}$, where p is the order of the NURBS basis functions. Then whenever our solution process needs a mesh, depending on the time, we use the control meshes belonging to either only the un-remeshed set or only the remeshed set (Figure 2). In the second choice, we perform knot insertion p times in the temporal representation of the surface at the right-most knot before the maximum value of the basis function corresponding to $t_c^{\beta+1}$, making that knot a new patch boundary. Then we do the mesh moving computation for the control meshes associated with the newly defined basis functions, not only the one at the new patch boundary, but also going back $(p-1)$ basis functions (Figure 3). We favor the second choice, because we believe that in many cases the need for remeshing is generated by a topological change, which we can avoid going over with a large step if we use the knot insertion process.

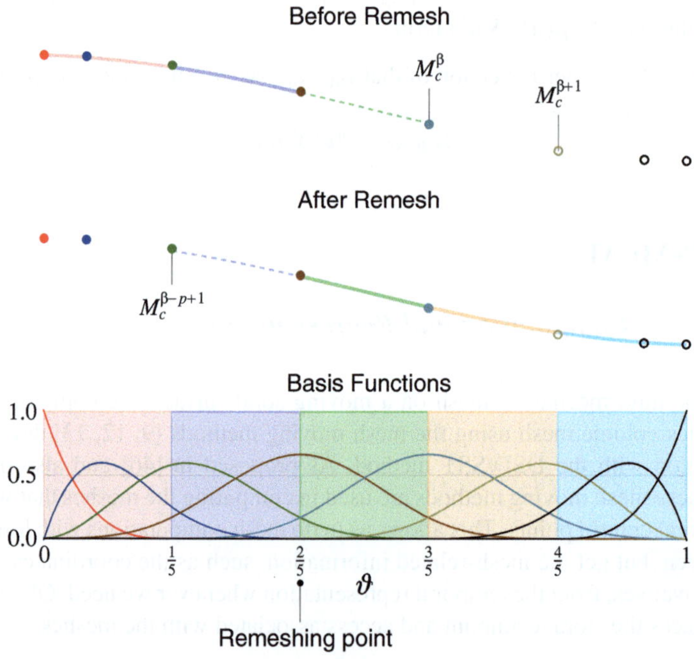

Fig. 2 Remeshing and trimming NURBS. A set of un-remeshed meshes (top). A set of remeshed meshes (middle). Common basis functions (bottom).

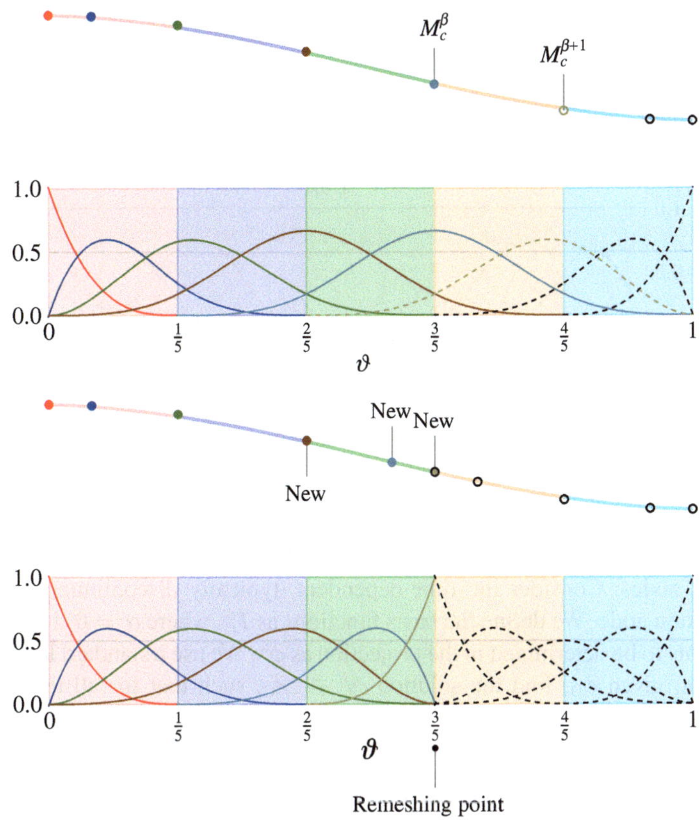

Fig. 3 Remeshing with knot insertion. For the set of un-remeshed meshes, there are p newly defined basis functions and the corresponding control points are marked "New." We carry out the mesh moving computations for those meshes.

5 ST-C Method

We describe, from [41], the version of the ST-C method used in extracting continuous temporal representation from computed data. This is essentially a post-processing method, and can also be seen as a data compression method. For the version used in direct computation of the solution with continuous temporal representation, see [41].

5.1 *Least-Squares Projection for Full Temporal Domain*

When we have the complete sequence of computed data, we can project that to a smooth representation, with basis functions that provide us that smooth representation, such as NURBS basis functions. As an example, Figure 4 shows the goal

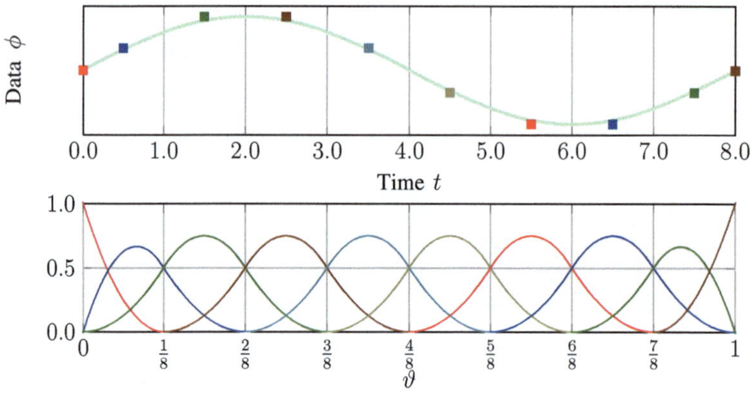

Fig. 4 Continuous solution (top) and its basis functions (bottom); ϑ is the parametric coordinate.

continuous data ϕ_C and its basis functions, where ϑ denotes the parametric temporal coordinate. The projection for each spatial node can be done independently from the other nodes. Consider the time-dependent, typically discontinuous computed data ϕ_D for a node. We define the basis functions as T_C^α, where $\alpha = 0, 1, \ldots$, and the coefficients to be determined in the projection as ϕ^α. We use a standard least-squares projection: given ϕ_D, find the solution $\phi_C \in \mathscr{S}_C$, such that for all test functions $w_C \in \mathscr{V}_C$:

$$\int_0^T w_C (\phi_C - \phi_D) \, dt = 0, \tag{15}$$

where T represents time period of the computation, and $\mathscr{S}_C$ and $\mathscr{V}_C$ are the solution and test function spaces constructed from the basis functions. This approach requires that we store all the computed data before the projection, and that would be a significant computer storage cost when the number of time steps is large.

5.2 *Successive-Projection Technique*

In ST-C with the successive-projection technique (ST-C-SPT), we extract the continuous solution shown in Figure 4 without storing all the computed data. We describe the technique here for the special case with quadratic B-splines. To explain the successive nature of the SPT, let us suppose that we have the continuous solution extracted up to $t_n = 4.0$, as shown in Figure 5. We assume that this continuous solution, which we will call $\overline{\phi}_C$, has already replaced ϕ_D up to $t_n = 4.0$. With that, we describe how we extract the continuous solution up to $t_{n+1} = 5.0$, as shown in Figure 6. With the newly computed data ϕ_D between $t_n = 4.0$ and $t_{n+1} = 5.0$,

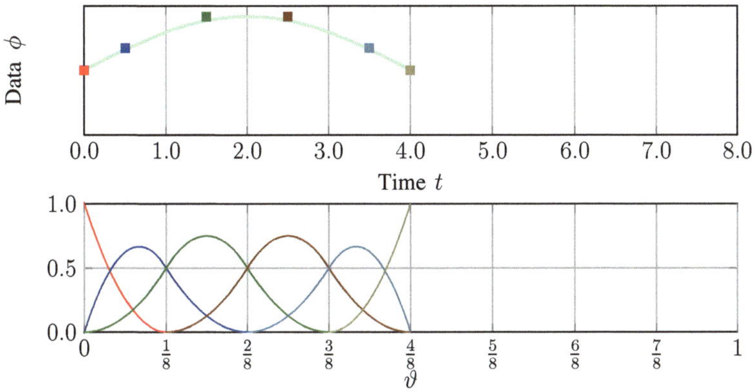

Fig. 5 Continuous solution up to $t_n = 4.0$ (top) and its basis functions (bottom).

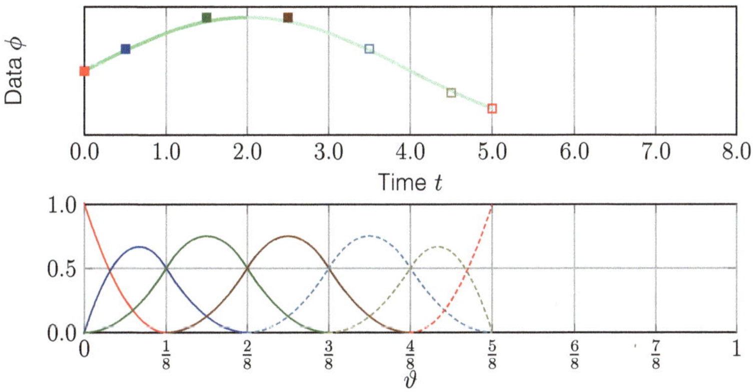

Fig. 6 Continuous solution up to $t_{n+1} = 5.0$ (top) and its basis functions (bottom). The bold part of the top curve indicates the part of the solution that does not change. The empty squares denote the temporal control values to be determined. The dashed lines denote the modified and new basis functions, which correspond to those control values.

we solve the following projection equation: given ϕ_{D} on $t \in (4.0, 5.0)$, $\overline{\phi}_{\mathrm{C}}$ on $t \in [2.0, 4.0]$, and $\phi_{\mathrm{C}}^{\alpha}$, $\alpha = 2, 3$, find $\phi_{\mathrm{C}} \in \mathscr{S}_{\mathrm{C}}$, such that $\forall w_{\mathrm{C}} \in \mathscr{V}_{\mathrm{C}}$:

$$\int_{2.0}^{4.0} w_{\mathrm{C}} \left(\phi_{\mathrm{C}} - \overline{\phi}_{\mathrm{C}}\right) \mathrm{d}t + \int_{4.0}^{5.0} w_{\mathrm{C}} \left(\phi_{\mathrm{C}} - \phi_{\mathrm{D}}\right) \mathrm{d}t = 0. \tag{16}$$

Equation (16) is essentially used for defining the coefficients $\phi_{\mathrm{C}}^{\alpha}$, $\alpha = 4, 5, 6$, which correspond to the basis functions T_{C}^{α}. How to deal with the initial part of the extraction, description of the ST-C-SPT for the general case (i.e., beyond quadratic B-splines), and comments on efficient implementation of the SPT can be found in [41].

6 ST-TC Method

6.1 TC

We consider two hypothetical cases of two bars to provide a context for TC. In the first case, shown in Figure 7, the bars are initially coinciding, with just one hole in the fluid mechanics domain. Then the red bar starts moving upward, creating a second hole. In the second case, shown in Figure 8, the bars are initially aligned with connected ends, again with a single hole in the domain. Then the red bar starts a flapping motion, up and down, creating a second hole in the domain, except when their ends become connected periodically during the flapping motion. When the red bar is in the upper position, the part of the domain below it is connected to the part of the domain above the blue bar. When the red bar is in the lower position, the part of the domain above it is connected to the part of the domain below the blue bar. These two cases are representatives of the typical TC challenges we expect to see in the classes of MBI problems we are targeting. Especially the first case is really not possible to treat in a consistent way without using an ST method.

6.2 Master–Slave System

We propose a very simple technique in the ST context. Having a constraint between nodes in a finite element formulation is quite common. These constraints reduce the number of unknowns, but in our implementation we delay that unknown elimination

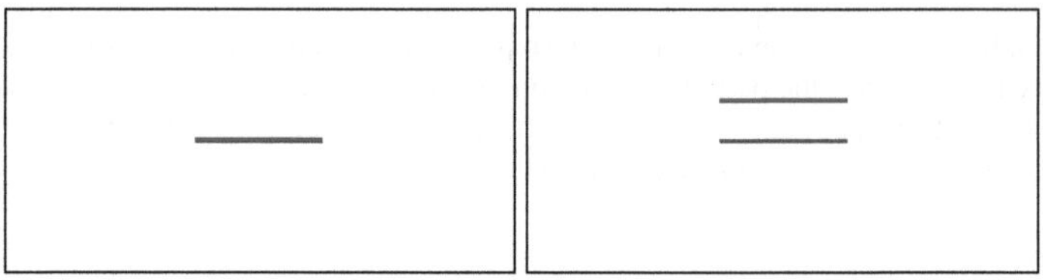

Fig. 7 Hypothetical case of two bars that are initially coinciding, with one hole in the domain (left). The red bar starts moving upward, creating a second hole in the domain (right).

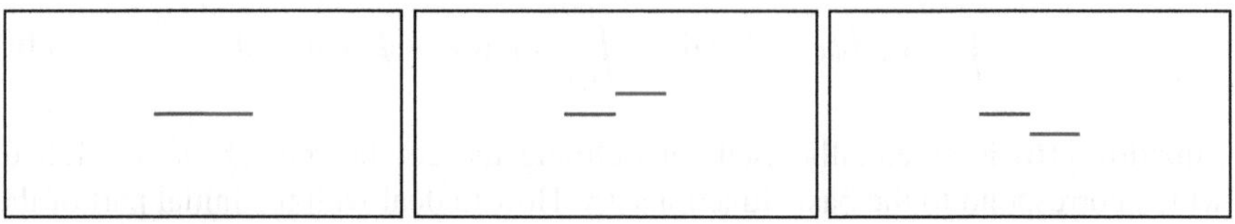

Fig. 8 Hypothetical case of two bars that are initially aligned with connected ends, with one hole in the domain (left). The red bar starts a flapping motion, up (middle) and down (right), creating a second hole in the domain, except when their ends become connected periodically.

until the iterative solution of the linear systems encountered at nonlinear iterations of a time step. The iterative solution of the linear systems is performed with reduced number of unknowns. The technique is easy to manage in a parallel-computing environment, especially if the preconditioner is simple enough. Typically we assign a master node to each slave node, and we use only the unknowns of the master nodes in iterative solution of the linear systems. We can use different master–slave relationships at different time levels. This is a practical alternative to, but less general than, using ST meshes that are unstructured in time. Still, we can use this concept to deal with the TC cases considered above, and the important point is that the connectivity of the "parent" mesh does not change. Consequently, the distribution model in the parallel-computing environment does not change during the computations.

With this technique, we need to implement one more functionality. We exclude certain elements from the integration of the finite element formulation. The exclusion principles are given below.

- Exclude all spatial elements with zero volume from the spatial integration.
- Exclude all ST elements with zero ST volume from the ST integration.
- We assume that checking if an ST element has zero ST volume is equivalent to checking if all the spatial elements associated with that ST element have zero volume. Therefore, for this purpose, we check the spatial-element volumes.
- To identify the spatial elements with zero volume, which should have zero Jacobian at all the integration points, instead of evaluating the Jacobians, we make the determination for a given spatial element from the master–slave relationship of its nodes. The method is explained more in [42].

7 ST-SI Method

In describing the ST-SI formulation, we will use the labels "Side A" and "Side B" to represent the two sides of the SI. In the ST-SI version of the formulation given by Eq. (5), we will have added boundary terms corresponding to the SI. The boundary terms for the two sides will be added separately, using test functions $\mathbf{w}_{\mathrm{A}}^h$ and q_{A}^h and $\mathbf{w}_{\mathrm{B}}^h$ and q_{B}^h. Here we give only the boundary terms for Side B:

$$
\begin{aligned}
&- \int_{(P_n)_{\mathrm{SI}}} q_{\mathrm{B}}^h \mathbf{n}_{\mathrm{B}} \cdot \mathbf{u}_{\mathrm{B}}^h \mathrm{d}P - \int_{(P_n)_{\mathrm{SI}}} \rho \mathbf{w}_{\mathrm{B}}^h \cdot \left(\mathbf{n}_{\mathrm{B}} \cdot \left(\mathbf{u}_{\mathrm{B}}^h - \mathbf{v}_{\mathrm{B}}^h\right)\right) \mathbf{u}_{\mathrm{B}}^h \mathrm{d}P \\
&+ \int_{(P_n)_{\mathrm{SI}}} q_{\mathrm{B}}^h \mathbf{n}_{\mathrm{B}} \cdot \frac{1}{2} \left(\mathbf{u}_{\mathrm{B}}^h + \mathbf{u}_{\mathrm{A}}^h\right) \mathrm{d}P \\
&+ \int_{(P_n)_{\mathrm{SI}}} \rho \mathbf{w}_{\mathrm{B}}^h \cdot \frac{1}{2} \left(\left(\mathbf{n}_{\mathrm{B}} \cdot \left(\mathbf{u}_{\mathrm{B}}^h - \mathbf{v}_{\mathrm{B}}^h\right) + \alpha^h\right) \mathbf{u}_{\mathrm{B}}^h - \left(\mathbf{n}_{\mathrm{A}} \cdot \left(\mathbf{u}_{\mathrm{A}}^h - \mathbf{v}_{\mathrm{A}}^h\right) + \alpha^h\right) \mathbf{u}_{\mathrm{A}}^h\right) \mathrm{d}P
\end{aligned}
$$

$$- \int_{(P_n)_{\mathrm{SI}}} \mathbf{w}_{\mathrm{B}}^h \cdot \frac{1}{2} \left(\mathbf{n}_{\mathrm{B}} \cdot \boldsymbol{\sigma}_{\mathrm{B}}^h - \mathbf{n}_{\mathrm{A}} \cdot \boldsymbol{\sigma}_{\mathrm{A}}^h \right) \mathrm{d}P - \int_{(P_n)_{\mathrm{SI}}} \mathbf{n}_{\mathrm{B}} \cdot \mu \boldsymbol{\varepsilon} \left(\mathbf{w}_{\mathrm{B}}^h \right) \cdot \left(\mathbf{u}_{\mathrm{B}}^h - \mathbf{u}_{\mathrm{A}}^h \right) \mathrm{d}P$$
$$+ \int_{(P_n)_{\mathrm{SI}}} \frac{\mu C}{h} \mathbf{w}_{\mathrm{B}}^h \cdot \left(\mathbf{u}_{\mathrm{B}}^h - \mathbf{u}_{\mathrm{A}}^h \right) \mathrm{d}P, \tag{17}$$

where

$$\alpha^h = \max \left(\left| \mathbf{n}_{\mathrm{B}} \cdot \left(\mathbf{u}_{\mathrm{B}}^h - \mathbf{v}_{\mathrm{B}}^h \right) \right|, \left| \mathbf{n}_{\mathrm{A}} \cdot \left(\mathbf{u}_{\mathrm{A}}^h - \mathbf{v}_{\mathrm{A}}^h \right) \right| \right), \tag{18}$$
$$h = \min(h_{\mathrm{B}}, h_{\mathrm{A}}), \tag{19}$$
$$h_{\mathrm{B}} = 2 \left(\sum_{\alpha=1}^{n_{\mathrm{ent}}} \sum_{a=1}^{n_{\mathrm{ens}}} \left| \mathbf{n}_{\mathrm{B}} \cdot \nabla N_a^\alpha \right| \right)^{-1} \text{ (for Side B)}, \tag{20}$$
$$h_{\mathrm{A}} = 2 \left(\sum_{\alpha=1}^{n_{\mathrm{ent}}} \sum_{a=1}^{n_{\mathrm{ens}}} \left| \mathbf{n}_{\mathrm{A}} \cdot \nabla N_a^\alpha \right| \right)^{-1} \text{ (for Side A)}. \tag{21}$$

Here, $(P_n)_{\mathrm{SI}}$ is the SI in the ST domain, $\mathbf{n}$ is the unit normal vector, and $\mathbf{v}$ is the mesh velocity. Side A counterpart of Eq. (17) can be written by simply interchanging subscripts A and B.

Remark 5. The first and third integrations set the volumetric flow rate at the boundary to $\mathbf{n}_{\mathrm{B}} \cdot \frac{1}{2} \left(\mathbf{u}_{\mathrm{B}}^h + \mathbf{u}_{\mathrm{A}}^h \right)$.

Remark 6. The second and fourth integrations set the advective flux at the boundary to the Lax–Friedrichs flux.

Remark 7. The fifth integration contains the average traction.

Remark 8. The sixth integration is the adjoint consistent term [47].

Remark 9. The seventh integration is a penalty term, where C is a nondimensional penalty constant, and usually $C = 1$ is large enough for stability.

For notation convenience, we introduce the symbols

$$\mathscr{F}_{\mathrm{B}}^h = \mathbf{n}_{\mathrm{B}} \cdot \left(\mathbf{u}_{\mathrm{B}}^h - \mathbf{v}_{\mathrm{B}}^h \right), \tag{22}$$
$$\mathscr{F}_{\mathrm{A}}^h = \mathbf{n}_{\mathrm{A}} \cdot \left(\mathbf{u}_{\mathrm{A}}^h - \mathbf{v}_{\mathrm{A}}^h \right). \tag{23}$$

With that, we replace the momentum flux as follows:

$$- \int_{(P_n)_{\mathrm{SI}}} q_{\mathrm{B}}^h \mathbf{n}_{\mathrm{B}} \cdot \mathbf{u}_{\mathrm{B}}^h \mathrm{d}P - \int_{(P_n)_{\mathrm{SI}}} \rho \mathbf{w}_{\mathrm{B}}^h \cdot \mathscr{F}_{\mathrm{B}}^h \mathbf{u}_{\mathrm{B}}^h \mathrm{d}P$$
$$+ \int_{(P_n)_{\mathrm{SI}}} q_{\mathrm{B}}^h \mathbf{n}_{\mathrm{B}} \cdot \frac{1}{2} \left(\mathbf{u}_{\mathrm{B}}^h + \mathbf{u}_{\mathrm{A}}^h \right) \mathrm{d}P$$

$$+ \int_{(P_n)_{\rm SI}} \rho \mathbf{w}_{\rm B}^h \cdot \frac{1}{2} \left(\left(\mathscr{F}_{\rm B}^h + \left| \mathscr{F}_{\rm B}^h \right| \right) \mathbf{u}_{\rm B}^h + \left(\mathscr{F}_{\rm B}^h - \left| \mathscr{F}_{\rm B}^h \right| \right) \mathbf{u}_{\rm A}^h \right) {\rm d}P$$

$$- \int_{(P_n)_{\rm SI}} \mathbf{w}_{\rm B}^h \cdot \frac{1}{2} \left(\mathbf{n}_{\rm B} \cdot \boldsymbol{\sigma}_{\rm B}^h - \mathbf{n}_{\rm A} \cdot \boldsymbol{\sigma}_{\rm A}^h \right) {\rm d}P - \int_{(P_n)_{\rm SI}} \mathbf{n}_{\rm B} \cdot \mu \boldsymbol{\varepsilon} \left(\mathbf{w}_{\rm B}^h \right) \cdot \left(\mathbf{u}_{\rm B}^h - \mathbf{u}_{\rm A}^h \right) {\rm d}P$$

$$+ \int_{(P_n)_{\rm SI}} \frac{\mu C}{h} \mathbf{w}_{\rm B}^h \cdot \left(\mathbf{u}_{\rm B}^h - \mathbf{u}_{\rm A}^h \right) {\rm d}P. \qquad (24)$$

Remark 10. Assuming that the discrete boundaries on the two sides of the interface are the same, $\mathbf{n}_{\rm B} = -\mathbf{n}_{\rm A}$. With that, we see the traction $\mathbf{h}_{\rm B}^h \equiv -\mathbf{h}_{\rm A}^h$ as

$$\mathbf{h}_{\rm B}^h = -\frac{1}{2} \left(p_{\rm B}^h + p_{\rm A}^h \right) \mathbf{n}_{\rm B} + \frac{1}{2} \mathbf{n}_{\rm B} \cdot \left(2\mu \boldsymbol{\varepsilon}(\mathbf{u}_{\rm B}^h) + 2\mu \boldsymbol{\varepsilon}(\mathbf{u}_{\rm A}^h) \right) - \frac{\mu C}{h} \left(\mathbf{u}_{\rm B}^h - \mathbf{u}_{\rm A}^h \right). \qquad (25)$$

8 Examples

The first two examples are from [48] for the ST-TC method and from [45] for the ST-C method. The third example is a 2D test computation based on the combination of ST-TC and ST-SI methods. For all computations presented, the core technology is the ST-VMS method. More examples for the ST-TC method can be found in [22, 42, 48]. Examples for the STNMUM can be found in [17–21] for flapping-wing aerodynamics, in [40] for separation aerodynamics of spacecraft, and in [21, 23] for wind-turbine aerodynamics. Examples for the ST-SI method can be found in [43].

8.1 Aortic-Valve Model with Coronary Arteries

We created a typical aortic-valve model, which has three leaflets with two outlets, corresponding to coronary arteries, and one main outlet, corresponding to the beginning of the aorta. Figure 9 shows the velocity magnitude when the valve is open and is about to close. For more on this computation, see [48].

8.2 Thermo-Fluid Analysis of a Ground Vehicle and Its Tires

First the thermo-fluid computation is carried out over the full domain, with a reasonable mesh refinement. The large amount of time-history data from that computation is stored using the ST-C-SPT method. This is followed by a higher-resolution computation over the local domain containing the tires we focus on. This gives us increased accuracy in the thermo-fluid analysis, including increased

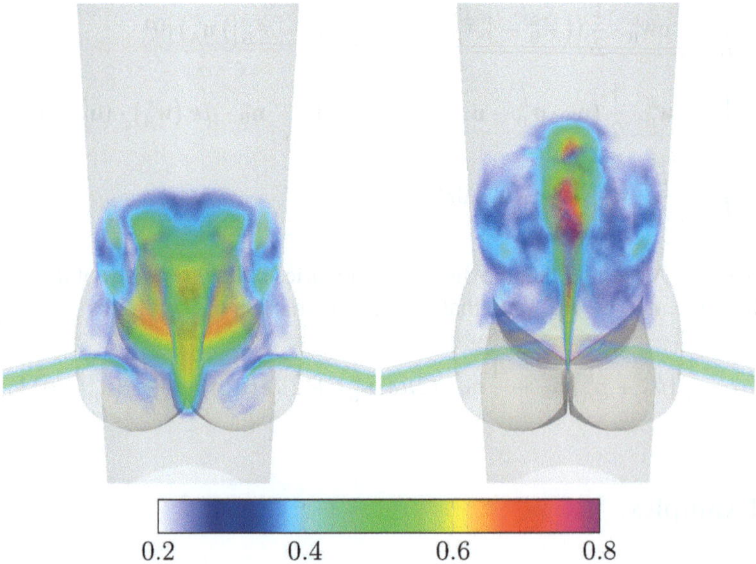

Fig. 9 Aortic-valve model with coronary arteries. Volume rendering of the velocity magnitude (m/s) at $t/T = 0.4$ and 0.5, where T is the cardiac period.

Fig. 10 Thermo-fluid analysis of a ground vehicle and its tires. Temperature volume rendering from the local computation. Colors from red to yellow indicate temperature from low to high.

accuracy in the heat transfer rates from the tires. Figure 10 shows the temperature patterns from the local computation. Figure 11 shows the Nusselt number from the global and local computations. For more on this computation, see [45].

Fig. 11 Time and circumferentially averaged Nusselt number from the global (left) and local (right) computations.

8.3 2D Model of Flow Past a Tire in Contact with the Road

This is an example of how the ST-TC and ST-SI methods can be used in combination for computational analysis of a challenging problem. The computation is challenging because the tire does not have circular symmetry and is in rotational contact with the road. The first challenge is addressed with the ST-SI method, which enables flow computations with spinning structures while maintaining the high-resolution representation of the boundary layers near the structure. The second challenge is addressed with the ST-TC method, which enables flow computations with contact between moving solid surfaces while maintaining the high-resolution representation of the boundary layers near the solid surfaces. Figure 12 shows the mesh and the velocity magnitude. More on this type of computations and the combined method and its implementation will be presented in a future paper.

9 Concluding Remarks

We have presented an overview of some of the new directions we have taken the ST methods since 2010 in bringing solution and analysis to different classes of computationally challenging engineering problems. The new directions include a) the VMS version of the DSD/SST method, which is called ST-VMS, b) ST methods based on using NURBS basis functions in temporal representation of the unknown variables and motion of the solid surfaces and fluid meshes, including the mesh update method STNMUM, c) ST techniques with continuous representation in time, which is called ST-C, d) ST interface-tracking with topology change, which is called ST-TC, and e) ST-VMS method for flow computations with slip interfaces, which is called ST-SI. We described the new directions and presented a few examples.

Fig. 12 2D model of flow past a tire in contact with the road. Velocity magnitude superimposed over the mesh. Colors from blue to red indicate values from low to high.

Acknowledgements This work was supported (first author) in part by Grant-in-Aid for Young Scientists (B) 24760144 from Japan Society for the Promotion of Science (JSPS); Grant-in-Aid for Scientific Research (S) 26220002 from the Ministry of Education, Culture, Sports, Science, and Technology of Japan (MEXT); Council for Science, Technology, and Innovation (CSTI), Cross-ministerial Strategic Innovation Promotion Program (SIP), "Innovative Combustion Technology" (Funding agency: JST); Japan Science and Technology Agency, CREST; and Rice–Waseda research agreement. This work was also supported (second author) in part by ARO Grant W911NF-12-1-0162. We thank Mr. Takashi Kuraishi (Waseda) for the computational thermo-fluid analysis of the ground vehicle and its tires.

References

1. Hughes, T.J.R., Liu, W.K., Zimmermann, T.K.: Lagrangian–Eulerian finite element formulation for incompressible viscous flows. Comput. Methods Appl. Mech. Eng. **29**, 329–349 (1981)
2. Miras, T., Schotte, J.-S., Ohayon, R.: Energy approach for static and linearized dynamic studies of elastic structures containing incompressible liquids with capillarity: a theoretical formulation. Comput. Mech. **50**, 729–741 (2012)
3. van Opstal, T.M., van Brummelen, E.H., de Borst, R., Lewis, M.R.: A finite-element/boundary-element method for large-displacement fluid-structure interaction. Comput. Mech. **50**, 779–788 (2012)
4. Bazilevs, Y., Takizawa, K., Tezduyar, T.E.: Computational Fluid-Structure Interaction: Methods and Applications. Wiley, New York (2013). ISBN 978-0470978771
5. Korobenko, A., Hsu, M.-C., Akkerman, I., Tippmann, J., Bazilevs, Y.: Structural mechanics modeling and FSI simulation of wind turbines. Math. Models Methods Appl. Sci. **23**, 249–272 (2013)
6. Hsu, M.-C., Akkerman, I., Bazilevs, Y.: Finite element simulation of wind turbine aerodynamics: validation study using NREL phase VI experiment. Wind Energy **17**, 461–481 (2014)

7. Tezduyar, T.E.: Stabilized finite element formulations for incompressible flow computations. Adv. Appl. Mech. **28**, 1–44 (1992) doi:10.1016/S0065-2156(08)70153-4
8. Tezduyar, T.E.: Computation of moving boundaries and interfaces and stabilization parameters. Int. J. Numer. Methods Fluids **43**, 555–575 (2003). doi:10.1002/fld.505
9. Tezduyar, T.E., Sathe, S.: Modeling of fluid-structure interactions with the space-time finite elements: solution techniques. Int. J. Numer. Methods Fluids **54**, 855–900 (2007). doi:10.1002/fld.1430
10. Takizawa, K., Tezduyar, T.E.: Multiscale space-time fluid-structure interaction techniques. Comput. Mech. **48**, 247–267 (2011). doi:10.1007/s00466-011-0571-z
11. Takizawa, K., Tezduyar, T.E.: Space-time fluid-structure interaction methods. Math. Models Methods Appl. Sci. **22**, 1230001 (2012). doi:10.1142/S0218202512300013
12. Tezduyar, T., Aliabadi, S., Behr, M., Johnson, A., Mittal, S.: Parallel finite-element computation of 3D flows. Computer **26**, 27–36 (1993). doi:10.1109/2.237441
13. Johnson, A.A., Tezduyar, T.E.: Mesh update strategies in parallel finite element computations of flow problems with moving boundaries and interfaces. Comput. Methods Appl. Mech. Eng. **119**, 73–94 (1994). doi:10.1016/0045-7825(94)00077-8
14. Tezduyar, T., Aliabadi, S., Behr, M., Johnson, A., Kalro, V., Litke, M.: Flow simulation and high performance computing. Comput. Mech. **18**, 397–412 (1996). doi:10.1007/BF00350249
15. Tezduyar, T.E.: Finite element methods for flow problems with moving boundaries and interfaces. Arch. Comput. Meth. Eng. **8**, 83–130 (2001). doi:10.1007/BF02897870
16. Takizawa, K., Tezduyar, T.E., Boben, J., Kostov, N., Boswell, C., Buscher, A.: Fluid-structure interaction modeling of clusters of spacecraft parachutes with modified geometric porosity. Comput. Mech. **52**, 1351–1364 (2013). doi:10.1007/s00466-013-0880-5
17. Takizawa, K., Henicke, B., Puntel, A., Kostov, N., Tezduyar, T.E.: Space-time techniques for computational aerodynamics modeling of flapping wings of an actual locust. Comput. Mech. **50**, 743–760 (2012). doi:10.1007/s00466-012-0759-x
18. Takizawa, K., Kostov, N., Puntel, A., Henicke, B., Tezduyar, T.E.: Space-time computational analysis of bio-inspired flapping-wing aerodynamics of a micro aerial vehicle. Comput. Mech. **50**, 761–778 (2012). doi:10.1007/s00466-012-0758-y
19. Takizawa, K., Henicke, B., Puntel, A., Kostov, N., Tezduyar, T.E.: Computer modeling techniques for flapping-wing aerodynamics of a locust. Comput. Fluids **85**, 125–134 (2013). doi:10.1016/j.compfluid.2012.11.008
20. Takizawa, K., Tezduyar, T.E., Kostov, N.: Sequentially-coupled space–time FSI analysis of bio-inspired flapping-wing aerodynamics of an MAV. Comput. Mech. **54**, 213–233 (2014). doi:10.1007/s00466-014-0980-x
21. Takizawa, K.: Computational engineering analysis with the new-generation space-time methods. Comput. Mech. **54**, 193–211 (2014). doi:10.1007/s00466-014-0999-z
22. Takizawa, K., Tezduyar, T.E., Buscher, A.: Space-time computational analysis of MAV flapping-wing aerodynamics with wing clapping. Comput. Mech. **55**, 1131–1141 (2015). doi:10.1007/s00466-014-1095-0
23. Takizawa, K., Tezduyar, T.E., McIntyre, S., Kostov, N., Kolesar, R., Habluetzel, C.: Space-time VMS computation of wind-turbine rotor and tower aerodynamics. Comput. Mech. **53**, 1–15 (2014). doi:10.1007/s00466-013-0888-x
24. Akin, J.E., Tezduyar, T.E., Ungor, M.: Computation of flow problems with the mixed interface-tracking/interface-capturing technique (MITICT). Comput. Fluids **36**, 2–11 (2007). doi:10.1016/j.compfluid.2005.07.008
25. Akkerman, I., Bazilevs, Y., Benson, D.J., Farthing, M.W., Kees, C.E.: Free-surface flow and fluid–object interaction modeling with emphasis on ship hydrodynamics. J. Appl. Mech. **79**, 010905 (2012)
26. Tezduyar, T.E., Takizawa, K., Moorman, C., Wright, S., Christopher, J.: Space-time finite element computation of complex fluid-structure interactions. Int. J. Numer. Methods Fluids **64**, 1201–1218 (2010). doi:10.1002/fld.2221
27. Wick, T.: Coupling of fully Eulerian and arbitrary Lagrangian–Eulerian methods for fluid-structure interaction computations. Comput. Mech. **52**, 1113–1124 (2013)

28. Wick, T.: Flapping and contact FSI computations with the fluid–solid interface-tracking/interface-capturing technique and mesh adaptivity. Comput. Mech. **53**, 29–43 (2014)
29. Hsu, M.-C., Kamensky, D., Bazilevs, Y., Sacks, M.S., Hughes, T.J.R.: Fluid-structure interaction analysis of bioprosthetic heart valves: significance of arterial wall deformation. Comput. Mech. **54**, 1055–1071 (2014). doi:10.1007/s00466-014-1059-4
30. Mittal, S., Tezduyar, T.E.: Parallel finite element simulation of 3D incompressible flows – Fluid-structure interactions. Int. J. Numer. Methods Fluids **21**, 933–953 (1995). doi:10.1002/fld.1650211011
31. Tezduyar, T.E., Aliabadi, S.K., Behr, M., Mittal, S.: Massively parallel finite element simulation of compressible and incompressible flows. Comput. Methods Appl. Mech. Eng. **119**, 157–177 (1994). doi:10.1016/0045-7825(94)00082-4
32. Takase, S., Kashiyama, K., Tanaka, S., Tezduyar, T.E.: Space-time SUPG finite element computation of shallow-water flows with moving shorelines. Comput. Mech. **48**, 293–306 (2011). doi:10.1007/s00466-011-0618-1
33. Kalro, V., Tezduyar, T.E.: A parallel 3D computational method for fluid–structure interactions in parachute systems. Comput. Methods Appl. Mech. Eng. **190**, 321–332 (2000). doi:10.1016/S0045-7825(00)00204-8
34. Brooks, A.N., Hughes, T.J.R.: Streamline upwind/Petrov-Galerkin formulations for convection dominated flows with particular emphasis on the incompressible Navier-Stokes equations. Comput. Methods Appl. Mech. Eng. **32**, 199–259 (1982)
35. Hughes, T.J.R.: Multiscale phenomena: Green's functions, the Dirichlet-to-Neumann formulation, subgrid scale models, bubbles, and the origins of stabilized methods. Comput. Methods Appl. Mech. Eng. **127**, 387–401 (1995)
36. Bazilevs, Y., Calo, V.M., Cottrell, J.A., Hughes, T.J.R., Reali, A., Scovazzi, G.: Variational multiscale residual-based turbulence modeling for large eddy simulation of incompressible flows. Comput. Methods Appl. Mech. Eng. **197**, 173–201 (2007)
37. Hughes, T.J.R., Cottrell, J.A., Bazilevs, Y.: Isogeometric analysis: CAD, finite elements, NURBS, exact geometry, and mesh refinement. Comput. Methods Appl. Mech. Eng. **194**, 4135–4195 (2005)
38. Bazilevs, Y., Calo, V.M., Hughes, T.J.R., Zhang, Y.: Isogeometric fluid–structure interaction: theory, algorithms, and computations. Comput. Mech. **43**, 3–37 (2008)
39. Takizawa, K., Wright, S., Moorman, C., Tezduyar, T.E.: Fluid-structure interaction modeling of parachute clusters. Int. J. Numer. Methods Fluids **65**, 286–307 (2011). doi: 10.1002/fld.2359
40. Takizawa, K., Montes, D., Fritze, M., McIntyre, S., Boben, J., Tezduyar, T.E.: Methods for FSI modeling of spacecraft parachute dynamics and cover separation. Math. Models Methods Appl. Sci. **23**, 307–338 (2013). doi:10.1142/S0218202513400058
41. Takizawa, K., Tezduyar, T.E.: Space-time computation techniques with continuous representation in time (ST-C). Comput. Mech. **53**, 91–99 (2014). doi:10.1007/s00466-013-0895-y
42. Takizawa, K., Tezduyar, T.E., Buscher, A., Asada, S.: Space-time interface-tracking with topology change (ST-TC). Comput. Mech. **54**, 955–971 (2014). doi:10.1007/s00466-013-0935-7
43. Takizawa, K., Tezduyar, T.E., Mochizuki, H., Hattori, H., Mei, S., Pan, L., Montel, K.: Space-time VMS method for flow computations with slip interfaces (ST-SI). Math. Models Methods Appl. Sci. **25**, 2377–2406 (2015). doi: 10.1142/S0218202515400126
44. Hsu, M.-C., Bazilevs, Y.: Fluid–structure interaction modeling of wind turbines: simulating the full machine. Comput. Mech. **50**, 821–833 (2012)
45. Takizawa, K., Tezduyar, T.E., Kuraishi, T.: Multiscale ST methods for thermo-fluid analysis of a ground vehicle and its tires. Math. Models Methods Appl. Sci. **25**, 2227–2255 (2015). doi: 10.1142/S0218202515400072
46. Takizawa, K., Henicke, B., Puntel, A., Spielman, T., Tezduyar, T.E.: Space-time computational techniques for the aerodynamics of flapping wings. J. Appl. Mech. **79**, 010903 (2012). doi:10.1115/1.4005073
47. Bazilevs, Y., Hughes, T.J.R.: Weak imposition of Dirichlet boundary conditions in fluid mechanics. Comput. Fluids **36**, 12–26 (2007)
48. Takizawa, K., Tezduyar, T.E., Buscher, A., Asada, S.: Space-time fluid mechanics computation of heart valve models. Comput. Mech. **54**, 973–986 (2014). doi:10.1007/s00466-014-1046-9

Part III
CFD: Phase-Field Modeling

Interfacial Instability of a Non-magnetized Drop in Ferrofluids Subjected to an Azimuthal Field: A Diffuse-Interface Approach

Ching-Yao Chen and Ting-Shiang Lin

Abstract We perform direct numerical simulations of a non-magnetic drop immersed in immiscible ferrofluids in a confined Hele-Shaw cell under an azimuthal field by a diffuse-interface method. The interface is unstable in such a condition because of the inward attraction of the ferrofluids induced by the magnetic field gradient. We focus on the fingering onset and pattern influenced by the coupling viscous effect with different viscously stable conditions, which is achieved by varying the viscosity contrast of the ferrofluids and non-magnetic drop. In a viscously stable condition, in which the viscosity of the ferrofluids is greater than the immersed drop, the fingering onset takes place earlier because of the higher mobility of drop. The fingering pattern is simpler with numerous straightly developed fingers. On the other hand, a viscously unstable interface of less viscous ferrofluids results in ramified fingering pattern associated with the secondary phenomena, e.g., competitions and tip-splits of fingers. However, the fingering onset is delayed because the drop is less mobile.

1 Introduction

Ferrofluids conveniently combine the fluidity of liquids and the magnetic properties of solids. They are stable colloidal suspensions of magnetic nano-particles suspended in a non magnetic carrier fluid, which has been a fascinating field of research for several decades [1, 2]. The pronounced superparamagnetic property of ferrofluids allows distant manipulation of their flow and interfacial behavior with external magnetic fields. Consequently, the phenomena of interfacial pattern formation in ferrofluids has received considerable attention. One striking example of pattern-forming systems in ferrofluids is the famous Rosensweig instability [3–7] in a uniform perpendicular field to an initially flat, ferrofluid free surface, which leads to the development of 3-dimensional hexagonal array of peaks. Another type of

C.-Y. Chen (✉) • T.-S. Lin
Department of Mechanical Engineering, National Chiao Tung University, Hsinchu, Taiwan 30010, R.O.C.
e-mail: chingyao@mail.nctu.edu.tw

Y. Bazilevs, K. Takizawa (eds.), *Advances in Computational Fluid-Structure Interaction and Flow Simulation*, Modeling and Simulation in Science, Engineering and Technology, DOI 10.1007/978-3-319-40827-9_14

remarkable patterns is associated with the so-called labyrinthine instability [8–13], where highly branched structures are formed when a ferrofluid droplet is trapped in the effectively 2-dimensional configuration of a Hele-Shaw cell [14] under a perpendicular uniform magnetic field. These two instabilities can co-exist to form a new hybrid instability if the confining upper plate of the Hele-Shaw cell is replaced by a free fluid surface [15]. Besides the two well-known interfacial instabilities under a uniform field, a non-uniform azimuthal field has also been demonstrated to stabilize a rotating heavier ferrofluid droplet surrounded by non-magnetic fluids, in which the interfacial instability is triggered by the centrifugal forces [16, 17]. This stabilizing effect is expected to reverse if an azimuthal field is applied to a non-magnetic drop immersed in ferrofluids, which is confirmed by a recent theoretical analysis [18]. In this paper we perform direct numerical simulations of this destabilizing effect by a highly accurate diffuse-interface method. In this context, our aim is to consider the coupling between capillary force and magnetic effects, and study how they give rise to still unexplored interfacial shapes and interesting dynamical behaviors.

2 Physical Problem and Governing Equations

We investigate the interfacial instability of a circular non-magnetized drop (fluid 2) with an initial radius R_0 surrounded by immiscible ferrofluids (fluid 1) in a Hele-Shaw cell, as the principle sketch shown in Fig. 1. The Hele-Shaw cell has gap spacing h. An azimuthal magnetic field $\mathbf{H}$ is generated by a wire carrying electric current strength I placed at the center of the drop. The viscosities of ferrofluids and the non-magnetic drop are denoted by η_1 and η_2, respectively. We focus on the magneto-induced motion, but allow the inner drop to be either more or less viscous than the outer ferrofluids. The governing equations of such an immiscible interface approached by diffuse-interface method are based on the model proposed by Cahn and Hilliard [19]. In the context of a Hele-Shaw cell system [20–25] incorporating with magnetic stress [16–18], the equations can be written as

$$\frac{\partial \rho}{\partial t} + \nabla \cdot (\rho \mathbf{u}) = 0, \tag{1}$$

$$\nabla p = -\frac{12\eta}{h^2}\mathbf{u} + \mu_0 M \nabla H - \epsilon \nabla \cdot [\rho(\nabla c)(\nabla c)^T], \tag{2}$$

$$\rho\left(\frac{\partial c}{\partial t} + \mathbf{u}\cdot\nabla c\right) = \alpha \nabla^2 \mu, \tag{3}$$

$$\mu = \frac{\partial f_0}{\partial c} - \frac{\epsilon}{\rho}\nabla \cdot (\rho \nabla c) - \frac{p}{\rho^2}\frac{d\rho}{dc}, \tag{4}$$

$$f_0 = f^* c^2 (1-c)^2. \tag{5}$$

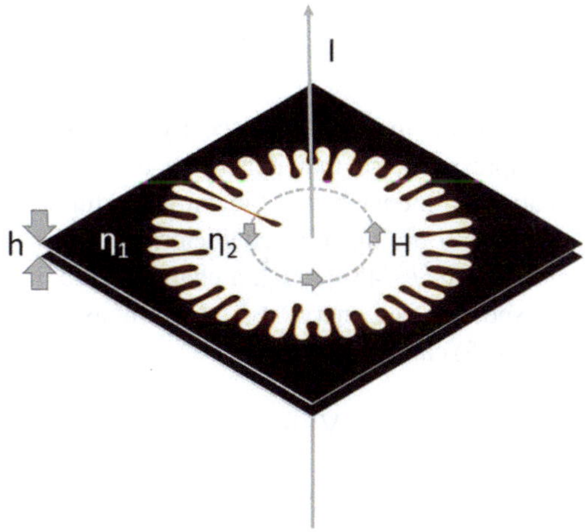

Fig. 1 A non-magnetized drop (bright color) inside ferrofluids (dark color) confined in a Hele-Shaw cell with width h subjected to an azimuthal field H generated by a current-carrying (strength I) wire.

Here, **u** denotes the fluid velocity vector and p the pressure. The viscosity and the density of the binary fluid system are represented by η and ρ, respectively. The phase-field variable is represented by c, so that $c = 1$ in the ferrofluid, and $c = 0$ in the drop. μ_0, M, and H are the permeability in vacuum, magnitudes of magnetization, and field strength, respectively. The coefficients of capillarity and mobility are denoted by ϵ and α. The chemical potential is denoted by μ, and f_0 is a free energy (or, the Helmholtz free energy) with a characteristic specific energy f^*. Equations (1)–(5) define the so-called Hele-Shaw-Cahn-Hilliard (HSCH) model [20–25], and result in a surface free energy given by

$$E = \rho \int \left(f_0 + \frac{\epsilon}{2} (\nabla c)^2 \right) dV, \tag{6}$$

where V is the volume of the fluid domain.

The magnetic field **H** generated by a wire with electric current I perpendicular to the cell is expressed by [16–18]

$$\mathbf{H} = \frac{I}{2\pi r} \mathbf{e}_\theta, \tag{7}$$

where r is the radial position, and $\mathbf{e}_\theta$ represents the unit vector along the azimuthal direction. A proportional magnetization strength of the ferrofluids to the field associated with a constant susceptibility χ can be approximated, so that

$$\mathbf{M}(c) = \chi \mathbf{H} c. \tag{8}$$

A correlation of viscosity (η) with the phase-field variable c is are required by the present approach. To take advantage of the highly accurate scheme previously developed for Hele-Shaw flows [26–28], we assume that

$$\eta(c) = \eta_1 e^{[R(1-c)]},\ R = \ln\left(\frac{\eta_2}{\eta_1}\right). \tag{9}$$

Here, R is the logarithmic viscosity contrast parameter. We further assume the flow incompressible, so that density ρ is constant. In order to render the governing equations dimensionless, the initial radius of the fluid drop R_0 and viscosity of ferrofluid η_1 are taken as characteristic scales. In conjunction with a characteristic velocity $\mu_0 h^2 I^2/48\eta_1\pi^2 R_0^3$, a characteristic pressure $\mu_0 I^2/4\pi^2 R_0^2$, and a characteristic specific free energy $f_0{}^*$, the dimensionless HSCH equations associated with Eq. (1)–(5) can be written as

$$\nabla \cdot \mathbf{u} = 0, \tag{10}$$

$$\nabla p = -\eta\mathbf{u} - \frac{\chi c}{r}\nabla(\frac{1}{r}) - \frac{C}{Mg}\nabla \cdot [(\nabla c)(\nabla c)^T], \tag{11}$$

$$\frac{\partial c}{\partial t} + \mathbf{u}\cdot\nabla c = \frac{1}{\mathrm{Pe}}\nabla^2\mu, \tag{12}$$

$$\mu = \frac{\partial f_0}{\partial c} - C\nabla^2 c, \tag{13}$$

$$f_0 = c^2(1-c)^2. \tag{14}$$

Dimensionless parameters, such as the Peclet number Pe, the Atwood number (normalized viscosity contrast) A, the Cahn number C, and magnetic strength Mg, are defined as

$$\mathrm{Pe} = \frac{\mu_0 h^2 I^2}{48\pi^2\alpha\eta_1 f^* R_0^2},\ A = \frac{e^R - 1}{e^R + 1},\ C = \frac{\epsilon}{f^* R_0^2},\ \mathrm{Mg} = \frac{\mu_0 I^2}{4\pi^2\rho f^* R_0^2}.$$

Moreover, the dimensionless free interfacial energy corresponding to Eq. (6) can be obtained accordingly as

$$E = \frac{1}{\mathrm{Mg}}\int\left(f_0 + \frac{C}{2}(\nabla c)^2\right)dV. \tag{15}$$

In order to solve the governing equations numerically, we recast the governing equations into the well-known streamfunction-vorticity formulation (ϕ, ω) [26], yielding

$$u = \frac{\partial\phi}{\partial y},\ v = -\frac{\partial\phi}{\partial x} \tag{16}$$

$$\nabla^2\phi = -\omega, \tag{17}$$

where

$$\omega = -R\left(u\frac{\partial c}{\partial y} - v\frac{\partial c}{\partial x}\right) - \frac{\chi}{\eta r^4}\left(y\frac{\partial c}{\partial x} - x\frac{\partial c}{\partial y}\right)$$
$$+ \frac{C}{\eta \mathrm{Mg}}\left[\frac{\partial c}{\partial x}\left(\frac{\partial^3 c}{\partial x^2 \partial y} + \frac{\partial^3 c}{\partial y^3}\right) - \frac{\partial c}{\partial y}\left(\frac{\partial^3 c}{\partial x \partial y^2} + \frac{\partial^3 c}{\partial x^3}\right)\right].$$

Boundary conditions are prescribed as follows:

$$x = \pm 1 \ : \phi = 0, \ \frac{\partial c}{\partial x} = 0, \ \frac{\partial^2 c}{\partial x^2} = 0, \tag{18}$$

$$y = \pm 1 \ : \ \phi = 0, \ \frac{\partial c}{\partial y} = 0, \ \frac{\partial^2 c}{\partial y^2} = 0. \tag{19}$$

To reproduce the extremely fine structures of the fingers, a highly accurate pseudospectral method is employed. As a result, the actual boundary conditions applied in the numerical code are $\partial\phi/\partial y = 0$ at $y = \pm 1$. However, at the present situation where no concentration gradient is generated on the boundaries, the above conditions automatically lead to $\phi = 0$. To ensure this condition, all the simulations are terminated when the inner fluid reaches a certain distance away from the computational boundaries. Both c and ϕ are expanded in a cosine series in the x-direction. In the y-direction, discretization is accomplished by sixth order compact finite differences. Time integration is fully explicit and utilizes a third order Runge–Kutta procedure. The evaluation of the nonlinearity at each time level is performed in a pseudospectral manner. The procedures reported in Refs. [23–25] are followed, in which $Pe = 3000$ and $C = 10^{-5}$ are applied to approach an immiscible interface. The simulations are terminated when the fastest penetrating fingers reach near the origin to avoid numerical instability. For a more detailed account about these numerical schemes the reader is referred to Refs. [23–28].

3 Numerical Results and Discussion

In the following simulations, we present the fingering patterns induced by the destabilizing magnetic field. Influences of the magnetic effects, e.g., the magnetic strength Mg and the magnetic susceptibility χ, coupled with the viscosity contrast A will be analyzed systematically. It is noticed that a positive/negative A represents more/less viscous non-magnetic drop than the surrounding ferrofluids, respectively. For the case of $A = 0$, the drop and ferrofluids are with the same viscosity.

Shown in Fig. 2 are three representative cases for $Mg = 2$ and $\chi = 1$ for various viscosity contrast $A = -0.462$, 0, and 0.462. The presence of magnetic field gradient triggered vigorous fingering instability. The negative radial field gradient attracts the ferrofluids to penetrate into the drop toward the origin, and

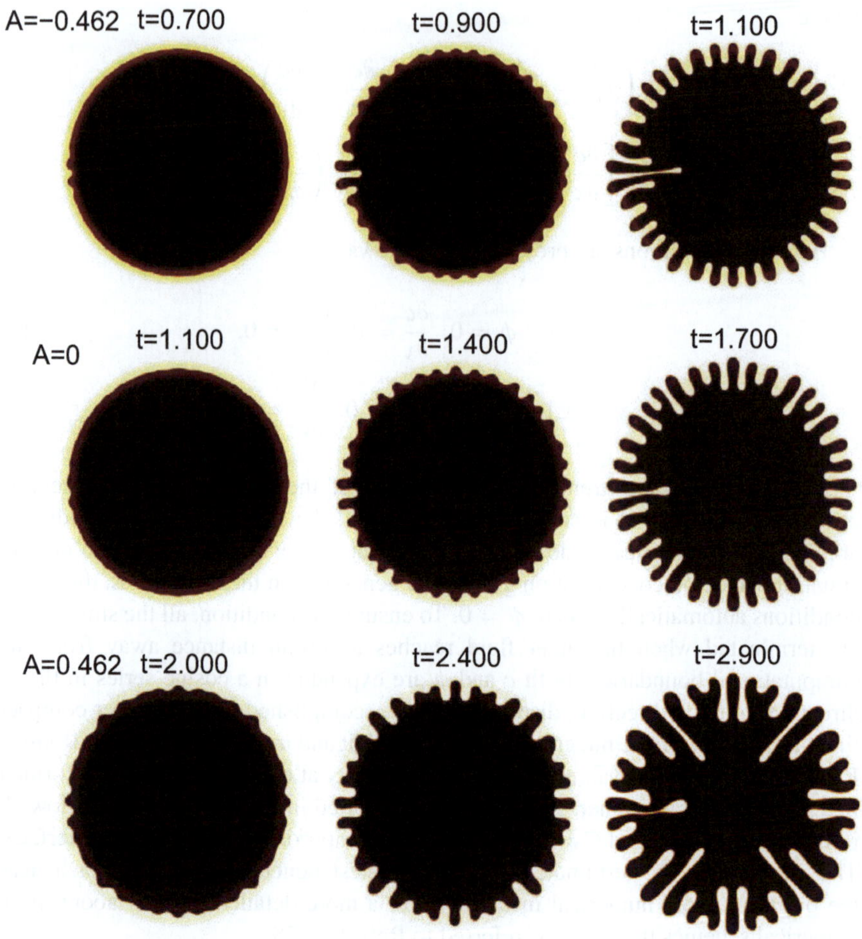

Fig. 2 Fingering patterns for $A = -0.462$ [top row], $A = 0$ [middle row], and $A = 0.462$ [bottom row]. The remaining parameters are: Mg=2 and $\chi = 1$. For better observation of the interfacial instability, the non-magnetic drop is shown by dark color in the present figures. The left, middle, and right columns show the patterns of (1) immediately after the emergences of fingers, (2) fully developed fingers, and (3) shortly before the most inward finger reaching the core, respectively.

squeeze partial mass of the drop to stretch outwardly. For easier identification, the penetrating ferrofluid fingers and stretching fingers of non-magnetic drop are denoted as inward fingers and outward fingers, respectively. For various Atwood numbers, the fingering patterns show apparent distinctions. For a negative Atwood number of $A = -0.462$ as the fingering pattern shown in the top row of Fig. 2, in which the viscosity of non-magnetic drop is smaller than the ferrofluids, the fingering pattern is very regular without strong finger competition, i.e., most of the fingers emerging compatibly. Both the outward and inward fingers develop in a nearly identical path to preserve circular inward and outward fingering fronts.

Nevertheless, because extremely significant field gradient near the origin as shown in Eq. (7), a few dominant fingers eventually emerge to reach the origin. Noted that the emergences of a few faster dominant fingers are similar to the fingering patterns of radial suction flow [24]. On the other hand, if the viscosity of the drop is increased to match the ferrofluids as the case of $A = 0$ shown in the middle row of Fig. 2, the fingering instability is enhanced. Some tips of outward fingers are seen to emerge secondary instability, e.g., fingertip split. The circular inward fingering front is no longer preserved, which indicates apparent fingering competition. If the viscosity of the drop is further increased to outmatch the ferrofluids, e.g., $A = 0.462$ as the pattern shown in the bottom row of Fig. 2, fingering competitions of the inward fingers and tip-split of the outward fingers are very prominent. The fingering patterns appear very different with variation of the viscosity contrast. These distinct fingering patterns can be realized by the underlining mechanism of well-developed viscous fingering phenomena. For the case of a negative $A = -0.462$, the interface is viscously stable, in which a more viscous ferrofluids penetrating (or displacing) a less viscous drop. The fingering pattern of viscously stable condition usually appears less fingering competition associated with a nearly circular fingering front as demonstrated in similar fingering patterns driven by centrifugal forces [23]. On the contrary, the less viscous ferrofluids penetrate inwardly into the more viscous drop in a positive $A = 0.462$, which is the typical instability of viscous fingering, so that the emergences of inward fingers appear very prominent with active secondary phenomena, e.g., fingering competitions and tip-split. As a result, the coupling effects of viscosity contrast with the de-stabilizing magnetic force can lead to entirely distinct patterns of interfacial instability. These coupling effects can be further enhanced by increasing the absolute magnitude of viscosity contrast to $A = -0.635$ and 0.635, as their fingering patterns shown in Fig. 3. The inward fingering competitions and outward tip-split are seen more prominent for these cases with stronger viscous influences.

As discussed in the previous paragraph, the fingering patters depend strongly on the viscous effect. It is also worthy to observe the onset and evolution of the fingering instability. The onset times of finger emergences for various Atwood numbers are shown in the left column of Figs. 2 and 3. It is interesting to discover that the onset of fingers occurs earlier for the condition, in which the ferrofluid is more viscous than the drop, e.g., onset at $t = 0.5, 0.7, 1.1, 2.0$, and 2.3 for $A = -0.635, -0.462, 0, 0.462$, and 0.635, respectively. The delay of fingering onset for less viscous ferrofluids penetrating into more viscous drop, in which the fingering pattern is more ramified as the fully evolved patterns shown in the middle and right columns of Figs. 2 and 3, suggests a contradict behavior in the measure of instability prominence. Considering the driven fluids are the ferrofluids, it is much easier for more viscous ferrofluids to displace the stationary less viscous drop whose mobility is higher. Consequently, the fingering onset occurs at earlier time. Nevertheless, since the interface is viscously stable, the fingering pattern appears simpler with primary straight inward fingers. On the contrary, it takes longer time to drive less viscous ferrofluids penetrating into the more viscous drop with less mobility. Afterward, the viscously unstable interface starts to evolve those

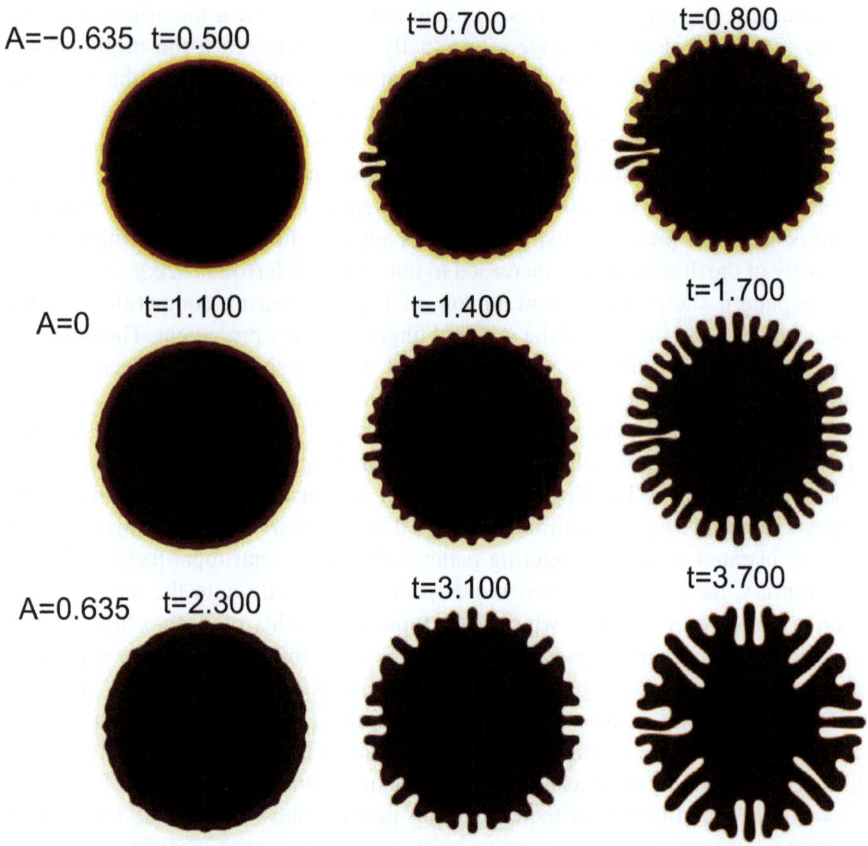

Fig. 3 Fingering patterns for $A = -0.635$ [top row], $A = 0$ [middle row], and $A = 0.635$ [bottom row]. The remaining parameters are: Mg=2 and $\chi = 1$.

secondary fingering phenomena and form more ramified pattern. As a result, the onset and pattern of fingering instability shows opposite trend with the viscosity contrast.

The effect of magnetic field is straightforward, such that stronger field strength shortens the onset of fingering instability, and results in a more vigorous fingering pattern. Quantitative measures will be presented in a latter paragraph. In Fig. 4, we show the patterns of $Mg = 4$ and $\chi = 0.5$ for $A = -0.635, 0$, and 0.635. It is noticed that they can be directly compared with the patterns shown in Fig. 3 with identical value of product for $\chi Mg = 2$. By the present dimensionless scaling, the main difference between Figs. 3 and 4 is the model dissipation. In general, even the evolution of the fingers is slower to allow longer simulating time period, the overall patterns show resemblances, e.g., same number of emerging fingers for $A = -0.635$ and similar fingering patterns for $A = 0.635$. These observations confirm that the model dissipation does not alter fingering patters significantly.

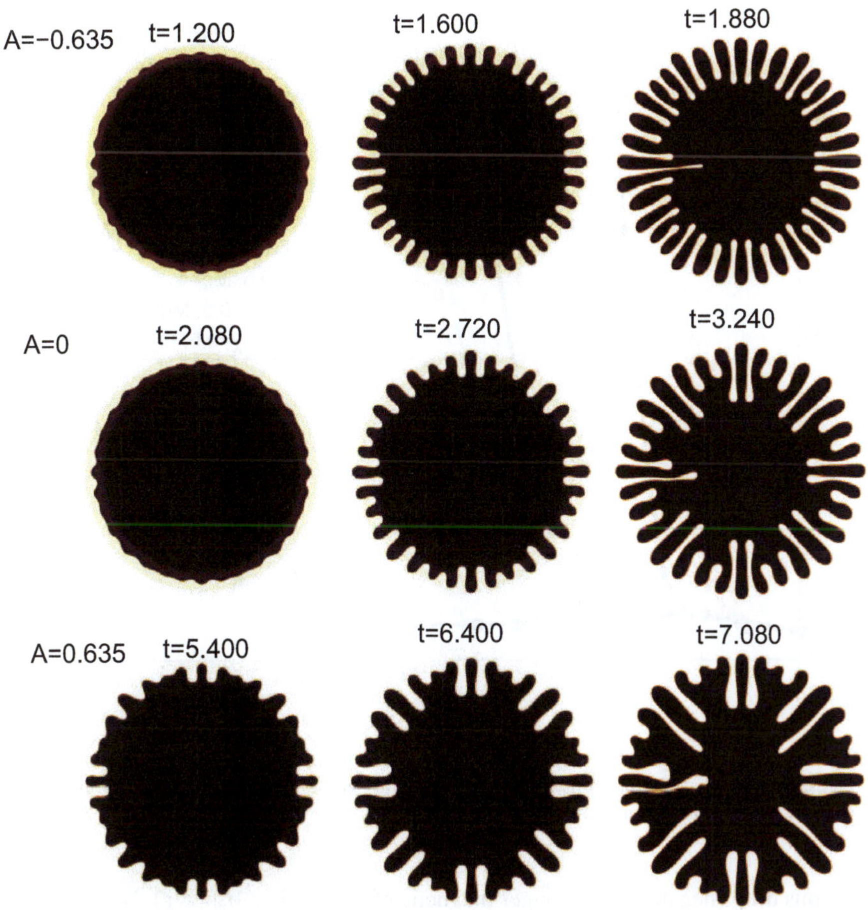

Fig. 4 Fingering patterns for $A = -0.635$ [top row], $A = 0$ [middle row], and $A = 0.635$ [bottom row]. The remaining parameters are: Mg=4 and $\chi = 0.5$.

To conclude this section, a quantitative measure of interfacial length L, which is capable to represent the prominence of the fingering instability, is presented for various control parameters. The interfacial length is approximated as [5]:

$$L_n(t) = \int_x \int_y \sqrt{(\frac{\partial c}{\partial x})^2 + (\frac{\partial c}{\partial y})^2 dxdy}. \tag{20}$$

The fingering onset can be clearly identified at the time when the interfacial length starts to deviate from the base value of circular shape as shown in Fig. 5. In addition, longer interfacial length generally indicates more vigorous fingering pattern. In line with the earlier discussion, earlier fingering onset occurs for the cases of less

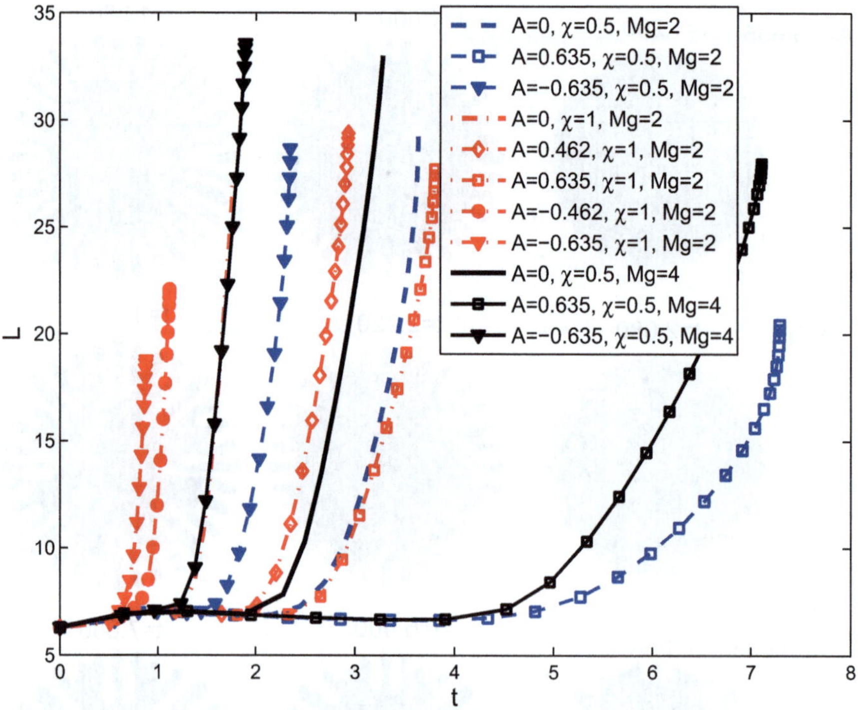

Fig. 5 Interfacial lengths for various parameters. Faster growth of interfacial length generally represents more unstable instability. Interfacial instability is in favor of the conditions with negative viscosity contrast and stronger magnetic effect.

viscous drop (negative A), stronger magnetic effect (higher Mg or χ). Nevertheless, the results for interfacial length appear contradictory. For $Mg = 2$ and $\chi = 0.5$, longer interfacial length at terminated simulating time is observed in the case of $A = -0.635$, which is viscously stable. Nevertheless, opposite results are found for case of $Mg = 2$ and $\chi = 1$, in which longer interfacial length appears for $A = 0.635$. This contradict behavior is resulted from the competitions of earlier fingering onset for a negative A and vigorous fingering pattern for a positive A. On the one hand, the interfacial length starts to growth after the onset time. On the other hand, the secondary fingering formation intensifies the growth of interfacial length. As a result, it leads to inconsistency to solely consider the interfacial length at the terminated time. Instead, it is more appropriate to consider the growth rate. All the cases clearly show higher growth rates of the interfacial lengths right before the terminated time for positive A, which indicates intensified secondary fingering behaviors in viscously unstable conditions.

4 Concluding Remarks

Direct numerical simulations of the interfacial instability of a non-magnetic drop surrounded by immiscible ferrofluids, confined in a Hele-Shaw cell, and subjected to an azimuthal field, are performed by a diffuse-interface method incorporating with highly accurate numerical schemes. The interface is unstable because of the inward attraction of the ferrofluids induced by the magnetic field gradient. The onset and pattern of interfacial fingering instability is analyzed to realize the coupling effects of the magnetic force and viscous contrast. If viscosity of the ferrofluids is greater than the immersed drop, in which the interface is viscously stable, the fingering onset takes place earlier because of the higher mobility of penetrated drop fluid. Nevertheless, the fingering pattern appears simpler with straightly emerging fingers. On the other hand, the viscously unstable interface, i.e., less viscous ferrofluids penetrating more viscous drop, results in more ramified pattern associated with the secondary fingering phenomena, e.g., competitions and tip-splits of fingers. However, the onset is delayed because less mobile drop fluid.

To quantify the prominence of fingering instability, development of the interfacial length is calculated as a global measure. In general, longer interfacial length indicates more unstable interface. We confirm the occurrences of earlier fingering onset and vigorous secondary phenomena both enhance growth of the interfacial length. Nevertheless, these two behaviors are favorable in opposite conditions, i.e., earlier onset and secondary phenomena in condition of more and less viscous ferrofluids, respectively, so that inconsistent evolutions of the interfacial length are observed. As a result, instead of the interfacial length, its growth rate is more appropriate for the consideration of the measure of instability. The earlier onset for cases of more viscous ferrofluids usually leads to greater growth rate in the early envelopment of fingering instability. On the contrary, the growth rate is dramatically enhanced at the later period when the secondary fingering phenomena are active for conditions of less viscous ferrofluids.

Acknowledgements support by R.O.C. MOST 104-2221-E-009-142-MY3 is acknowledged.

References

1. Rosensweig, R.E.: Ferrohydrodynamics. Cambridge University Press, Cambridge (1985) and references therein
2. Blums, E., Cebers, A., Maiorov, M.M.: Magnetic Fluids. De Gruyter, New York (1997) and references therein
3. Cowley, M.D., Rosensweig, R.E.: J. Fluid Mech. **30**, 671 (1967)
4. Friedrichs, R., Engel, A.: Phys. Rev. E **64**, 021406 (2001)
5. Chen, C.-Y., Cheng, Z.-Y.: Phys. Fluids E **20**, 054105 (2008)
6. Chen, C.-Y., Li, C.-S.: Phys. Fluids E **22**, 054105 (2010)
7. Chen, C.-Y., Wu, W.-L., Miranda, J.A.: Phys. Rev. E **82**, 056321 (2010)
8. Tsebers, A.O., Maiorov, M.M.: Magnetohydrodynamics (N.Y.) **16**, 21 (1980)

9. Langer, S.A., Goldstein, R.E., Jackson, D.P.: Phys. Rev. A **46**, 4894 (1992)
10. Jackson, D.P., Goldstein, R.E., Cebers, A.O.: Phys. Rev. E **50**, 298 (1994)
11. Pacitto, G., Flament, C., Bacri, J.-C., Widom, M.: Phys. Rev. E **62**, 7941 (2000)
12. Chen, C.-Y., Wu, H.-J.: Phys. Fluids **17**, 042101 (2005)
13. Wen, C.-Y., Chen, C.-Y., Kuan, D.-C.: Phys. Fluids **19**, 084101 (2007)
14. Saffman, P.G., Taylor, G.I.: Proc. R. Soc. Lond. Ser. A **245**, 312 (1958)
15. Chen, C.-Y., Tsai, W.-K., Miranda, J.A.: Phys. Rev. E **77**, 056306 (2008)
16. Miranda, J.A.: Phys. Rev. E **62**, 2985 (2000)
17. Chen, C.-Y., Wu, S.-Y., Miranda, J.A.: Phys. Rev. E **75**, 036310 (2007)
18. Dias, E., Miranda, J.A.: Phys. Rev. E **91**, 023020 (2015)
19. Cahn, J.W., Hilliard, J.E.: J. Chem. Phys. **28**, 258 (1958)
20. Lowengrub, J., Truskinovsky, L.: Proc. R. Soc. Lond. A **454**, 3617 (1998)
21. Lee, H.-G., Lowengrub, J., Goodman, J.: Phys. Fluids **14**, 492 (2002)
22. Lee, H.-G., Lowengrub, J., Goodman, J.: Phys. Fluids **14**, 514 (2002)
23. Chen, C.-Y., Huang, Y.-S., Miranda, J.A.: Phys. Rev E **84**, 046302 (2011)
24. Chen, C.-Y., Huang, Y.-S., Miranda, J. A.: Phys. Rev E **89**, 053006 (2014)
25. Huang, Y.-S., Chen, C.-Y.: Comput. Mech. **55**(2), 407–420 (2015)
26. Chen, C.-Y., Meiburg, E.: J. Fluid Mech. **371**, 233 (1998)
27. Ruith, M., Meiburg, E.: J. Fluid Mech. **420**, 225 (2000)
28. Meiburg, E., Chen, C.-Y.: SPE J. **5**, 2 (2000)

Numerical Analysis of Backward Erosion of Soils by Solving the Darcy–Brinkman Equations

Kazunori Fujisawa

Abstract The backward erosion of soils, which is induced by seepage flows, is numerically simulated. To the end, the following three aspects need to be computed: Water flow fields, onset and speed of erosion and boundary tracking between the soil and the water phases. The authors employ the Darcy–Brinkman equations as the governing equations for the water flow fields around the soils, which easily enable the simultaneous analysis of the seepage flows in the porous media and the water flows in the fluid domain. The onset and the speed of the seepage-induced erosion is predicted by an empirical formula constructed from the flow velocity and the pressure gradient of the seepage water. The boundary tracking scheme based on the phase-field equation is applied for tracking the soil boundary changing with the erosion. This paper shows that the combination of the above three aspects achieves the stable computation of the seepage-induced backward erosion of soils.

1 Introduction

When erosion occurs around soil surface where a seepage water flow comes out, the erosion develops in the opposite direction to the water flow. This type of erosion is known as backward erosion, which leads to piping of soils. Recently, the damages and failures of soil structures, such as levees and small embankment dams for irrigation reservoirs, have occurred more frequently because of a greater chance of severe typhoons and localized heavy rains. The piping phenomenon, induced by the soil erosion due to the seepage flows, is known as a primary cause of embankment breaks. Actually, Foster et al. [1] statistically investigated the failures and the incidents involving embankment dams around the world, and reported that the soil piping accounted for approximately 45 % of these incidents. Therefore, the soil erosion is considered to be a major threat to the structures made of earth materials and the objective of this paper is to develop a numerical method to compute the seepage-induced erosion, especially the backward erosion.

K. Fujisawa (✉)
Kyoto University, Kitashirakawa Oiwake-cho, Sakyo-ku, Kyoto 606-8502, Japan
e-mail: fujik@kais.kyoto-u.ac.jp

Y. Bazilevs, K. Takizawa (eds.), *Advances in Computational Fluid-Structure Interaction and Flow Simulation*, Modeling and Simulation in Science, Engineering and Technology, DOI 10.1007/978-3-319-40827-9_15

To this end, this article begins with the simultaneous computation of seepage flows in porous media and regular flows in fluid domains, because the erosion of soils is affected by water flows inside and outside the soils, and these two flow fields need to be grasped for predicting how the erosion develops. As described in the next section, the authors employ the Darcy–Brinkman equations as the governing equations for this problem and propose a numerical method to simultaneously solve the Navier–Stokes flow in the fluid domains and the Darcy flow in the porous media. Combining the boundary tracking with the simultaneous analysis of the Darcy and the Navier–Stokes flows, the computation of the seepage-induced erosion is carried out. In order to determine the moving speed of the soil boundary due to the erosion, an empirical formula predicting the discharge rate of the soils is utilized. The numerical method proposed by Sun and Beckermann [4] is applied to tracking the soil boundary changing with the erosion. The method enables the sharp interface tracking based on the phase-field equation. The numerical results presented in the end of this article show that the combination of the above numerical methods achieves the stable computation of the seepage-induced erosion.

2 Governing Equations

2.1 Water Flow Field

The Darcy–Brinkman equations are adopted as the governing equations for the coupled analysis of the Navier–Stokes and the Darcy flows:

$$\frac{\partial u_i}{\partial x_i} = 0 \tag{1}$$

$$\frac{\partial u_i}{\partial t} + \frac{\partial}{\partial x_j}\left(\frac{u_i u_j}{\lambda}\right) = -\frac{\lambda}{\rho}\frac{\partial p}{\partial x_i} + \nu\frac{\partial u_i}{\partial x_j \partial x_j} - \frac{\lambda g}{k}u_i \tag{2}$$

where u_i, p, ρ, ν, λ, k, g, t, and x_i denote the flow velocity, the piezometric pressure, the density of water, the kinematic viscosity of water, the porosity, the hydraulic conductivity, the gravitational acceleration, time, and Cartesian coordinates, respectively.

Eq. (2) can describe the Navier–Stokes equations in the fluid phase by giving $\lambda = 1.0$ and $1/k = 0$, and can approximate the Darcy's law in the Darcy phase when the hydraulic conductivity is sufficiently small. Therefore, the Darcy–Brinkman equations allow us to simulate the Darcy and the Navier–Stokes flows without employing the different governing equations between the fluid and the Darcy phases.

2.2 Boundary Tracking

The governing equation based on the phase-field equation is used for boundary tracking. This method regards the zero contour of the phase-field variable ϕ as the interface between two different phases, and the phase-field variable ϕ has a hyperbolic tangent profile across the interface. The usual phase-field method allows us to track the interface by solving the following equation derived from the general interface advection equation:

$$\frac{\partial \phi}{\partial t} + a|\nabla \phi| = b\left[\nabla^2 \phi + \frac{\phi(1-\phi^2)}{W}\right] \tag{3}$$

where ϕ, a, b, and W denote the phase-field variable, the moving velocity normal to the interface, the curvature coefficient, and a measure of the width of the hyperbolic tangent profile, respectively. However, this equation provides not only the normal interface motion but also the curvature-driven motion which is not necessary for the alteration of the soil-water interface caused by the seepage-induced erosion. Hence, the following equation developed by Sun and Beckermann [4] below is employed, which can avoid the curvature-driven interface motion.

$$\frac{\partial \phi}{\partial t} + a|\nabla \phi| = b\left[\nabla^2 \phi + \frac{\phi(1-\phi^2)}{W} - |\nabla \phi|\nabla \cdot \left(\frac{\nabla \phi}{|\nabla \phi|}\right)\right] \tag{4}$$

In the right-hand side of Eq. (4), the curvature term included in the right-hand side of Eq. (3), $\nabla^2 \phi + \phi\left(1-\phi^2\right)/W$, is cancelled out by the counter term, $|\nabla \phi|\nabla \cdot (\nabla \phi/|\nabla \phi|)$, whereby the interface motion is driven only by the velocity normal to the interface, denoted by a.

3 Numerical Method

The numerical method developed by the authors, which achieves the stable computation of the Darcy–Brinkman equations, is explained. The detailed numerical procedures for the boundary tracking by Eq. (4) are well described in [4]. The method presented in this section is based on the one proposed by Kim and Choi [3], which can solve the Navier–Stokes equations for incompressible fluids by the finite volume method with unstructured grids. Their method is characterized by the grid system shown in Fig. (1). The velocity and the pressure are stored at the center of the finite volume cells and the flux U is additionally computed at the mid-point of each cell face, which has the following definition:

$$U = (u_i)_{face}\, n_i \tag{5}$$

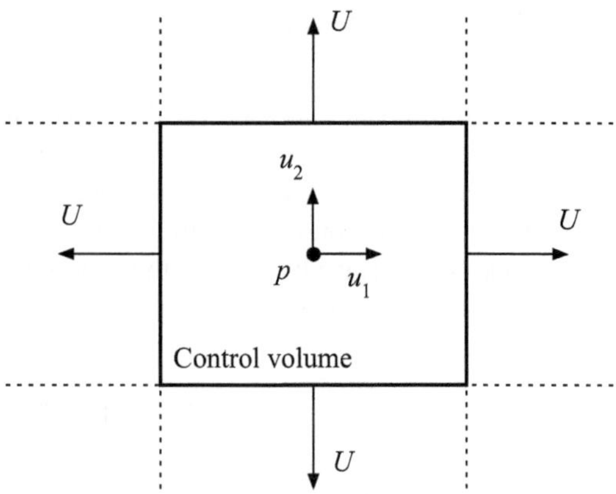

Fig. 1 Finite volume cell and variables: the velocity and the pressure are placed at the center of the cell, and the flux is defined at each of the cell edges

where $(u_i)_{face}$ and n_i denote the flow velocity and outward normal unit vector on the cell face, respectively. Applying a fractional step method and the Crank–Nicolson method to the time integration of Eq.(2), and spatially integrating it over the finite volume cells, the following equations are obtained:

$$\frac{u_i' - u_i^m}{\Delta t} = -\frac{\lambda g}{2k}\left(u_i' + u_i^m\right) \tag{6}$$

$$\frac{\hat{u}_i - u_i'}{\Delta t} = -\frac{1}{A}\oint_l \frac{1}{2\lambda}\left(\hat{u}_i U^m + u_i^m \hat{u}_j n_j\right) dl - \frac{\lambda}{A\rho}\oint_l p^m n_i dl + \frac{\nu}{2A}\oint_l \frac{\partial}{\partial n}\left(\hat{u}_i + u_i^m\right) dl \tag{7}$$

$$\frac{u_i^* - \hat{u}_i}{\Delta t} = -\frac{\lambda}{A\rho}\oint_l p^m n_i dl \tag{8}$$

$$\oint_l \frac{\lambda}{\rho}\frac{\partial p^{m+1}}{\partial n} dl = \frac{1}{\Delta t}\oint_l U^* dl \tag{9}$$

$$\frac{u_i^{m+1} - u_i^*}{\Delta t} = -\frac{\lambda}{A\rho}\oint_l p^{m+1} n_i dl \tag{10}$$

$$\frac{U^{m+1} - U^*}{\Delta t} = -\frac{\lambda}{\rho}\frac{\partial p^{m+1}}{\partial n} \tag{11}$$

where A, l, n_i, and Δt denote the area of the cell, the length of the cell faces, the outward normal unit vector at the cell faces, and the time step size, respectively, and the superscript m implies the number of time steps. u_i', $\hat{u}_i$, and u_i^* are the intermediate velocities between u_i^m and u_i^{m+1}. U^* is also the intermediate flux defined at the cell edges. The numerical procedures for solving the Darcy–Brinkman equations are to conduct computation of Eqs. (6) to (11) in the above order. Eqs. (7) and (9) result in the linear systems for $\hat{u}_i$ and p^{m+1}, which require the inversion of the matrices.

In order to compute the integrals appearing in Eqs. (6) to (11), the velocities, the pressure, and their directional derivatives need to be evaluated at the mid-point of each cell face. The values of the velocities and the pressure are interpolated from those at the centers of neighboring cells. Then, the manner of interpolating these variables plays an important role in the stable computation at the interface between the porous and the fluid domains. The simple linear interpolation of the variables may induce the physically unrealistic oscillations at the interface of the two different domains. In order to avoid the oscillations, the interpolation scheme described by the following equations is applied to the rectangular finite volume cells:

$$p_f = \frac{\frac{k_a}{\delta_a} p_a + \frac{k_b}{\delta_b} p_b}{\frac{k_a}{\delta_a} + \frac{k_b}{\delta_b}} \tag{12}$$

$$\left.\frac{\partial p}{\partial n}\right|_a = \frac{p_b - p_a}{\delta_a + \delta_b}, \quad \left.\frac{\partial p}{\partial n}\right|_b = \frac{p_a - p_b}{\delta_a + \delta_b} \tag{13}$$

$$u_{i,f} = \frac{\frac{k_b}{\delta_a} u_{i,a} + \frac{k_a}{\delta_b} u_{i,b}}{\frac{k_b}{\delta_a} + \frac{k_a}{\delta_b}} \tag{14}$$

$$\left.\frac{\partial u_i}{\partial n}\right|_a = \frac{u_{i,f} - u_{i,a}}{\delta_a}, \quad \left.\frac{\partial u_i}{\partial n}\right|_b = \frac{u_{i,f} - u_{i,b}}{\delta_b} \tag{15}$$

where p_f and $u_{i,f}$ denote the values of the pressure and the velocity at the interface, respectively, and δ is the distance from the cell center to the interface. The subscripts a and b mean the indexes for neighboring cells (See Fig. 2 or 3). Eq. (12) means that when a cell face is located on the interface between the porous and the fluid domains, the value of the pressure stored in the cell of the fluid domain is given to the interface (See Fig. 2 and note $k_b = \infty$). On the other hand, the velocity is interpolated in an opposite manner by Eq. (14), i.e., the velocity of the porous domain is given to the interface (See Fig. 3). While the cell face is located in either domain, the pressure and the velocity are linearly interpolated onto the cell face.

4 Numerical Result

The moving speed of the interface denoted by a in Eq. (4) corresponds to the discharge rate of soils. The interface speed a can be estimated by the following formula:

$$a = \frac{u_n}{\lambda} - \left\{ f + (1 - \lambda) \frac{\partial p}{\partial X} \right\} \frac{k}{\lambda^2 \rho g} \tag{16}$$

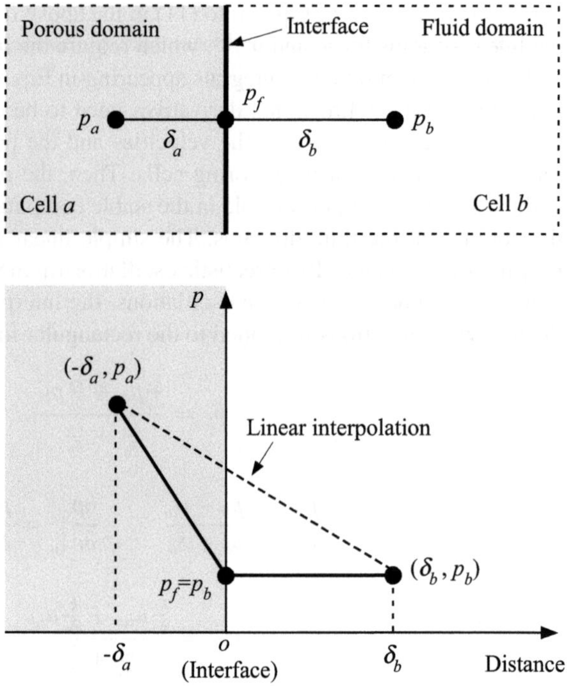

Fig. 2 Interpolation of pressure between the porous and the fluid domains

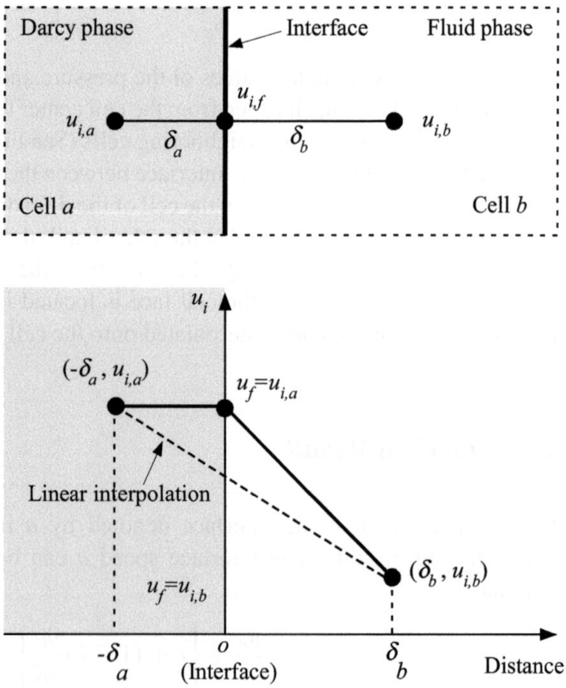

Fig. 3 Interpolation of pressure between the porous and the fluid domains

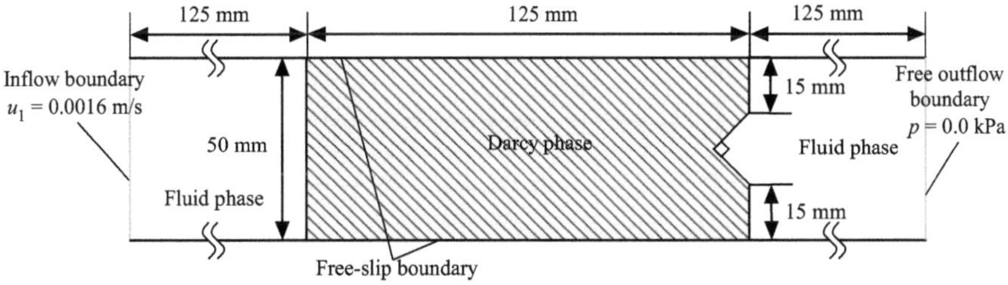

Fig. 4 Geometry and boundary conditions for numerical analysis of backward erosion

where u_n, X, and f denote the outward normal seepage flow velocity at the interface, the coordinate normal to the interface, and the maximum resisting force exerted onto the soil particles. Eq. (16) is derived from the experiments by Fujisawa et al. [2] and is based on the equilibrium of the forces exerted onto the soil particles in a direction perpendicular to the interface. Eq. (16) assumes that the effective stress of soils vanishes when the soil particles discharge, so it is applicable for the seepage erosion occurring on the soil surface. The profile of the boundary between the porous and the fluid domains can be updated by solving Eq. (4) at each time step after the values of a are calculated from the numerical solutions of the Darcy–Brinkman equations.

The numerical results of the backward erosion are presented herein. Fig. 4 shows the geometry and the boundary conditions. The porous domain of a 125 mm long soil block was installed at the middle of the computational domain and the rightward water flow was induced by the imposed boundary conditions. The right side of the porous domain was made concave, which intended to accelerate the seepage flow to the exit and to concentrate the seepage-induced erosion to the center of the porous domain. The other region was occupied by water, i.e., the fluid domain.

As for the boundary conditions, the horizontal velocity of 0.0016 m/s was given to the left side of the computational domain and the free outflow boundary condition was set on the right side. The free-slip condition was imposed on both the upside and the downside. The hydraulic conductivity and the porosity of the porous domain were assumed to be 1.0×10^{-3} m/s and 0.4, respectively. After the initial flow velocity and the initial water pressure were set to zero, the numerical computation was carried out until the penetration of the soil block occurred.

Figure 5 shows the profile of the computed interface between the porous and the fluid domains changing due to the erosion. As seen in the figure, the boundary moves in a direction opposite to the seepage flow and the backward seepage erosion can be observed. Fig 6 shows the contour plots of the horizontal flow velocity 10 seconds after the erosion started and after the penetration, respectively. These results show that the flow velocity in the region where the soil is eroded became even greater than the other region, which concentrated the water flow into the eroded region and developed the seepage-induced erosion straightly upstream. After the soil block was penetrated, the erosion no longer proceeded because the water flow almost fully concentrated to the connected fluid domain (Fig. 7).

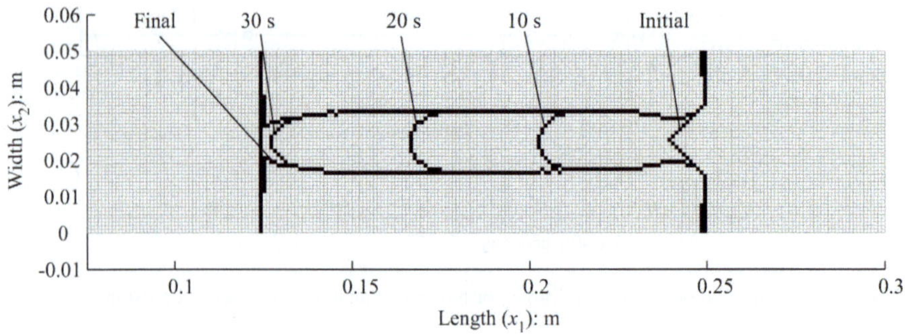

Fig. 5 The profile of the interface altering with elapsed time due to the backward erosion

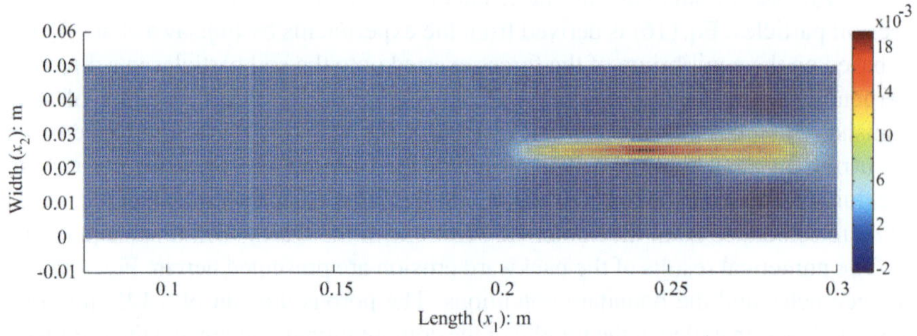

Fig. 6 Horizontal velocity 10 s after the start of erosion (Unit: m/s)

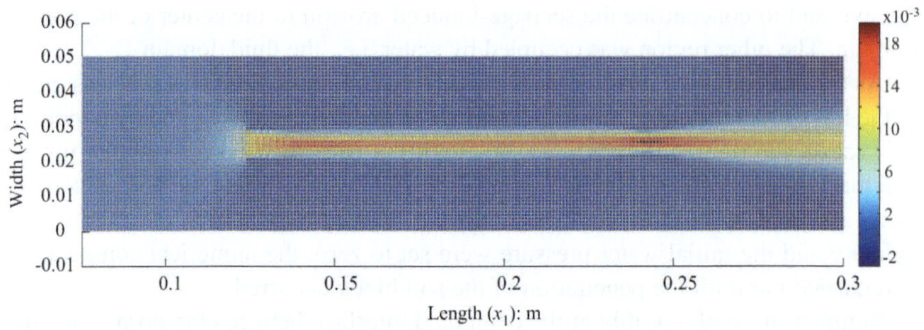

Fig. 7 Final horizontal velocity after penetration (Unit: m/s)

5 Conclusions

This article has proposed a numerical method for the computation of the soil erosion induced by seepage flows. This method is built from the three parts, i.e., the simultaneous analysis of the Darcy and the Navier–Stokes flows, the estimation of the erosion rate, and the computation of the boundary tracking. The authors employed the Darcy–Brinkman equations as the governing equations to achieve the simultaneous analysis of the Darcy and the Navier–Stokes flows. The erosion rate was estimated by the experimental formula of [2] and the tracking of the interface between the Darcy and the fluid phases was conducted by solving the phase-field equation modified by Sun and Beckermann [4]. The backward seepage erosion was computed by the proposed method, which has revealed that it can numerically predict the typical behavior of the seepage-induced erosion, which straightly develops upstream.

References

1. Foster, M., Fell, R., Spannagle, M.: The statistics of embankment dam failures and accidents. Can. Geotech. J. **37**, 1000–1024 (2000)
2. Fujisawa, K., Nishimura, S., Nakatani, A., Murakami, A.: Velocity of sand particles transported by upward seepage flow during sand boil. IDRE J. **80**(5), 409–416 (2012)
3. Kim, D., Choi, H.: A second-order time-accurate finite volume method for unsteady incompressible flow on hybrid unstructured grids. J. Comput. Phys. **162**, 411–428 (2000)
4. Sun, Y., Beckermann, C.: Sharp interface tracking using the phase-field equation. J. Comput. Phys. **220**, 626–653 (2007)

A Diffuse Interface Model for Incompressible Two-Phase Flow with Large Density Ratios

Yu Xie, Olga Wodo, and Baskar Ganapathysubramanian

Abstract In this chapter, we explore numerical simulations of incompressible and immiscible two-phase flows. The description of the fluid–fluid interface is introduced via a diffuse interface approach. The two-phase fluid system is represented by a coupled Cahn–Hilliard Navier–Stokes set of equations. We discuss challenges and approaches to solving this coupled set of equations using a stabilized finite element formulation, especially in the case of a large density ratio between the two fluids. Specific features that enabled efficient solution of the equations include: (i) a conservative form of the convective term in the Cahn–Hilliard equation which ensures mass conservation of both fluid components; (ii) a continuous formula to compute the interfacial surface tension which results in lower requirement on the spatial resolution of the interface; and (iii) a four-step fractional scheme to decouple pressure from velocity in the Navier–Stokes equation. These are integrated with standard streamline-upwind Petrov–Galerkin stabilization to avoid spurious oscillations. We perform numerical tests to determine the minimal resolution of spatial discretization. Finally, we illustrate the accuracy of the framework using the analytical results of Prosperetti for a damped oscillating interface between two fluids with a density contrast.

1 Introduction

Simulation of multiphase flows involving interface evolution has long been an interesting and challenging topic. Accurate and efficient methods of computing the position of the interface between immiscible fluid components, particularly in cases of surface tension driven fluid flow, are of particular interests. There are broadly

Y. Xie • B. Ganapathysubramanian (✉)
Department of Mechanical Engineering, Iowa State University, Ames, IA, USA
e-mail: yuxie@iastate.edu; baskarg@iastate.edu

O. Wodo
Materials Design and Innovation Department, Mechanical and Aerospace Department, University at Buffalo, Buffalo, NY, USA
e-mail: olgawodo@buffalo.edu

Y. Bazilevs, K. Takizawa (eds.), *Advances in Computational Fluid-Structure Interaction and Flow Simulation*, Modeling and Simulation in Science, Engineering and Technology, DOI 10.1007/978-3-319-40827-9_16

two families of methods of simulating such immiscible, multiphase fluid systems. The first class of methods track/capture the interface between two components. Examples of this class include marker (or front-tracking) methods ([19]), which interpolate the interface with a set of connected marker points; volume of fluid methods ([10]), which represent the fluid portion in a natural way with a color function; level set methods ([16]), which capture the interface as the zero level curve of a continuous function; and many of their variants based on this idea.

Instead of explicitly tracking the sharp interface, an alternative approach is based on the idea of smearing the interface and calculating its evolution by embedding it into a field. This class of methods are called diffuse interface methods. The idea of diffuse interface methods has been widely used and is reviewed in [1]. Diffuse interface methods represent the interface using the phase field variable that changes smoothly across the interface that has non-zero thickness. Various interfacial properties, like the surface tension, can be directly computed from the phase field without reconstructing the interface first.

Numerical algorithm using diffuse interface approaches has been the focus of several investigations. Hohenberg and Halperin [11] derived an abstract model for simulating interfacial flows of incompressible fluids with matched densities by coupling hydraulic equation with the Cahn–Hilliard equation, which is well known as model H. Lowengrub et al. [14] pointed out that binary fluids with incomprehensible components may be actually compressible, and they derived a quasi-incompressible formula for the flows of binary mixture with a density contrast. Yue et al. [23] proposed a 2D fully adaptive finite element model for simulating interfacial dynamics in incompressible viscoelastic fluids. A 3D numerical scheme was later developed [25] based on the 2D model. Ceniceros et al [5] decoupled the discrete model H system with a semi-implicit time discretization, and solved the linear system with a multigrid method combining a mesh refinement algorithm. Ding [8] derived a finite volume scheme for simulating binary mixture flows with large density ratios. Caginalp [3] mathematically verified the existence of sharp interface limit for the phase field model. Numerical investigations of the convergence of phase field model to sharp interface limit can be found in [14] and [23]. Yue et al. [24] and [22] later extended the study of sharp interface limit on the moving contact line problem.

Regardless of impressive advancement in the diffuse interface approaches, there are several important questions on the application of diffuse interface model, especially for surface tension mediated flows and for large density contrast. These include designing: (i) thermodynamically consistent models for two-phase flows ([9, 18]), (ii) methods that ensure mass conservation of fluid components regardless of the density ratio, (iii) approaches to accurately incorporate surface tension ([12]), as well as (iv) questions related to the spatial resolution requirements of the diffuse interface and designing scalable numerical scheme for such models. These questions motivate the developments in this chapter.

This chapter describes an efficient and robust framework that utilizes a stabilized finite element formulation for the simulation of incompressible and immiscible two-phase flows with a large density contrast. Our contributions include the following:

(1) a semi-coupled, mass-conserved Cahn–Hilliard Navier–Stokes equation system with an accurate continuous approximation of the surface tension for two-phase flows; (2) a four-step fractional scheme combined with a stabilized finite element method. The ensuing decoupling of pressure from the momentum equation enables scaling up of the framework; (3) comparison of the method with analytical results for two-phase flow under large density ratios; and (4) grid resolution studies of the combined Cahn–Hilliard Navier–Stokes system, which provides a guidance for grid size selection to ensure physically meaningful results.

This chapter is organized as follows: In section 2, we introduce the governing equations, discuss evaluation of surface tension, and strategy for dealing with large density ratios. Then the numerical schemes are discussed in section 3. Validations of the method and further analysis are discussed in section 4.

2 Governing Equations

Consider a finite domain, Ω, occupied by two immiscible fluids (*fluid*$_+$ and *fluid*$_-$) with density (ρ_+, ρ_-) and viscosity (η_+, η_-). A phase field variable ϕ is defined over the domain that essentially serves as an indicator function to the fluids. That is, $\phi(x) = 1$, when x lies in *fluid*$_+$, and $\phi(x) = -1$ when x lies in *fluid*$_-$. Alternatively, ϕ can also be considered to be a measure of the volume fraction of the immiscible fluid components. At the fluid–fluid interface the value of ϕ smoothly varies from -1 to 1 over a finite distance (hence, the term "diffuse interface"). The fluid properties can now be written in terms of the individual properties and the phase field variable as

$$\rho(\phi) = \rho_- \frac{1-\phi}{2} + \rho_+ \frac{1+\phi}{2}, \quad \eta(\phi) = \eta_- \frac{1-\phi}{2} + \eta_+ \frac{1+\phi}{2}, \tag{1}$$

The motion of the two fluids is governed by the transient incompressible Navier–Stokes equation with an additional surface tension $\boldsymbol{f}_s$ (that depends on the interface, and hence is described by the phase field function ϕ)

$$\rho \left[\frac{\partial \boldsymbol{u}}{\partial t} + (\nabla \cdot \boldsymbol{u})\, \boldsymbol{u} \right] = \nabla \cdot \boldsymbol{\tau} + \rho \boldsymbol{g} + \boldsymbol{f}_s, \tag{2}$$

$$\nabla \cdot \boldsymbol{u} = 0, \tag{3}$$

where $\boldsymbol{g}$ is the gravitational acceleration, $\boldsymbol{\tau}$ is the stress tensor given as

$$\boldsymbol{\tau} = -p\boldsymbol{I} + 2\eta \boldsymbol{v}(\boldsymbol{u}), \quad \boldsymbol{v}(\boldsymbol{u}) = \frac{1}{2}\left[\nabla \boldsymbol{u} + (\nabla \boldsymbol{u})^T\right], \tag{4}$$

where $\boldsymbol{I}$ is the identity tensor. The boundary conditions for Eqn. 3 are given as

$$\boldsymbol{u} = \boldsymbol{g}(\boldsymbol{x}), \quad \boldsymbol{x} \in \Gamma_1, \quad \text{and} \quad \boldsymbol{\tau} \cdot \boldsymbol{n} = \boldsymbol{h}(\boldsymbol{x}), \quad \boldsymbol{x} \in \Gamma_2, \tag{5}$$

where Γ_1 and Γ_2 are the essential boundary (or Dirichlet boundary) and natural boundary (or Neumann boundary), respectively, and $\Gamma_1 \cup \Gamma_2 = \partial\Omega$, $\Gamma_1 \cap \Gamma_2 = \emptyset$.

The diffuse interface model assumes a non-zero thickness of the interface, as a smooth transition between the two fluid components. Evolution of the interface is described by the Cahn–Hilliard equation [4], which is derived by minimizing the free energy of the system given by the Ginzburg–Landau form $H(\phi) = \int_\Omega \left[f(\phi) + \frac{1}{2}\epsilon^2|\nabla\phi|^2\right] \mathrm{d}\Omega$, where ϵ is the constant (interfacial parameter) that encodes the interfacial thickness (or the relative energy required to form an interface). The total energy consists of the bulk energy density $f(\phi)$ and interfacial energy $\frac{1}{2}\epsilon^2|\nabla\phi|^2$. Function $f(\phi)$ is written as a double-well function, $f(\phi) = \frac{1}{4}(1-\phi^2)^2$ with two minima for $\phi \in [-1, 1]$, corresponding to the two equilibrium states of the binary fluid components. The advective Cahn–Hilliard equation is represented as

$$\frac{\partial \phi}{\partial t} + \nabla \cdot (\phi \boldsymbol{u}) = \nabla \cdot [M \nabla \mu(\phi)], \tag{6}$$

$$\mu(\phi) = \frac{\delta H(\phi)}{\delta \phi} = \phi^3 - \phi - \epsilon^2 \Delta \phi, \tag{7}$$

where $\boldsymbol{u}$ is the velocity, M is the mobility, and $\mu(\phi)$ is the chemical potential. Here we set M as a constant over the whole domain.[1]

The governing equation (Eqn 7) is supplemented by the following boundary condition:

$$\boldsymbol{n} \cdot \nabla \mu = 0 \quad \boldsymbol{x} \in \partial\Omega, \quad \boldsymbol{n} \cdot \nabla \phi = \frac{1}{\sqrt{2}\epsilon} \cos(\theta_S)(1 - \phi^2) \quad \boldsymbol{x} \in \partial\Omega, \tag{8}$$

where θ_S is the static contact angle. The boundary condition for μ acts as the no-flux condition for the fluid components. The condition for ϕ determines the contact angle of the liquid(s) with the substrate [13].

The Navier–Stokes equation and Cahn–Hilliard equation are coupled by the convective term $\nabla \cdot (\phi \boldsymbol{u})$ in the Cahn–Hilliard equation and by the surface tension $\boldsymbol{f}_s = \boldsymbol{f}_s(\phi)$ in the Navier–Stokes equation. We choose the conservative form of the advection term [15] to ensure numerical mass conservation.

The surface tension $\boldsymbol{f}_s$ in the momentum equation (Eqn 3) is a numerical approximation of the continuum surface force formula [2], which converts the

---

[1] [20] also suggests a degenerate form of M, $M = D(1 - \phi^2)$, where D is the diffusivity. We obtain identical results for the results presented here, with the constant mobility case converging faster (iterations per time step).

surface tension into a continuous body force as a function of ϕ. Kim [12] discussed the various forms of the surface function. We choose the following form of $\boldsymbol{f}_s$ (which is an approximation of the force term also used in level set method [6]):

$$\boldsymbol{f}_s = -\sigma \nabla \cdot \left(\frac{\nabla \phi}{|\nabla \phi|} \right) \epsilon \alpha |\nabla \phi|^2 \frac{\nabla \phi}{|\nabla \phi|}, \tag{9}$$

where $\nabla \cdot \left(\frac{\nabla \phi}{|\nabla \phi|} \right)$ represents the interface curvature κ, $\epsilon \alpha |\nabla \phi|^2$ approximates the Dirac delta function $\delta(\phi)$, $\frac{\nabla \phi}{|\nabla \phi|}$ is the normalized gradient of ϕ, and $\alpha = \frac{2\sqrt{2}}{3}$. With this choice, the curvature κ can be easily solved from a Poisson type of equation, without introducing higher order derivatives of ϕ. This enables computationally efficient FEM solutions using linear basis functions.

We write the governing equations in dimensionless form by scaling the variables using the characteristic velocity magnitude U, length L, density ρ_+:

$$\boldsymbol{u}^* - \frac{\boldsymbol{u}}{U}, \quad \boldsymbol{x}^* = \boldsymbol{x}/L, \quad t^* = \frac{t}{L/U}, \quad p^* = \frac{p}{\rho_+ U^2} \tag{10}$$

The dimensionless governing equations are (after dropping the superscript "*" in the sequel)

$$\tilde{\rho} \left(\frac{\partial \boldsymbol{u}}{\partial t} + \boldsymbol{u} \cdot \nabla \boldsymbol{u} \right) + \nabla p - \frac{1}{Re} \nabla \cdot \left[\tilde{\eta} \left(\nabla \boldsymbol{u} + (\nabla \boldsymbol{u})^T \right) \right]$$

$$= \frac{\tilde{\rho} \boldsymbol{g}_0}{Fr} - \frac{2\sqrt{2}}{3} \frac{Cn}{We} \nabla \cdot \left(\frac{\nabla \phi}{|\nabla \phi|} \right) |\nabla \phi|^2 \frac{\nabla \phi}{|\nabla \phi|}, \tag{11}$$

$$\nabla \cdot \boldsymbol{u} = 0, \tag{12}$$

$$\frac{\partial \phi}{\partial t} + \nabla \cdot (\phi \boldsymbol{u}) = \frac{1}{Pe} \nabla \cdot [M \nabla \mu(\phi)], \tag{13}$$

$$\mu(\phi) = \phi^3 - \phi - Cn^2 \Delta \phi, \tag{14}$$

where $\tilde{\rho} = \rho_- / \rho_+$ is the density ratio, $\tilde{\eta} = \eta_- / \eta_+$ is the viscosity ratio, and $\boldsymbol{g}_0$ indicates the gravitational direction unit vector. The dimensionless numbers are: Reynolds number, *Re*, defined as $Re = \rho_+ UL/\mu_+$, representing the ratio between inertia force and viscous force; Froude number *Fr*, defined as $Fr = U^2/gL$, with g the magnitude of gravitational acceleration, is the ratio of inertia to external forces; Weber number, *We*, defined as $We = \rho_+ U^2 L/\sigma$, representing the ratio between inertial force and surface tension; Cahn number, *Cn*, defined as $Cn = \epsilon/L$, representing the relative interface thickness; and *Pe* is the Péclet number, $Pe = LU/(M\mu)$, where M and μ are the characteristic mobility and chemical potential. For most numerical simulations, the Péclet number is set to be proportional to the inverse of Cahn number ([5, 8]), i.e., $Pe = O(1/Cn)$. We take $Pe = 1/Cn.$ for the results presented in this chapter.

3 Numerical Schemes

The phase field and chemical potential at time step t^{n+1} are obtained by solving Cahn–Hilliard equation with a Crank–Nicolson scheme

$$\frac{\phi^{n+1}-\phi^{n}}{\Delta t}+\frac{1}{2}\nabla\cdot(\phi^{n}\boldsymbol{u}^{n})+\frac{1}{2}\nabla\cdot\left(\phi^{n+1}\boldsymbol{u}^{n+1}\right)=\frac{1}{2}\frac{1}{Pe}\nabla\cdot\nabla\mu^{n}+\frac{1}{2}\frac{1}{Pe}\nabla\cdot\nabla\mu^{n+1}, \tag{15}$$

$$\mu^{n+1}=(\phi^{n+1})^{3}-\phi^{n+1}-Cn^{2}\nabla\cdot\nabla\phi^{n+1}. \tag{16}$$

The curvature κ^{n+1} is then calculated using the solved field function ϕ^{n+1}:

$$\kappa^{n+1}=\nabla\cdot\left(\frac{\nabla\phi^{n+1}}{|\nabla\phi^{n+1}|}\right). \tag{17}$$

Velocity and pressure are solved by a four-step fractional step method ([7]), where pressure is decoupled from the momentum equation and solved by a Poisson type equation. The four-step fractional method incorporated with surface tension and varying density/viscosity is given by:

$$\tilde{\rho}^{n+1}\left(\frac{\hat{\boldsymbol{u}}-\boldsymbol{u}^{n}}{\Delta t}+\frac{1}{2}\left(\hat{\boldsymbol{u}}\cdot\nabla\hat{\boldsymbol{u}}+\boldsymbol{u}^{n}\cdot\nabla\boldsymbol{u}^{n}\right)\right)+\nabla p^{n}-\frac{1}{2}\frac{1}{Re}\nabla\cdot\left[\tilde{\eta}\left(\nabla\boldsymbol{u}^{n}+(\nabla\boldsymbol{u}^{n})^{T}\right)\right]$$
$$-\frac{1}{2}\frac{1}{Re}\nabla\cdot\left[\tilde{\eta}\left(\nabla\boldsymbol{u}^{n+1}+(\nabla\boldsymbol{u}^{n+1})^{T}\right)\right]=\frac{\tilde{\rho}^{n+1}}{Fr}\boldsymbol{g}_{0}-\frac{2\sqrt{2}}{3}\frac{Cn}{We}\kappa^{n+1}|\nabla\phi^{n+1}|\nabla\phi^{n+1}, \tag{18}$$

$$\tilde{\rho}^{n+1}\frac{\boldsymbol{u}^{*}-\hat{\boldsymbol{u}}}{\Delta t}=\nabla p^{n}, \tag{19}$$

$$\nabla\cdot\nabla p^{n+1}=\frac{\tilde{\rho}^{n+1}}{\Delta t}\nabla\cdot\boldsymbol{u}^{*}, \tag{20}$$

$$\tilde{\rho}^{n+1}\frac{\boldsymbol{u}^{n+1}-\boldsymbol{u}^{*}}{\Delta t}=-\nabla p^{n+1}. \tag{21}$$

In the above equations, $\hat{\boldsymbol{u}}$ and $\boldsymbol{u}^{*}$ are the intermediate velocities obtained based on pressure p^{n} from the previous time step. Pressure p^{n+1} is updated via Eqn. 20 with the intermediate velocities. Notice that the intermediate velocities do not satisfy the incompressible condition, i.e., $\nabla\cdot\boldsymbol{u}^{*}\neq 0$. So the last step corrects the velocity with the latest pressure field p^{n+1} to guarantee the incompressibility condition. We apply a Crank–Nicolson representation for the convective term and diffusive term in the momentum equation.

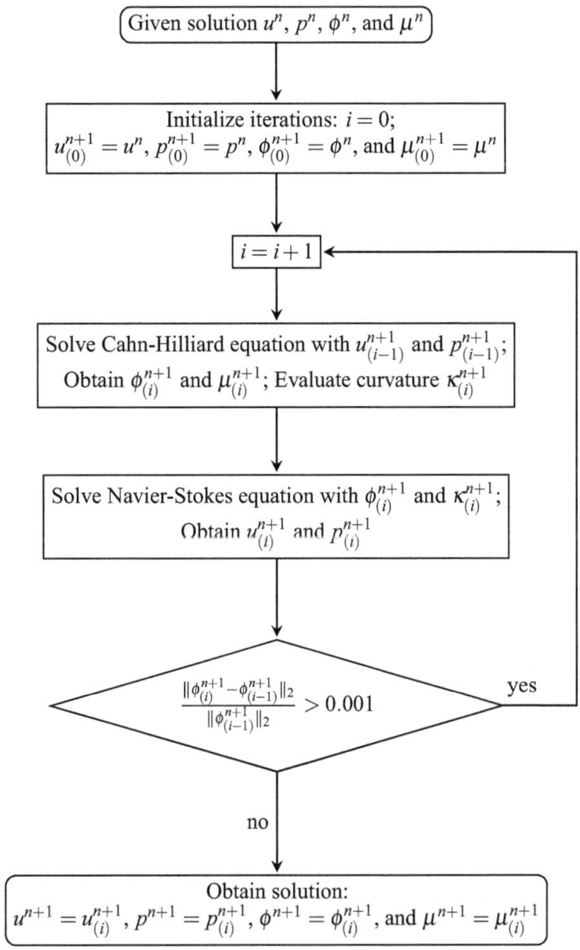

Fig. 1 Flowchart of the iteration solver for the semi-coupled Cahn–Hilliard Navier–Stokes equation system.

Eqns. 15–21 form a coupled system of ϕ^{n+1}, μ^{n+1}, $\boldsymbol{u}^{n+1}$, and p^{n+1}. Instead of directly solving the whole system in one step, we perform an iterative algorithm to solve a semi-decoupled equation system, shown by the flowchart in figure 1. In all simulations shown, the convergence is achieved within two or three iterations. A streamline-upwind Petrov–Galerkin stabilized finite element framework is used to solve this set of equations.

Evaluation of density and viscosity given by Eqn. 1 is valid with the assumption that ϕ is a scalar with value in $[-1, 1]$. However, the numerical scheme does not restrict ϕ strictly between -1 and 1. Notice that Eqn (1) gives a negative ρ when a lower bound of ϕ is reached, i.e., $\phi < -(1 + \tilde{\rho})/(1 - \tilde{\rho})$. To ensure positivity of density and viscosity, a normalization procedure is applied on ϕ *only* for computing ρ and η [9],

$$\hat{\phi} = \begin{cases} \phi & |\phi| <= 1 \\ \text{sign}(\phi) & \text{otherwise}, \end{cases} \tag{22}$$

and $\hat{\phi}$ is used for evaluating density and viscosity

$$\rho = \rho_- \frac{1-\hat{\phi}}{2} + \rho_+ \frac{1+\hat{\phi}}{2}, \quad \eta = \eta_- \frac{1-\hat{\phi}}{2} + \eta_+ \frac{1+\hat{\phi}}{2}. \tag{23}$$

Notice that Eqn. 22 is employed only for calculating the physical properties, ρ and η, appearing in Navier–Stokes equation to avoid numerical instability. The phase field ϕ obtained from Cahn–Hilliard equation is actually not modified by Eqn. 22. Therefore applying Eqn. 22 will not compromise mass conservation of fluid components.

4 Numerical Results

4.1 Convergence Tests

In this sub-section, we test the numerical method by using a *manufactured solution* suggested by Dong and Shen [9]

$$\begin{aligned} u &= \sin(\pi x)\cos(\pi y)\sin t, \quad & v &= -\cos(\pi x)\sin(\pi y)\sin t, \\ p &= \sin(\pi x)\sin(\pi y)\cos t, \quad & \phi &= \cos(\pi x)\cos(\pi y)\sin t. \end{aligned} \tag{24}$$

The above expressions satisfy the continuity equation (Eqn. 13). However, they do not satisfy the momentum equation (Eqn. 12) and Cahn–Hilliard equation (Eqn. 14). To remedy this issue, an artificial source term to the right-hand sides of both equation (Eqn. 12) and (Eqn. 14) is added.

The simulation is performed in a rectangular domain, $\Omega = [0, 2] \times [-1, 1]$ with Dirichlet conditions for velocity components u and v (set using the analytical solution (24)) applied on all boundaries. Quadrilateral elements with bilinear basis functions are used. The simulations are run with the following parameters:

$$Re = 100, \quad Cn = 0.1, \quad Pe = 10, \quad We = 100, \quad \text{and} \quad Fr = 0. \tag{25}$$

The time step Δt is fixed at 1×10^{-4}.

Figure 2 (left) shows mesh convergence results for velocity and phase field parameter. The relative L2 error is computed:

$$\text{error}_a = \frac{\|\boldsymbol{a}_{numeric} - \boldsymbol{a}_{analytical}\|_2}{\|\boldsymbol{a}_{analytical}\|_2} \tag{26}$$

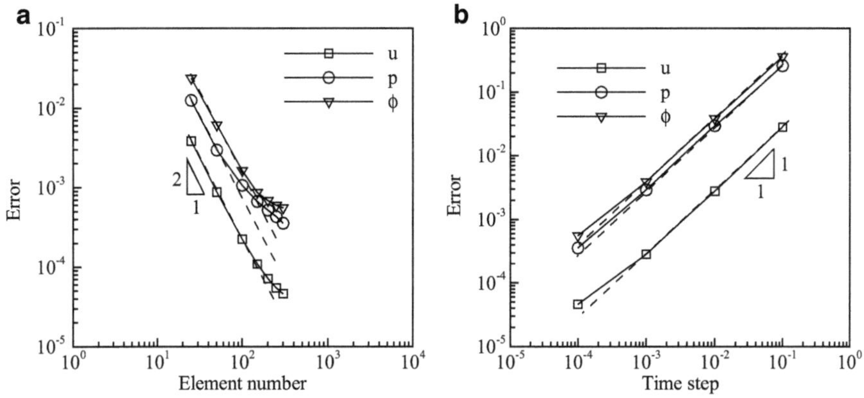

Fig. 2 Temporal and spatial convergence tests. Dashed lines reflect the estimated convergence rates. (a) Convergence of element numbers on one edge; (b) Convergence of time step Δt.

for $a = \{u, p, \phi\}$. The log–log slopes for all variables are around 2, indicating consistent second order spatial convergence rate. Decrease in the slope for large element size is probably due to the saturation from the temporal error due to constant Δt.

Figure 2 (right) shows the temporal convergence results. The spatial grid is fixed at 300×300 linear quadrilateral elements. The slope for all three variables is close to 1. Note that the error is evaluated at $t = 1$ (the last time step).

4.2 Damped Oscillation of a Capillary Wave

We study the damped oscillation of an interface between two incompressible viscous fluids with the lighter fluid placed on top of the heavier one. The interface is initially perturbed by sinusoidal function with a small amplitude H_0 and wave number k. The analytical solution [17] is derived for two fluids having the same kinematic viscosity ν, but different density. It describes the evolution of the amplitude of the interface, $H(t)$, and is given in [17]. This serves as a good benchmark test for the computational framework.

The motion of interface is simulated in a rectangular domain. The initial shape of the interface is set using sinusoidal function, $y_c = 0.01 \cos(2\pi x)$, where the wave length of interface oscillation is chosen as the characteristic length. The computational domain is chosen as $[0, 0.5] \times [-1, 1]$. In this way, the top and bottom boundaries are far enough from the interface to eliminate end effects. No-slip boundary condition for velocity components is applied on top and bottom boundaries. No-flux condition is applied on the two side walls. Reynolds number Re, Weber number We, and Froude number Fr are chosen as 100, 1, and 1, respectively.

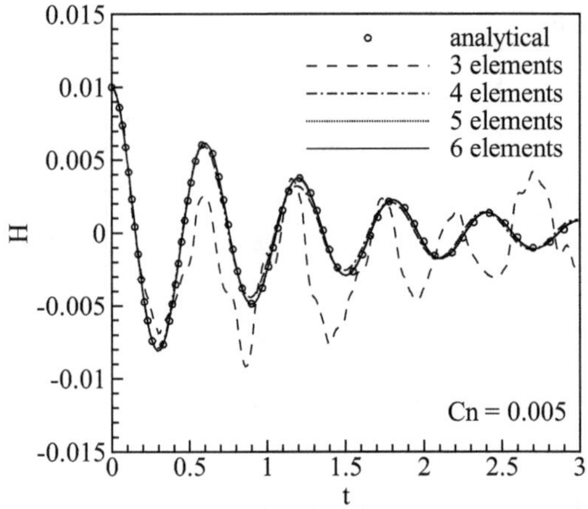

Fig. 3 Evolution of interface oscillation magnitude with various number of elements per interface. The Cahn number *Cn* is fixed at 0.005 for this test.

Table 1 Relative L∞ and relative L2 errors for the interface discretization tests.

elements per interface	relative L∞ error	relative L2 error
3	0.677	0.983
4	0.0729	0.110
5	0.0266	0.0423
6	0.0132	0.0203

Spatial discretization plays an important role in determining the accuracy of diffusive interface model. Wodo et al. [20] showed that at least four elements across the interface (where ϕ is approximately between -0.9 and 0.9) are necessary for accurate results of the Cahn–Hilliard model. However, this analysis was performed without the convective term or coupling with the Navier–Stokes equation. We extend the analysis in [20] to investigate how the diffuse interface model is affected by the element size. We fix the time step increment Δt to a small value, i.e., $\Delta t = 1 \times 10^{-4}$. Density and viscosity ratios are both set to unity. Cahn number *Cn* is selected as 0.005, which results in Péclet number $Pe = 200$. Below, we report results for 3, 4, 5, and 6 elements per interface.

Figure 3 shows the evolution of amplitude $H(t)$ for different spatial resolutions overlaid with the analytical result. The plot clearly shows that 3 elements per interface is not enough to capture the physical properties. Table 1 details the relative errors for all 4 cases. Our analysis clearly show that the simulations with at least 4 elements per interface have significant smaller errors than case with 3 elements per interface. This result is in agreement with previous recommendation [20], where no fluid flow was accounted for, but both phase separation and coarsening were studied.

An important highlight of the presented numerical scheme is its capability to simulate flows behavior of fluids with large density and viscosity ratios. Here, we consider three density ratio: 10, 100, and 1000. We choose a discretization corresponding to six elements per interface and time step $\Delta t = 1 \times 10^{-4}$. Results are shown in figure 4. As expected the oscillation for larger density ratio reduces the damping effect thus has longer oscillating period and slower damping rate. These features are accurately captured by the proposed model.

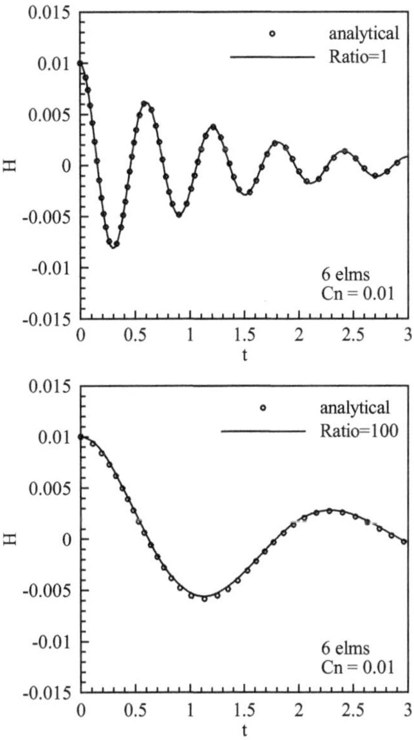

Fig. 4 Evolution of interface oscillation with different density and viscosity ratios. (a) Same density/viscosity ratio; (b) $\rho_+ : \rho_- = \mu_+ : \mu_- = 10 : 1$; (c) $\rho_+ : \rho_- = \mu_+ : \mu_- = 100 : 1$; (d) $\rho_+ : \rho_- = \mu_+ : \mu_- = 1000 : 1$;

5 Conclusions

In this chapter, we develop a finite element framework to model two-phase flows using a diffuse interface model. A four-step fractional temporal discretization is employed to decoupled the pressure and momentum equations. The coupling terms between the Navier–Stokes equation and Cahn–Hilliard equation are carefully investigated to guarantee that the scheme respects conservation of mass and accurately captures the effects of surface tension, especially for fluids with a large density contrast. Coupling between the two equations is treated iteratively. Validations are then performed to test the accuracy of the numerical scheme. We use a manufactured solution to illustrate order of convergence. Finally, we report the excellent agreement with the analytical solution for the damped oscillation of an interface between two incompressible fluids. We report the agreement for the whole range of density ratios. This framework has utility for modeling two-phase flows involving surface tension effects which exhibit a large density contrast (e.g., air-water, air-solvent systems). Specifically, we intend to leverage this framework to explore the effects of geometrical patterning on spreading behavior during manufacturing of polymer blend thin films [21].

Acknowledgements The authors acknowledge partial funding from KAUST CRG, NSF 1236839, and NSF 1149365. Computing support from NSF XSEDE via TG-CTS110007 is gratefully acknowledged.

References

1. Anderson, D.M., McFadden, G.B., Wheeler, A.A.: Diffuse-interface methods in fluid mechanics. Annu. Rev. Fluid Mech. **30**(1), 139–165 (1998)
2. Brackbill, J.U., Kothe, D.B., Zemach, C.: A continuum method for modeling surface tension. J. Comput. Phys. **100**(2), 335–354 (1992)
3. Caginalp, G., Chen, X.: Convergence of the phase field model to its sharp interface limits. Eur. J. Appl. Math. **9**(4), 417–445 (1998)
4. Cahn, J.W., Hilliard, J.E.: Free energy of a nonuniform system. i. interfacial free energy. J. Chem. Phys. **28**(2), 258–267 (1958)
5. Ceniceros, H.D., Nós, R.L., Roma, A.M.: Three-dimensional, fully adaptive simulations of phase-field fluid models. J. Comput. Phys. **229**(17), 6135–6155 (2010)
6. Chang, Y.-C., Hou, T.Y., Merriman, B., Osher, S.: A level set formulation of Eulerian interface capturing methods for incompressible fluid flows. J. Comput. Phys. **124**(2), 449–464 (1996)
7. Choi, H.G., Choi, H., Yoo, J.Y.: A fractional four-step finite element formulation of the unsteady incompressible Navier-Stokes equations using SUPG and linear equal-order element methods. Comput. Methods Appl. Mech. Eng. **143**(3), 333–348 (1997)
8. Ding, H., Spelt, P.D.M., Shu, C.: Diffuse interface model for incompressible two-phase flows with large density ratios. J. Comput. Phys. **226**(2), 2078–2095 (2007)
9. Dong, S., Shen, J.: A time-stepping scheme involving constant coefficient matrices for phase-field simulations of two-phase incompressible flows with large density ratios. J. Comput. Phys. **231**(17), 5788–5804 (2012)
10. Gueyffier, D., Li, J., Nadim, A., Scardovelli, R., Zaleski, S.: Volume-of-fluid interface tracking with smoothed surface stress methods for three-dimensional flows. J. Comput. Phys. **152**(2), 423–456 (1999)
11. Hohenberg, P.C., Halperin, B.I.: Theory of dynamic critical phenomena. Rev. Mod. Phys. **49**(3), 435 (1977)
12. Kim, J.: A continuous surface tension force formulation for diffuse-interface models. J. Comput. Phys. **204**(2), 784–804 (2005)
13. Li, X., Lowengrub, J., Rätz, A., Voigt, A.: Solving PDEs in complex geometries: a diffuse domain approach. Commun. Math. Sci. **7**(1), 81 (2009)
14. Lowengrub, J., Truskinovsky, L.: Quasi-incompressible Cahn–Hilliard fluids and topological transitions. Proc. R. Soc. London Ser. A **454**(1978), 2617–2654 (1998)
15. Minjeaud, S.: An unconditionally stable uncoupled scheme for a triphasic Cahn–Hilliard/Navier–Stokes model. Numer. Methods Partial Differ. Equ., **29**(2), 584–618 (2013)
16. Osher, S., Sethian, J.A.: Fronts propagating with curvature-dependent speed: algorithms based on Hamilton-Jacobi formulations. J. Comput. Phys. **79**(1), 12–49 (1988)
17. Prosperetti, A.: Motion of two superposed viscous fluids. Phys. Fluids **24**(7), 1217 (1981)
18. Simsek, G., van der Zee, K.G., van Brummelen, E.H.: Stable–time scheme for energy dissipative quasi–incompressible two–phase diffuse–interface flows. In: Book of Abstracts, European Conference on Numerical Mathematics and Advanced Applications p. 159 (2015)

19. Tryggvason, G., Bunner, B., Esmaeeli, A., Juric, D., Al-Rawahi, N., Tauber, W., Han, J., Nas, S., Jan, Y.-J.: A front-tracking method for the computations of multiphase flow. J. Comput. Phys. **169**(2), 708–759 (2001)
20. Wodo, O., Ganapathysubramanian, B.: Computationally efficient solution to the Cahn-Hilliard equation: adaptive implicit time schemes, mesh sensitivity analysis and the 3d isoperimetric problem. J. Comput. Phys. **230**(15), 6037–6060 (2011)
21. Xie, Y., Wodo, O., Ganapathysubramanian, B.: A diffuse interface model for incompressible two-phase flows: stabilized FEM, large density ratios, and the 3d patterned substrate wetting problem. Comput. Fluids (submitted) doi:10.1016/j.compfluid.2016.04.011
22. Yue, P., Feng, J.J.: Can diffuse-interface models quantitatively describe moving contact lines? Eur. Phys. J. E Spec. Top. **197**(1), 37–46 (2011)
23. Yue, P., Zhou, C., Feng, J.J., Ollivier-Gooch, C.F., Hu, H.H.: Phase-field simulations of interfacial dynamics in viscoelastic fluids using finite elements with adaptive meshing. J. Comput. Phys. **219**(1), 47–67 (2006)
24. Yue, P., Zhou, C., Feng, J.J.: Sharp-interface limit of the Cahn–Hilliard model for moving contact lines. J. Fluid Mech. **645**, 279–294 (2010)
25. Zhou, C., Yue, P., Feng, J.J., Ollivier-Gooch, C.F., Hu, H.H.: 3d phase-field simulations of interfacial dynamics in Newtonian and viscoelastic fluids. J. Comput. Phys. **229**(2), 498–511 (2010)

Isogeometric Phase-Field Simulation of Boiling

Ju Liu and Thomas J.R. Hughes

Abstract In this work we consider the Navier–Stokes–Korteweg equations, a diffuse-interface model describing liquid–vapor phase transitions. A numerical scheme for this model is constructed based on functional entropy variables and a new time integration concept. The fully discrete scheme is unconditionally stable in entropy and second-order time-accurate. Isogeometric analysis is utilized for spatial discretization. The boiling problem is numerically investigated by making proper assumptions on transport parameters and boundary conditions. Compared with traditional multiphase solvers, the dependence on empirical data is significantly reduced, and this modeling approach provides a unified predictive tool for both nucleate and film boiling. Both two- and three-dimensional simulation results are provided.

1 Introduction

Boiling is a thermally induced phase transition process in which new liquid–vapor interfaces are generated in a bulk liquid region [2]. It is an extremely effective mechanism in energy transfer and is widely used in energy conversion facilities. Despite its importance in industry, the fundamental mechanism of boiling is still not well understood [2]. A predictive model for boiling is highly desired for engineering designs. Film boiling is regarded as most amenable to modeling, since its governing mechanism is principally the Rayleigh–Taylor instability. However, existing simulations all start with a preexisting perturbed flat interface as the initial condition [8]. In other words, none of those methods captured the film generation process. On the other side, very few simulations of nucleate boiling have been performed because more physical mechanisms are involved in this phenomenon.

J. Liu (✉) • T.J.R. Hughes
Institute for Computational Engineering and Sciences, The University of Texas at Austin,
201 East 24th Street, 1 University Station C0200, Austin, TX 78712, USA
e-mail: jliu@ices.utexas.edu; hughes@ices.utexas.edu

Y. Bazilevs, K. Takizawa (eds.), *Advances in Computational Fluid-Structure Interaction and Flow Simulation*, Modeling and Simulation in Science, Engineering and Technology, DOI 10.1007/978-3-319-40827-9_17

Traditional interface-tracking and interface-capturing methods are designed based on geometrical information of existing interfaces. This is perhaps the reason why these methods become intractable for phase transition phenomena. Phase-field or diffuse-interface methods were proposed as an alternative interface-capturing method, that use thermodynamic state variables to distinguish different phases [1]. The solid mathematical and thermodynamic foundations of phase-field models allow them to describe these complicated phenomena without resorting to modeling "tricks." The initial instantiation of phase-field methods is the Navier–Stokes–Korteweg equations, which are constructed based on the van der Waals theory [1, 4]. In the past decades, this theory has been developed further [3], and a rational thermomechanical framework for the Navier–Stokes–Korteweg equations has been presented very recently [10].

For phase-field problems, the non-convexity of the entropy function precludes the possibility of directly applying many existing robust numerical methodologies [6, 11]. To overcome the challenges posed by the non-convexity of the entropy, first, functional entropy variables are introduced to construct an entropy-stable spatial discretization [9, 10]. Second, to develop a stable temporal scheme, we adopt the methodology based on special quadrature rules [5, 9]. This time integration concept can be viewed as a second-order modification to the mid-point rule. The modifications are designed so that the temporal approximation is provably entropy dissipative. Since this temporal scheme does not require convexity for the entropy function, it is anticipated to be applicable to many more general problems.

2 The Navier–Stokes–Korteweg Equations

We consider a fixed, connected, and bounded domain $\Omega \subset \mathbb{R}^3$. The time interval of interest is denoted $(0, T)$, with $T > 0$. The dimensionless Navier–Stokes–Korteweg equations are considered in the space-time domain $\Omega \times (0, T)$ as

$$\frac{\partial \rho}{\partial t} + \nabla \cdot (\rho \mathbf{u}) = 0, \tag{1}$$

$$\frac{\partial (\rho \mathbf{u})}{\partial t} + \nabla \cdot (\rho \mathbf{u} \otimes \mathbf{u}) + \nabla p - \nabla \cdot \boldsymbol{\tau} - \nabla \cdot \boldsymbol{\varsigma} = \rho \mathbf{b}, \tag{2}$$

$$\frac{\partial (\rho E)}{\partial t} + \nabla \cdot ((\rho E + p)\mathbf{u} - (\boldsymbol{\tau} + \boldsymbol{\varsigma})\mathbf{u}) + \nabla \cdot \mathbf{q} + \nabla \cdot \boldsymbol{\Pi} = \rho \mathbf{b} \cdot \mathbf{u} + \rho r. \tag{3}$$

In the above equations, ρ is the density, $\mathbf{u}$ is the velocity, E is the total energy, p is the thermodynamic pressure, $\boldsymbol{\tau}$ is the viscous stress, $\boldsymbol{\varsigma}$ is the Korteweg stress, $\mathbf{q}$ is the heat flux; $\boldsymbol{\Pi}$ is the interstitial working flux [3, 10], $\mathbf{b}$ is the prescribed body force per unit mass, and r is the heat source per unit mass. The constitutive relations are

$$p = \frac{8\theta\rho}{27(1-\rho)} - \rho^2, \qquad \boldsymbol{\tau} = \frac{1}{\mathrm{Re}}\left(\nabla\mathbf{u} + \nabla\mathbf{u}^T - \frac{2}{3}\nabla\cdot\mathbf{u}\mathbf{I}\right), \qquad \mathbf{q} = -\kappa\nabla\theta,$$

$$\boldsymbol{\varsigma} = \frac{1}{\mathrm{We}}\left(\left(\rho\Delta\rho + \frac{1}{2}|\nabla\rho|^2\right)\mathbf{I} - \nabla\rho\otimes\nabla\rho\right), \qquad \iota_{loc} = -\rho + \frac{8}{27(\gamma-1)}\theta,$$

$$\iota = \iota_{loc} + \frac{1}{2\mathrm{We}\rho}|\nabla\rho|^2, \qquad E = \iota + \frac{1}{2}|\mathbf{u}|^2, \qquad \boldsymbol{\Pi} = \frac{1}{\mathrm{We}}\rho\nabla\cdot\mathbf{u}\nabla\rho,$$

wherein θ is the temperature, κ is the conductivity, Re is the Reynolds number, We is the Weber number, γ is the heat capacity ratio, and ι is the internal energy density per unit mass. The mathematical entropy function H and the local Helmholtz free energy Ψ_{loc} are defined as

$$H := \frac{8}{27}\rho\ln\left(\frac{\rho}{1-\rho}\right) - \frac{8}{27(1-\gamma)}\rho\ln(\theta),$$

$$\Psi_{loc}(\rho,\theta) := -\rho + \frac{8}{27(\gamma-1)}\theta + \frac{8}{27}\theta\ln\left(\frac{\rho}{1-\rho}\right) - \frac{8}{27(1-\gamma)}\theta\ln\theta.$$

In three dimensions, the vector of conservation variables is

$$\mathbf{U}^T = [U_1, U_2, U_3, U_4, U_5] := [\rho, \rho u_1, \rho u_2, \rho u_3, \rho E].$$

Due to the appearance of the gradient-squared term, the mathematical entropy function H is no longer just an algebraic function of the conservation variables, but rather it is a functional of the conservation variables. We define the entropy variables $\mathbf{V}^T = [V_1, V_2, V_3, V_4, V_5]$ as the functional derivatives of H with respect to $\mathbf{U}$:

$$V_i[\delta v_i] = \frac{\delta H}{\delta U_i}[\delta v_i], \quad i = 1,\ldots,5,$$

wherein $\delta\mathbf{v}^T = [\delta v_1, \delta v_2, \delta v_3, \delta v_4, \delta v_5]$ are the test functions. The entropy variables $\mathbf{V}$ can be written explicitly as

$$V_1[\delta v_1] = \frac{1}{\theta}\left(\nu_{loc} - \frac{|\mathbf{u}|^2}{2}\right)\delta v_1 + \frac{1}{\mathrm{We}}\frac{1}{\theta}\nabla\rho\cdot\nabla\delta v_1,$$

$$V_i[\delta v_i] = \frac{u_{i-1}}{\theta}\delta v_i, \quad i = 2,3,4, \quad V_5[\delta v_5] = -\frac{1}{\theta}\delta v_5,$$

wherein

$$\nu_{loc} = -2\rho + \frac{8}{27}\theta\ln\left(\frac{\rho}{1-\rho}\right) - \frac{8}{27(\gamma-1)}\theta\left(\ln(\theta) - 1\right) + \frac{8\theta}{27(1-\rho)}$$

is the local electrochemical potential. Inspired from the form of V_1, we introduce a new independent variable V as

$$V := \frac{1}{\theta}\left(\nu_{loc} - \frac{|\mathbf{u}|^2}{2}\right) - \frac{1}{\mathrm{We}}\nabla\cdot\left(\frac{\nabla\rho}{\theta}\right).$$

The fundamental thermodynamic relation between p and ν_{loc} allows us to express p in terms of V as

$$p = \rho V\theta - \rho\Psi_{loc} + \frac{\rho|\mathbf{u}|^2}{2} + \frac{1}{\mathrm{We}}\rho\theta\nabla\cdot\left(\frac{\nabla\rho}{\theta}\right). \tag{4}$$

Making use of the relation (4), the original strong-form problem (1)–(3) can be rewritten as

$$\frac{\partial\rho}{\partial t} + \nabla\cdot(\rho\mathbf{u}) = 0, \tag{5}$$

$$\frac{\partial(\rho\mathbf{u})}{\partial t} + \nabla\cdot(\rho\mathbf{u}\otimes\mathbf{u}) + \nabla\left(\rho V\theta + \frac{\rho|\mathbf{u}|^2}{2} + \frac{1}{\mathrm{We}}\rho\theta\nabla\cdot\left(\frac{\nabla\rho}{\theta}\right)\right)$$
$$-\left(V\theta + \frac{|\mathbf{u}|^2}{2} + \frac{1}{\mathrm{We}}\theta\nabla\cdot\left(\frac{\nabla\rho}{\theta}\right)\right)\nabla\rho - H\nabla\theta - \nabla\cdot\boldsymbol{\tau} - \nabla\cdot\boldsymbol{\varsigma} = \rho\mathbf{b}, \tag{6}$$

$$\frac{\partial(\rho E)}{\partial t} + \nabla\cdot\left(\left(\rho V\theta - \theta H + \frac{1}{2\mathrm{We}}|\nabla\rho|^2 + \rho|\mathbf{u}|^2 + \frac{1}{\mathrm{We}}\rho\theta\nabla\cdot\left(\frac{\nabla\rho}{\theta}\right)\right)\mathbf{u}\right)$$
$$-\nabla\cdot((\boldsymbol{\tau}+\boldsymbol{\varsigma})\mathbf{u}) + \nabla\cdot\mathbf{q} + \nabla\cdot\boldsymbol{\Pi} = \rho\mathbf{b}\cdot\mathbf{u} + \rho r, \tag{7}$$

$$V = \frac{1}{\theta}\left(\nu_{loc} - \frac{|\mathbf{u}|^2}{2}\right) - \frac{1}{\mathrm{We}}\nabla\cdot\left(\frac{\nabla\rho}{\theta}\right). \tag{8}$$

The new strong-form problem (5)–(8) is an equivalent statement of the original Navier–Stokes–Korteweg equations (1)–(3).

3 The Fully Discrete Scheme

The time interval $(0, T)$ is divided into N_{ts} subintervals (t_n, t_{n+1}), $n = 0, \cdots, N_{ts}-1$, of size $\Delta t_n = t_{n+1} - t_n$. We use the notation

$$\mathbf{Y}_n^h := \left[\rho_n^h, \frac{u_{1,n}^h}{\theta_n^h}, \frac{u_{2,n}^h}{\theta_n^h}, \frac{u_{3,n}^h}{\theta_n^h}, \frac{-1}{\theta_n^h}, V_n^h\right]^T$$

to represent the fully discrete solutions at the time level n. We define the jump of density, linear momentum, and total energy over each time step as

$$
\begin{aligned}
[\![\rho_n^h]\!] &:= \rho_{n+1}^h - \rho_n^h, \quad [\![\rho_n^h \mathbf{u}_n^h]\!] := \rho_{n+1}^h \mathbf{u}_{n+1}^h - \rho_n^h \mathbf{u}_n^h, \\
\left[\rho_n^h E(\rho_n^h, \mathbf{u}_n^h, \theta_n^h)\right] &:= (\rho \Psi_{loc})(\rho_{n+\frac{1}{2}}^h, \theta_{n+1}^h) - (\rho \Psi_{loc})(\rho_{n+\frac{1}{2}}^h, \theta_n^h) \\
&+ (\rho \Psi_{loc})(\rho_{n+1}^h, \theta_{n+\frac{1}{2}}^h) - (\rho \Psi_{loc})(\rho_n^h, \theta_{n+\frac{1}{2}}^h) \\
&- \theta_{n+\frac{1}{2}}^h \left(H(\rho_{n+1}^h, \theta_{n+1}^h) - H(\rho_n^h, \theta_n^h)\right) \\
&- \frac{\theta_{n+1}^h - \theta_n^h}{2} \left(H(\rho_{n+\frac{1}{2}}^h, \theta_{n+1}^h) + H(\rho_{n+\frac{1}{2}}^h, \theta_n^h)\right) \\
&+ \frac{(\theta_{n+1}^h - \theta_n^h)^3}{12} \frac{\partial^2 H}{\partial \theta^2}(\rho_{n+\frac{1}{2}}^h, \theta_{n+1}^h) \\
&+ \frac{1}{2}\left(\rho_{n+1}^h |\mathbf{u}_{n+1}^h|^2 - \rho_n^h |\mathbf{u}_n^h|^2\right) + \frac{1}{2\mathrm{We}}\left(|\nabla \rho_{n+1}^h|^2 - |\nabla \rho_n^h|^2\right).
\end{aligned}
$$

With the jump operators defined above, the fully discrete scheme can be stated as follows. In each time step, given $\mathbf{Y}_n^h$ and the time step Δt_n, find $\mathbf{Y}_{n+1}^h$ such that for all $w_1^h \in \mathcal{V}^h$, $\mathbf{w}^h = (w_2^h; w_3^h; w_4^h)^T \in \left(\mathcal{V}^h\right)^3$, $w_5^h \in \mathcal{V}^h$, and $w_6^h \in \mathcal{V}^h$,

$$
\mathbf{B}^M(w_1^h; \mathbf{Y}_{n+1}^h) := \left(w_1^h, \frac{[\![\rho_n^h]\!]}{\Delta t_n}\right)_\Omega - \left(\nabla w_1^h, \rho_{n+\frac{1}{2}}^h \mathbf{u}_{n+\frac{1}{2}}^h\right)_\Omega = 0, \tag{9}
$$

$$
\begin{aligned}
\mathbf{B}^U(\mathbf{w}^h; \mathbf{Y}_{n+1}^h) &:= \left(\mathbf{w}^h, \frac{[\![\rho_n^h \mathbf{u}_n^h]\!]}{\Delta t_n}\right)_\Omega - \left(\nabla \mathbf{w}^h, \rho_{n+\frac{1}{2}}^h \mathbf{u}_{n+\frac{1}{2}}^h \otimes \mathbf{u}_{n+\frac{1}{2}}^h\right)_\Omega \\
&- \left(\nabla \cdot \mathbf{w}^h, \rho_{n+\frac{1}{2}}^h V_{n+\frac{1}{2}}^h \theta_{n+\frac{1}{2}}^h + \frac{1}{2}\rho_{n+\frac{1}{2}}^h |\mathbf{u}_{n+\frac{1}{2}}^h|^2 + \frac{1}{\mathrm{We}} \rho_{n+\frac{1}{2}}^h \theta_{n+\frac{1}{2}}^h \nabla \cdot \left(\frac{\nabla \rho_{n+\frac{1}{2}}^h}{\theta_{n+\frac{1}{2}}^h}\right)\right)_\Omega \\
&- \left(\mathbf{w}^h, \left(V_{n+\frac{1}{2}}^h \theta_{n+\frac{1}{2}}^h + \frac{|\mathbf{u}_{n+\frac{1}{2}}^h|^2}{2} + \frac{1}{\mathrm{We}} \theta_{n+\frac{1}{2}}^h \nabla \cdot \left(\frac{\nabla \rho_{n+\frac{1}{2}}^h}{\theta_{n+\frac{1}{2}}^h}\right)\right) \nabla \rho_{n+\frac{1}{2}}^h\right)_\Omega \\
&- \left(\mathbf{w}^h, H_{n+\frac{1}{2}}^h \nabla \theta_{n+\frac{1}{2}}^h\right)_\Omega + \left(\nabla \mathbf{w}^h, \boldsymbol{\tau}_{n+\frac{1}{2}}^h + \boldsymbol{\varsigma}_{n+\frac{1}{2}}^h\right)_\Omega - \left(\mathbf{w}^h, \rho_{n+\frac{1}{2}}^h \mathbf{b}\right)_\Omega = \mathbf{0},
\end{aligned} \tag{10}
$$

$$
\begin{aligned}
\mathbf{B}^E(w_5^h; \mathbf{Y}_{n+1}^h) &:= \left(w_5^h, \frac{\left[\rho_n^h E(\rho_n^h, \mathbf{u}_n^h, \theta_n^h)\right]}{\Delta t_n}\right)_\Omega - \Bigg(\nabla w_5^h, \Bigg(\rho_{n+\frac{1}{2}}^h V_{n+\frac{1}{2}}^h \theta_{n+\frac{1}{2}}^h \\
&- \theta_{n+\frac{1}{2}}^h H_{n+\frac{1}{2}}^h + \frac{|\nabla \rho_{n+\frac{1}{2}}^h|^2}{2\mathrm{We}} + \frac{\rho_{n+\frac{1}{2}}^h \theta_{n+\frac{1}{2}}^h}{\mathrm{We}} \nabla \cdot \left(\frac{\nabla \rho_{n+\frac{1}{2}}^h}{\theta_{n+\frac{1}{2}}^h}\right) + \rho_{n+\frac{1}{2}}^h |\mathbf{u}_{n+\frac{1}{2}}^h|^2\Bigg) \mathbf{u}_{n+\frac{1}{2}}^h\Bigg)_\Omega \\
&+ \left(\nabla w_5^h, \boldsymbol{\tau}_{n+\frac{1}{2}}^h \mathbf{u}_{n+\frac{1}{2}}^h\right)_\Omega + \left(\nabla w_5^h, \boldsymbol{\varsigma}_{n+\frac{1}{2}}^h \mathbf{u}_{n+\frac{1}{2}}^h\right)_\Omega - \left(\nabla w_5^h, \mathbf{q}_{n+\frac{1}{2}}^h + \boldsymbol{\Pi}_{n+\frac{1}{2}}^h\right)_\Omega \\
&- \left(w_5^h, \rho_{n+\frac{1}{2}}^h \mathbf{b} \cdot \mathbf{u}_{n+\frac{1}{2}}^h\right)_\Omega - \left(w_5^h, \rho_{n+\frac{1}{2}}^h r\right)_\Omega = 0,
\end{aligned} \tag{11}
$$

$$\mathbf{B}^A(w_6^h;\mathbf{Y}_{n+1}^h) := \left(w_6^h, V_{n+\frac{1}{2}}^h - \frac{1}{2\theta_{n+\frac{1}{2}}^h}\left(\left(\nu_{loc}(\rho_n^h,\theta_{n+\frac{1}{2}}^h) + \nu_{loc}(\rho_{n+1}^h,\theta_{n+\frac{1}{2}}^h)\right)\right.\right.$$

$$\left.\left. + \frac{[\![\rho_n^h]\!]^2}{12}\frac{\partial^2\nu_{loc}}{\partial\rho^2}(\rho_n^h,\theta_{n+\frac{1}{2}}^h)\right) - \frac{\mathbf{u}_n^h\cdot\mathbf{u}_{n+1}^h}{2\theta_{n+\frac{1}{2}}^h}\right)_{\Omega} - \left(\nabla w_6^h, \frac{\nabla\rho_{n+\frac{1}{2}}^h}{\mathrm{We}\theta_{n+\frac{1}{2}}^h}\right)_{\Omega} = 0. \qquad (12)$$

In our work, Non-Uniform Rational B-Splines (NURBS) basis functions are used to define $\mathcal{V}^h$ as well as the computational domain. Consequently, this approach may be considered as isogeometric analysis method [7]. The main results of the fully discrete scheme (9)–(12) are stated in the following two theorems:

Theorem 1. *The solutions of the fully discrete scheme (9)–(12) satisfy*

$$\int_\Omega \left(\frac{H(\rho_{n+1}^h,\theta_{n+1}^h) - H(\rho_n^h,\theta_n^h)}{\Delta t_n} + \nabla\cdot\left(H(\rho_{n+\frac{1}{2}}^h,\theta_{n+\frac{1}{2}}^h)\mathbf{u}_{n+\frac{1}{2}}^h\right) - \nabla\cdot\left(\frac{\mathbf{q}_{n+\frac{1}{2}}^h}{\theta_{n+\frac{1}{2}}^h}\right)\right.$$

$$\left. + \frac{\rho_{n+\frac{1}{2}}^h r}{\theta_{n+\frac{1}{2}}^h}\right) dV_{\mathbf{x}} = -\int_\Omega \frac{1}{\theta_{n+\frac{1}{2}}^h}\boldsymbol{\tau}_{n+\frac{1}{2}}^h : \nabla\mathbf{u}_{n+\frac{1}{2}}^h dV_{\mathbf{x}} - \int_\Omega \frac{\kappa|\nabla\theta_{n+\frac{1}{2}}^h|^2}{\left(\theta_{n+\frac{1}{2}}^h\right)^2} dV_{\mathbf{x}}$$

$$-\int_\Omega \frac{1}{\theta_{n+\frac{1}{2}}^h\Delta t_n}\left(\frac{[\![\rho_n^h]\!]^4}{24}\frac{\partial^3\nu_{loc}}{\partial\rho^3}(\rho_{n+\xi_1}^h,\theta_{n+\frac{1}{2}}^h) - \frac{[\![\theta_n^h]\!]^4}{24}\frac{\partial^3 H}{\partial\theta^3}(\rho_{n+\frac{1}{2}}^h,\theta_{n+\xi_2}^h)\right) dV_{\mathbf{x}} \le 0.$$

Theorem 2. *The local truncation error in time* $\boldsymbol{\Theta}(t) = \left(\Theta_\rho(t);\boldsymbol{\Theta}_{\mathbf{u}}^T(t);\Theta_E(t)\right)^T$ *can be bounded by* $|\boldsymbol{\Theta}(t_n)| \le K\Delta t_n^2\mathbf{1}_5$ *for all* $t_n \in [0,T]$, *where K is a constant independent of* Δt_n *and* $\mathbf{1}_5 = (1;1;1;1;1)^T$.

The proofs of the above two theorems can be found in [10]. Theorem 1 states that the method is unconditionally entropy stable, because $\partial^3\nu_{loc}/\partial\rho^3 \ge 0$ and $\partial^3 H/\partial\theta^3 \le 0$, which follow from properties of the van der Waals fluid. Theorem 2 establishes the second-order time-accuracy of the method.

4 Boiling

To obtain successful boiling simulations, there are several additional modeling considerations. First, the transport parameters need to be density dependent in order to differentiate the properties of the liquid and vapor phases. In our simulations, the dimensionless viscosity coefficient and the dimensionless conductivity are modeled as

$$\bar{\mu} = C_\mu^{boil}\rho, \qquad \kappa = C_\kappa^{boil}\rho,$$

with C_μ^{boil} and C_κ^{boil} being constants independent of ρ. Second, the gravity effect need to be taken into account to generate buoyancy. The dimensionless body force $\mathbf{b}$ is chosen as $\mathbf{b} = (0; 0; -0.025)^T$ for the three-dimensional case and $\mathbf{b} = (0; -0.025)^T$ for the two-dimensional case. Third, the ninety-degree contact angle boundary condition is used for the density variable, and the slip boundary condition is applied to the velocity. To specify the boundary condition for $Y_5 = -1/\theta$, the boundary $\partial\Omega$ is divided into three non-overlapping parts:

$$\partial\Omega = \Gamma_t \cup \Gamma_b \cup \Gamma_v, \qquad \Gamma_t = \{\mathbf{x} \in \partial\Omega | \mathbf{n}(\mathbf{x}) \cdot \mathbf{b} < 0\},$$
$$\Gamma_b = \{\mathbf{x} \in \partial\Omega | \mathbf{n}(\mathbf{x}) \cdot \mathbf{b} > 0\}, \quad \Gamma_v = \{\mathbf{x} \in \partial\Omega | \mathbf{n}(\mathbf{x}) \cdot \mathbf{b} = 0\}.$$

With the above partition, the boundary condition for Y_5 is

$$Y_5 = -\frac{1}{0.950} + \delta_{Y_{5,h}}(\mathbf{x}), \quad \text{on } \Gamma_b \times (0, T),$$
$$Y_5 = -\frac{1}{0.775} + \delta_{Y_{5,c}}(\mathbf{x}), \quad \text{on } \Gamma_t \times (0, T),$$
$$-\mathbf{q} \cdot \mathbf{n} = 0, \quad \text{on } \Gamma_v \times (0, T),$$

wherein $\delta_{Y_{5,h}}(\mathbf{x})$ and $\delta_{Y_{5,c}}(\mathbf{x})$ are small scalar perturbation functions that mimic the uneven temperature distribution on the solid surface. The initial conditions represent a static free surface, with liquid in the bottom region and vapor in the top region. It is worth emphasizing that, in contrast to the existing boiling models, there is no artificial manipulation used to serve as boiling onset in this model; the initial liquid and vapor densities are uniform with no perturbations.

4.1 Two-Dimensional Nucleate Boiling

In this example, we simulate boiling flows in a two-dimensional rectangular domain $\Omega = (0, 1) \times (0, 0.5)$. The material parameters are chosen as We $= 8.401 \times 10^6$, $\gamma = 1.333$, $C_\mu^{boil} = 1.150 \times 10^{-4}$, and $C_\kappa^{boil} = 1.725 \times 10^{-5}$. The initial conditions for this problem are

$$\rho_0(\mathbf{x}) = 0.3660 - 0.2971 \tanh\left(\frac{x_2 - 0.35}{2}\sqrt{\text{We}}\right),$$
$$\mathbf{u}_0(\mathbf{x}) = \mathbf{0},$$
$$\theta_0(\mathbf{x}) = 0.775.$$

The perturbations for the temperature on the boundary $\delta_{Y_{5,h}}(\mathbf{x})$ and $\delta_{Y_{5,c}}(\mathbf{x})$ are uniform random distributions and satisfy

$$\delta_{Y_{5,h}}(\mathbf{x}) \in [-5.0 \times 10^{-2}, 5.0 \times 10^{-2}], \quad \delta_{Y_{5,c}}(\mathbf{x}) \in [-5.0 \times 10^{-3}, 5.0 \times 10^{-3}].$$

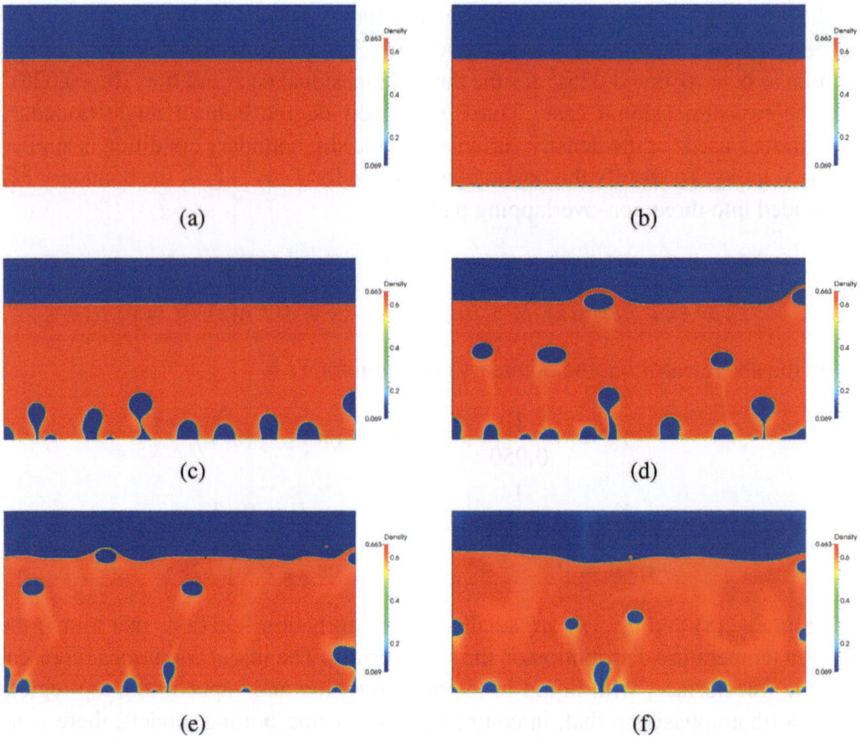

Fig. 1 Two-dimensional nucleate boiling simulation: Density profiles at (a) $t = 0.0$, (b) $t = 1.25$, (c) $t = 18.75$, (d) $t = 31.25$, (e) $t = 62.5$, and (f) $t = 100.0$.

The spatial mesh consists of 2048×1024 quadratic NURBS elements. The problem is integrated up to the final time $T = 100.0$ with time step fixed as $\Delta t = 5.0 \times 10^{-4}$. In Figure 1, snapshots of the density are depicted at different time steps. It can be observed that tiny vapor bubbles are generated at discrete sites of the heated wall surface during the initial times. The increase of bubble size leads to the increase of buoyancy. At about $t = 18.75$, the first three bubbles get detached from the bottom. More bubbles are generated on the bottom surface subsequently. Interestingly, small droplets can be observed at $t = 62.5$ and $t = 100.0$ as a result of the breakage of the liquid film when the vapor bubbles reach the free surface. There are 30 bubbles formed in the time interval of $0 < t < 100$.

4.2 *Two-Dimensional Film Boiling*

In the second example, the same two-dimensional problem considered in the preceding section is simulated again with a different parameter C_{μ}^{boil}. Here, the parameter C_{μ}^{boil} is chosen to be 4.600×10^{-4}, which is four times larger than that of the previous example. Since the fluid motion in this example is slower,

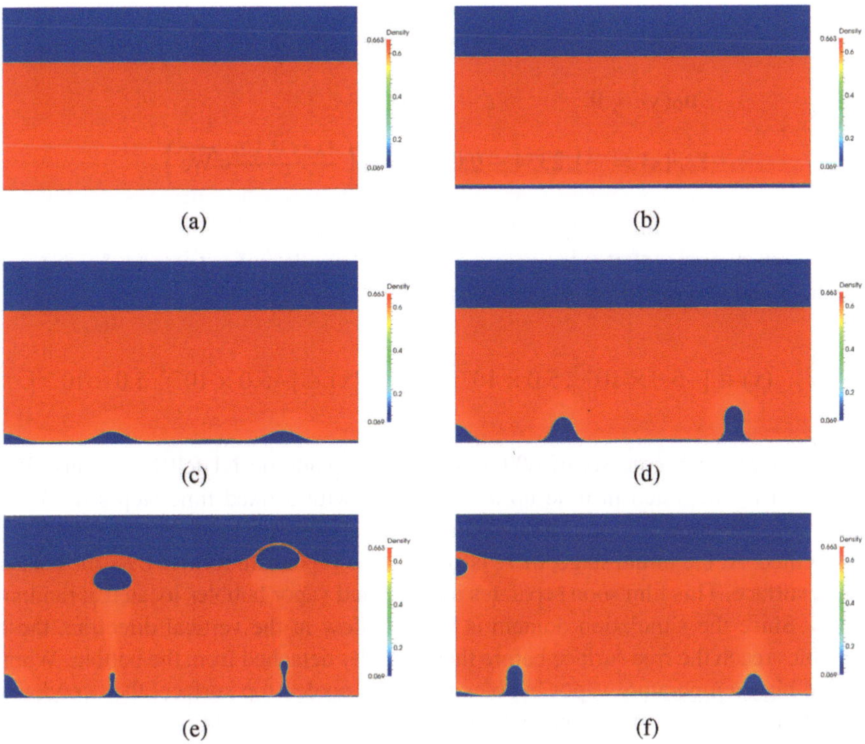

Fig. 2 Two-dimensional film boiling simulation: Density profiles at (a) $t = 0.0$, (b) $t = 100.0$, (c) $t = 175.0$, (d) $t = 200.0$, (e) $t = 225.0$, and (f) $t = 500.0$. The generation of the thin vapor film is visible at $t = 100.0$.

the simulation is integrated in time up to $T = 500.0$. All the other conditions are identical to those of the previous case. In Figure 2, snapshots of the density at different time steps are depicted. A thin vapor film is gradually generated at the bottom during the early stage of the simulation. As time evolves, the interface becomes unstable and there are vapor bubbles formed. From $t = 200.0$ to $t = 225.0$, the first two vapor bubbles pinch off from the vapor film and rise upward in ellipsoidal shapes. This process repeats itself periodically. By final time $t = 500.0$, there are seven bubbles detached from the vapor film.

4.3 Three-Dimensional Boiling

As the last example, we simulate the Navier–Stokes–Korteweg equations in a three-dimensional domain $\Omega = (0, 1) \times (0, 0.5) \times (0, 0.25)$. The material properties are chosen as $\mathrm{We} = 6.533 \times 10^5$, $\gamma = 1.333$, $C_\mu^{boil} = 1.289 \times 10^{-4}$, and $C_\kappa^{boil} = 7.732 \times 10^{-5}$. The initial conditions for this three-dimensional problem are

$$\rho_0(\mathbf{x}) = 0.33565 - 0.26675 \tanh\left(\frac{x_3 - 0.15}{2}\sqrt{\mathrm{We}}\right),$$

$$\mathbf{u}_0(\mathbf{x}) = \mathbf{0},$$

$$Y_{5,0}(\mathbf{x}) = -1.2334 - 0.0569 \tanh\left(\frac{x_3 - 0.15}{2}\sqrt{\mathrm{We}}\right).$$

The perturbations of the temperature on the boundary $\delta_{Y_{5,h}}(\mathbf{x})$ and $\delta_{Y_{5,c}}(\mathbf{x})$ are uniform random distributions and satisfy

$$\delta_{Y_{5,h}}(\mathbf{x}) \in [-5.0 \times 10^{-2}, 5.0 \times 10^{-2}], \quad \delta_{Y_{5,c}}(\mathbf{x}) \in [-5.0 \times 10^{-3}, 5.0 \times 10^{-3}].$$

The spatial mesh consists of $600 \times 300 \times 150$ quadratic NURBS elements. The problem is integrated in time up to $T = 20.0$ with a fixed time step size $\Delta t = 2.0 \times 10^{-3}$. In Figure 3, snapshots of density isosurfaces and velocity streamlines are presented. At the initial stage, there is an unstable vapor film formed over the heated wall surface. This film soon separates into isolated vapor bubbles located at random sites. Since the simulation domain is very shallow in the vertical direction, these bubbles reach the free surface before they get fully detached from the bottom. When these high-temperature vapor bubbles reach the cooled top surface, they condense into liquid droplets instantaneously (see Fig. 3 (e)). At $t = 20.0$, a second round of vapor bubbles is clearly generated on the bottom and the liquid droplets on the top surface merge together.

5 Conclusion

In this work, we presented theoretical and numerical methodologies for the study of boiling, capable of describing complicated phase transition phenomena without resorting to empirical assumptions. Our algorithm is provably entropy-stable and second-order accurate in time. It provides a unified predictive framework for nucleate and film boiling in two and three dimensions. In the future, the presented methodologies will be applied to the study of other important phase transition phenomena, such as cavitation, spray and mist formation.

Acknowledgements This work was partially supported by the Office of Naval Research under contract number N00014-08-1-0992.

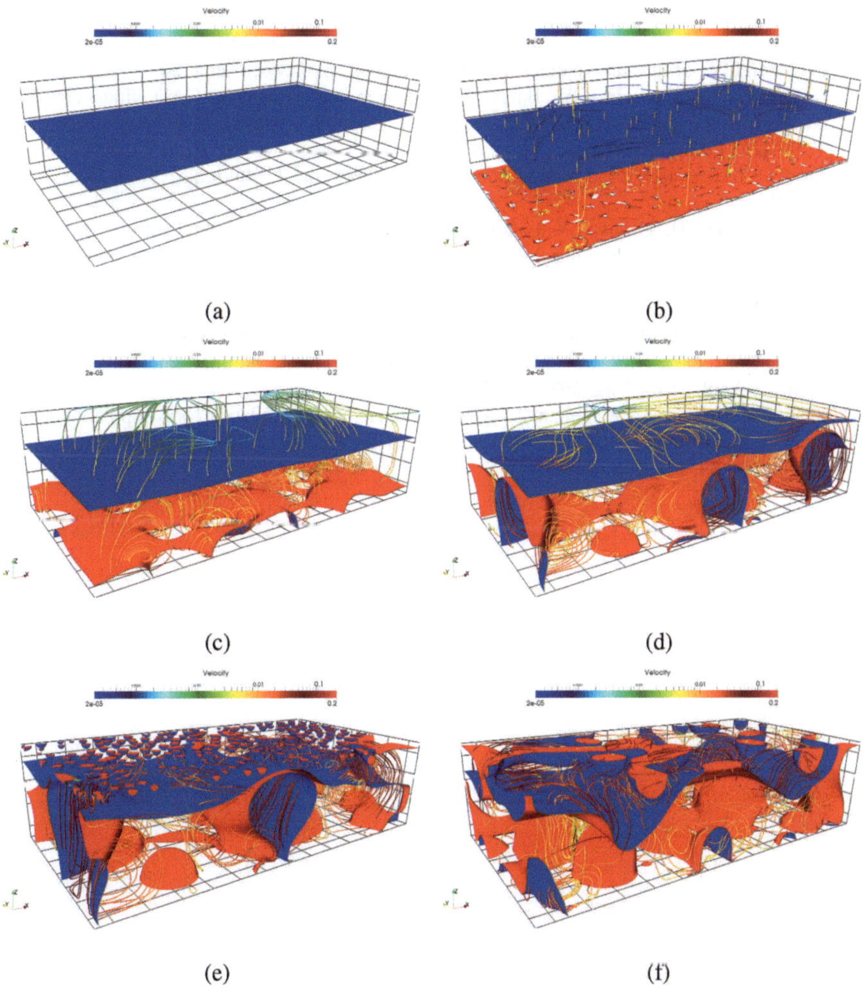

Fig. 3 Three-dimensional boiling simulation: Density isosurfaces and velocity streamlines at (a) $t = 0.0$, (b) $t = 0.6$, (c) $t = 5.0$, (d) $t = 11.0$, (e) $t = 14.0$, and (f) $t = 20.0$.

References

1. Anderson, D.M., McFadden, G.B., Wheeler, A.A.: Diffuse-interface methods in fluid mechanics. Annu. Rev. Fluid Mech. **30**, 139–165 (1998)
2. Dhir, V.K.: Boiling heat transfer. Annu. Rev. Fluid Mech. **30**, 365–401 (1998)
3. Dunn, J.E., Serrin, J.: On the thermomechanics of interstitial working. Arch. Ration. Mech. Anal. **88**, 95–133 (1985)
4. Gomez, H., Hughes, T.J.R., Nogueira, X., Calo, V.: Isogeometric analysis of the Navier-Stokes-Korteweg equations. Comput. Methods Appl. Mech. Eng. **199**, 1828–1840 (2010)
5. Gomez, H., Hughes, T.J.R.: Provably unconditionally stable, second-order time-accurate, mixed variational methods for phase-field models. J. Comput. Phys. **230**, 5310–5327 (2011)

6. Hughes, T.J.R., Franca, L.P., Mallet, M.: A new finite element formulation for computational fluid dynamics: I. Symmetric forms of the compressible Euler and Navier-Stokes equations and the second law of thermodynamics. Comput. Methods Appl. Mech. Eng. **54**, 223–234 (1986)
7. Hughes, T.J.R., Cottrell, J.A., Bazilevs, Y.: Isogeometric analysis: CAD, finite elements, NURBS, exact geometry and mesh refinement. Comput. Methods Appl. Mech. Eng. **194**, 4135–4195 (2005)
8. Juric, D., Tryggvason, G.: Computation of boiling flows. Int. J. Multiphase Flow **24**, 387–410 (1998)
9. Liu, J., Gomez, H., Evans, J.A., Landis, C.M., Hughes, T.J.R.: Functional entropy variables: a new methodology for deriving thermodynamically consistent algorithms for complex fluids, with particular reference to the isothermal Navier-Stokes-Korteweg equations. J. Comput. Phys. **248**, 47–86 (2013)
10. Liu, J., Landis, C.M., Gomez, H., Hughes, T.J.R.: Liquid-vapor phase transition: thermomechanical theory, entropy stable numerical formulation, and boiling simulations. Comput. Methods Appl. Mech. Eng. **297**, 476–553 (2015)
11. Shakib, F., Hughes, T.J.R., Johan, Z.: A new finite element formulation for computational fluid dynamics: X. The compressible Euler and Navier-Stokes equations. Comput. Methods Appl. Mech. Eng. **89**, 141–219 (1991)

Part IV
Computer Science and HPC Aspects

How to Generate Effective Block Jacobi Preconditioners for Solving Large Sparse Linear Systems

Yao Zhu and Ahmed H. Sameh

Abstract We present a procedure for creating effective block Jacobi preconditioners for solving large sparse linear systems using Krylov subspace methods. Such preconditioners are constructed using two techniques. The first is a reordering scheme based on weighted graph partitioning which strengthens the block diagonal structure of the coefficient matrix while simultaneously accommodating load balancing on distributed memory architectures. The second technique is the factorization of the resulting diagonal blocks, or the factorization of slightly perturbed diagonal blocks to guard against possible singularity. Focusing on a set of linear systems arising in several computational fluid dynamics applications, we demonstrate the effectiveness of our enhanced block Jacobi preconditioners. Compared to a well-known sparse direct linear system solver, our parallel solver (block Jacobi preconditioned Krylov subspace method) proves to be equally robust and achieves appreciable speed improvements on a distributed memory parallel computing platform.

1 Introduction

Block Jacobi preconditioners enjoy the perfect coarse-grain parallelism when they are used in conjunction with Krylov subspace methods for solving large sparse linear systems of equations. In general, the existence of a nonsingular block Jacobi preconditioner is questionable at best. Further, even if one exists, the effectiveness of such a preconditioner is equally questionable. In this paper, we present a procedure for generating *enhanced block Jacobi preconditioners*, and experimentally demonstrate: (1) their effectiveness in realizing robust Krylov subspace solvers of large

Y. Zhu • A.H. Sameh (✉)
Department of Computer Science, Purdue University, 305 N. University Street, West Lafayette, IN 47907, USA
e-mail: yaozhu@purdue.edu; sameh@cs.purdue.edu

Y. Bazilevs, K. Takizawa (eds.), *Advances in Computational Fluid-Structure Interaction and Flow Simulation*, Modeling and Simulation in Science, Engineering and Technology, DOI 10.1007/978-3-319-40827-9_18

sparse linear systems that arise from computational fluid dynamics applications; and (2) the parallel speed improvement realized by our solver over a state-of-the-art direct sparse linear system solver (PARDISO) on a distributed memory parallel architecture.

In Section 2, we outline the classical procedures for preconditioned Krylov subspace methods for solving large sparse linear systems of equations. In Section 3, we present an algorithm for creating effective block Jacobi preconditioners. Specifically, we describe: (i) a reordering scheme of the sparse coefficient matrix involving weighted graph partitioning, and (ii) a perturbed factorization scheme for constructing the enhanced block Jacobi preconditioners. In Section 4, we present numerical experiments that demonstrate the robustness and parallel scalability of our resulting block Jacobi preconditioned BiCGstab solver.

2 Background

We consider the problem of solving the sparse linear system of equations

$$Ax = f \tag{1}$$

on a parallel computing platform using a Krylov subspace method (e.g., BiCGstab), where $A \in \mathbb{R}^{n\times n}$ is nonsingular. In order to improve the robustness of Krylov subspace methods, it is critical to construct an effective preconditioner M to the linear system (1) [2]. A *right-preconditioned* linear system is given by

$$AM^{-1}y = f \tag{2}$$

where

$$y = Mx$$

The choice of the preconditioner M is mainly constrained by two objectives

1. Either the eigenvalues of AM^{-1} are well clustered around 1, or at least the condition number $\kappa(AM^{-1}) \ll \kappa(A)$, and
2. Solving systems involving the preconditioner, i.e., solving

$$Mz = r \tag{3}$$

is much easier than solving $Ax = f$ and much more amenable for efficient implementation on parallel architectures..

In this paper, we focus on techniques for generating effective *block Jacobi preconditioners* for solving sparse linear systems. A block Jacobi preconditioner M is a block diagonal matrix consisting of diagonal blocks $M_1, \ldots, M_p$, where p is the

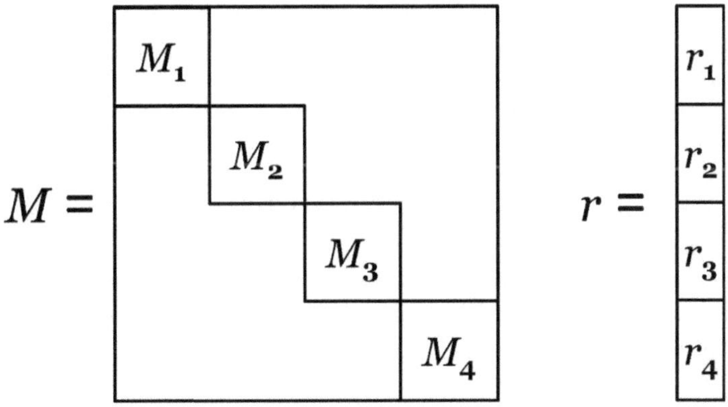

Fig. 1 A block Jacobi preconditioner and the correspondingly partitioned RHS, with $p = 4$.

number of blocks, e.g., see, Figure 1 for $p = 4$. The block diagonal structure of M leads to perfect coarse-grain parallelism when solving the preconditioning step (3), which consists of p independent sub-systems

$$M_k z_k = r_k \tag{4}$$

for $k = 1, \ldots, p$.

3 Enhanced Block Jacobi Preconditioners

In this section, we describe two techniques for generating effective block Jacobi preconditioners. The first technique is a reordering scheme that insures that the constructed block Jacobi preconditioner M encapsulates as many of the heaviest elements of the coefficient matrix A as possible without sacrificing load balancing on a distributed memory parallel architecture. The second consists of obtaining an LU-factorization of M or of a slightly perturbed M should one (or more) of the diagonal blocks be singular or close to singular. We refer to the resulting block Jacobi preconditioner as the *Enhanced Block Jacobi* (E-BJacobi) preconditioner.

3.1 Load Balancing

Given that the p independent sub-systems (4) can be solved in perfect parallelism (one sub-system per a multicore node), the load balancing of the work involved in the LU-factorization of each M_k (or a slightly perturbed M_k) as well as the following triangular solves using the nonsingular factors L_k and U_k becomes a critical issue.

If the work is not load-balanced, the performance of the entire solver will be adversely affected. Let W_k denote the work pertaining to the k-th block for solving the sub-system (4), and the average work $\overline{W} = (W_1 + \cdots + W_p)/p$. We enforce load balancing by constraining the *imbalance ratio*

$$\frac{\max_{1,\ldots,p} W_k}{\overline{W}} \leq 1 + \delta \tag{5}$$

where the parameter δ specifies the tolerance of load imbalance. As a coarse (but reasonable) approximation of W_k, we use the *number-of-nonzeros* (*nnz*) of M_k as an estimate of W_k.

3.2 Weighted Graph Partitioning Reordering

It has been demonstrated in [7] that the *maximum product traversal* permutation improves the convergence of iterative solution methods based on matrix splitting and ILU-preconditioned Krylov subspace schemes. Thus, in the following, we assume that the coefficient matrix A has already been column-permuted (e.g., via using the HSL MC64 subroutine) to satisfy the maximum product traversal property [7]. Our target is to derive a symmetric reordering[1] P such that the block Jacobi preconditioner $M = \mathrm{diag}(M_1, \ldots, M_p)$ extracted from the reordered matrix PAP^{T} will encapsulate as many of the heaviest elements of A as possible, while satisfying the load balancing constraint (5). This is essentially the *weighted graph partitioning* problem [8, 9] defined below.

Definition 1 (*p*-**GPES**). Given an undirected and weighted graph $\mathscr{G} = (\mathscr{V}, \mathscr{E})$ with weight function $w(i,j) : \mathscr{E} \to \mathbb{R}^+$, then the problem of p-way graph partitioning by edge separators (p-GPES) is given by: partition $\mathscr{V}$ into p disjoint subsets $\mathscr{V} = \mathscr{V}_1 \cup \cdots \cup \mathscr{V}_p$, such that the total weight of edges crossing different subsets is minimized while keeping the volumes of the subsets $\mathrm{vol}(\mathscr{V}_k), k = 1, \ldots, p$ balanced.

Let the coefficient matrix A be of order n. We define its associated graph $\mathscr{G}[A] = (\mathscr{V}[A], \mathscr{E}[A])$ with the weight function as follows:

$$\begin{aligned} \mathscr{V}[A] &= \{1, \ldots, n\} \\ \mathscr{E}[A] &= \{(i,j) : |A|_{ij} + |A|_{ji} \neq 0\} \\ w(i,j) &= \frac{|A|_{ij} + |A|_{ji}}{2} \end{aligned} \tag{6}$$

[1]Note that a symmetric reordering preserves the maximum product traversal property.

From a p-GPES result of $\mathscr{G}[A]$, we define a block Jacobi preconditioner M of which the k-th diagonal block is given by

$$M_k = A(\mathscr{V}_k, \mathscr{V}_k) \tag{7}$$

for $k = 1, \ldots, p$, where $A(\mathscr{V}_k, \mathscr{V}_k)$ is that submatrix of A with row and column indices in $\mathscr{V}_k$.

Ideally when applying the p-GPES to $\mathscr{G}[A]$, we should define the volume function $\mathrm{vol}(\mathscr{V}_k) = W_k$, i.e., the work involved in factorization and triangular solve of M_k. We use the number of nonzeros (*nnz*) of M_k as an estimate of W_k. Thus, we define the volume function of $\mathscr{V}_k$ as follows:

$$\mathrm{vol}(\mathscr{V}_k) = \sum_{i \in \mathscr{V}_k} \Big| \{j \neq i : |A|_{ij} + |A|_{ji} \neq 0\} \Big|$$

Note that a more accurate estimate of W_k depends on the amount of fill-ins incurred during the factorization of M_k (or a slightly perturbed M_k). Such a more accurate estimate, however, is hard to predict without performing actual fill-reduce reordering and symbolic factorization [12].

A p-GPES result induces a family of reorderings under which the indices in the same subset $\mathscr{V}_k$ are numbered contiguously. We call a reordering of such a kind the *Weighted Graph Partitioning Reordering* (WGPR).

Definition 2 (WGPR). Let P be a reordering matrix. We call P a weighted graph partitioning reordering that respects the graph partitioning result $\mathscr{V} = \mathscr{V}_1 \cup \cdots \cup \mathscr{V}_p$, if for any $k \neq l$ either $\pi_P(i) < \pi_P(j)$ for all $i \in \mathscr{V}_k$ and all $j \in \mathscr{V}_l$; or $\pi_P(i) > \pi_P(j)$ for all $i \in \mathscr{V}_k$ and all $j \in \mathscr{V}_l$, where π_P is the permutation represented by the reordering matrix P.

In a recent study, [15], it is proposed that one should also incorporate matrix values into graph partitioning for constructing effective block Jacobi preconditioners. Unlike our reordering WGPR, this study, however, considers only preconditioning symmetric positive definite systems. The approach in [15] is based on recursive spectral bisection in which each step requires solving a generalized eigenvalue problem that could be very time consuming on its own. In contrast, our formulation based on p-GPES admits very efficient algorithms and implementations [8, 9].

3.3 Perturbed Factorization of Diagonal Blocks

We solve the sub-system (4) using a direct sparse solver [5, 10, 14]. It is possible, however, that a certain diagonal block M_k is singular or near singular. This is handled through diagonal *pivot perturbation*, e.g., as implemented in PARDISO [14]. During the process of the LU-factorization of M_k, if the diagonal pivot $|L_k(i, i)| < \epsilon \|M_k\|_\infty$, then we set

$$L_k(i,i) = \text{sgn}(L_k(i,i)) \cdot \epsilon \|M_k\|_\infty \tag{8}$$

where sgn(·) is the sign function, and ϵ is a user specified parameter. Such perturbed factorization results in the nonsingular factors L_k and U_k such that

$$\widetilde{P}_k M_k \widetilde{Q}_k^{\mathrm{T}} + E_k = L_k U_k$$

where the perturbation matrix E_k is possibly of low rank with $\|E_k\|_\infty \ll \|M\|_\infty$, and $\widetilde{P}_k$ ($\widetilde{Q}_k$) being row (column) permutations. We define

$$\widetilde{M}_k = M_k + \widetilde{P}_k^{\mathrm{T}} E_k \widetilde{Q}_k \tag{9}$$

3.4 *The Parallel Implementation of E-BJacobi*

In our parallel implementation of E-BJacobi, we use the state-of-the-art parallel graph partitioning tool ParMETIS[2] [9] to solve the p-GPES problem for deriving a reordering matrix P via WGPR. For the LU-factorization of M_k with pivot perturbation, we use the state-of-the-art parallel sparse direct solver PARDISO[3] [14]. Let $\widetilde{M}_k$ be M_k with possible perturbation as given in (9). The E-BJacobi preconditioner is then defined to be

$$\widetilde{M} = \text{diag}(\widetilde{M}_1, \ldots, \widetilde{M}_k) \tag{10}$$

We summarize the algorithm using E-BJacobi to precondition the BiCGstab method for solving a large sparse linear system in parallel in Algorithm 1. We point out that the column-permutation in the HSL subroutine MC64 is the only major serial step of the entire E-BJacobi scheme.

4 Numerical Experiments

Our parallel implementation of E-BJacobi is written in FORTRAN 90 and uses Intel's Math Kernel Library (MKL). All the numerical experiments in this section are conducted on a distributed memory platform consisting of 8 nodes with each node containing 24 cores (Intel Xeon E5-4617 processor) with 64 GB memory. Each node is assigned at most 1 MPI process, with 16 OpenMP threads as well as 16 MKL threads. For a given input coefficient matrix, we generate the RHS of the linear system by taking the all-one vector as the solution of the sparse linear system.

[2]http://glaros.dtc.umn.edu/gkhome/metis/parmetis/overview

[3]http://www.pardiso-project.org/

Algorithm 1 E-BJacobi $(A, f, p, \rho, maxit)$

Require: Nonsingular coefficient matrix $A \in \mathbb{R}^{n \times n}$; the RHS f; the number of blocks in the E-BJacobi preconditioner p; the stopping criterion ρ and the maximum number of iterations *maxit* of the BiCGstab method.

1: Apply MC64 column-permutation Q to A for maximum product traversal.
2: Construct the undirected weighted $\mathcal{G}[AQ^{\mathrm{T}}]$ according to (6).
3: Solve the p-GPES problem on $\mathcal{G}[AQ^{\mathrm{T}}]$ in parallel using ParMETIS. Let an induced WGPR be P.
4: Apply the WGPR reordering P to get the reordered matrix $PAQ^{\mathrm{T}}P^{\mathrm{T}}$.
5: Extract from $PAQ^{\mathrm{T}}P^{\mathrm{T}}$ the block Jacobi preconditioner $M = \mathrm{diag}(M_1, \ldots, M_p)$, where the k-th block is given by (7).
6: Each diagonal block M_k $(k = 1, \ldots, p)$ is assigned to the k-th multicore node. Each of the p nodes factors M_k with pivot perturbation according to (8) using PARDISO in parallel. The E-BJacobi preconditioner is then given by $\widetilde{M}$ of (10).
7: Solve the reordered linear system $(PAQ^{\mathrm{T}}P^{\mathrm{T}})(PQx) = Pf$ using BiCGstab with the E-BJacobi preconditioner $\widetilde{M}$, on a cluster of multicore nodes, given the parameters ρ and *maxit*.
8: Either an approximate solution x is found, or FAILURE is reported.

4.1 Robustness

In this section, we demonstrate the robustness of our solver E-BJacobi using the WGPR reordering by comparing it with: (i) block Jacobi preconditioner after only column-permutations using MC64 (referred to as MC64-BJacobi in what follows), and (ii) ILUTP preconditioners [13]. Following [3, 4], we apply the MC64 column-permutation and the symmetric RCM reordering before forming a particular ILUTP preconditioner. In our case, we select the relative numerical drop tolerance, *droptol*, to 10^{-2}, with no limit on the resulting fill-in.

We chose, from the University of Florida sparse matrix collection [6], 24 matrices all arising from different applications in computational fluid dynamics, as summarized in Table 1. In Table 2, we report the number of BiCGstab iterations preconditioned by E-BJacobi, MC64-BJacobi, and ILUTP, respectively, with the stopping criterion of BiCGstab (relative residual) being set to $\rho = 10^{-4}$. From Table 2, we see that E-BJacobi improves the success rate to 83% as compared to the 67% of MC64-BJacobi. If we strengthen the stopping criterion to be $\rho = 10^{-10}$, then as shown in Table 3[4], the success rate of E-BJacobi (75%) is more than 20% higher than that of MC64-BJacobi (54%). And for those systems for which both E-BJacobi and MC64-BJacobi converge, E-BJacobi significantly reduces the number of BiCGstab iterations, especially for the linear systems `cfd2`, `af23560`, `poisson3Db`, `venkat25`, and `venkat50`.

[4]Given the poor performance of ILUTP in Table 2 when $\rho = 10^{-4}$, we did not include ILUTP for $\rho = 10^{-10}$.

Table 1 Linear systems for robustness experiments.

Linear system	n	nnz	Application
StocF-1465	1,465,137	21,005,389	Flow in porous media
shallow_water1	81,920	327,680	Weather shallow water equations
shallow_water2	81,920	327,680	Weather shallow water equations
cfd1	70,656	1,825,580	Pressure equation
cfd2	123,440	3,085,406	Pressure equation
lung2	109,460	492,564	Coupled temperature and water vapor transport in a lung
atmosmodd	1,270,432	8,814,880	Atmospheric modeling
atmosmodj	1,270,432	8,814,880	Atmospheric modeling
atmosmodl	1,489,752	10,319,760	Atmospheric modeling
af23560	23,560	460,598	Flow over airfoils
poisson3Db	85,623	2,374,949	3D Poisson problem
venkat01	62,424	1,717,792	Unstructured 2D Euler solver
venkat25	62,424	1,717,763	Unstructured 2D Euler solver
venkat50	62,424	1,717,777	Unstructured 2D Euler solver
Ill_Stokes	20,896	191,368	Ill-conditioned matrix from a Stokes problem
ns3Da	20,414	1,679,599	3D Navier–Stokes
parabolic_fem	525,825	3,674,625	Parabolic FEM problem
raefsky3	21,200	1,488,768	Fluid-structure interaction turbulence problem
rma10	46,835	2,329,092	3D CFD model
water_tank	60,740	2,035,281	3D fluid flow of water in a tank
bbmat	38,744	1,771,722	2D airfoil exact Jacobian
invextr1_new	30,412	1,793,881	POLYFLOW flat die
mixtank_new	29,957	1,990,919	POLYFLOW mixing tank
rim	22,560	1,014,951	Fluid mechanics problem

In Figure 2(a), we plot the sparsity pattern with magnitude of Ill_Stokes after only the MC64 column-permutation, and in Figure 2(b) the sparsity pattern after applying both MC64 column-permutation and WGPR reordering with $p = 8$. Clearly, the block diagonal structure in Figure 2(b) is much more dominant than that in Figure 2(a). The dominant block diagonal structure resulting from the WGPR reordering explains the robustness of E-BJacobi. Note also that the 8 diagonal blocks in Figure 2(b) are well load-balanced.

Table 2 Number of BiCGstab iterations preconditioned by E-BJacobi, MC64-BJacobi, and ILUTP, respectively. For E-BJacobi and MC64-BJacobi, we set the number of diagonal blocks $p = 8$. For ILUTP, we set the relative *droptol* $= 10^{-2}$. The stopping criterion of BiCGstab (achieved relative residual) set as $\rho = 10^{-4}$, with the maximum number of BiCGstab iterations set as *maxit* $= 500$. Here, "-" indicates a failure to converge.

Linear system	E-BJacobi	MC64-BJacobi	ILUTP
`StocF-1465`	28	**17.5**	-
`shallow_water1`	2	2.5	1
`shallow_water2`	**3.5**	4	1.5
`cfd1`	50.5	85.5	-
`cfd2`	56.5	**37**	-
`lung2`	2	2	1
`atmosmodd`	25.5	18	**11**
`atmosmodj`	23.5	16.5	**8.5**
`atmosmodl`	13.5	28.5	**5**
`af23560`	**37**	133	108.5
`poisson3Db`	**19**	95.5	-
`venkat01`	4.5	7.5	3.5
`venkat25`	**13**	22.5	-
`venkat50`	**14**	23	-
`Ill_Stokes`	**47.5**	-	-
`ns3Da`	**58**	-	-
`parabolic_fem`	**93**	-	46.5
`raefsky3`	0.5	0.5	-
`rma10`	**55**	-	-
`water_tank`	**32**	38	-
`bbmat`	-	-	-
`invextr1_new`	-	-	-
`mixtank_new`	-	-	-
`rim`	-	-	-
Success rate	83%	67%	38%

4.2 *Parallel Scalability*

We demonstrate the parallel scalability of E-BJacobi on the linear systems in Table 4. The coefficient matrix `Nastran_2` is the stiffness matrix of a Nastran benchmark for car body dynamics [11]. For each linear system, we profile the total time of the entire E-BJacobi algorithm as outlined in Algorithm 1. The total time ($T_{\texttt{E-BJacobi}}$) includes the times of: MC64 column-permutation ($T_{\texttt{MC64}}$), weighted graph partitioning ($T_{\texttt{WGPR}}$), setting up the E-BJacobi preconditioner ($T_{\texttt{setup}}$), and the preconditioned BiCGstab iterations ($T_{\texttt{BiCGstab}}$). As preprocessing steps, the times of $T_{\texttt{MC64}}$, $T_{\texttt{WGPR}}$, and $T_{\texttt{setup}}$ can be amortized when solving for a sequence of right-hand sides with the same coefficient matrix. Nevertheless in the following we still use $T_{\texttt{E-BJacobi}}$ to measure the parallel scalability of E-BJacobi when solving for just one RHS. We fix the stopping criterion of BiCGstab to $\rho = 10^{-4}$, and the maximum number of iterations to *maxit* $= 500$. For comparison, we also solve each linear system directly using PARDISO (under the default parameter setting) on one node with 16 threads, and denote the time PARDISO consumes by $T_{\texttt{PARDISO}}$.

Table 3 The first and second columns list the number of BiCGstab iterations as preconditioned by E-BJacobi and MC64-BJacobi, respectively, with the number of diagonal blocks $p = 8$, the stopping criterion of BiCGstab $\rho = 10^{-10}$, and the maximum number of iterations *maxit* = 500. Here, "-" indicates a failure to reach the stopping criterion. In the last column, we report the relative residual achieved by the sparse direct solver PARDISO (under the default parameter setting).

Linear system	E-BJacobi	MC64-BJacobi	PARDISO rel. res.
StocF-1465	-	-	1.9×10^{-14}
shallow_water1	6	6.5	4.0×10^{-16}
shallow_water2	8.5	10	8.4×10^{-16}
cfd1	**128**	152.5	1.4×10^{-15}
cfd2	**293.5**	411.5	6.9×10^{-16}
lung2	5.5	6	6.6×10^{-16}
atmosmodd	45.5	**37**	8.5×10^{-15}
atmosmodj	53.5	**35**	7.1×10^{-15}
atmosmodl	**34.5**	63.5	5.6×10^{-15}
af23560	**74.5**	193.5	5.4×10^{-14}
poisson3Db	**42**	224.5	4.1×10^{-15}
venkat01	**13.5**	20.5	9.4×10^{-16}
venkat25	**46**	92	8.4×10^{-16}
venkat50	**52.5**	122.5	8.3×10^{-16}
Ill_Stokes	**130.5**	-	5.5×10^{-15}
ns3Da	**120.5**	-	9.0×10^{-15}
parabolic_fem	**176**	-	2.2×10^{-11}
raefsky3	**197.5**	-	6.1×10^{-16}
rma10	**158.5**	-	2.1×10^{-15}
water_tank	-	-	9.3×10^{-16}
bbmat	-	-	1.8×10^{-12}
invextr1_new	-	-	1.8×10^{-12}
mixtank_new	-	-	2.9×10^{-4}
rim	-	-	1.1×10^{-2}
Success rate	75%	54%	92%

In Figure 3(a)–3(c) we plot the speed improvement of E-BJacobi vs. PARDISO ($T_{\texttt{PARDISO}}/T_{\texttt{E-BJacobi}}$) for different number of nodes (same as the number of blocks) p=2, 4, and 8. From Figure 3(a) we see that on atmosmodl E-BJacobi achieves almost superlinear speed improvement over PARDISO. Even though the effectiveness of the block Jacobi preconditioner decreases as the number of diagonal blocks increases, the observed superlinear speed improvement may be due to the significant reduction in time consumed by the factorization and triangular solves on each node. In contrast, the speed improvements realized for StocF-1465 and $\texttt{Nastran}_2$ are much more modest. On $\texttt{Nastran}_2$ the parallel efficiency $T_{\texttt{PARDISO}}/(p \cdot T_{\texttt{E-BJacobi}})$ is about 0.5 for 8 nodes. The main cause of such modest

Table 4 Linear systems for parallel scalability experiments.

Linear system	n	nnz	nnz/n	Application
`atmosmodl`	1,489,752	10,319,760	7	Atmospheric modeling
`StocF-1465`	1,465,137	21,005,389	14	Flow in porous media
`Nastran_2`	7,234,582	417,297,732	58	Car body dynamics

parallel performance on $\texttt{Nastran}_2$ is the sequential nature of the MC64 column-permutation. In Table 5, we report the percentage of $T_{\texttt{MC64}}/T_{\texttt{E-BJacobi}}$ for the number of blocks p=2, 4, 8. We see that for `atmosmodl` and `StocF-1465`, the time of MC64 column-permutation $T_{\texttt{MC64}}$ is no more than 10% of the total time $T_{\texttt{E-BJacobi}}$, which implies that the sequential nature of MC64 is not an impediment to parallel scalability for these two linear systems. However, for $\texttt{Nastran}_2$ the percentage of $T_{\texttt{MC64}}/T_{\texttt{E-BJacobi}}$ is almost 40% when p=8, which explains the deterioration of parallel performance as p increases. We suspect the dramatic increase of $T_{\texttt{MC64}}$ on $\texttt{Nastran}_2$ is due to its much higher density (i.e., nnz/n) than that of `atmosmodl` and `StocF-1465`, as shown in Table 4.

5 Conclusion

In this paper, we present a procedure for constructing effective block Jacobi preconditioners. Two primary techniques for constructing such preconditioners include: (a) a reordering scheme based on weighted graph partitioning (WGPR) which leads to a dominant block diagonal structure, and hence of the E-BJacobi preconditioner, by encapsulating as many of the heaviest elements of the coefficient matrix as possible, while simultaneously maintaining load balancing on a distributed memory parallel architecture, and (b) factorization with diagonal pivot perturbation, which ensures the actual E-BJacobi preconditioner $\widetilde{M}$ is nonsingular. Our experiments on linear systems arising from applications in computational fluid dynamics demonstrate the robustness of E-BJacobi preconditioners. Compared to solving sparse linear systems using a direct method, the parallel implementation of E-BJacobi achieves appreciable speed improvement on a distributed memory platform. We observe that the sequential nature of the MC64 column-permutation will compromise the parallel performance of E-BJacobi when the density of the coefficient matrix is high. In this respect, for future work, we propose to study parallel algorithms for maximum product traversal (e.g., [1]).

Acknowledgements This work was supported by the Army Research Office, ARO grant number 7W911NF-11-1-0401.

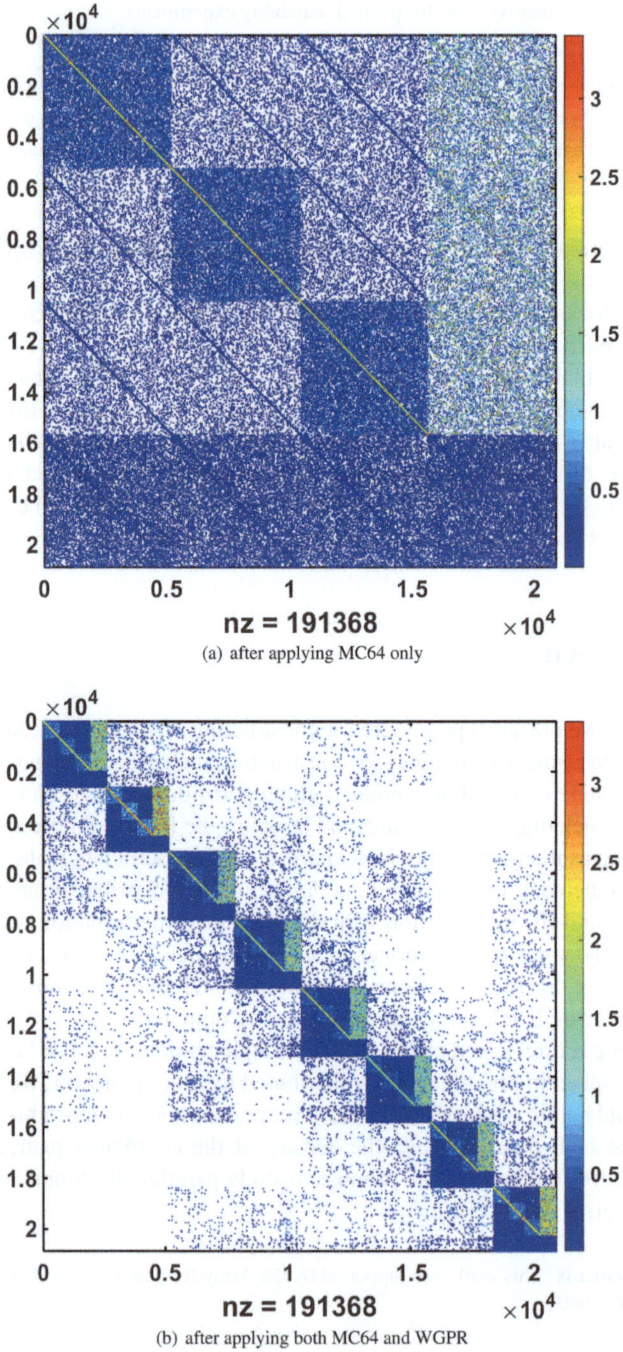

(a) after applying MC64 only

(b) after applying both MC64 and WGPR

Fig. 2 Sparsity pattern of `Ill_Stokes`. The nonzeros are colored based on their magnitudes.

Fig. 3 E-BJacobi speed improvement vs. PARDISO.

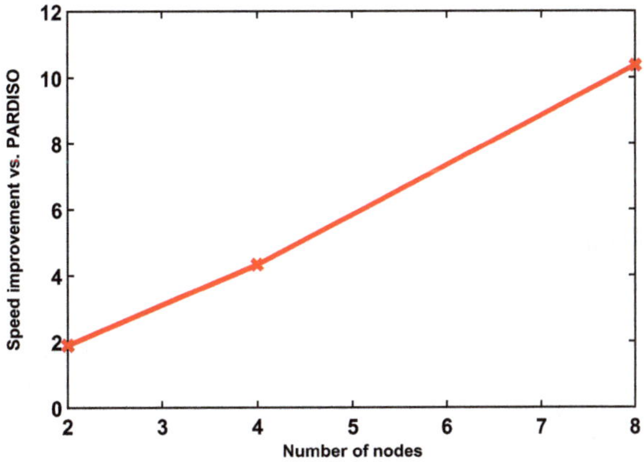

(a) `atmosmodl`

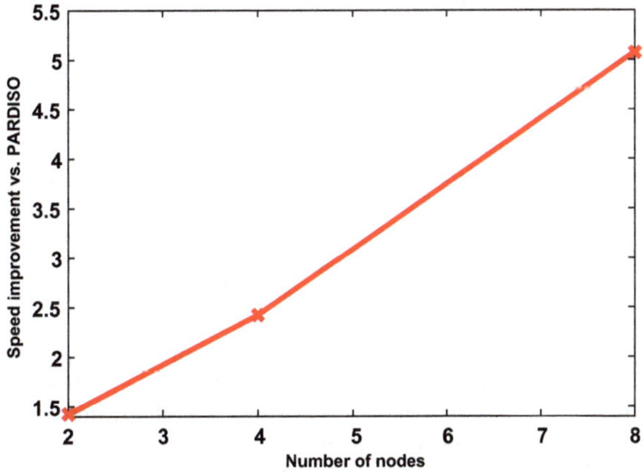

(b) `StocF-1465`

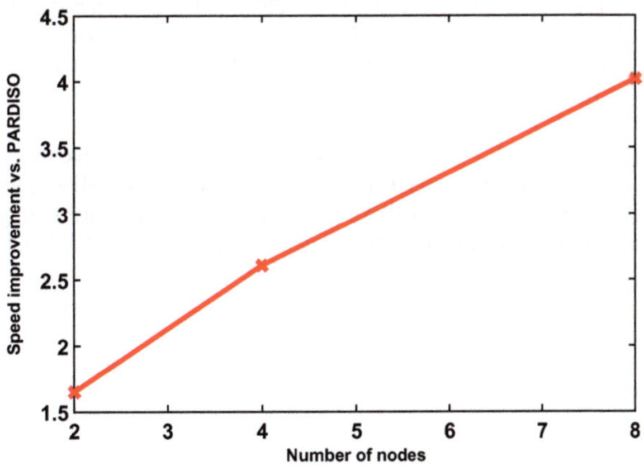

(c) `Nastran_2`

Table 5 $T_{\texttt{MC64}}/T_{\texttt{E-BJacobi}}$ for different number of blocks p.

Linear system	$p = 2$	$p = 4$	$p = 8$
atmosmodl	1.2%	2.8%	6.9%
StocF-1465	1.8%	3.0%	6.4%
Nastran_2	16.6%	27.3%	39.5%

References

1. Amestoy, P.R., Duff, I.S., Ruiz, D., Ucar, B.: Towards parallel bipartite matching algorithms. In: Scheduling for Large-Scale Systems (2009) http://perso.ens-lyon.fr/bora.ucar/papers/knoxville.pdf And the website of the workshop is at: http://graal.ens-lyon.fr/~abenoit/knoxville/
2. Benzi, M.: Preconditioning techniques for large linear systems: a survey. J. Comput. Phys. **182**(2), 418–477 (2002)
3. Benzi, M., Szyld, D.B., van Duin, A.: Orderings for incomplete factorization preconditionings of nonsymmetric problems. SIAM J. Sci. Comput. **20**, 1652–1670 (1999)
4. Benzi, M., Haws, J.C., Tuma, M.: Preconditioning highly indefinite and nonsymmetric matrices. SIAM J. Sci. Comput. **22**(4), 1333–1353 (2000)
5. Davis, T.A.: Algorithm 832: UMFPACK V4.3 - an unsymmetric-pattern multifrontal method. ACM Trans. Math. Softw. **30**(2), 196–199 (2004)
6. Davis, T.A., Hu, Y.: The University of Florida sparse matrix collection. ACM Trans. Math. Softw. **38**(1), 1–25 (2011)
7. Duff, I.S., Koster, J.: On algorithms for permuting large entries to the diagonal of a sparse matrix. SIAM J. Matrix Anal. Appl. **22**, 973–996 (2001)
8. Karpis, G., Kumar, V.: A fast and high quality multilevel scheme for partitioning irregular graphs. SIAM J. Sci. Comput. **20**(1), 359–392 (1998)
9. Karypis, G., Kumar, V.: A parallel algorithm for multilevel graph partitioning and sparse matrix ordering. J. Parallel Distrib. Comput. **48**(1), 71–95 (1998)
10. Li, X.S.: An overview of SuperLU: algorithms, implementation, and user interface. ACM Trans. Math. Softw. **31**(3), 302–325 (2005)
11. Mehrmann, V., Schroder, C.: Nonlinear eigenvalue and frequency problems in industrial practice. J. Math. Ind. 1:7 (2011) http://link.springer.com/article/10.1186/2190-5983-1-7
12. Moulitsas, I., Karypis, G.: Algorithms for graph partitioning and fill reducing ordering for domain decomposition methods. Technical Report, RAL, Oxfordshire and ENSEEIHT-IRIT, Toulouse (2006)
13. Saad, Y.: Iterative Methods for Sparse Linear Systems, 2nd edn. SIAM, Philadelphia (2003)
14. Schenk, O., Gärtner, K.: Solving unsymmetric sparse systems of linear equations with PARDISO. Futur. Gener. Comput. Syst. **20**(3), 475–487 (2004)
15. Vecharynski, E., Saad, Y., Sosonkina, M.: Graph partitioning using matrix values for preconditioning symmetric positive definite systems. SIAM J. Sci. Comput. **36**(1), A63–A87 (2014)

Parallel Analysis System for Fluid–Structure Interaction with Free-Surfaces Using ADVENTURE_Solid and LexADV_EMPS

Naoto Mitsume, Tomonori Yamada, Shinobu Yoshimura, and Kohei Murotani

Abstract In this chapter, we present a parallel analysis system for fluid–structure interaction (FSI) analysis with free-surfaces. It is based on a method that uses the moving-particle semi-implicit/simulation (MPS) method for flow computations and the finite element (FE) method for structural computations. The MPS-FE method is an efficient and robust approach for FSI problems involving free-surface flow. To develop the system presented herein, we use two existing open-source software modules: ADVENTURE_Solid, a large-scale FE solver for structural computations; and LexADV_EMPS, a library for large-scale MPS computations for free-surface flow. The explicitly represented polygon (ERP) wall boundary model employed in LexADV_EMPS is accurate and stable, and it expresses wall boundaries as a set of arbitrarily shaped triangular polygons with appropriately imposed boundary conditions. Thus, when the ERP is used, in both the fluid and the structure computations, the fluid–structure interfaces are matched, and therefore, preprocessing of the data for FSI analysis is greatly facilitated. We demonstrate the applicability of the developed system by solving a dam-break problem with an elastic obstacle.

1 Introduction

As seen with the Great East Japan Earthquake and Tsunami on March 11, 2011, a large tsunami can inflict devastating damage. This has led to an increased awareness of the vital importance of designing safe, tsunami-resistant structures, including the buildings and machinery of electric power, energy, and chemical plants, and evacuation facilities in coastal regions. The damage done by a tsunami is caused not only by the direct force of the wave, but also by such forces as, for example, elastic

N. Mitsume • S. Yoshimura (✉) • K. Murotani
Department of Systems Innovation, School of Engineering, The University of Tokyo, 7-3-1 Hongo, Bunkyo-ku, Tokyo 113-8654, Japan
e-mail: mitsume@save.sys.t.u-tokyo.ac.jp; yoshi@sys.t.u-tokyo.ac.jp

T. Yamada
RACE, The University of Tokyo, 5-1-5 Kashiwanoha, Kashiwa-shi, Chiba 277-0882, Japan

Y. Bazilevs, K. Takizawa (eds.), *Advances in Computational Fluid-Structure Interaction and Flow Simulation*, Modeling and Simulation in Science, Engineering and Technology, DOI 10.1007/978-3-319-40827-9_19

and plastic deformation, collisions with floating objects, and buoyancy [26]; these all must be considered when developing tsunami-resistant designs, but it is difficult to predict them when using conventional processes based on empirical knowledge. Because of this, numerical simulations can be a powerful and important part of the design process.

There have been many researches into prevention and mitigation of water-related disasters, such as tsunamis, floods, and storm surges. In these studies, shallow water equations are solved to predict the areas that will be inundated and the variation in water level in coastal areas [21], and the Navier–Stokes equations are solved to evaluate the wave forces exerted on affected structures [3]. However, few studies have used numerical simulation to evaluate water-related damage from various fluid–structure interactions (FSIs). A versatile analysis system has not yet been developed for this, because the phenomena involve free-surfaces and moving boundaries, which are difficult to deal with in a robust way.

In order to be able to better analyze potential damage due to these factors, we developed the MPS-FE method [13], which is an efficient and robust method for FSI problems with free-surfaces and moving boundaries; it uses a mesh-free particle method, the moving-particle semi-implicit/simulation (MPS) method [7] for free-surface flow computations, and the finite element (FE) method for structural computations. By taking advantage of both methods, the MPS-FE method has good robustness when dealing with free-surfaces and moving boundaries in flow computations, and it gives highly reliable solutions to various constitutive equations. Subsequently, we proposed an improved MPS-FE method [12] that used a polygon wall boundary method [5] to express the wall boundaries as a set of arbitrarily shaped triangular polygons instead of as conventional wall particles. In that paper [12], we indicated that preprocessing of the data was greatly facilitated, because in both the fluid and structure computations, the fluid–structure interfaces are matched. The improved MPS-FE method is thus more accurate and more stable than other existing methods [8, 13, 23]. We have proposed a stable and efficient model, the explicitly represented polygon (ERP) wall boundary model [14], which is based on polygon wall boundary models and other models that can treat arbitrarily shaped boundaries [15, 16].

To develop a large-scale parallel algorithm and system for analysis, it is important to solve an actual engineering problem. In this chapter, we present a parallel analysis system for FSI with free-surfaces; it is based on the MPS-FE method and uses existing modules for large-scale parallel computations: ADVENTURE_Solid [1, 25], a large-scale parallel FE solver for the structural computations; LexADV_EMPS [9, 17], a domain decomposition and interdomain communication library for MPS computations with free-surface flow; and HDDM_EMPS, a large-scale parallel solver that uses LexADV_EMPS [9, 17]. To exchange physical values between these solvers, the MPSFE_Coupler module was newly developed and used. In the following sections, we will explain this system, and then we will demonstrate its applicability by solving a dam-break problem with an elastic obstacle.

2 Large-Scale Parallel FE Solver

The domain decomposition method (DDM) is a well-known and widely used approach for parallel FE computations. In the DDM, the area to be analyzed is split into a number of non-overlapping subdomains, and the unknowns in the interiors of the subdomains are eliminated; this transforms the global linear system into a Schur complement system on the unknowns associated with the interfaces between the subdomains. ADVENTURE_ Solid, the large-scale parallel FE solver used in our system, employs the hierarchical domain decomposition method [22] (HDDM) and the balancing domain decomposition (BDD) preconditioner [10]. ADVENTURE_ Solid is one of the modules developed by the ADVENTURE project [1], which has the aim of developing a computational mechanics system for large-scale analysis and design; development of this project is led by Shinobu Yoshimura. This project is open source and includes many modules, such as fluid, thermal, and magnetic solvers, a mesh generator, an IO data formatter, and a post-processor. ADVENTURE_ Solid decomposes the area to be analyzed into a two-level hierarchy of subdomains; the top level is called the *part*, and the second is called the *subdomain*. This is performed by a module for domain decomposition, ADVENTURE_ Metis, which is based on ParMETIS [19], a library for the parallel computation of minimal-cut partitions of graphs and meshes. Adventure_ Solid can obtain excellent parallel efficiency and convergence rates, even when solving very large-scale problems, such as those with over 100 million degrees of freedom.We note that ADVENTURE_ Solid has been used for simulations of various engineering applications involving coupled systems [2, 6, 11, 24, 27].

3 Large-Scale Parallel MPS Solver

There have been several studies on the use of distributed-memory parallel computation for mesh-free particle methods. In contrast to the domain decomposition approach of the finite element method, it is extremely difficult for mesh-free particle methods to obtain good parallel efficiency by using static domain decomposition because the particles move in a Lagrangian fashion. Murotani et al. [17, 18] proposed a dynamic domain decomposition approach for the explicit MPS method [20]. This approach uses ParMETIS so that the number of particles in each domain is equalized and communication between domains is minimized; this results in a good load balance. In addition, an open-source library for dynamic domain decomposition and interdomain communication, LexADV_EMPS [9], has been developed. In the system proposed in this paper, for the free-surface flow computations, we use a large-scale parallel MPS solver that is based on LexADV_EMPS and HDDM_EMPS.

4 Parallel Analysis System for FSI with Free-Surfaces

4.1 Model of Wall Boundary Used in MPS Computations

As mentioned in the Introduction, we propose an accurate and practicable system for FSI analysis, and we do this by ensuring that the fluid–structure interface is consistent. We accomplished this by using the polygon wall boundary model [5], which represents wall boundaries as a set of arbitrarily shaped triangular polygons [12]. However, the existing polygon wall boundary model has problems in terms of accuracy and stability, so we proposed the ERP as a more accurate and stable model [14].

We are currently developing a library for the ERP model for large-scale parallel MPS computations. An example of a simulation of a breaking dam and three rectangular columns is shown in Fig. 1; the HDDM_EMPS with the ERP model was used, and colors indicate the domain of each particle. The wall boundaries were represented by 44 polygons. Six message-passing interface (MPI) processes (flat MPI) were used, and the calculation area was split into six domains. As can be seen in Fig. 1, the domains are dynamically decomposed as the particles move. Although the conventional approach of using particles for MPS computations requires several layers at a wall boundary, the ERP model does not use particles, and in areas where a boundary is flat, it can be discretized by course polygons, with no loss of accuracy. Because of this, when the wall boundaries are relatively flat, as in Fig. 1, the computational cost can be significantly reduced.

4.2 Fluid–Structure Interaction Model for the MPS-FE Method Using the ERP Model

In the ERP model, the force exerted on a polygon wall by a particle is defined as the reaction to the force exerted on the particle by the walls. The force on particle *i*, f_i^{wall}, in the ERP model is (Fig. 2)

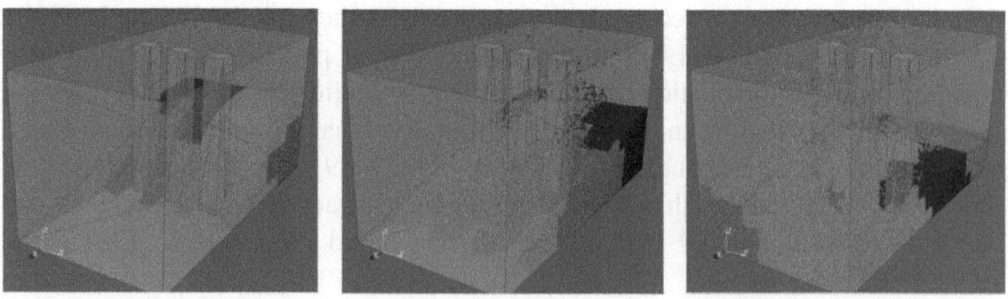

Fig. 1 Dynamic domain decomposition by LexADV_EMPS in a dam-break simulation

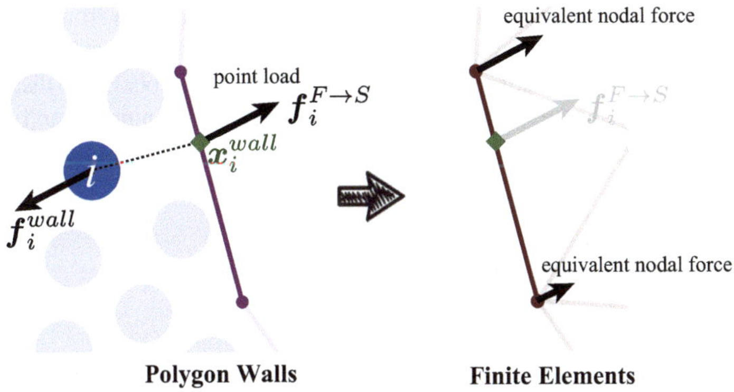

Fig. 2 Force exerted on polygon wall by particle i, and corresponding equivalent nodal forces on finite element nodes

$$f_i^{wall} = m_i\left(-\frac{1}{\rho^0}\left\langle\nabla p\right\rangle_i^{wall} + \nu\left\langle\nabla^2 \boldsymbol{v}\right\rangle_i^{wall}\right), \tag{1}$$

where m_i is the mass of particle i, ρ^0 is the constant fluid density, and ν is the kinetic viscosity. Here, $\langle\nabla p\rangle_i^{wall}$ and $\langle\nabla^2 \boldsymbol{v}\rangle_i^{wall}$ are the contributions from the wall of the pressure gradient and the velocity Laplacian, respectively, when the pressure Neumann and slip/no-slip boundary conditions are imposed. The force that particle i exerts on the polygon wall, that is, the force from the fluid to the structure, $\boldsymbol{f}_i^{F\to S}$, is regarded as the reaction force of $\boldsymbol{f}_i^{wall}$,

$$\boldsymbol{f}_i^{F\to S} = -\boldsymbol{f}_i^{wall}, \tag{2}$$

acting on the nearest point $\boldsymbol{x}_i^{wall}$ as the point load. With this assumption, the total distributed load, $\boldsymbol{q}^{F\to S}$, can be written as the summation of the point loads on each of the particles:

$$\boldsymbol{q}^{F\to S} = \sum_i \delta\left(\boldsymbol{x} - \boldsymbol{x}_i^{wall}\right)\boldsymbol{f}_i^{F\to S}, \tag{3}$$

where δ is the Dirac delta function. By Eq. (3), the equivalent nodal forces on the FE node j at the fluid–structure interface Γ_I, $\boldsymbol{f}_j^{node}$, can be written as follows:

$$\boldsymbol{f}_j^{node} = \int_{\Gamma_I} \boldsymbol{q}^{F\to S} N_j^g(\boldsymbol{x})d\Gamma \tag{4}$$

$$= -\sum_i \boldsymbol{f}_i^{wall} N_j^g(\boldsymbol{x}_i^{wall}), \tag{5}$$

where N_j^g is the global shape function of node j.

On the other hand, the process of expressing the contribution from the structure to the fluid is now simplified significantly, since the interfaces are matched in the fluid and structure computations, due to the use of the ERP model. The displacement (current position), velocity, and acceleration for each polygon node can be updated using their value on the corresponding FE node; no interpolation is necessary.

4.3 Coupler for Exchanging Physical Values Between Two Existing Solvers

The existing solvers introduced above were connected by the FSI model discussed in Sec. 4.2 and a partitioned coupling approach [4]. We needed to exchange physical values between different parallel solvers, so we developed MPSFE_ Coupler, i.e., a coupler for ADVENTURE_ Solid and HDDM_ EMPS.

A schematic view of the developed FSI system is shown in Fig. 3. As indicated in Fig. 3, since the polygon wall data for the wall boundary can be obtained by a surface extraction of the finite element data, the preprocessing is considerably simplified. The data exchanged by MPSFE_ Coupler (the sets of node and corresponding part numbers for each polygon, the subdomain numbers, and the local node numbers for the subdomain of the FE nodes) can be obtained easily. Other studies that performed FSI simulations using the mesh-free particle method and the finite element method [8, 13, 23] used particles to express the wall boundaries; as a result, they had difficulties when dealing with complicated surface shapes. In addition, we pointed out in [12] that in the existing studies, spurious injections or dissipations

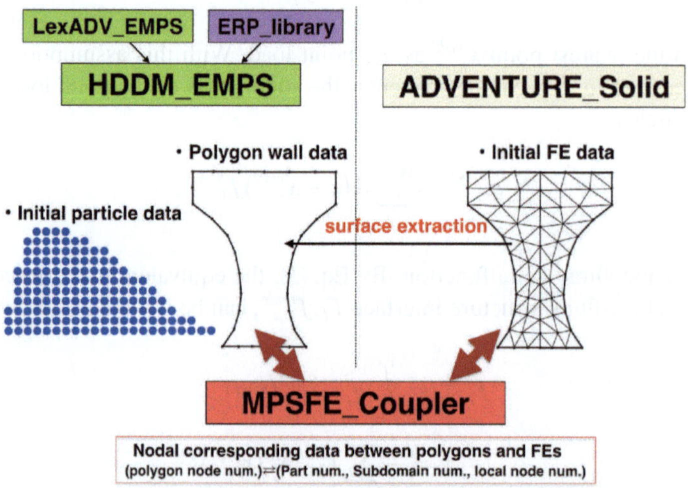

Fig. 3 Schematic view of the large-scale parallel FSI system using ADVENTURE_ Solid and HDDM_ EMPS

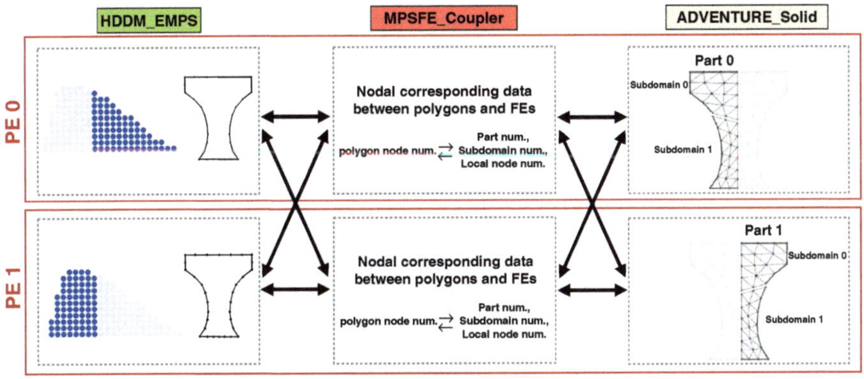

Fig. 4 Image of data distribution and communication in the FSI system

of momentum and energy occur on the interface, and this makes the computation unstable, since it makes it difficult to maintain a consistent force at the fluid–structure interface. In contrast to the existing studies, the approach adopted in this study does not have these problems, because the fluid–structure interface is matched by using the ERP model and the forces are consistent in the FSI model introduced in Sec. 4.2.

The distribution of data between the solvers and the communications by MPSFE_Coupler for two MPI processes is shown in Fig. 4. In HDDM_EMPS with the ERP model in the present version, each process has one domain containing particles and all processes have complete data for the polygon walls. On the other hand, in ADVENTURE_Solid, each process has one part that contains subdomains. MPSFE_Coupler exchanges the physical values between the polygons and the FE nodes.

5 Example of FSI Simulation

To demonstrate the applicability of our system for large-scale parallel FSI analysis, we solved a dam-break problem with an elastic obstacle, as shown in Fig. 5. The water column was a Newtonian fluid, discretized with 105,000 particles set in a cubic grid with 0.02[m] spacing. The fluid density was 1.0×10^3[kg/m^3], and the kinetic viscosity was 1.0×10^{-6}[m^2/s]. The obstacle was a linear elastic body discretized with 17,915 second-order tetrahedron elements and 27,328 FE nodes; there was a surface constraint on the bottom. The density was 1.2×10^3[kg/m^3], Young's modulus was 1.0×10^6[kg/s^{2}m], and Poisson's ratio was 0.4. A total of 3,592 polygon walls and 7,186 polygon nodes were obtained by surface extraction from the FE data; these were used as the wall boundary on the fluid–structure interface in the free-surface flow computation. The nodal corresponding data for

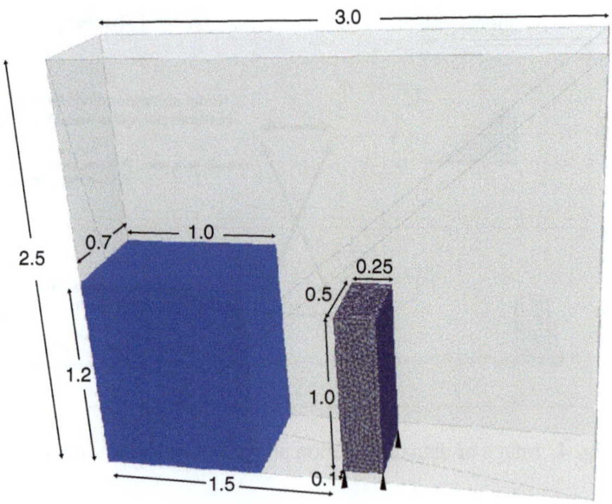

Fig. 5 Initial configuration of dam-break problem with an elastic obstacle (units: [m])

MPSFE_Coupler were obtained simultaneously. The outer cuboid wall boundary was expressed by ten polygon walls, and these were used only the fluid computations. In both the fluid and structure computations, the gravitational acceleration was $9.8[\mathrm{m/s^2}]$, and the time step was $1.0 \times 10^4[\mathrm{s}]$. The conventional serial staggered method with a linear predictor [4] was used as the coupling method for the partitions.

The simulation was conducted with six MPI processes on a single PC with six-core CPU (Intel Core i7-3970X, 3.50[GHz]). The particles and FEs were split into six domains in HDDM_EMPS and into six parts in ADVENTURE_Solid; each part had 70 subdomains. The computation times for each time step varied between 0.8[s] and 7.1[s]; the time required depended on the number of particles close to the polygon walls. Obviously, this was due to calculations of the distance between polygon walls and particles, and this algorithm should be made more efficient.

The results of this simulation are shown in Fig. 6; the velocity is indicated by color. As can be seen in Fig. 6, the wave force due to the dam break deforms the elastic obstacle, and the deformation of the obstacle affects the motion of the fluid. Although complicated free-surface and boundary motions occur, our system was able to perform the simulation robustly. This simulation example is unrealistic in terms of actual disaster prevention and mitigation, but we emphasize that the results indicate that our system can be used to simulate various situations, and it can include the effects of forces due to elastic and plastic deformation, uplift by buoyancy, and other factors. In addition, this system simplifies the preprocessing phase, and so it is especially practical for resistant and resilient designs that require the consideration of a variety of potential situations.

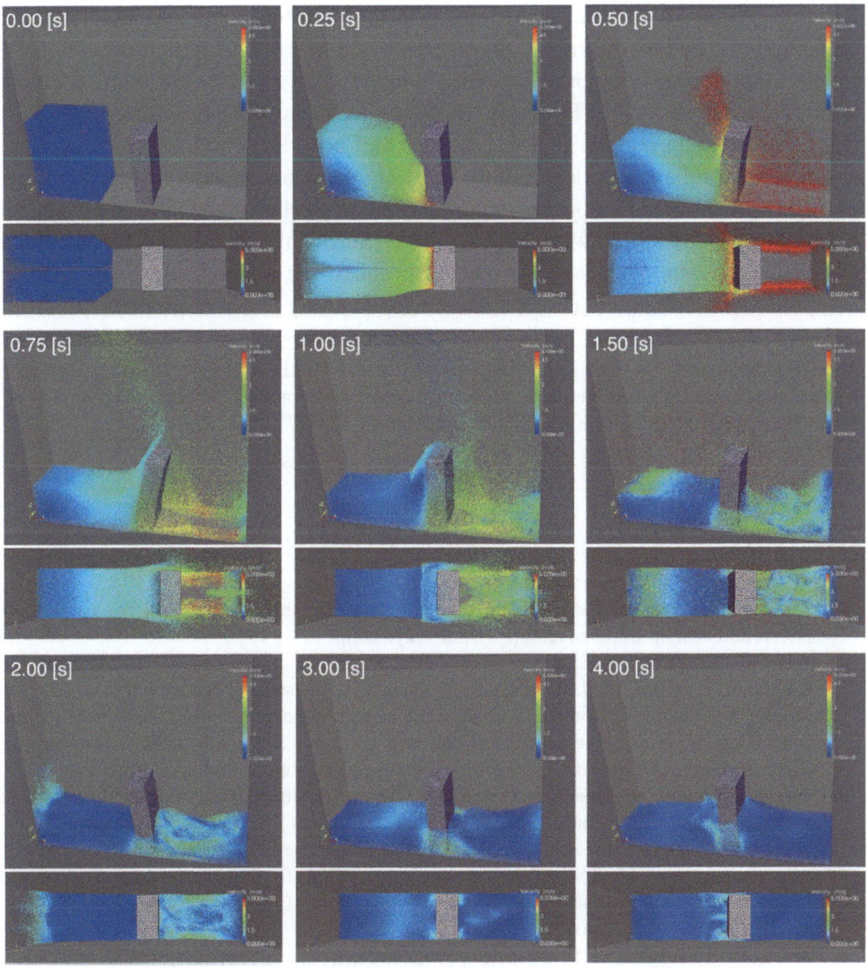

Fig. 6 Simulation results for dam-break problem with an elastic obstacle; velocity contours are shown by color (min: 0.0[m/s], max: 5.0[m/s])

6 Conclusions

In this chapter, we presented a parallel analysis system for FSI with free-surfaces; it is based on the MPS-FE method and uses ADVENTURE_Solid, a parallel FEM solver for structural computations with HDDM and the BDD preconditioner, and LexADV_EMPS, a domain decomposition and interdomain communication library for the MPS computation of free-surface flows. We incorporated the ERP wall boundary model, which is a polygon-based model, to improve the accuracy and stability of our FSI model, and we developed a large-scale parallel FSI analysis system that uses a coupler to exchange values between the solvers. In this system, the fluid–structure interfaces are matched in both the fluid and the structure computations, and therefore, the preprocess of the data is facilitated considerably. Finally, we used our proposed system to analyze the FSI of a dam-break problem with an elastic obstacle. The results indicate that the system is sufficiently robust when dealing with complicated free-surface and boundary motions, and therefore, it will be a useful and practical tool for the design of resistant and resilient structure.

Although we only demonstrated our method on a relatively small-scale problem with simple boundaries, we intend to more thoroughly investigate the accuracy and stability of the proposed method, to evaluate the parallel efficiency and the computational cost, and to demonstrate its applicability to actual problems.

Acknowledgements This study was supported by the JSTCREST project "Development of a Numerical Library Based on Hierarchical Domain Decomposition for Post Petascale Simulation." It was also supported by JSPS KAKENHI Grant Number 15J09772.

References

1. ADVENTURE Project: Development of computational mechanics system for large scale analysis and design. http://adventure.sys.t.u-tokyo.ac.jp/
2. Akiba, H., Ohyama, T., Shibata, Y., Yuyama, K., Katai, Y., Takeuchi, R., Hoshino, T., Yoshimura, S., Noguchi, H., Gupta, M., Gunnels, J.A., Austel, V., Sabharwal, Y., Garg, R., Kato, S., Kawakami, T., Todokoro, S., Ikeda, J.: Large scale drop impact analysis of mobile phone using ADVC on blue gene/l. In: Proceedings of the 2006 ACM/IEEE Conference on Supercomputing, SC '06. ACM, New York (2006). (2006 Gordon Bell Award finalist)
3. Asai, M., Fujimoto, K., Tanabe, S., Beppu, M.: Slip and no-slip boundary treatment for particle simulation model with incompatible step-shaped boundaries by using a virtual maker. Trans. Jpn. Soc. Comput. Eng. Sci. (20120012) (2012). (in Japanese) **2013**, 20130011
4. Farhat, C., Lesoinne, M.: Two efficient staggered algorithms for the serial and parallel solution of three-dimensional nonlinear transient aeroelastic problems. Comput. Methods Appl. Mech. Eng. **182**(3), 499–515 (2000)
5. Harada, T., Koshizuka, S., Shimazaki, K.: Improvement of wall boundary calculation model for MPS method. Trans. Jpn. Soc. Comput. Eng. Sci. (20080006) (2008). (in Japanese) **2008**, 20080006
6. Kataoka, S., Minami, S., Kawai, H., Yamada, T., Yoshimura, S.: A parallel iterative partitioned coupling analysis system for large-scale acoustic fluid–structure interactions. Comput. Mech. **53**(6), 1299–1310 (2014)
7. Koshizuka, S., Oka, Y.: Moving-particle semi-implicit method for fragmentation of incompressible fluid. Nucl. Sci. Eng. **123**(3), 421–434 (1996)
8. Lee, C.J.K., Noguchi, H., Koshizuka, S.: Fluid–shell structure interaction analysis by coupled particle and finite element method. Comput. Struct. **85**(11), 688–697 (2007)
9. LexADV: http://adventure.sys.t.u-tokyo.ac.jp/lexadv/
10. Mandel, J.: Balancing domain decomposition. Commun. Numer. Methods Eng. **9**(3), 233–241 (1993)
11. Minami, S., Kawai, H., Yoshimura, S.: Parallel BDD-based monolithic approach for acoustic fluid-structure interaction. Comput. Mech. **50**(6), 707–718 (2012)
12. Mitsume, N., Yoshimura, S., Murotani, K., Yamada, T.: Improved MPS-FE fluid-structure interaction coupled method with MPS polygon wall boundary model. Comput. Model. Eng. Sci. **101**(4), 229–247 (2014)

13. Mitsume, N., Yoshimura, S., Murotani, K., Yamada, T.: MPS-FEM partitioned coupling approach for fluid-structure interaction with free surface flow. Int. J. Comput. Methods **11**(4), 1350,101(16 pp.) (2014)
14. Mitsume, N., Yoshimura, S., Murotani, K., Yamada, T.: Explicitly represented polygon wall boundary model for explicit MPS method. Comput. Part. Mech. **2**(1), 73–89 (2015)
15. Monaghan, J., Kos, A., Issa, N.: Fluid motion generated by impact. J. Waterw. Port Coast. Ocean Eng. **129**(6), 250–259 (2003)
16. Morris, J.P., Fox, P.J., Zhu, Y.: Modeling low Reynolds number incompressible flows using SPH. J. Comput. Phys. **136**(1), 214–226 (1997)
17. Murotani, K., Oochi, M., Fujisawa, T., Koshizuka, S., Yoshimura, S.: Distributed memory parallel algorithm for explicit MPS using ParMETIS. Trans. Jpn. Soc. Comput. Eng. Sci. (20120012) (2012). (in Japanese) **2012**, 20120012
18. Murotani, K., Koshizuka, S., Tamai, T., Shibata, K., Mitsume, N., Shinobu, Y., Tanaka, S., Hasegawa, K., Nagai, E., Fujisawa, T.: Development of hierarchical domain decomposition explicit MPS method and application to large-scale tsunami analysis with floating objects. J. Adv. Simul. Sci. Eng. **1**(1), 16–35 (2014)
19. ParMETIS - Parallel Graph Partitioning and Fill-reducing Matrix Ordering. http://glaros.dtc.umn.edu/gkhome/metis/parmetis/overview
20. Shakibaeinia, A., Jin, Y.C.: A weakly compressible MPS method for modeling of open-boundary free-surface flow. Int. J. Numer. Methods Fluids **63**(10), 1208–1232 (2010)
21. Westerink, J.J., Luettich, R.A., Feyen, J.C., Atkinson, J.H., Dawson, C., Roberts, H.J., Powell, M.D., Dunion, J.P., Kubatko, E.J., Pourtaheri, H.: A basin-to channel-scale unstructured grid hurricane storm surge model applied to Southern Louisiana. Mon. Weather Rev. **136**(3), 833–864 (2008)
22. Yagawa, G., Shioya, R.: Parallel finite elements on a massively parallel computer with domain decomposition. Comput. Syst. Eng. **4**(4), 495–503 (1993)
23. Yang, Q., Jones, V., McCue, L.: Free-surface flow interactions with deformable structures using an SPH–FEM model. Ocean Eng. **55**, 136–147 (2012)
24. Yoshimura, S.: Virtual demonstration tests of large-scale and complex artifacts using an open source parallel CAE system, adventure. Solid State Phenom. **110**, 133–142 (2005)
25. Yoshimura, S., Shioya, R., Noguchi, H., Miyamura, T.: Advanced general-purpose computational mechanics system for large-scale analysis and design. J. Comput. Appl. Math. **149**(1), 279–296 (2002)
26. Yoshimura, S., Nakamura, H., Kanayama, H., Aoki, T., Himeno, T., Sakai, Y., Koshizuka, S., Isobe, D., Fujii, H.: Understanding the mechanism of tsunami-induced damage to machines and structures based on a discipline of mechanics. In: Report of the JSME Research Committee on the Great East Japan Earthquake Disaster, Chap. 4, pp. 24–57. The Japan Society of Mechanical Engineers, Sapporo (2014)
27. Yoshimura, S., Kobayashi, K., Akiba, H., Suzuki, S., Ogino, M.: Seismic response analysis of full-scale boiling water reactor using three-dimensional finite element method. J. Nucl. Sci. Technol. **52**(4), 546–567 (2015)

A Review on Fast Quasi-Newton and Accelerated Fixed-Point Iterations for Partitioned Fluid–Structure Interaction Simulation

David Blom, Florian Lindner, Miriam Mehl, Klaudius Scheufele, Benjamin Uekermann, and Alexander van Zuijlen

Abstract The partitioned simulation of fluid–structure interactions offers great flexibility in terms of exchanging flow and structure solver and using existing established codes. However, it often suffers from slow convergence and limited parallel scalability. Quasi-Newton or accelerated fixed-point iterations are a very efficient way to solve the convergence issue. At the same time, they stabilize and speed up not only the standard staggered fluid–structure coupling iterations, but also the variant with simultaneous execution of flow and structure solver that is fairly inefficient if no acceleration methods for the underlying fixed-point iteration are used. In this chapter, we present a review on combinations of iteration patterns (parallel and staggered) and of quasi-Newton methods and compare their suitability in terms of convergence speed, robustness, and parallel scalability. Some of these variants use the so-called manifold mapping that yields an additional speedup by using an approach that can be interpreted as a generalization of the multi-level idea.

D. Blom • A. van Zuijlen
Faculty of Aerospace Engineering, Delft University of Technology,
P. O. Box 5058, 2600 GB Delft, The Netherlands
e-mail: d.s.blom@tudelft.nl; a.h.vanzuijlen@tudelft.nl

F. Lindner • M. Mehl (✉) • K. Scheufele
Institute for Parallel and Distributed Systems, Universität Stuttgart,
Universitätsstraße 38, Stuttgart, Germany
e-mail: florian.lindner@ipvs.uni-stuttgart.de; miriam.mehl@ipvs.uni-stuttgart.de; klaudius.scheufele@ipvs.uni-stuttgart.de

B. Uekermann
Department of Computer Science, Technische Universität München,
Boltzmannstraße 3, Garching, Germany
e-mail: uekerman@in.tum.de

Y. Bazilevs, K. Takizawa (eds.), *Advances in Computational Fluid-Structure Interaction and Flow Simulation*, Modeling and Simulation in Science, Engineering and Technology, DOI 10.1007/978-3-319-40827-9_20

1 Introduction

The simulation of fluid–structure interactions (FSI) is a prominent example of multi-physics simulations where multiple physical fields are involved. FSI alone or in combination with further physical fields is, for example, needed in aero-elasticity [15], arterial flow simulation [10], airbag deployment [21], parachute systems [23], wind turbines [3], tidal power plants, and air conditioning devices. Typically, the simulations are required to be finished in a limited time but, at the same time, with a high accuracy.

In addition, many of these coupled multi-physics simulations are still evolving, i.e. they work with models that might have to be enhanced by further physical fields, use a discretization approach or equation solvers that might have to be replaced by a different method, or focus on the fast development of a new simulation environment. In these cases, the partitioned coupling approach going back to [16] using existing single-field solvers and combining them in a modular way to a coupled simulation environment is favourable over the monolithic approach. The latter allows the development and implementation of highly efficient solvers [1, 19, 24], but requires the explicit or implicit assembly of a large overall system of equations and, thus, a reimplementation of the whole solver. For a modular, partitioned approach, concepts for its efficient and robust realization are extremely valuable, but have to fulfil the important requirement to use only little information on the solvers, in other words, to be able to deal with black-box solvers exposing only input and output values of the respective single-field simulation.

In the remainder of this paper, we present state-of-the-art and new combinations of partitioned iteration patterns with quasi-Newton variants that we developed during the last 3 years [4–6, 20, 26] and compare the resulting convergence speed, robustness, and parallel scalability. For some cases, we add an additional improvement by using the so-called manifold mapping, a generalization of the multi-level idea.

2 Quasi-Newton or Fixed-Point Acceleration Schemes

Before presenting quasi-Newton and fixed-point acceleration methods, we shortly recapitulate the introduction of fixed-point equations realizing the coupling at the wet surface between fluid and structure for different execution orders of the flow and structure solvers.

The serial implicit coupling scheme (S-System) is used in most fluid–structure coupling approaches. The solvers are executed in a staggered way, i.e. the flow solver (F) first computes stresses or forces from wet surface displacements and velocities (x_d). These forces are communicated to the structure solver (S) that afterwards computes new wet surface displacements or velocities. The corresponding interface equation reads

$$x_d \stackrel{!}{=} S \circ F(x_d)\,. \tag{1}$$

The parallel implicit coupling scheme (V-System) was developed due to substantial drawbacks of the above serial implicit coupling scheme regarding efficient parallelization. There is a significant mismatch of work load between the structure and the fluid solver, which does not allow for an efficient parallelization using the S-system. The only way to overcome these limitations in parallel efficiency is to evaluate the fluid and the structure solver in parallel. Hereby, the V-system uses the original input/output relation for both solvers but the boundary values are exchanged after each simultaneous solving of flow and structure. This leads to the vectorial fixed-point equation

$$\begin{pmatrix} x_f \\ x_d \end{pmatrix} \stackrel{!}{=} \begin{pmatrix} 0 & F \\ S & 0 \end{pmatrix} \begin{pmatrix} x_f \\ x_d \end{pmatrix}. \tag{2}$$

The vectorial system results in two independent instances of the S-system if solved by a pure fixed-point iteration, but quasi-Newton solvers turn out to be powerful enough such that one iteration of the V-system is comparable to one iteration of the S-system (cf. [8, 26]).

We describe two quasi-Newton schemes, which can be applied to both the aforementioned fixed-point equations (1) and (2). For the sake of clarity, we introduce a unified notation for the two types of fixed-point equations presented above:

$$H(x) = x \;\Leftrightarrow\; R(x) := H(x) - x = 0 \tag{3}$$

with $H = S \circ F$ and $x = x_d$ or $H = \begin{pmatrix} 0 & F \\ S & 0 \end{pmatrix}$ and $x = \begin{pmatrix} x_f \\ x_d \end{pmatrix}$.

A pure fixed-point iteration for both systems, the S-system and the V-system, tends to be unstable and, in particular, is two times slower for the V-system than for the S-system [22]. Therefore, we accelerate the fixed-point iteration by a subsequent Newton step

$$x^{k+1} = \underbrace{H(x^k)}_{=:\,\tilde{x}^k} - J^{-1}_{I-H^{-1}} (\underbrace{H(x^k) - x^k}_{=\,\tilde{x}^k - H^{-1}(\tilde{x}^k)}) \tag{4}$$

with $\tilde{R} := I - H^{-1}$ mapping $\tilde{x}^k$ to the residual $R(x^k) = \tilde{R}(\tilde{x}^k) = H(x^k) - x^k$.

This Newton step requires the Jacobian and the solution of the respective linear system (inversion of the Jacobian). For our black-box approach, the Jacobian is inaccessible and we use an approximation

$$\widehat{J_{\tilde{R}}^{-1}}(\tilde{x}^k) \approx J_{\tilde{R}}^{-1}(\tilde{x}^k).$$

For this purpose, we collect input–output data throughout our iterations within a time step and generate the following matrices:

$$W_k = \left[\Delta\tilde{x}_0^k, \Delta\tilde{x}_1^k, \cdots, \Delta\tilde{x}_{k-1}^k\right], \text{ with } \Delta\tilde{x}_i^k = \tilde{x}^k - \tilde{x}^i ,$$
$$V_k = \left[\Delta R_0^k, \Delta R_1^k, \cdots, \Delta R_{k-1}^k\right], \text{ with } \Delta R_i^k = R(x^k) - R(x^i) .$$

This yields the secant equation

$$\widehat{J_{\tilde{R}}}^{-1}(\tilde{x}^k) V_k = W_k . \tag{5}$$

for the approximate inverse Jacobian of $\tilde{R}$. As k is in general much smaller than the number of degrees of freedom at the coupling interface, W_k and V_k are tall and skinny matrices and the system (5) is underdetermined. In the following, we present two approaches to regularize this system based on different additional norm minimization conditions for $\widehat{J_{\tilde{R}}}^{-1}(\tilde{x}^k)$.

The interface quasi-Newton least squares (IQN-LS) method presented in [12] uses the norm minimization

$$\left\| \widehat{J_{\tilde{R}}}^{-1}(\tilde{x}^k) \right\|_F \to \min \tag{6}$$

with the Frobenius norm $\|\cdot\|_F$. With this, we get the approximate inverse Jacobian

$$\widehat{J_{\tilde{R}}}^{-1}(x^k) = W_k \underbrace{(V_k^T V_k)^{-1} V_k^T}_{= V_k^\dagger}$$

with the pseudo-inverse $V_k^\dagger$ of V_k and the update formula

$$x^{k+1} = \tilde{x}^k + W_k \underbrace{V_k^\dagger \left(-R(x^k)\right)}_{=: \alpha} .$$

We do not have to explicitly compute the inverse Jacobian, but can restrict ourselves to compute only the vector α. This can be realized very efficiently by solving the least squares problem

$$\min_{\alpha \in \mathbb{R}^k} \|V_k \alpha + R(x^k)\|_2 \text{ with the Euclidian norm } \|\cdot\|_2.$$

The convergence properties of the IQN-LS method can be greatly improved, if the input/output informations from previous time steps are incorporated into the secant equation, i.e. into W_k and V_k. However, the optimal number R of reused time steps is highly problem dependent and cannot be determined analytically. Thus, in practice, R has to be determined based on experiences and in a costly try-and-error process. Also, linear dependencies and contradicting information within the accumulated difference matrices need to be handled properly. The alternative quasi-Newton approach presented in the next paragraph provides an automatic implicit incorporation of information from passed time steps and, thus, avoids these drawbacks of the IQN-LS method. However, it requires to explicitly compute $\widehat{J_{\tilde{R}}}^{-1}$ instead of only the short vector $\alpha \in \mathbb{R}^k$.[1]

The interface quasi-Newton multiple vector Jacobian (IQN-MVJ) method presented here is a newly developed quasi-Newton fluid–structure coupling approach. It combines the idea of approximating the Newton iteration (4) based on the secant equation (5) with the ideas presented in [7] for Jacobian approximations in the context of a block-iterative Newton method. To implicitly use information from previous time steps, the IQN-MVJ method uses a norm minimization for the difference between the current time step's inverse Jacobian approximation and the last time step's approximation:

$$\left\| \widehat{J_{\tilde{R}}}^{-1}(\tilde{x}^k) - \widehat{J_{\tilde{R}prev}}^{-1} \right\|_F \rightarrow \min \,, \tag{7}$$

where $\widehat{J_{\tilde{R}prev}}^{-1}$ denotes the inverse Jacobian approximation of the previous time step. Thus, our approximations always stay as close as possible to the approximation from the last time step. This automatically guaranties that we profit from past information without having to explicitely use old W and V matrices again. We get the approximate inverse Jacobian

$$\widehat{J_{\tilde{R}}}^{-1}(\tilde{x}^k) = \widehat{J_{\tilde{R}prev}}^{-1} + \left(W_k - \widehat{J_{\tilde{R}prev}}^{-1} V_k \right) V_k^{\dagger}$$

and the update formula

$$x^{k+1} = \tilde{x}^k + \left(\widehat{J_{\tilde{R}prev}}^{-1} + \left(W_k - \widehat{J_{\tilde{R}prev}}^{-1} V_k \right) V_k^{\dagger} \right) \left(-R(x^k) \right) .$$

There is a close relation of the IQN-MVJ update scheme to the Broyden method: The Broyden method also minimizes distances between successively computed Jacobian approximations within a Newton iteration. However, Broyden minimizes the distance of approximations between two successive iterations, whereas we minimize the distance between Jacobian approximations in two successive time steps.

[1]Note that the update formula for x^{k+1} also shows that skipping the fixed-point iteration step (computing $\tilde{x}^k = H(x^k)$) before using a quasi-Newton step would have led to linearly dependent columns in W_k: We then would always correct x^k to x^{k+1} by adding multiples of differences $x^k - x^i$ from previous iterations as we would have to use $W_k = (\Delta x_0^k, \Delta x_1^k, \ldots, \Delta x_{k-1}^k)$ with $\Delta x_i^k = x^k - x^i$ in this case. Using induction over the iterations, we see that all columns of W_k would be in the space spanned by x^0 and $x^1 - x^0$.

3 Manifold Mapping

Manifold mapping is a surrogate-based optimization technique, i.e. the approximation quality a low-fidelity model is iteratively improved. The goal of a surrogate-based optimization technique is to decrease the computational time of an optimization process. For FSI simulations the coupling represents the optimization problem.

In the following subsection, the basic terminology is introduced and the manifold mapping is explained. The reader is referred to Echeverría and Hemker [14] for the theoretical basis of the technique.

Fine and Coarse Models. Two types of models are distinguished: a fine model and a coarse model. It is assumed that the fine model is accurate, but requires a high computational cost to evaluate. The coarse model, on the contrary, is considered to be computationally less costly, but also less accurate in comparison to the fine model. The fine model response for FSI problems is the interface residual as already used above, i.e.

$$R(x) = H(x) - x$$

with the fixed-point mapping H either from the S- or from the V-system. The optimization problem is the minimization of the residual norm, i.e. we have to solve

$$\mathbf{x}^{\star}_{fine} = \arg\min_{x} \|R(x)\| . \tag{8}$$

The coarse model response is denoted by $c(z)$, where z represents the control variable of the coarse model, e.g. a coarser representation of x. Correspondingly, $c(z)$ is a coarse representation of the fine model residual $R(x)$. The coarse model optimization problem is defined as

$$\mathbf{x}^{\star}_{coarse} = \arg\min_{x} \left\| p^{-1}(c(p(x))) - q_k \right\| \tag{9}$$

with a suitable mapping p from the fine model design space to the coarse model design space. In this work, we use radial basis function interpolation for the mapping p. The design specification q_k of the coarse model is iteratively updated during the minimization process of the fine model. k represents the iteration counter of the manifold mapping algorithm. In the remainder of this work, we use the short notation

$$R_c(x) := p^{-1}(c(p(x))).$$

The manifold mapping algorithm (IQN-ILS-MM) [13] is based on the observation that the solution of the fine model optimization problem can be computed based on the coarse model and a suitable mapping S between fine and coarse model response:

$$\mathbf{x}^{\star}_{fine} = \arg\min_{x} ||SR_c(x)|| \quad (10)$$

The mapping S has the properties that $R_c(x^{\star}_{fine})$ is mapped to $R(x^{\star}_{fine})$ and that the tangent plane for R_c at $R_c(x^{\star}_{fine})$ is mapped to the tangent plane for R at $R(x^{\star}_{fine})$. Such a mapping is called a manifold mapping and can be defined as an affine mapping

$$S(R_c(x)) = R(x^{\star}_{fine}) + \bar{S}(R_c(x) - R_c(x^{\star}_{fine})) \text{ with } \bar{S} = J_R(x)J^{\dagger}_{R_c}(x) \quad (11)$$

where J_R is the Jacobian of the fine model and $J^{\dagger}_{R_c}$ the pseudo-inverse of the Jacobian of the coarse model. J_R and $J^{\dagger}_{R_c}$ are, however, unavailable for black-box solvers used in the FSI problem such that we have to use approximations and solve the respective approximative version of (10) several times in an iterative manner. Using the definition (11), we can rewrite the optimization problem and get

$$x^{\star}_{fine} = \arg\min_{x} \left\| R_c(x) - R_c(x_k) + \bar{S}^{\dagger} R(x_k) \right\| \quad (12)$$

where $\bar{S}^{\dagger}$ is the pseudo-inverse of $\bar{S}$. Instead of approximating J_R and J_{R_c}, we directly approximate the mapping $\bar{S}^{\dagger}$ by a sequence $\bar{S}^{\dagger}_k$ similar to the approximation of the inverse Jacobians in the quasi-Newton approaches presented in the previous section. That is, we use the matrices

$$F_k = \left[\Delta R^k_{k-1}, \cdots, \Delta R^k_0\right] \text{ with } \Delta R^k_j = R(x^k) - R(x^j) \text{ and} \quad (13)$$

$$C_k = \left[\Delta R^k_{c,k-1}, \cdots, \Delta R^k_{c,0}\right] \text{ with } \Delta R^k_{c,j} = R_c(x^k) - R_c(x^j). \quad (14)$$

The approximation of the pseudo-inverse of the manifold mapping function $\bar{S}^{\dagger}$ is given by

$$\bar{S}^{\dagger}_k = C_k F^{\dagger}_k \quad (15)$$

as described in [13][2]. This results in the iteration

$$x_{k+1} = \arg\min_{x} \left\| R_c(x) - R_c(x_k) + \bar{S}^{\dagger}_k R(x_k) \right\|.$$

Each iteration requires the evaluation of a fine-model response $R(x_k)$. The manifold mapping procedure is started with an extrapolation step of the solutions of the previous time steps in order to initialize the algorithm with a good initial guess. The number n of degrees of freedom on the fluid–structure interface is in general

[2]For the computation of $C_k F^{\dagger}_k$, we can either use a QR decomposition of F_k as we do in our quasi-Newton approaches to compute $V^{\dagger}_k$ or singular value decompositions of C_k and F_k as described in [5]

much larger than the number of columns of C_k and F_k. Therefore, the computational cost of the manifold mapping technique is limited. In order to reduce the number of coarse model optimizations, the matrices C_k and F_k can be combined with the information from r previous time steps as in the IQN-ILS(r) approach [12].

Each coarse optimization problem requires one execution of the fine model to compute $R(x^k)$. The coarse optimization problem can be solved with a coupling scheme of the users choice. We use the quasi-Newton methods described in the previous section. Note that the coupling scheme needs to meet the design specification $\mathbf{q}_k$, which can be included in the formulation of the residual for the FSI problem.

4 Comparison of the Methods — Numerical Results

In this section, we present results for the coupling approaches presented in the previous section and two FSI scenarios, a two-dimensional flow around a cylinder flap and a three-dimensional flow through a flexible tube.

The cylinder flap test case is referred to as FSI3 benchmark scenario in [25] and simulates a two-dimensional flow around a fixed cylinder with attached flexible cantilever, placed in the middle of the flow channel with a small vertical offset to foster oscillations. The geometry and simulation results are shown in Fig. 1 (a). The computational domain is $2.5\,\mathrm{m} \times 0.41\,\mathrm{m}$ in size. The fluid flow is driven by a parabolic velocity profile v with mean inflow velocity $\bar{v}$ $=0.2\,\mathrm{m\,s^{-1}}$ at the left boundary and free outflow at the right boundary. An incompressible fluid with density $\rho_f = 1 \times 10^3\,\mathrm{kg\,m^{-3}}$, a dynamic viscosity of $\nu_f = 1 \times 10^{-3}\,\mathrm{m^2\,s^{-1}}$, and a Reynolds number $Re = 200$ is considered. The flexible cantilever is modeled using the Saint-Venant–Kirchhoff material model with a Young's modulus of E $=1.4 \times 10^6\,\mathrm{N\,m^{-2}}$, a Poisson's ratio of $\nu_s = 0.4$, and a density of $\rho_s =$

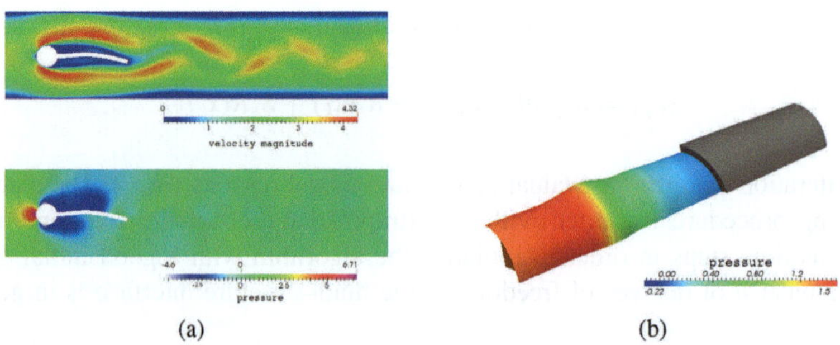

Fig. 1 Geometry and results for the simulated scenarios. (a) Two-dimensional flow around fixed cantilever. Velocity magnitude and pressure contours in the fluid domain at $t = 0.697\,s$. (b) Wave propagation in a three-dimensional elastic tube. Geometry and pressure contours on the fluid–structure interface at $t = 3.0 \cdot 10^{-1}\,s$

$1 \times 10^3\,\mathrm{kg\,m^{-3}}$. The flow solver uses a mesh with 23 924 cells, the solid mesh consists of 328 cells. A time frame of T =2 s is simulated in steps of 1×10^{-3} s after the flow is fully developed without structure movement. The dynamic movement of the cantilever converges to a periodic oscillation with an average period of 128 time steps, i. e., 0.128 s.

The flexible tube example [2, 11, 17] simulates a wave propagating in a straight elastic tube with a length of 0.05 m, a wall thickness of 0.001 m, and an inner diameter of 0.01 m. Both ends of the tube are fixed. From $t = 0$ s until $t = 0.003$ s, the pressure at the left boundary is set to 1333.2 Pa. After this time, the inlet pressure is set to zero. At the right boundary, the pressure is zero at all times. The fluid has a density of 10^3 kg/m^3 and a dynamic viscosity of $3.0 \cdot 10^{-3}$ Pa · s. The structure is assumed to be elastic and compressible with a density of $1.2 \cdot 10^3$ kg/m^3, the Young's modulus $3.0 \cdot 10^5$ N/m^2, and the Poisson's ratio 0.3. The flow solver uses a mesh with 20 800 cells, whereas the structure mesh has 6 400 cells. The initial solution of the displacement is computed based on an extrapolation from previous time steps for each numerical method under consideration.

All numerical experiments were conducted using the coupling library preCICE and the simulation toolbox OpenFOAM. We give a brief summary of both:

preCICE[3] is a library for flexible numerical coupling of single-physics solvers. It is developed at the Technische Universität München and the University of Stuttgart. preCICE uses a partitioned black-box coupling approach and offers a wide variety of runtime configurable aspects of numerical coupling like serial and parallel as well as explicit and implicit coupling schemes. For more information, refer to [18].

OpenFOAM[4] was used for the fluid and the structure simulations based on a second order finite volume discretization of the incompressible Navier–Stokes equation for the fluid. Instead of the standard pressure implicit with splitting of operator (PISO) algorithm, it uses a coupled solution algorithm as described in [9]. For more information, refer to, e.g., [4].

Convergence Criterion. To test convergence, we use a criterion that checks both forces and displacements based on the relative residuals

$$\frac{\|x_d^k - \tilde{x}_d^k\|_2}{\|x_d^k\|_2} \overset{!}{<} \epsilon_{rel} \quad \wedge \quad \frac{\|x_f^k - \tilde{x}_f^k\|_2}{\|x_f^k\|_2} \overset{!}{<} \epsilon_{rel} \text{ with } \epsilon_{rel} = 10^{-5}$$

[3]http://www5.in.tum.de/wiki/index.php/PreCICE_Webpage

[4]http://www.openfoam.org/

Table 1 Two-dimensional flow around a fixed cylinder flap (FSI3 benchmark). Averaged iteration counts per time step for the S-system (1) and the V-system (2), the IQN-ILS method, the IQN-ILS mapping with acceleration based on manifold mapping and the parameter-free IQN-MVJ method.

		Reuse 0		Reuse 4		Reuse 8	
System	Method	n_f	n_c	n_f	n_c	n_f	n_c
S	IQN-ILS	14.8	0.0	5.4	0.0	5.0	0.0
S	IQN-ILS-MM	6.4	48.5	6.0	34.9	5.9	34.4
S	IQN-MVJ	6.1	0.0	—	—	—	—
V	IQN-ILS	26.2	0.0	7.7	0.0	6.5	0.0
V	IQN-ILS-MM	8.9	98.9	4.8	30.9	5.1	34.3
V	IQN-MVJ	7.6	0.0	—	—	—	—

where

$$\tilde{x}_d^k = \begin{cases} S \circ F(x_d^k) & \text{for the S} - \text{system,} \\ S(x_f^k) & \text{for the V} - \text{system} \end{cases} \quad \text{and} \quad \tilde{x}_f^k = \begin{cases} F \circ S(x_f^k) & \text{for the S} - \text{system,} \\ F(x_d^k) & \text{for the V} - \text{system.} \end{cases}$$

S-System versus V-System. Switching our coupling pattern from the S-system (1) to the V-system (2) induces more parallelism. More specifically, the time for one iteration is reduced by a factor of two if we have parallelized both solvers to appropriate numbers of cores such that their runtime per time step is equal. For the total time step, we thus benefit from the enhanced parallelism of the V-system if the number of iterations required per time step is less than twice the number of iterations required for the S-system. Tables 1 and 2 show that the iteration counts using the IQN-ILS method for both systems are comparable for the FSI3 test case (Table 1) if an appropriate number of previous time steps is reused. The results for the flexible tube (Table 2) show a disadvantage of the V-system in terms of convergence. For the more sophisticated IQN method accelerated with the manifold mapping (IQN-ILS-MM) and the robust parameter-free variant IQN-MVJ, the V-system yields iteration counts (of the fine model) comparable to those of the S-system for both scenarios. Note that these good iteration counts for the V-system are only due to the efficiency of the quasi-Newton approach. The pure fixed-point iteration (if converging) for the V-systems obviously would require exactly twice as many iterations as for the S-system.

Convergence Boost by Manifold Mapping. For IQN-ILS-MM, we use coarse models of the fluid with 2 600 cells (tube) / 5 981 cells (FSI3), respectively, and 800 cells (tube) / 82 cells (FSI3) for the structure. In addition, viscous effects are ignored for the coarse fluid model. Table 2 shows the iteration counts. The manifold mapping technique substantially reduces the number of fine model iterations for the S-system as well as for the V-system. Note that the number of fine model evaluations for the flexible tube was reduced to only a slightly higher number than the best result

Table 2 Three-dimensional flow through a flexible tube. Averaged iteration counts per time step for the S-system (1) and the V-system (2), the IQN-ILS method, the IQN-ILS mapping with acceleration based on manifold mapping and the parameter-free IQN-MVJ method.

System	Method	Reuse 0		Reuse 8		Reuse 16		Reuse 24	
		n_f	n_c	n_f	n_c	n_f	n_c	n_f	n_c
S	IQN-ILS	17.0	0.0	6.8	0.0	5.9	0.0	5.8	0.0
S	IQN-ILS-MM	6.9	55.4	5.4	34.0	5.5	32.0	5.9	34.9
S	IQN-MVJ	7.19	0.0	—	—	—	—	—	—
V	IQN-ILS	30.4	0.0	11.3	0.0	9.8	0.0	9.2	0.0
V	IQN-ILS-MM	6.9	95.0	4.8	41.0	4.6	35.4	4.5	32.9
V	IQN-MVJ	8.68	0.0	—	—	—	—	—	—

for the S-system, whereas the number of iterations for the IQN-ILS method alone applied to the V-system is almost twice as high as for the S-system.

Parameter-Free Robustness with MVJ. Lines three and six in Tables 1 and 2 show the averaged iteration numbers for the coupling using the Jacobian updates based on the minimization of the distance to the Jacobian of the previous time step. For both sequential and the parallel coupling, this method yields good convergence with no reuse of previous time step information. Thus, we consider the MVJ method as a very robust, generally applicable, and efficient coupling method which could, in a further step, also be combined with the manifold mapping algorithm to further increase its efficiency.

5 Conclusions

An overview of state-of-the-art coupling schemes for the partitioned simulation of FSIs has been given. By considering the fluid and structure solvers as black-boxes, the convergence of the coupling iterations is often hampered by slow convergence. Quasi-Newton methods and the multi-level technique manifold mapping have been applied to solve this convergence problem. The standard staggered execution of fluid–structure coupling iterations is accelerated by those methods, as well as a parallel or simultaneous execution of the flow and structure solvers. The IQN-MVJ method incorporates the information from previous time steps implicitly, avoiding a manual tuning of the number of time steps reused. The manifold mapping technique results in the least amount of coupling iterations per time step and is not sensitive to the number of time steps reused. Optimal combinations of the presented methods represent a very powerful, flexible, and robust toolbox for partitioned fluid–structure simulations and presumably also other multi-field simulations.

References

1. Barker, A.T., Cai, X.C.: Scalable parallel methods for monolithic coupling in fluid–structure interaction with application to blood flow modeling. J. Comput. Phys. **229**(3), 642–659 (2010)
2. Bathe, K.J., Ledezma, G.A.: Benchmark problems for incompressible fluid flows with structural interactions. Comput. Struct. **85**(11–14), 628–644 (2007)
3. Bazilevs, Y., Korobenko, A., Deng, X., Yan, J., Kinzel, M., Dabiri, J.: Fluid-structure interaction modeling of vertical-axis wind turbines. J. Appl. Mech. **81**(8), 081,006 (2014)
4. Blom, D.S., van Zuijlen, A.H., Bijl, H.: Acceleration of strongly coupled fluid-structure interaction with manifold mapping. In: Oñate, E., Oliver, X., Huerta, A. (eds.) Proceedings of the 11th World Congress on Computational Mechanics, 5th European Congress on Computational Mechanics, 6th European Congress on Computational Fluid Dynamics, pp. 4484–4495 (2014)
5. Blom, D.S., Uekermann, B., Mehl, M., van Zuijlen, A., Bijl, H.: Multi-level acceleration of parallel coupled partitioned fluid-structure interaction with manifold mapping. In: Mehl, M., Bischoff, M., Schäfer, M. (eds.) Recent Trends in Computational Engineering - CE 2014. Lecture Notes in Computational Science and Engineering. Springer, Berlin, Heidelberg (2015)
6. Blom, D.S., van Zuijlen, A.H., Bijl, H.: Multi-level acceleration with manifold mapping of strongly coupled partitioned fluid-structure interaction (2015, submitted)
7. Bogaers, A.E.J., Kok, S., Reddy, B.D., Franz, T.: Quasi-Newton methods for implicit black-box FSI coupling. Comput. Methods Appl. Mech. Eng. **279**, 113–132 (2014)
8. Bungartz, H.J., Lindner, F., Mehl, M., Uekermann, B.: A plug-and-play coupling approach for parallel multi-field simulations. In: Computational Mechanics, pp. 1–11 (2014). doi:10.1007/s00466-014-1113-2. http://dx.doi.org/10.1007/s00466-014-1113-2
9. Darwish, M., Sraj, I., Moukalled, F.: A coupled finite volume solver for the solution of incompressible flows on unstructured grids. J. Comput. Phys. **228**(1), 180–201 (2009). doi:10.1016/j.jcp.2008.08.027
10. Degroote, J.: Partitioned simulation of fluid-structure interaction. Arch. Comput. Meth. Eng. **20**(3), 185–238 (2013)
11. Degroote, J., Bruggeman, P., Haelterman, R., Vierendeels, J.: Stability of a coupling technique for partitioned solvers in FSI applications. Comput. Struct. **86**, 2224–2234 (2008)
12. Degroote, J., Bathe, K.J., Vierendeels, J.: Performance of a new partitioned procedure versus a monolithic procedure in fluid-structure interaction. Comput. Struct. **87**(11–12), 793–801 (2009)
13. Echeverría, D., Hemker, P.W.: Space mapping and defect correction. Comput. Methods Appl. Math. **5**(2), 107–136 (2005)
14. Echeverría, D., Hemker, P.W.: Manifold mapping: a two-level optimization technique. Comput. Vis. Sci. **11**(4–6), 193–206 (2008). doi:10.1007/s00791-008-0096-y. http://dx.doi.org/10.1007/s00791-008-0096-y
15. Farhat, C., Lesoinne, M.: Two efficient staggered algorithms for the serial and parallel solution of three-dimensional nonlinear transient aeroelastic problems. Comput. Methods Appl. Mech. Eng. **182**(3–4), 499–515 (2000)
16. Felippa, C., Park, K., Farhat, C.: Partitioned analysis of coupled mechanical systems. Comput. Methods Appl. Mech. Eng. **190**(24–25), 3247–3270 (2001)
17. Fernández, M.Á., Moubachir, M.: A Newton method using exact Jacobians for solving fluid-structure coupling. Comput. Struct. **83**(2–3), 127–142 (2005)
18. Gatzhammer, B.: Efficient and flexible partitioned simulation of fluid-structure interactions. Dissertation, Institut für Informatik, Technische Universität München, München (2015)
19. Gee, M., Küttler, U., Wall, W.: Truly monolithic algebraic multigrid for fluid-structure interaction. Int. J. Numer. Methods Eng. **85**(8), 987–1016 (2011). doi:10.1002/nme.3001
20. Lindner, F., Mehl, M., Scheufele, K., Uekermann, B.: A comparison of various quasi-Newton schemes for partitioned fluid-structure interaction. In: Proceedings of 6th International Conference on Computational Methods for Coupled Problems in Science and Engineering, Venice, pp. 1–12 (2015)

21. Marklund, P.O., Nilsson, L.: Simulation of airbag inflation processes using a coupled fluid structure approach. Comput. Mech. **29**(4–5), 289–297 (2002)
22. Mehl, M., Uekermann, B., Bijl, H., Blom, D.S., Gatzhammer, B., van Zuijlen, A.H.: Parallel coupling numerics for partitioned fluid-structure interaction simulations. SIAM Sci. Comput. (2013, submitted)
23. Stein, K., Benney, R., Kalro, V., Tezduyar, T.E., Leonard, J., Accorsi, M.: Parachute fluid-structure interactions: 3-D computation. Comput. Methods Appl. Mech. Eng. **190**(3–4), 373–386 (2000)
24. Tezduyar, T.E., Sathe, S., Keedy, R., Stein, K.: Space–time finite element techniques for computation of fluid–structure interactions. Comput. Methods Appl. Mech. Eng. **195**(17), 2002–2027 (2006)
25. Turek, S., Hron, J.: Proposal for numerical benchmarking of fluid-structure interaction between an elastic object and laminar incompressible flow. In: Bungartz, H.J., Schäfer, M. (eds.) Fluid-Structure Interaction. Modelling, Simulation, Optimisation, vol. 53, pp. 371–385. Springer, Berlin, Heidelberg (2006)
26. Uekermann, B., Bungartz, H.J., Gatzhammer, B., Mehl, M.: A parallel, black-box coupling algorithm for fluid-structure interaction. In: Proceedings of 5th International Conference on Computational Methods for Coupled Problems in Science and Engineering, Ibiza, pp. 1–12 (2013)

Rhino 3D to Abaqus: A T-Spline Based Isogeometric Analysis Software Framework

Yicong Lai, Lei Liu, Yongjie Jessica Zhang, Joshua Chen, Eugene Fang, and Jim Lua

Abstract This chapter presents a novel software framework for T-spline based isogeometric analysis (IGA), interfacing between computer aided design (CAD) and finite element analysis environment for "design-through-analysis" concept. The framework is built on *Rhinoceros* 3D (Rhino) and *SIMULIA* Abaqus (Abaqus). This framework has three primary steps: creating CAD surface in Rhino with T-spline Plugin, converting surface T-spline representations into volumetric T-spline, and performing analysis with T-spline models in Abaqus through its user element subroutine. Both 2D and 3D examples are provided in the end to demonstrate our T-spline based IGA software framework.

1 Introduction

Non-uniform rational B-spline (NURBS) is the standard mathematical representation for geometry in most commercial computer aided design (CAD) software. Finite element analysis (FEA) has been widely used in most commercial computer aided engineering (CAE) software. Geometry designed in CAD software must undergo an approximating discretization process known as meshing, so that it can be used for analysis in CAE software. A seminal effort to integrate CAD and CAE, termed isogeometric analysis (IGA) [1], gained great attention as a new "design-through-analysis" methodology. It enabled direct analysis of the designed geometry without meshing.

T-spline [2] was introduced as a generalization of NURBS allowing local refinement. The introduction of the *Autodesk* T-Splines Plugin for Rhino in 2004 (http://www.tsplines.com/) provides designers the tool to create and manipulate T-spline geometries. T-spline basis functions have been incorporated into IGA [3], further generalizing the "design-through-analysis" framework. Analysis-suitable

Y. Lai • L. Liu • Y.J. Zhang (✉) • J. Chen
Department of Mechanical Engineering, Carnegie Mellon University, Pittsburgh, PA 15213, USA
e-mail: jessicaz@andrew.cmu.edu

E. Fang • J. Lua
Global Engineering and Materials, Inc, Princeton, NJ 08540, USA

Y. Bazilevs, K. Takizawa (eds.), *Advances in Computational Fluid-Structure Interaction and Flow Simulation*, Modeling and Simulation in Science, Engineering and Technology, DOI 10.1007/978-3-319-40827-9_21

T-splines in IGA were characterized [4], and various methods were developed to construct volumetric T-splines for IGA [5–9]. Open-source NURBS based IGA implementations [10, 11] were introduced, but they are primarily used in research. Recent efforts have increased accessibility of T-spline based IGA to industry. Data structures for T-spline based IGA were developed [12, 13]. Moreover, NURBS based IGA was implemented in Abaqus [14]. Although the improvements of IGA are vast, it is still at the beginning stage in industry. So far, T-spline based IGA implementations have not yet been available in many commercial software like Abaqus.

In this chapter, we present a novel CAD-CAE software framework for T-spline based IGA. Since Rhino and Abaqus are built for engineering design and analysis specifically, here we develop a T-spline based IGA software framework upon them. Users can benefit from the strengths of both software. This framework allows:

- Boundary value problem (BVP) specification on T-spline geometry;
- Volumetric T-spline construction from surface representation;
- Efficient and compact trivariate T-spline data structure; and
- Abaqus T-spline IGA user element subroutine based on Bézier extraction.

The reminder of this chapter is organized as follows. Section 1 overviews the framework and the pipeline. Section 3 describes surface T-spline geometry creation and BVP specification. Section 4 presents T-spline data structure. Section 5 explains how volumetric T-splines are constructed from surface T-splines. Section 6 presents our T-spline Abaqus user subroutine. Finally, section 7 shows numerical results and draws conclusions.

2 Software Framework and Pipeline Overview

To integrate Rhino with Abaqus, the framework uses Rhino with the T-spline plugin to create and manipulate T-spline geometries, and uses Abaqus for analysis through its user element subroutines (UEL/UELMAT). The framework incorporates two self-developed plugins as shown in Fig. 1, the Rhino plugin in the grey and blue blocks and the Abaqus plugin in the red block. This framework requires the following software environment:

- *Rhinoceros* 3D, Version 5 or newer;
- *Autodesk* T-splines Plugin for Rhino, Version 3.4 or newer;
- *SIMULIA* Abaqus Unified FEA, Version 6.0 or newer; and
- *Intel* FORTRAN, Version 11 or newer.

As shown in Fig. 1, surface geometry is created in Rhino first. With the Rhino plugin, BVP problem is specified and the geometry is pre-processed, ending up with Abaqus input files. Depending on whether a 2D or 3D geometry is needed, users can choose whether to construct volumetric T-splines. For a 2D case, a Rhino .iga file is directly converted into Abaqus .inp file and .bezier file. For a 3D case, .TSM files

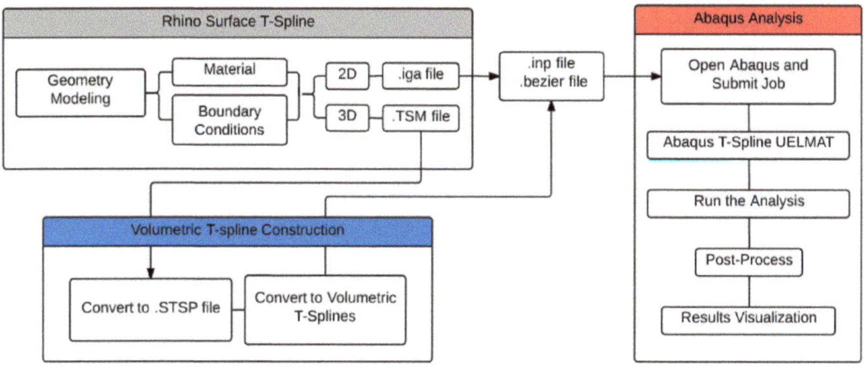

Fig. 1 Rhino 3D to Abaqus Software Pipeline Overview.

are saved and converted to our .STSP file for volumetric T-spline construction, and then Abaqus .inp file and .bezier file are generated for analysis. When the analysis is completed through the Abaqus plugin, a post-processing function is called in the Abaqus plugin to generate the .odb file for visualization. Based on the results, the user can go back to Rhino to refine or modify the geometry.

3 Geometry Generation and BVP Specification

Our software framework supports both surface and volumetric T-splines. Since Rhino only supports surface modeling, surface T-splines must undergo a conversion process to become volumetric T-splines. Currently, we support two types of volumetric T-spline geometries. The first type is a 3D geometry created by sweeping a surface T-spline patch, and the second type is the geometry with genus-zero cube topology which can be created using the conformal parametric mapping method[7]. Fig. 2 shows three T-spline surface models designed in Rhino. (a) and (b) show 2D open surfaces which can be used in analysis directly or converted to volumetric T-spline via sweeping. (c) shows a structure with genus-zero topology and six patches, which can be converted into volumetric T-spline via parametric mapping.

In Rhino, all the boundary conditions are applied on relevant control points. In the pre-processing step, the Rhino plugin extracts the information of these control points and writes to Abaqus .inp file. To apply boundary conditions, we use the T-Spline plugin selection node sets [15] to choose desired control points. The platform is currently limited to specifying Dirichlet boundary conditions, but extending to other boundary conditions is straightforward. For the material properties, simple isotropic material properties are supported currently. The users can specify the Young's Modulus E and the Poisson's Ratio ν.

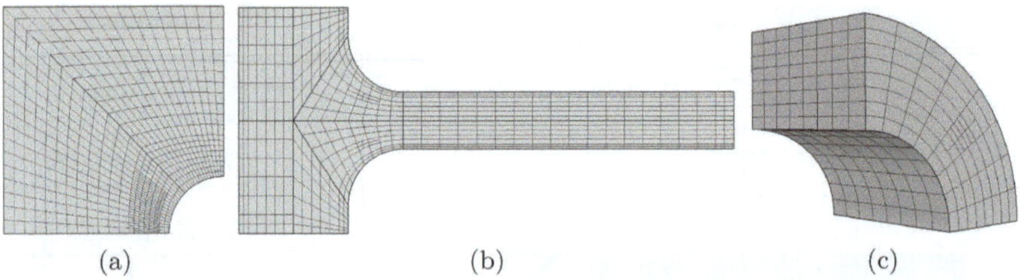

Fig. 2 (a) A 2D quarter of a plate with a hole; (b) a 2D notch structure prepared for sweeping; and (c) a quarter cylinder with genus-zero cube topology.

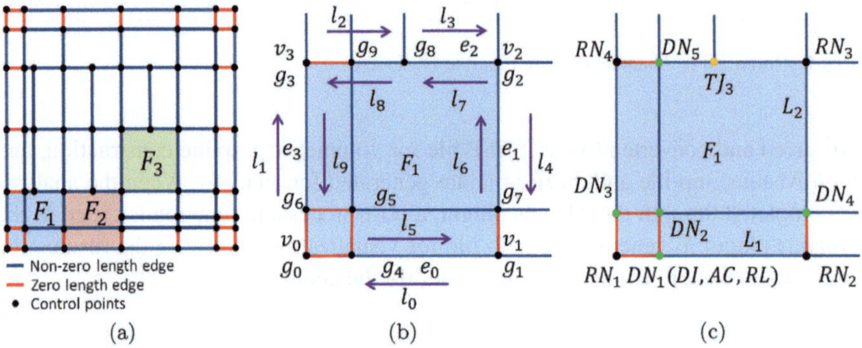

Fig. 3 (a) A T-mesh with three representative faces/elements, the corner face F_1, the edge face F_2, and the interior face F_3; (b) a corner face in *.TSM* data format. Abbreviations are: *l-link, v-vertex, g-grip, F-face, e-edge*; and (c) a corner element in *.STSP* data format. Abbreviations are: *AC-Associate Corner, DI-Duplicated Node Index, DN-Duplicated Node, F- Element, L-Parametric Length, RL-Relative Location, RN-Regular Node, TJ-T-Junction.*

4 T-Spline Data Structure

In this section, we first explain the commercial *.TSM* data structure used in Rhino. Then, we introduce our *.STSP/.VTSP* data structure and explain the conversion between them.

4.1 TSM Data Format

The *.TSM* file originated from "half edge" data structure[13] which contains parametric and physical information of a T-spline control mesh (T-mesh). Fig. 3(a) shows three representative faces or elements of a T-mesh. F_1 is a corner face, F_2 is an edge face, and F_3 is an interior face. Fig. 3(b) takes a corner face as an example to show detailed components in a *.TSM* face, including links, vertices, grips, faces, and edges.

Link {previous, next, opposite, vertex, face, edge, type}. *Links* tie the topology of a surface together. Every edge, face, or vertex refers to a link to determine its location. The first 3 values indicate the id of the previous, next, and opposite links. For example, in Fig. 3(b), the previous, next, and opposite links of l_5 are l_9, l_6, and l_0, respectively. Each link has an associated vertex, face, and edge which are the three values following the opposite link. For example, l_5 originates from v_0 and belongs to F_1, associating to e_0. The type indicates what kind of node the link starts from, either a regular node (Type 0), a T-junction (Type 1), or an L-junction (Type 2). For example, the type of l_5 is 0 since it starts from a regular node while the type of l_8 is 1 since it starts from a T-junction. Note that the T-junction can break one edge into multiple links. Here, the T-junction g_8 breaks the edge e_2 into l_7 and l_8.

Vertex {link, direction}. A *vertex* is the origin of a link. The direction that the link points away from the vertex is stored as one of the North (N), South (S), East (E), and West (W). For example, v_0, v_1, v_2, and v_3 are four vertices of the corner face F_1 in Fig. 3(b).

Grip {x, y, z, weight}. A *grip*, equivalent to a control point, is defined by its x, y, z coordinates and weight. A grip may be parent of one or more other grips. This kind of grips is also known as *compound grips*. For example, in Fig. 3(b), g_0 is a compound grip which is the parent of g_4, g_5, and g_6.

Face {start link, flag}. A *face* is used to represent an element in the T-mesh, and it is defined by a start link and a flag. The connectivity of a face is given by the counter-clockwise cycle of links at its borders, starting from its start link. The *flag* of a face stores various properties of this face like if the face is hidden or not.

Edge {link, interval}. An *edge* connects vertices in the T-mesh. It is defined by a link and an interval. Its link is one of the two links running along it, while the "interval" is its parametric length.

4.2 STSP/VTSP Data Format and Conversion

Let us first define four types of control points for our T-spline data structure, including the regular node, the duplicated node, the T-junction node, and the extraordinary node. A *regular node* is a control point with valence 4. A *duplicated node* is a duplication of its neighbor, see the green dots in Fig. 3(c). A *T-junction* is a control point that is analogical to the hanging nodes in classical finite elements, see the orange dot. For T-junction nodes, we record the edge it locates as well as its index. An *extraordinary node* is an interior control point with valence other than 4, and not a T-junction. In our data structure, regular node, T-junctions, and extraordinary nodes are used to describe the connectivity of elements. Regular nodes are often the parents of duplicated nodes. A corner regular node is the parent of three duplicated nodes while a boundary regular node is the parent of one duplicated node. For example, in Fig. 3(c), RN_1 is the parent of DN_1, DN_2, and DN_3, and RN_2 is the parent of DN_4. According to the algorithm introduced in [16], extraordinary nodes cannot be the parents of duplicated nodes.

The *.STSP/.VTSP* data structure is efficient in storing surface and volumetric T-spline information. This data structure is designed especially for the convenience of extracting knot vectors. It is compact, concise, and easy to interpret and fit into the commercial software. Our data format contains two types of data, the control point data and the element data. The former contains the basic information of T-spline control points, and the latter contains connectivity, parametric length, T-junctions, and duplicated nodes in the element.

1. **Control Point {x, y, z, weight, type}**. A *control point* is defined by its x, y, z coordinates, weight, and type. For the type, we use 0 to represent the regular node, 1 for duplicated node, 2 for T-junction node, and 3 for extraordinary node.
2. **Element {Corner ID$_{1-4}$, Parametric Length$_{1-2}$, T-junction ID$_{1-4}$, Number of Duplicated Nodes, (Duplicated Node ID, Associated Corner, Relative Location)$_{1-totalnumber}$}**. The first four values are the indices of the four corners of this element. The next two values specify the parametric length of the edges. The following four values are the T-junction indices for each edge. If there is a T-junction on one edge, we record its index at the corresponding position, otherwise we put "-1" there. For example, in Fig. 3(c), there is a T-junction on the third edge. Assuming its index is TJ_3, we record the T-junctions as "-1 -1 TJ_3 -1". Following the T-junctions, the next value indicates the total number of duplicated nodes, which is 5 for a corner element and 2 for an edge element. The remaining values specify the relative location of the duplicated nodes with respect to the associated corner of this element. A duplicated node has three consecutive values, the global index, the corner in the T-mesh to which it belongs, and the relative location of this node to the corner.

.TSM to .STSP Conversion. To obtain a logical order of the duplicated nodes, we introduce an *STSP* plane to help extract the T-spline information from the *.TSM* patch in Rhino. Fig. 4(a) shows the *STSP* plane with the center point labeled as 8. This plane is a designed pattern to store the relative location of duplicated nodes with respect to their associated corners. The edge interval length between a duplicated node and its associated corner is zero. This information is further used in extracting knot vectors. For open surface topologically equivalent to a unit square, there are four types of corner elements. Thus, the *STSP* plane has four subsquares, each of which can be applied to one certain type of corner element or edge element to store the duplicated nodes information. For example, the green square in Fig. 4(a) can be applied to the corner element in Fig. 4(b) while the orange square can be applied to the corner element in Fig. 4(c). In Fig. 4(b), there are three nodes that have duplicated nodes, g_0, g_1, and g_3. Since g_0 has three duplicated nodes g_4, g_5, and g_6, we first put g_0 at the center of the *STSP* plane which is position 8. In this way, g_4, g_5, and g_6, are located at positions 3, 4, and 5 in the *STSP* plane, respectively. This indicates g_4 is associated with the corner 0 at the relative location of 3 in the *STSP* plane. Thus, we record the relative location of this duplicated node as "4 0 3". Similarly for g_5 and g_6, we record them as "5 0 4" and "6 0 5," respectively. We can apply the same rule to the remaining g_1 and g_3 with only one duplicated node each.

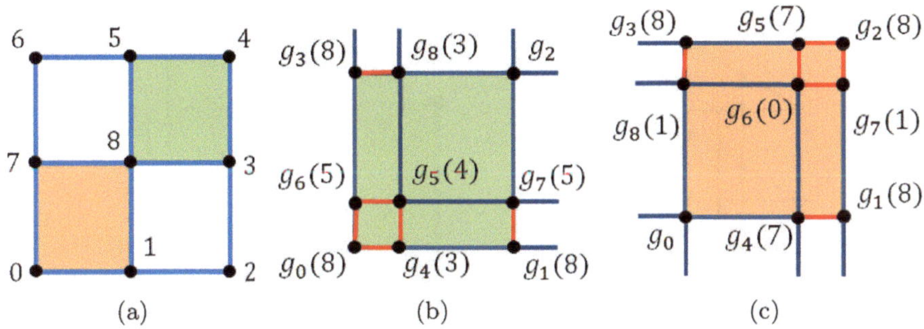

Fig. 4 (a) The *STSP* plane; (b) a corner element applied to the green square in (a); and (c) another corner element applied to the orange square in (a). The red edges have zero length.

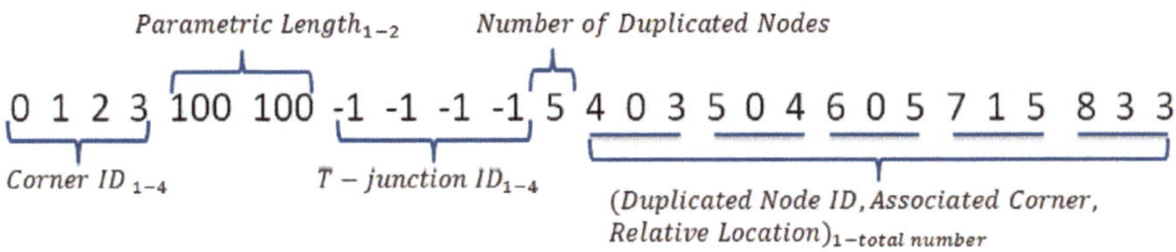

Fig. 5 Sample element data line for the T-mesh element in Fig. 4(a).

We put g_1 and g_3 at the center of STSP plane and g_7 and g_8 can be recorded as "7 1 5" and "8 3 3," respectively. Similarly, Fig. 4(c) shows the relative location of all duplicated nodes in another corner element.

To convert the *.TSM* data to the *.STSP* data, we first follow the face link loop to find all the vertices and grips of this element. The four grips at the corners determine the connectivity of this element. They are also the first four values in our data format. With the links, we extract the edge interval associated with the links and record them as parametric length. Then we loop through each grip to order the information of duplicated nodes. We count the total number of duplicated nodes and then record the relative position using the *STSP* plane. Suppose this element has an edge length of 100 in both parametric directions and has no T-junction, its representation " 0 1 2 3 100 100 -1 -1 -1 -1 5 4 0 3 5 0 4 6 0 5 7 1 5 8 3 3 " is shown in Fig. 5.

After *.TSM* to *.STSP* data format conversion, we obtain all the information of the T-mesh. If only a 2D geometry is required for analysis, we directly extract the knot vectors and calculate the T-spline surface. If there are extraordinary nodes in the T-mesh. We use the interval duplication method [16] to deal with it. If 3D geometry is needed for analysis, we convert *.TSM* file to *.VTSP* file.

.TSM to .VTSP Conversion. We extend the designed pattern from *STSP* plane to *VTSP* cube in 3D case. Fig. 6(a) shows the *VTSP* cube and its labels. In 3D, we put each corner of this element to the center of the cube and for each duplicated node we record its relative location in the cube using the same way as in 2D case. As the example shown in Fig. 6(b), g_6 is associated with corner 1 in the element. After putting corner 1 to the center of the cube which is position 17, g_6

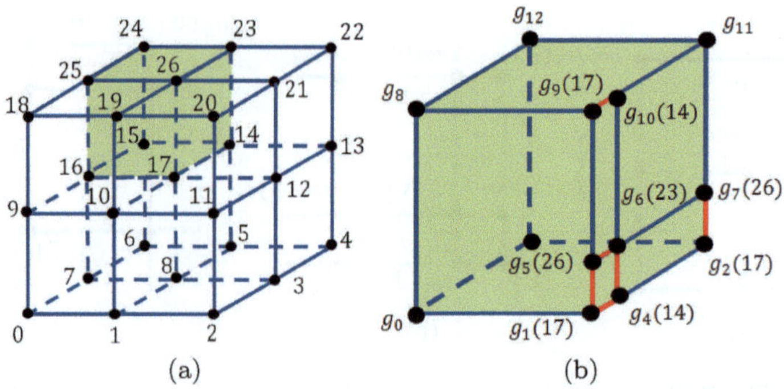

Fig. 6 (a) The *VTSP* cube; and (b) a corner element applied to the green cube in (a). The red edges have zero length.

is located at position 23 automatically. Thus we record the duplicated node g_6 as "6 1 23". Other duplicated nodes can be recorded in the same way. With everything recorded, the information is then used in knot vector extraction and volumetric T-spline construction.

5 Volumetric T-Spline Construction

After data structure conversion, we perform volumetric T-spline construction, using the parametric mapping algorithm [7]. The algorithm first identifies the eight corner nodes on the boundary surface and then maps the surface patch to a unit cube. The interior domain is adaptively subdivided using octree subdivision until each face or edge contains at most one T-junction. Two boundary layers are inserted between the input boundary and subdivision results using the pillowing technique to preserve the input boundary representation. Then, templates are applied around extraordinary nodes to build a valid T-mesh. Finally, the knot vectors are extracted and volumetric T-splines are constructed.

In our approach, instead of inserting two boundary layers, we pillow the 6 patches one by one, resulting in one boundary layer. The reason is that for CAD model with sharp edges, the original method may generate elements with bad quality and introduce unnecessary extraordinary nodes. Our modified method improves the quality and no new extraordinary node is introduced. We use a 16×16 cube to show the difference between these two methods. Fig. 7(a) shows the original input boundary representation. Fig. 7(b) shows the constructed T-splines in Bézier representations using the original algorithm, and Fig. 7(c) shows the results in Bézier representations using the modified algorithm. Note that in the modified method, the one layer inserted is used as the zero-length layer, so they are not extracted for Bézier representations. Finally, the Bézier operators [12] are calculated, which are used for IGA in Abaqus.

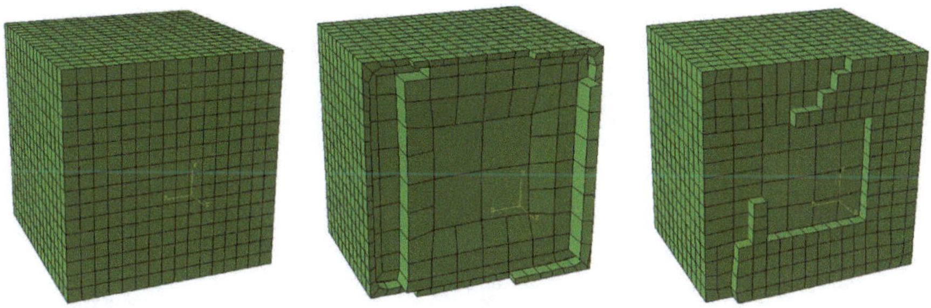

(a) (b) (c)

Fig. 7 Comparison between the original method and the modified method. (a) The boundary representation of the cube; (b) the constructed volumetric T-splines using the original method [7]; and (c) the T-splines using our modified method.

6 Abaqus IGA and Visualization

With the surface or constructed volumetric T-spline, the plugin generates an Abaqus .inp file and a .bezier file. The Abaqus .inp file contains T-spline control point coordinates, element connectivities, and the BVP information. The .bezier file is our self-defined file format containing information of T-spline control point weight and Bézier operators. Both files are read into the UELMAT through the User EXTERNAL DataBase (UEXTERNALDB) in Abaqus. The T-spline based UELMAT uses Bézier operator to compute the T-spline basis functions, we then build the stiffness matrix and force vector for the analysis[12].

Abaqus does not support the visualization of user-defined elements. So post-processing is necessary to project the results to a finite element mesh for visualization. In addition to Abaqus, other software like ParaView can also be used for visualization. For linear elastic problems, the displacement can be extracted from the .fil file generated by Abaqus. Then, the nodal values of each element are computed based on the analysis results.

7 Numerical Results and Conclusion

Using the three geometries in Fig. 2, linear elastic problems with a Young's modulus of 10,000 and a Poisson ratio of 0.3 are solved here, see Fig. 8. The first example shows the displacement and normal strain of a quarter of plate with a hole, which was designed directly in Rhino. The quarter plate is constrained from moving along the horizontal direction at the right edge and vertical direction at the bottom edge. The second example shows the displacement and normal strain of a notch structure under bending, which was created by sweeping the geometry in Fig. 2(b). The generated 2D notch model has one extraordinary node. After dealing with it

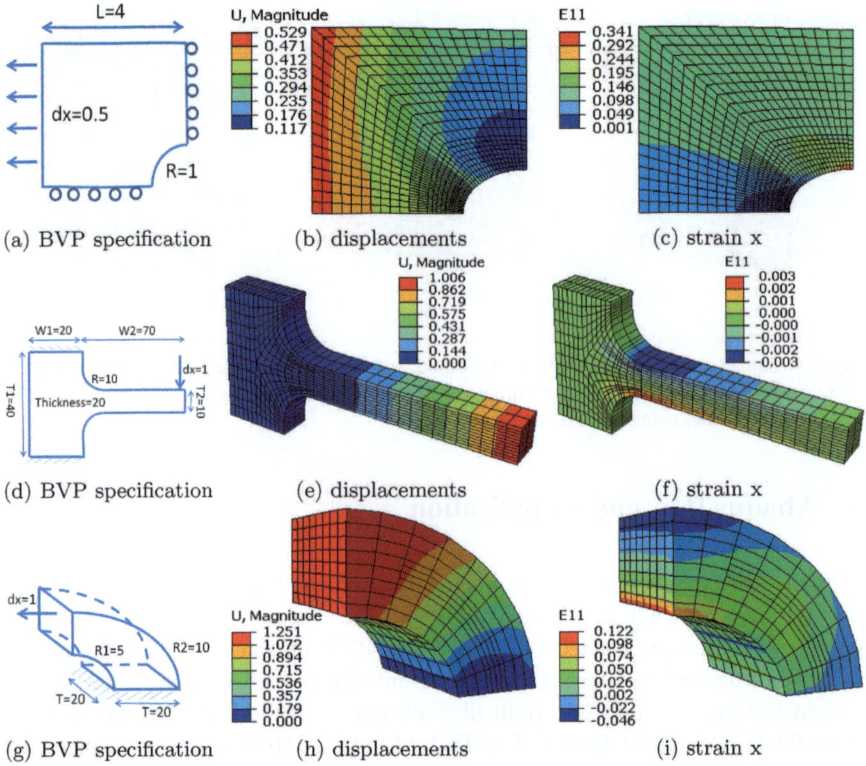

Fig. 8 Analysis results. (a-c) 2D plate with a hole; (d-f) 3D notch structure; and (g-i) 3D quarter cylinder.

with the interval duplication method [16], we sweep the geometry to obtain the volumetric T-spline. There is no further requirements for dealing the generated partial extraordinary nodes and we can directly obtain the Bézier extraction matrix. The geometry is fixed at the top and bottom while a load is applied at the tip of the notch structure. The third example shows the deformation and strain of a quarter cylinder, which was created using the conformal parametric mapping method [7]. The cylinder is fixed at one end and the load is applied at the other end.

In conclusion, this chapter presented a novel integrated CAD-CAE T-spline IGA software framework based on Rhino and Abaqus. The framework can solve both surface and volumetric T-spline problems. In particular, this framework realizes the transformation from Rhino surface T-splines to volumetric T-splines based on efficient *STSP/VTSP* data structures. Three examples were given to demonstrate our software framework. Generalizing this platform to arbitrary topology geometry is possible [6]. We are planning to implement this algorithm, enabling a more general set of geometries in the future.

Acknowledgements This project was supported by an ONR SBIR Phase I project. In addition, the research at CMU was supported in part by the PECASE Award N00014-14-1-0234.

References

1. Hughes, T.J.R., Cottrell, J.A., Bazilevs, Y.: Isogeometric analysis: CAD, finite elements, NURBS, exact geometry, and mesh refinement. Comput. Methods Appl. Mech. Eng. **194**(39), 4135–4195 (2005)
2. Sederberg, T.W., Zheng, J., Bakenov, A., Nasri, A.: T-splines and T-NURCCs. ACM Trans. Graph. **22**(3), 477–484 (2003)
3. Bazilevs, Y., Calo, V.M., Cottrell, J.A., Evans, J.A., Hughes, T.J.R., Lipton, S., Scott, M.A., Sederberg, T.W.: Isogeometric analysis using T-splines. Comput. Methods Appl. Mech. Eng. **199**(5), 229–263 (2010)
4. Li, X., Zheng, J., Sederberg, T.W., Hughes, T.J.R., Scott, M.A.: On linear independence of T-spline blending functions. Comput. Aided Geom. Des. **29**(1), 63–76 (2012)
5. Zhang, Y., Wang, W., Hughes, T.J.R.: Solid T-spline construction from boundary representations for genus-zero geometry. Comput. Methods Appl. Mech. Eng. **249–252**, 185–197 (2012)
6. Wang, W., Zhang, Y., Liu, L., Hughes, T.J.R.: Trivariate solid T-spline construction from boundary triangulations with arbitrary genus topology. Comput. Aided Des. **45**(2), 351–360 (2013)
7. Zhang, Y., Wang, W., Hughes, T.J.R.: Conformal solid T-spline construction from boundary T-spline representations. Comput. Mech. **51**(6), 1051–1059 (2013)
8. Wang, W., Zhang, Y., Xu, G., Hughes, T.J.R.: Converting an unstructured quadrilateral/hexahedral mesh to a rational T-spline. Comput. Mech. **50**(1), 65–84 (2012)
9. Liu, L., Zhang, Y., Hughes, T.J.R., Scott, M.A., Sederberg, T.W.: Volumetric T-spline construction using Boolean operations. Eng. Comput. **30**(4), 425–439 (2014)
10. Vuong, A.V., Heinrich, C., Simeon, B.: ISOGAT: A 2D tutorial MATLAB code for isogeometric analysis. Comput. Aided Geom. Des. **27**(8), 644–655 (2010)
11. De Falco, C., Reali, A., Vázquez, R.: GeoPDEs: a research tool for isogeometric analysis of PDEs. Adv. Eng. Softw. **42**(12), 1020–1034 (2011).
12. Scott, M.A., Borden, M.J., Verhoosel, C.V., Sederberg, T.W., Hughes, T.J.R.: Isogeometric finite element data structures based on Bézier Extraction of T-splines. Int. J. Numer. Methods Eng. **88**(2), 126–156 (2011)
13. Asche, C., Berkhahn, V.: Efficient data structures for T-spline modeling. EG-ICE 2012 International Workshop: Intelligent Computing in Engineering, Technische Universität München, Germany (2012)
14. Elguedj, T., Duval, A., Maurin, F., Al-Akhras, H.: Abaqus user element implementation of NURBS based isogeometric analysis. In: 6th European Congress on Computational Methods in Applied Sciences and Engineering, Vienna (2012)
15. Scott, M.A., Hughes, T.J.R., Sederberg, T.W., Sederberg, M.T.: An integrated approach to engineering design and analysis using the Autodesk T-spline plugin for Rhino3d, ICES REPORT 14-33. The Institute for Computational Engineering and Sciences, The University of Texas at Austin (2014)
16. L. Liu, Y. Zhang, X. Wei, Handling extraordinary nodes with weighted T-spline basis functions. 24th International Meshing Roundtable. Procedia Engineering 124, 161–173 (2015)

A Multithreaded Recursive and Nonrecursive Parallel Sparse Direct Solver

Ercan Selcuk Bolukbasi and Murat Manguoglu

Abstract Sparse linear system of equations often arises after discretization of the partial differential equations (PDEs) such as computational fluid dynamics, material science, and structural engineering. There are, however, sparse linear systems that are not governed by PDEs, some examples of such applications are circuit simulations, power network analysis, and social network analysis. For solution of sparse linear systems one can choose using either a direct or an iterative method. Direct solvers are based on some factorization of the coefficient matrix such as the LU, QR, or singular value decompositions and are known to be robust. Classical preconditioned iterative solvers, on the other hand, are not as robust as direct solvers and finding an effective preconditioner is often problem dependent. Due to their sequential nature, direct solvers often have limited parallel scalability. In this chapter, we present a new parallel recursive sparse direct solver that is based on the sparse DS factorization. We implement our algorithm using MIT's Cilk programming language which is also a part of the Intel C++ compiler. We show the scalability and robustness of our algorithm and compare it to Pardiso direct solver.

1 Introduction

The chip producers can no longer efficiently increase the clock frequency of a processor. The Moore's law which originally stated that the number of transistors doubled every 2 years [18], is, however, still valid. The Moore's law could be translated as the number of cores double every 2 years today. As a result of this paradigm shift, researchers have been working on parallelizing the existing sequential algorithms in order to effectively use all the cores of the processors. A more innovative approach is to design a completely parallel algorithm with parallelism in mind.

Solution of sparse linear systems is required by many applications in science and engineering. Often, the linear solution step is the main performance bottleneck.

E.S. Bolukbasi • M. Manguoglu (✉)
Department of Computer Engineering, Middle East Technical University, Ankara, Turkey
e-mail: ercan@ceng.metu.edu.tr; manguoglu@ceng.metu.edu.tr

Y. Bazilevs, K. Takizawa (eds.), *Advances in Computational Fluid-Structure Interaction and Flow Simulation*, Modeling and Simulation in Science, Engineering and Technology, DOI 10.1007/978-3-319-40827-9_22

There are two main classes of linear system solvers, namely direct and iterative methods. While iterative solvers are not as robust as direct solvers [19], iterative solvers are considered to scale better on parallel computers. There are many parallel sparse direct solver implementations some of the most well known of these are SuperLU [5, 15], MUMPS [1], PARDISO [23–26], and WSMP [7, 8]. All of these direct solvers, however, are based on some form of the classical LU factorization, performed in parallel. After factorization, the system is solved using the computed factors via forward and backward sweeps. For most sparse systems there are some dependencies between unknowns, which limit the parallelism in factorization phase and, due to fill in, this is even more pronounced during the triangular solution phase.

To alleviate these drawbacks of the existing direct solvers we have developed a new general parallel sparse direct solver [2] based on the sparse DS factorization [3]. The idea of the DS factorization is first introduced in [20–22] for banded linear systems which is called the SPIKE algorithm due the structure of the S matrix. A recursive banded DS factorization is introduced in [21] which applies recursion on the S matrix. Our approach, on the other hand, is to apply the recursion on the smaller reduced system. A generalization of the banded DS factorization to sparse linear systems and its hybrid (direct/iterative) implementation, in which the reduced system is solved iteratively, is given in [16, 17].

Given a banded or sparse linear system of equations and number of blocks,

$$Ax = f \tag{1}$$

The DS decomposition factors A into two matrices D and S such that,

$$A = DS \tag{2}$$

where D is just the block diagonal of A. Hence the splitting $A = D + R$ where R is the remaining block off-diagonal entries of A. There is no computation to obtain D. The S matrix, on the other hand, is obtained by $S = D^{-1}A$ or taking advantage of the fact that the block diagonals of S is identity,

$$S = D^{-1}(D + R). \tag{3}$$

We obtain $S = I + G$ which involves solving independent systems in parallel to obtain $G = D^{-1}R$. After obtaining the DS factorization, if we multiply both sides of the Equation (1) with D^{-1} from left,

$$D^{-1}Ax = D^{-1}f, \tag{4}$$

and obtain a new system

$$Sx = g \tag{5}$$

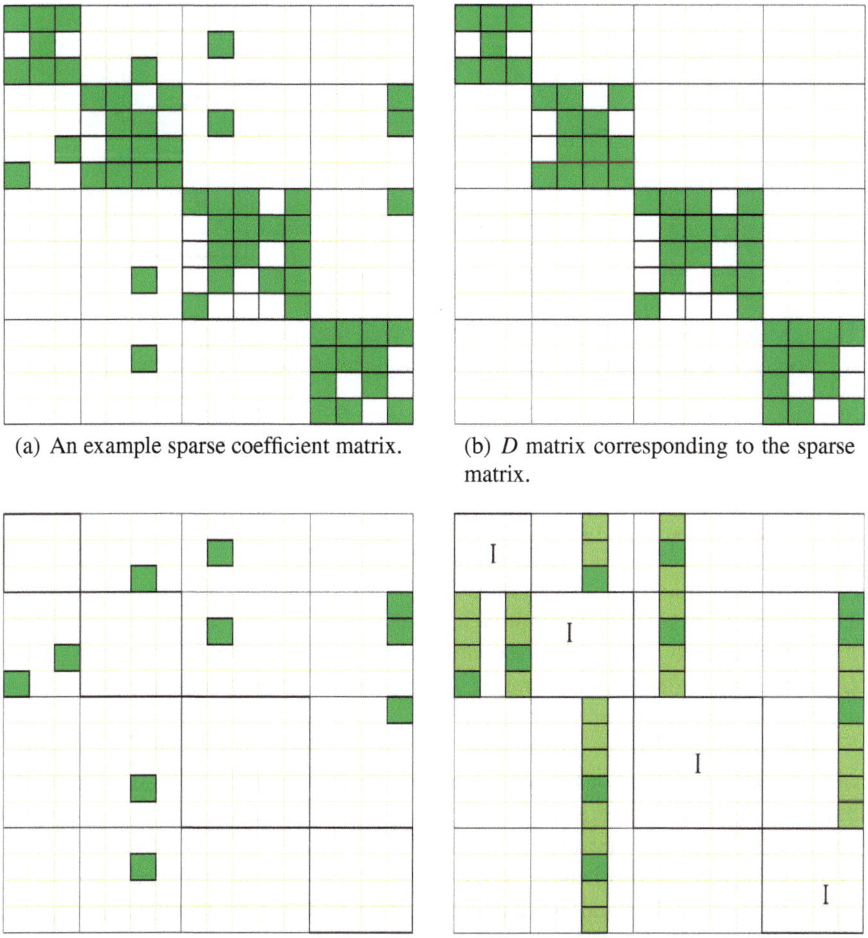

(a) An example sparse coefficient matrix.

(b) D matrix corresponding to the sparse matrix.

(c) R matrix such that $A = D + R$ for the small example.

(d) S matrix for the smaller example, light color indicates the matrix entries can be relatively small if the matrix is diagonally or block diagonally dominant.

Fig. 1 The coefficient matrix and the corresponding D, R, and S matrices using 4 partitions.

where $g = D^{-1}f$. The new system in Equation (5) contains a smaller subsystem which did not exist in the original system of equations. The reduced system is obtained by identifying the indices, c, of the columns which contain nonzeros in R matrix. Then, the reduced system is formed simply by selecting the rows and columns, c, from the S matrix (i.e., $S_{(c,c)}$).

For a small example the sparsity structure of A, D, R, S , and $S_{(c,c)}$ is given in Figures 1(a), 1(b), 1(c), 1(d), 2 respectively.

The reduced system could be formed

$$S_{(c,c)}x_{(c)} = g_{(c)}. \tag{6}$$

Fig. 2 Extracted reduced system $S_{(c,c)}$.

The smaller reduced subsystem can be solved independently and the complete solution can be retrieved also in parallel using one of the following two methods:

$$x = g - G_{(:,c)}x_{(c)} \tag{7}$$

or equivalently

$$x = g - D^{-1}(R_{(:,c)}x_{(c)}). \tag{8}$$

The difference between these approaches is that the first one requires G matrix to be formed completely, while in the second approach we only need to compute a partial G just enough to form $S_{(c,c)}$. However, it involves additional triangular solves which are independent and completely parallel.

Note that the size of the reduced system is highly dependent on the initial structure of the matrix. In fact sparse matrix partitioning tools that are designed to minimize the communication volume in parallel sparse matrix vector multiplication and load balance can be used in sparse DS factorization. The objective of the partitioning in DS factorization is to decrease the size of the reduced system and, hence, to improve the parallel scalability. Furthermore, for the factorization to exist, D must be nonsingular. To achieve this, one can apply a nonsymmetric permutation to strengthen the diagonal entries.

The rest of the chapter is organized as follows. In Section 2, we introduce the new recursive sparse solver and its variations. Programming and computing environments are described and numerical results on sparse linear systems obtained from the University of Florida Sparse Matrix Collection are given in Section 3. Finally, we conclude and summarize the results in Section 4.

2 The Recursive Sparse DS Factorization

Before we apply the recursive algorithm on the linear system, we apply symmetric and nonsymmetric permutations as mentioned before. In the following pseudocode the linear systems $Ax = f$ are assumed to be the permuted system.

Algorithm 1 The Algorithm

1: **procedure** RDS(A, x, f, t) ▷ to solve $Ax = f$ with t number of threads
2: $D + R \leftarrow A$
3: Identify nonzero columns of R and store their indices in c
4: (a) $[G, g] \leftarrow D^{-1}[R, f]$
5: (b) $g \leftarrow D^{-1}f$
6: $S \leftarrow I + G$
7: **if** $t \geq 4$ **then**
8: RDS($S_{(c,c)}, x_{(c)}, g_{(c)}, t/2$)
9: **else**
10: $x_{(c)} \leftarrow S^{-1}_{(c,c)} g_{(c)}$
11: **end if**
12: (a) $x \leftarrow g - G_{(:,c)} x_{(c)}$
13: (b) $x \leftarrow g - D^{-1}(R_{(:,c)} x_{(c)})$
14: **end procedure**

The pseudocode of the recursive DS factorization is given as follows:

Two options are indicated with (a) and (b). They are mathematically equivalent but computationally not. If we choose option (a), the G matrix needs to be formed explicitly which is expensive since linear systems with multiple right-hand sides need to be solved in parallel where D is the coefficient matrix (Line 4). Obtaining the final solution in option (a) is just a matrix vector multiplication (Line 12). Option b, on the other hand, requires a matrix vector multiplication followed by a parallel triangular solve with a single right-hand side (Line 13) and G is no longer need to be computed (Line 5). In our implementation, in all variations, we are (sequentially) using Pardiso direct solver for each of the diagonal blocks in D. Even if we choose option (b), we still need to solve the reduced system. Note that the reduced system is formed using only certain entries from G. The system we form to solve for G has R matrix as the right-hand side. This allows us to use the feature that is provided in Pardiso to allow one to compute only certain entries of the solution vector if the right-hand side is sparse. Therefore, in order to keep the computational costs lower, if we choose option (b) we use the sparse right-hand side feature of Pardiso and compute just some entries of G that is required to form the reduced system. Before factorization, we reorder and partition the initial matrix once, and since the smaller reduced system maintains a similar sparsity structure as the reordered original system we do not need to repartition at every recursive call.

3 Numerical Results

We implement our algorithms using the Intel Cilk Plus which is an extension of C and C++ languages [11]. In our work it is used to ensure efficient multithreading with recursion. Intel MKL is a library of mathematical subroutines such as BLAS, LAPACK, ScaLAPACK, and sparse linear system solvers [12]. In our implementations, all BLAS and Sparse BLAS operations are called from the available routines in Intel MKL version 10.3 [10]. As the direct solver we use Pardiso version 4.1.2.

For symmetric and nonsymmetric permutations we use METIS and HSL MC64, respectively. METIS is a collection of algorithms for graph partitioning, finite element mesh partitioning, and fill-reducing ordering for sparse matrices [13, 14]. It is used in our work to gather the nonzero values in the diagonal blocks as much as possible by setting it to minimize the communication volume. METIS finds a symmetric permutation of matrices and works on undirected graphs. In order to obtain a partitioning for nonsymmetric matrices, we use a symmetric matrix that matches to the structure of our nonsymmetric matrix (i.e., using $(|A|^T + |A|)$). Version 5.1.0 of METIS is used in our work. Some of the matrices that are used in our experiments have zero values on their main diagonals. Since having even one zero value in the main diagonal means that our matrix is indefinite and the diagonal blocks could be singular, we apply a nonsymmetric permutation. HSL MC64 is a collection of Fortran codes to find a column permutation vector to ensure that the matrix will have only nonzero entries on its main diagonal [9]. The permutation vector created by HSL MC64 is used if the matrix is indefinite.

In addition to two recursive variations of the DS factorization based sparse solver, we have also implemented two nonrecursive variations where the reduced system is directly solved via Pardiso. For comparison there are many available parallel sparse direct solvers, in this chapter we compared our results with multithreaded Pardiso direct solver. For many problem types, Pardiso has been shown to be one of most efficient direct solvers available today [6]. Furthermore, Pardiso is provided as a part of Intel MKL.

In summary, we implemented 4 variations of the DS factorization based sparse solver:

- Nonrecursive DS factorization using the sparse right-hand side feature of PARDISO in its computations (DS-NR-SP)
- Nonrecursive DS factorization without using the sparse right-hand side feature of PARDISO (DS-NR-NS)
- Recursive DS factorization using the sparse right-hand side feature of PARDISO (DS-RE-SP)
- Recursive DS factorization without using the sparse right-hand side feature of PARDISO (DS-RE-NS)

In our naming convention RE, NR, SP, and NS stand for recursive algorithm, nonrecursive algorithm, using the sparse right-hand side feature (i.e., not computing the G matrix explicitly) and not using the sparse right-hand side feature (i.e., computing the G matrix), respectively.

For all runs using the sparse DS factorization based solver, we set the number of partitions to be equal to the number of threads. The matrices used for testing are retrieved from University of Florida Sparse Matrix collection [4]. The properties of matrices are given in Table 1. We use a right-hand side vector that consists of all ones.

Table 1 Properties of test matrices, n, nnz, and dd stand for the number of rows and columns and the number of nonzeros, respectively.

#	Matrix Name	n	nnz	Problem domain
1	**ASIC_320k**	321,821	1,931,828	Sandia, circuit simulation
2	**ASIC_680ks**	682,712	1,693,767	Circuit simulation
3	**crashbasis**	160,000	1,750,416	Optimization
4	**ecology2**	999,999	4,995,991	2D/3D
5	**Freescale1**	3,428,755	17,052,626	Circuit simulation
6	**hvdc2**	189,860	1,339,638	Power network
7	**Kaufhold**	160,000	1,750,416	Counter-example
8	**Lin**	256,000	1,766,400	Structural
9	**majorbasis**	160,000	1,750,416	Optimization
10	**Raj1**	263,743	1,300,261	Circuit simulation
11	**rajat21**	411,676	1,876,011	Circuit simulation
12	**scircuit**	170,998	958,936	Circuit simulation
13	**stomach**	213,360	3,021,648	2D/3D
14	**torso3**	259,156	4,429,042	2D/3D
15	**transient**	178,866	961,368	Circuit simulation
16	**xenon2**	157,464	3,866,688	Materials

For all numerical experiments, we use a single node of the Nar cluster. Nar is the High Performance Computing Facility of Middle East Technical University Department of Computer Engineering. A single node of Nar contains 2 x Intel Xeon E5430 Quad-Core CPU (2.66 GHz, 12 MP L2 Cache, 1333 MHz FSB) and 16 GB Memory. Nar uses an open source Linux distribution developed by Fermi National Accelerator Laboratory (Fermilab) and European Organization for Nuclear Research (CERN), Scientific Linux v5.2 64bit, as its operating system. Since each node has 8 cores, we run the algorithms using up to 8 threads.

In Table 2, the speed improvement of the recursive DS factorization algorithm (RDS) and multithreaded Pardiso is compared to single threaded Pardiso runs. Timings include the total time to reorder, factorize, and solve the given linear system. We ran all the variations of the RDS algorithm. In the table, due to limited space, all RDS runs presented are using DS-RE-SP variation of the algorithm except for three cases. For matrix #6 we are using DS-RE-NS, for matrices #10 and #15 we are using DS-NR-SP variations of the algorithm since DS-RE-SP does not give a comparable relative residual to multithreaded Pardiso. Also note that the recursive versions of the solver is defined when the number of partitions is equal to or greater than 4. Hence, we are using the nonrecursive DS-NR-SP for all cases if the number of threads is 2. In Table 3, 2-norm of the final relative residuals are given. Based on the results of the runs, the proposed algorithm is faster than Pardiso using 8 threads for 10 cases out of 16 obtaining comparable relative residuals. We note that for the cases where RDS performs worse than multithreaded Pardiso, sequential solution time is very short and hence we could not expect much improvement to begin with.

Table 2 Speedup of multithreaded Pardiso and RDS algorithms compared to the Sequential Pardiso time for all test matrices using 2,4, and 8 threads.

#	Sequential	PARDISO			RDS		
	Time (s)	2	4	8	2	4	8
1	13,88	1,92	2,74	3,39	1,93	2,71	3,52
2	104,30	1,99	3,91	7,18	1,99	3,84	7,04
3	3,74	1,89	3,34	3,60	1,96	3,90	7,33
4	7,49	1,89	3,39	4,57	1,92	3,82	7,57
5	7,64	1,84	3,03	3,37	1,91	3,78	6,42
6	0,45	1,29	1,55	1,96	-	5,63	11,25
7	0,04	1,33	2,00	2,00	0,31	0,40	0,27
8	81,50	1,90	3,37	3,35	0,93	2,59	6,17
9	14,50	1,90	3,42	4,41	1,99	3,61	5,27
10	0,84	1,83	2,80	4,42	1,65	2,47	4,00
11	1,05	1,88	3,00	1,67	1,78	2,69	3,75
12	1,02	1,06	1,52	1,28	0,68	0,92	0,97
13	10,73	1,95	3,55	5,09	1,96	3,78	5,80
14	59,32	1,89	3,54	3,16	1,96	3,66	3,11
15	0,43	1,79	2,69	3,91	1,48	1,95	2,39
16	15,95	1,91	3,56	4,68	1,95	3,68	5,86

Table 3 Relative residual norms for RDS and Pardiso using 2, 4, and 8 threads.

#	PARDISO			RDS		
	2	4	8	2	4	8
1	1,41E-10	9,03E-11	5,31E-11	1,12E-15	1,47E-15	1,16E-15
2	9,45E-08	9,45E-08	9,45E-08	6,19E-10	8,40E-10	8,13E-10
3	2,20E-15	2,20E-15	2,19E-15	2,20E-15	2,01E-14	2,44E-14
4	1,31E-16	1,31E-16	1,29E-16	1,32E-16	2,04E-16	1,71E-16
5	1,79E-10	1,79E-10	1,85E-10	4,47E-15	7,11E-15	1,03E-15
6	2,80E-09	2,80E-09	2,85E-09	-	9,51E-11	1,24E-10
7	9,72E-15	9,72E-15	9,72E-15	1,26E-16	3,14E-04	6,51E-16
8	8,81E-16	8,07E-16	7,16E-16	5,63E-16	3,70E-13	6,13E-13
9	1,71E-15	1,71E-15	1,70E-15	1,09E-16	9,61E-05	2,91E-11
10	6,16E-10	4,50E-10	6,47E-10	4,67E-07	1,21E-08	1,04E-08
11	7,97E-06	8,38E-06	1,16E-05	4,93E-07	3,18E-04	9,96E-05
12	2,03E-09	1,60E-09	1,59E-09	4,63E-15	1,73E-15	1,44E-15
13	7,71E-16	7,64E-16	7,52E-16	7,65E-16	3,20E-15	3,27E-15
14	1,02E-15	1,01E-15	9,93E-16	4,80E-15	2,11E-15	8,59E-16
15	1,42E-06	1,56E-06	1,68E-06	1,96E-10	1,80E-10	1,47E-10
16	4,39E-12	4,37E-12	4,35E-12	9,41E-12	2,21E-11	5,23E-11

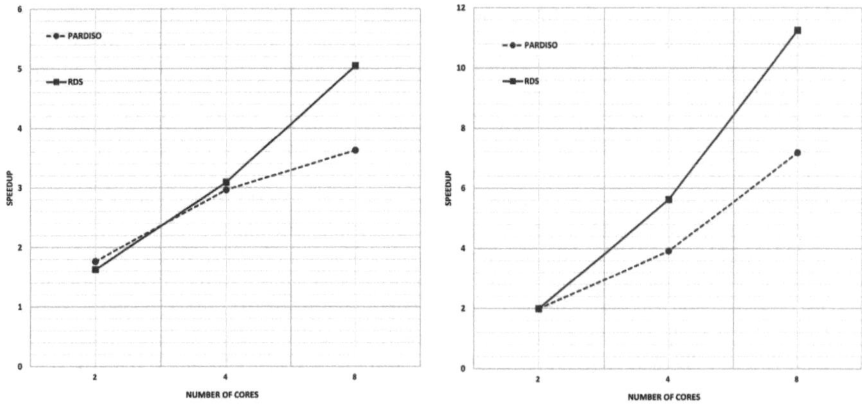

(a) Average (b) Best

Fig. 3 The average and the best speed improvements obtained using RDS and Pardiso compared to the sequential time with Pardiso.

In Figure 3(a), the average speed improvement of RDS is compared against the average speed improvement for Pardiso for all problems. The improvement of RDS more pronounced as the number of threads (i.e., cores) is increased from 4 to 8.

In Figure 3(b), we plot the best speed improvement for both RDS and Pardiso. Again, the improvement is more pronounced as the number of threads increase.

4 Conclusions

We present a recursive sparse DS factorization based direct solver. The results show that on a multicore environment, the scalability of the proposed algorithm is better than the classical LU factorization based solvers in most examples. The improvement is more pronounced if the number of cores is large.

Acknowledgements The HPC resources were provided by the Department of Computer Engineering, Middle East Technical University. This work has been supported by TUBITAK Career Award EEAG111E238, METU BAP-08-11-2011-128, and Turkish Academy of Sciences Distinguished Young Scientist Award TUBA-GEBIP/2012-19.

References

1. Amestoy, P., Duff, I., L'Excellent, J.Y., Koster, J.: Mumps: a general purpose distributed memory sparse solver. In: Sørevik, T., Manne, F., Gebremedhin, A., Moe, R. (eds.) Applied Parallel Computing. New Paradigms for HPC in Industry and Academia. Lecture Notes in Computer Science, vol. 1947, pp. 121–130. Springer, Berlin, Heidelberg (2001)
2. Bolukbasi, E.: A new multi-threaded and recursive direct algorithm for parallel solution of sparse linear systems. Master's thesis, Middle East Technical University, Ankara (2013)

3. Bolukbasi, E.S., Manguoglu, M.: A new multi-threaded and recursive direct algorithm for parallel solution of sparse linear systems. 5th International Conference of the ERCIM (European Research Consortium for Informatics and Mathematics) Working Group on Computing & Statistics (2012)
4. Davis, T.A., Hu, Y.: The university of Florida sparse matrix collection. ACM Trans. Math. Softw. **38**(1), 1:1–1:25 (2011). doi:10.1145/2049662.2049663. http://doi.acm.org/10.1145/2049662.2049663
5. Demmel, J.W., Eisenstat, S.C., Gilbert, J.R., Li, X.S., Liu, J.W.H.: A supernodal approach to sparse partial pivoting. SIAM J. Matrix Anal. Appl. **20**(3), 720–755 (1999)
6. Gould, N.I.M., Scott, J.A., Hu, Y.: A numerical evaluation of sparse direct solvers for the solution of large sparse symmetric linear systems of equations. ACM Trans. Math. Softw. **33**(2) (2007). doi:10.1145/1236463.1236465. http://doi.acm.org/10.1145/1236463.1236465
7. Gupta, A.: WSMP: Watson sparse matrix package (part-i: direct solution of symmetric sparse systems). IBM TJ Watson Research Center, Yorktown Heights, NY. Technical Report, RC, vol. 21886 (2000)
8. Gupta, A.: Wsmp: Watson sparse matrix package (part-ii: direct solution of general sparse systems). Technical Report, Technical Report RC 21888 (98472). IBM TJ Watson Research Center, Yorktown Heights, NY (2000). http://www.cs.umn.edu/~agupta/doc/wssmp-paper.ps
9. HSL: A collection of Fortran codes for large scale scientific computation (2011). http://www.hsl.rl.ac.uk
10. Intel. Intel MKL 10.3 release notes (2012). http://software.intel.com/en-us/articles/intel-mkl-103-release-notes
11. Intel: Intel Cilk plus (2013). http://software.intel.com/en-us/intel-cilk-plus
12. Intel: Intel math kernel library 11.0 (2013). http://software.intel.com/en-us/intel-mkl
13. Karypis, G.: Metis - serial graph partitioning and fill-reducing matrix ordering (2013). http://glaros.dtc.umn.edu/gkhome/metis/metis/overview
14. Karypis, G., Kumar, V.: A fast and high quality multilevel scheme for partitioning irregular graphs. SIAM J. Sci. Comput. **20**(1), 359–392 (1998). doi:10.1137/S1064827595287997. http://epubs.siam.org/doi/abs/10.1137/S1064827595287997
15. Li, X.S.: An overview of SuperLU: Algorithms, implementation, and user interface. ACM Trans. Math. Softw. **31**(3), 302–325 (2005)
16. Manguoglu, M.: A domain-decomposing parallel sparse linear system solver. J. Comput. Appl. Math. **236**(3), 319–325 (2011)
17. Manguoglu, M.: Parallel solution of sparse linear systems. In: High-Performance Scientific Computing, pp. 171–184. Springer, London (2012)
18. Moore, G.E.: Cramming more components onto integrated circuits. IEEE Solid State Circuits Soc. Newsl. **11**(5), 33–35 (2006) [Reprinted from Electronics **38**(8), 114 ff. (1965)]. doi: 10.1109/N-SSC.2006.4785860
19. Saad, Y.: Iterative Methods for Sparse Linear Systems, 2nd edn. SIAM, Philadelphia (2003)
20. Sameh, A.H., Kuck, D.J.: On stable parallel linear system solvers. J. ACM **25**(1), 81–91 (1978)
21. Sameh, A.H., Polizzi, E.: A parallel hybrid banded system solver: the spike algorithm. Parallel Comput. **32**(2), 177–194 (2006)
22. Sameh, A.H., Polizzi, E.: Spike: a parallel environment for solving banded linear systems. Comput. Fluids **36**(1), 113–120 (2007)
23. Schenk, O., Gärtner, K.: Solving unsymmetric sparse systems of linear equations with pardiso. Futur. Gener. Comput. Syst. **20**(3), 475–487 (2004)
24. Schenk, O., Gärtner, K.: On fast factorization pivoting methods for sparse symmetric indefinite systems. Electron. Trans. Numer. Anal. **23**, 158–179 (2006)
25. Schenk, O., Wächter, A., Hagemann, M.: Matching-based preprocessing algorithms to the solution of saddle-point problems in large-scale nonconvex interior-point optimization. Comput. Optim. Appl. **36**(2–3), 321–341 (2007)
26. Schenk, O., Bollhöfer, M., Römer, R.A.: On large-scale diagonalization techniques for the Anderson model of localization. SIAM Rev. **50**(1), 91–112 (2008)

Part V
Mathematical Methods

Macroscopic First Order Models of Multicomponent Human Crowds with Behavioral Dynamics

N. Bellomo, S. Berrone, L. Gibelli, and A.B. Pieri

Abstract This paper presents a new approach to the behavioral dynamics of human crowds. Macroscopic first order models are derived based on mass conservation at the macroscopic scale, while methods of the kinetic theory are used to model the decisional process by which walkers select their velocity direction. The present approach is applied to describe the dynamics of a homogeneous crowd in venues with complex geometries. Numerical results are obtained using a finite volume method on unstructured grids. Our results visualize the predictive ability of the model. Solutions for heterogeneous crowd can be obtained by the same technique where crowd heterogeneity is modeled by dividing the whole system into subsystems identified by different features.

1 Plan of the Paper

The modeling of crowd dynamics can be developed, at the three scales, namely *microscopic* (individual based), *macroscopic* (corresponding to the dynamics of mean averaged quantities), and to the intermediate *mesoscopic* (corresponding to the dynamics of a probability distribution function over the microscopic scale state of individuals), see the book [11]. The latter approach is such that interactions are modeled at the micro-scale, while mean quantities, such as local number density and linear momentum, are obtained by velocity weighted moments of the aforesaid probability distribution.

A critical analysis of the advantages and drawbacks of the different scales selected for the modeling approach is discussed in various papers, [3, 7, 9] where it is stated that the present state of the art does not yet allow well-defined hallmarks

N. Bellomo (✉)
King Abdulaziz University, Jeddah, Saudi Arabia
Politecnico di Torino, Torino, Italy
e-mail: nicola.bellomo@polito.it

S. Berrone • L. Gibelli • A.B. Pieri
Department of Mathematical Sciences, Politecnico di Torino, Torino, Italy
e-mail: stefano.berrone@polito.it; livio.gibelli@polito.it; alexandre.pieri@polito.it

Y. Bazilevs, K. Takizawa (eds.), *Advances in Computational Fluid-Structure Interaction and Flow Simulation*, Modeling and Simulation in Science, Engineering and Technology, DOI 10.1007/978-3-319-40827-9_23

to support an optimal choice. A detailed investigation has been carried out to understand the complex dynamics at the microscopic scale, see, among others, [13] and [15]. These models can contribute to implement both meso-scale models [5] and hybrid models [1], where the state of the system is defined in probability over the velocity direction and deterministically over the velocity.

Macroscopic, hydrodynamic, models are of great interest in that they are far less computationally demanding than those at the other two scales. This requirement is particularly important when dealing with complex flows such as coupling pedestrian flows to vehicular traffic networks [10]. However, macroscopic models suffer a number of drawbacks. Firstly, the heterogeneous behavior of walkers gets lost in the averaging process needed by their derivations. An additional difficulty, well documented in the paper by Hughes [18], consists in modeling the process by which walkers select their velocity, namely direction and speed, in a crowd. The nonlocal feature of interactions is a further issue to be taken into account in the modeling approach, as walkers are not classical particles and modify their velocity before encountering a wall. Finally, a challenging problem is the study of propagation of anomalous behaviors, which might be induced by panic conditions [16], that can induce large deviations in the collective dynamics. Due to all the aforementioned motivation the term *social dynamics* has been introduced in [2] to enhance the heterogeneous behavior of walkers, who might modify their strategy induced by interaction not only with other individuals, but also with the specific features of the environment where they walk.

The present paper aims at tackling the aforesaid drawbacks in the case of macroscopic first order models. These are simply obtained by mass conservation equations, which involve local density and mean velocity closed by phenomenological models relating the local mean velocity to local density distribution. Although this approach is a simple way of looking at the dynamics, substantial developments are needed with respect to the existing literature, to achieve a realistic modeling of the decision process by which individuals select the velocity direction and adjust their velocity to the local density conditions, as well as to deal with the heterogeneous behavior of the crowd.

The contents of the paper are presented as follows: Section 2 defines the mathematical structure underlying the modeling approach, which consists in a system of mass conservation equations for a crowd subdivided into various populations, which can be called after [6], functional subsystems. More specifically two structures are derived which correspond to homogeneous and to heterogeneous populations. Section 3 shows how specific models can be derived according to the aforesaid structure. More specifically, two classes of models are proposed corresponding to the structures defined in Section 2. Section 4 presents numerical results obtained by solving the simple case of a homogeneous crowd in a complex domain. Section 5 looks ahead to research perspectives.

2 Mathematical Structures

Let us consider the dynamics of a crowd in a domain Σ, which may include internal obstacles and inlet/outlet segments on the boundary $\partial\Sigma$. This section searches for appropriate mathematical structures, which can provide the conceptual basis for the derivation of first order macroscopic models. It is assumed that the state of the system is described by local density and mean velocity to be viewed as dependent variables of the differential system. However, since we deal with first order models, it is also necessary looking for a functional relation linking the mean velocity to the density.

Dimensionless quantities are used according to the following definitions:

- $\rho := n/n_M$ is the ratio between the number density n of individuals per unit area and the number n_M corresponding to the highest admissible packing density.
- $\boldsymbol{\xi} := \boldsymbol{v}/v_M$ is the dimensionless local mean velocity obtained by referring the dimensional velocity $\boldsymbol{v}$ to the highest limit of the mean velocity v_M, which can be reached by a walker in a low density limit in optimal environmental conditions.

Density and velocity depend on time and space coordinates, that is: $\rho = \rho(t, \mathbf{x}) =: \rho(t, x, y)$ and $\boldsymbol{\xi} = \boldsymbol{\xi}(t, \mathbf{x}) =: \boldsymbol{\xi}(t, x, y)$. The independent variables time t and space $\mathbf{x}$ are also dimensionless with respect to a characteristic dimension ℓ of Σ and ℓ/v_M, respectively. Moreover, following [2], we introduce a number of dimensionless parameters that account for some specific features of the crowd already discussed in previous papers based on the kinetic theory approach:

- $\alpha \in [0, 1]$ models the quality of the environment where $\alpha = 1$ stands for the optimal quality of the environment, which allows to reach high velocity, while $\alpha = 0$ stands for the worst quality, which prevents the motion;
- $\beta \geq 0$ models the attraction toward the direction of the highest density gradient, where $\beta = 0$ stands for the highest search of the less congested areas, while increasing value of β denotes increasing attraction to follow what the other do.

These parameters are used in this section only at a formal level, while their physical meaning will be made precise in the next section devoted to derivation of specific models.

In this present paper, two models are considered. One corresponds to a crowd with homogeneous walking ability, while the other to a crowd subdivided into different populations featured by different walking abilities.

- *Homogeneous crowd:* Let us first consider the derivation of the mass conservation equation for a homogeneous crowd, where all individuals have the same walking ability. Classically, the said equation writes as follows:

$$\partial_t \rho + \nabla_{\mathbf{x}} (\rho\, \boldsymbol{\xi}) = 0 \,. \tag{1}$$

The closure of the equation can be obtained by modeling the dependence of $\boldsymbol{\xi}$ on ρ by a phenomenological relation of the type $\boldsymbol{\xi} = \boldsymbol{\xi}[\rho](\alpha, \beta)$, so that the conservation equation formally writes as follows:

$$\partial_t \rho + \nabla_{\mathbf{x}} (\rho\, \boldsymbol{\xi}[\rho](\alpha, \beta)) = 0\,, \tag{2}$$

where square brackets denote that functional relations must be used, rather than functions, to link the local mean velocity to the local density. In the present study, walkers select their preferred directions based not only on the local density, but also on density gradients. Therefore, specific models can be obtained by some heuristic interpretations of physical reality leading to $\boldsymbol{\xi} = \boldsymbol{\xi}[\rho]$ and inserting such model into Eq.(2), which is the mathematical structure to be used for the modeling.

- *Heterogeneous crowd:* Let us now consider a more general framework, where walkers are subdivided into a number n of populations corresponding to different levels of expressing their walking abilities. Therefore, the state of the system is defined by a set of dimensionless number densities

$$\rho_i = \rho_i(t, \mathbf{x})\,, \qquad \rho(t, \mathbf{x}) = \sum_{i=1}^{n} \rho_i(t, \mathbf{x})\,, \tag{3}$$

The subscript corresponds to a discrete variable, modeling the walking ability of each population $i = 1, \ldots, n$, being $i = 1$ and $i = n$ to the lowest and highest motility, respectively.

The new structure simply needs the following modification:

$$\partial_t \rho_i + \nabla_{\mathbf{x}} (\rho_i \boldsymbol{\xi}_i[\rho](\alpha, \beta)) = 0, \qquad i = 1, \ldots, n\,, \tag{4}$$

where the modeling of the mean velocity differs for each population $\boldsymbol{\xi}_i = \boldsymbol{\xi}_i[\rho]$. Therefore, the structure consists in a system of n nonlinear PDEs.

Remark 2.1. Structures defined by Eqs. (2) and (4) constitute the underlying framework to derive specific models. In all cases, the derivation depends on the environment, where the crowd moves, not only its quality, but also its shape. In fact, the model requires implementing nonlocal interaction with the walls. The formal structures proposed in this section need two types of parameters, namely α, β, corresponding to two specific features of the dynamics.

3 Derivation of Models

Let us now consider the derivation of specific models consistent with the frameworks presented in the preceding section. Therefore, two different classes of models are derived in the following subsections according to conceivable requirements of applications. More in detail, we consider both homogeneous and heterogeneous populations in absence of social interactions, that would allow individuals to shift from one subpopulation to the other.

Let us first consider the simple case of a homogeneous crowd, where all individuals behave in the same manner. Derivation of models requires simply to describe analytically the dependence of $\boldsymbol{\xi}$ on the local density distribution. Polar coordinates are used for the velocity variable, so that

$$\boldsymbol{\xi} = \xi\,(\cos\theta\,\mathbf{i} + \sin\theta\,\mathbf{j}) := \xi\,\boldsymbol{\omega}, \tag{5}$$

where θ is the angle between $\boldsymbol{\xi}$ and the x-axis of an orthogonal system in the plane, $\mathbf{i}$ and $\mathbf{j}$ are the unit vector of the aforesaid axes, ξ is the velocity modulus of $\boldsymbol{\xi}$ occasionally called *speed*.

The idea that walkers adjust their dynamics according to a decision process, which can be modeled by theoretical tools of game theory [21], has been developed in a sequel of papers [1, 2, 4, 5] devoted to crowd dynamics.

In detail, the modeling proposed in this paper takes advantage of the approach to individual based interaction proposed in [2], where the decision process is supposed to act according to the following sequence:

1. Walkers move along the direction $\boldsymbol{\omega}$, forming an angle θ with respect to the x-axis, selected according to a decision process which account the following trends: search of the exit, avoid walls, search of less congested areas, and attraction toward the density gradients. Details are given in the following.
2. Once the direction $\boldsymbol{\omega}$ has been selected, walkers adjust their velocity modulus to the so-called local *perceived density* ρ_θ^* along $\boldsymbol{\omega}$. This quantity is defined, according to [8], as follows:

$$\rho_\theta^* = \rho_\theta^*[\rho] = \rho + \frac{\partial_\theta \rho}{\sqrt{1 + (\partial_\theta \rho)^2}}\left[(1-\rho)\,H(\partial_\theta\rho) + \rho\,H(-\partial_\theta\rho)\right], \tag{6}$$

where ∂_θ denotes the derivative along the direction θ, while $H(\cdot)$ is the Heaviside function $H(\cdot \geq 0) = 1$, and $H(\cdot < 0) = 0$. Therefore, positive gradients increase the perceived density up to the limit $\rho = 1$, while negative gradients decrease it down to the limit $\rho = 0$ in a way that

$$\partial_\theta\rho \to \infty \Rightarrow \rho_\theta^* \to 1\,, \quad \partial_\theta\rho = 0 \Rightarrow \rho_\theta^* = \rho\,, \quad \partial_\theta\rho \to -\infty \Rightarrow \rho_\theta^* \to 0\,. \tag{7}$$

3. The relation $\xi = \xi(\rho_\theta^*;\alpha)$ between ξ and ρ_θ^* depends on the quality of the environment. In the attempt of reproducing empirical data [12, 17, 19, 20, 22, 23], such a relation is modeled by is modeled by a polynomial of ρ_θ^* fulfilling the following constraints: $\xi(0) = \alpha$; $\xi'(0) = 0$; $\xi(1) = 0$; $\xi'(1) = 0$, where prime denotes derivative with respect to ρ_θ^*.

The first step of the above process leads to the following expression of the preferred direction:

$$\boldsymbol{\omega}(\rho, \mathbf{x}; \beta) = \frac{(1-\rho)\boldsymbol{\nu}_T + \rho\,[\beta\boldsymbol{\nu}_S + (1-\beta)\boldsymbol{\nu}_V]\,[1 - \exp(-d/d_w)]}{\|(1-\rho)\boldsymbol{\nu}_T + \rho\,[\beta\boldsymbol{\nu}_S + (1-\beta)\boldsymbol{\nu}_V]\,[1 - \exp(-d/d_w)]\|}, \tag{8}$$

where

$$\boldsymbol{\nu}_S = -\boldsymbol{\nu}_V = \frac{\nabla\rho}{||\nabla\rho||}, \tag{9}$$

while the parameter β may account for panic conditions [2].

In Equation (8), the role of the nonlocal interactions with the wall and obstacles has been modeled by adapting the hallmarks of [2] to this present case. More specifically, we define the boundary parameter d_w and the cell distance from the wall, d. Then, the contribution of velocity components $\boldsymbol{\nu}_S$ and $\boldsymbol{\nu}_V$ is assumed to decrease exponentially while approaching the wall and becomes naught at the boundary.

The second and third step yield the following polynomial expression:

$$\xi = \xi[\rho] = \alpha(1 - 3\,\rho_\theta^{*2} + 2\,\rho_\theta^{*3})[\rho]. \tag{10}$$

These two quantities can be inserted into the framework Eq. (1), which writes as follows:

$$\partial_t\rho + \nabla_{\mathbf{x}}\Big(\alpha\,\rho(1 - 3\,\rho_\theta^{*2} + 2\,\rho_\theta^{*3})\,\boldsymbol{\omega}(\rho, \mathbf{x}; \beta)\Big) = 0\,. \tag{11}$$

Let us now consider the case of a heterogeneous crowd. The reference mathematical structure is now given by Eq. (4), which consists in a system of n PDEs. Hence, the modeling problem consists in modeling the mean velocity $\boldsymbol{\xi}_i$ for each population of walkers, which, as an example, can correspond to slow, mean, and fast individuals. Hereinafter, the model is presented in the case of a system in absence of boundaries, referring the technical generalization to the approach proposed above. By assuming that walkers exploit the quality of the environment depending also on the quality of their walking ability, the following model is obtained:

$$\boldsymbol{\xi}_i = \boldsymbol{\xi}_i[\rho] = \alpha\,\frac{i}{n}\,(1 - 3\,\rho_\theta^{*2} + 2\,\rho_\theta^{*3})[\rho] \tag{12}$$

which inserted into the structure yields

$$\partial_t\rho_i + \nabla_{\mathbf{x}}\left(\rho_i\,\alpha\,\frac{i}{n}\,(1 - 3\,\rho_\theta^{*2} + 2\,\rho_\theta^{*3})[\rho]\,\boldsymbol{\omega}_i(\rho, \mathbf{x}; \beta)\right) = 0\,, \quad i = 1, \ldots, n\,, \tag{13}$$

where the preferred direction $\boldsymbol{\omega}_i$ is now computed for all components of the fluid viewed as a mixture.

The models proposed in this section assume that the parameter β is a constant. This assumption can be improved by letting β depend on ρ. In fact, high density induce individuals to look of less congested area. Then one can assume heuristically that β decreases with the density. This matter is further discussed in the last section.

4 Applications and Simulations

4.1 Computational Method

Equation (11) is integrated using finite volumes [14]. This method has the advantage to be globally mass conserving and to be easily extended to complex geometries. In the following we denote Σ the computational domain. We then introduce a tessellation of Σ into N polygons $\mathscr{T} = \bigcup_{i=1}^{N} P_i$ where $P_i \cap P_j = \varnothing$ for two disconnected polygons and $P_i \cap P_j = \{e, v\}$ if two adjacent polygons (i,j) share a common edge e or a common vertex v. The effective hexagonal grid is generated using *polymesher* [24]. For a given polygon $P \in \mathscr{T}$, the integral form of Equation (11) reads

$$\iint_P \frac{\partial \rho}{\partial t} = -\sum_{e\in\mathscr{E}} \int_e \left(\alpha\, \rho(1 - 3\,\rho_\theta^{*2} + 2\,\rho_\theta^{*3})\,\boldsymbol{\omega}(\rho, \mathbf{x}; \beta)\right) \cdot \boldsymbol{n}_e ds, \tag{14}$$

where $\mathscr{E}$ denotes the set of all edges of polygon P and $\boldsymbol{n}_e$ the outward-pointing unit normal to edge e. Integrals involved in Equation (14) are approximated using the midpoint rule.

Quantities defined by Equation (6) and Equations (8)–(10) must then be computed at edge centers $e_{1/2}$ while the left-hand side of (14) requires density computed at cell center G. Finally, we need to solve

$$\frac{\partial \rho_{G(P)}}{\partial t} \approx -\frac{1}{\mathscr{A}(P)} \sum_{e\in\mathscr{E}} \rho(e_{1/2}) l(e) \left(\alpha(e_{1/2})\, \xi_F(e_{1/2})\, \boldsymbol{\omega}(\rho(e_{1/2}), \boldsymbol{e}_{1/2}; \beta)\right) \cdot \boldsymbol{n}_e, \tag{15}$$

where $\mathscr{A}(P)$ is the area of polygon P and $l(e)$ stands for the length of edge e. To compute the preferred direction $\boldsymbol{\omega}$, the perceived density ρ_θ^*, and the mean speed $\boldsymbol{\xi}$ we need to compute density gradients at edge centers which in turn involve density at cell centers.

The definition of the target direction ν_T which enter in Equation (8) is an open choice. In the present study, we find suitable to use the normalized Darcy velocity $-\nabla\phi_D/\|\nabla\phi_D\|$ for porous media as local target direction which has the strong advantage to provide admissible (compatible with domain boundaries) velocity directions. In practice, a Poisson equation for the pressure field ϕ_D is solved once for all at the beginning. High and low pressure are imposed at entrances and exits, respectively, while a nonpenetrability condition is used at the solid walls. As a result, the preferred direction is a trade-off between hydrodynamic velocity $\boldsymbol{v}_T$ and behavioral velocity $\boldsymbol{v}_S$.

4.2 Numerical Results

We test our computational model on two different geometries both presented in Figure 1. These geometries have been chosen to be representative of common situations. One represents a convergent–divergent corridor with length 10m and with throat at 5m. This configuration has been selected to underline the capability of the present numerical approach to deal with curvilinear boundaries. In this case, people go from left to right, toward the exit. The second represents a square room with two exits at upper-left and upper-right corners with room entrance located at the bottom side. People initially move northward and then divide into two groups which exit the room from both sides. For the first configuration, two different dynamics are studied, namely $\beta = 0.5$ and $\beta = 0$. Results are presented in Figure 2. Numerical results for the second configuration are reported in Figure 3 and correspond to the

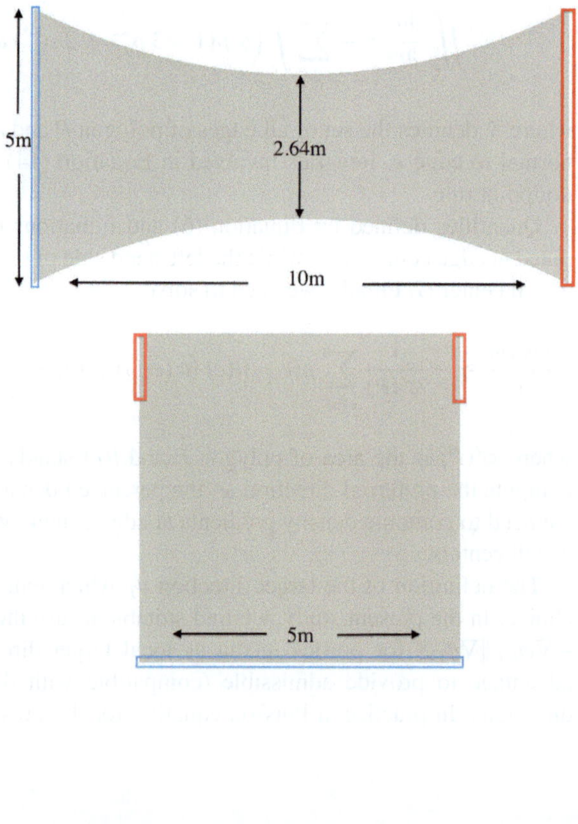

Fig. 1 Geometries for the two test cases. Bottom panel: two-exits room. Top panel: convergent corridor. Red indicates exits whereas blue indicates entrance.

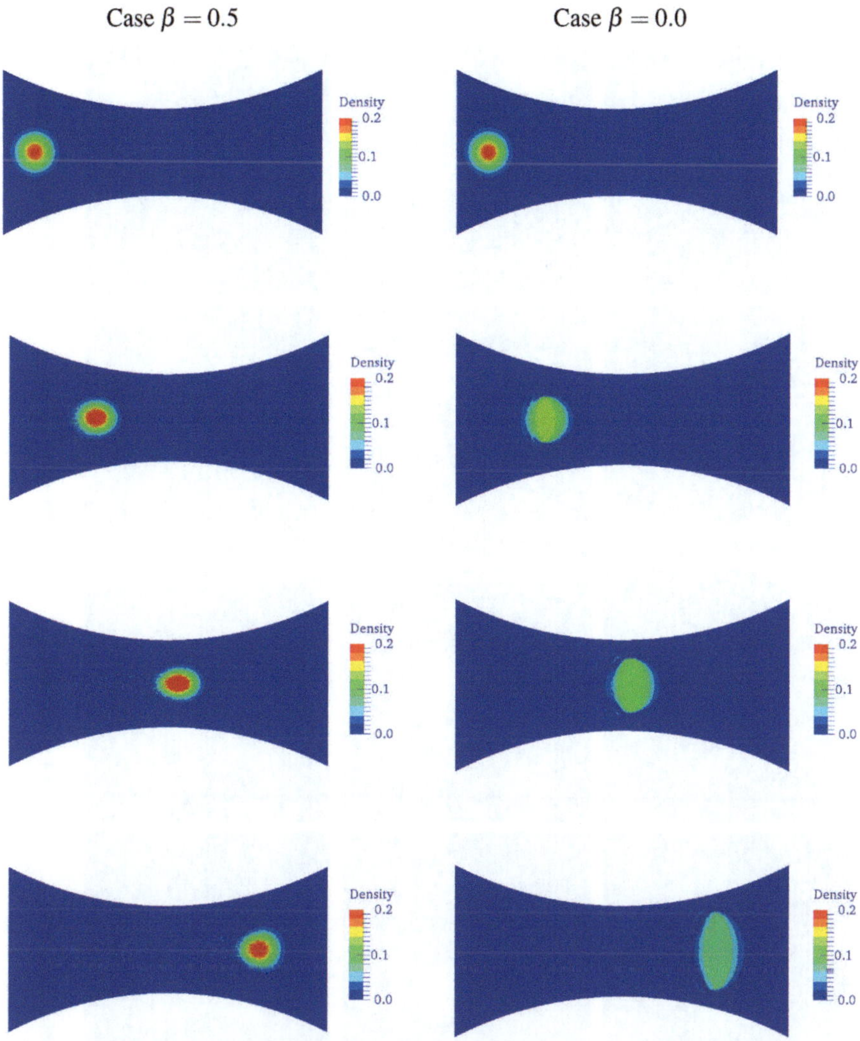

Fig. 2 Crowd density distribution for the convergent corridor case. Left panels: $\beta = 0.5$. Right panels: $\beta = 0.0$. From top to bottom: t=0; t=2; t=5; t=8.

cases $\beta = 0.5$, $\beta = 0.25$, and $\beta = 0$. Numerical experiments indicate that values of $\beta > 0.5$ produce regions with density passing the threshold $\rho = 1$, which denotes the onset of an incident and localized injury of walkers.

Setting parameter $\beta = 1/2$ we expect the preferred velocity to be exactly the hydrodynamics velocity $\boldsymbol{v}_T$. When $\beta = 0$, people in the crowd try to avoid high density regions and include in their decisional process a direction $\boldsymbol{v}_V$ opposite to density gradients. These behaviors are correctly reproduced by the macroscopic model. In particular, when the parameter β approaches zero, the crowd progressively

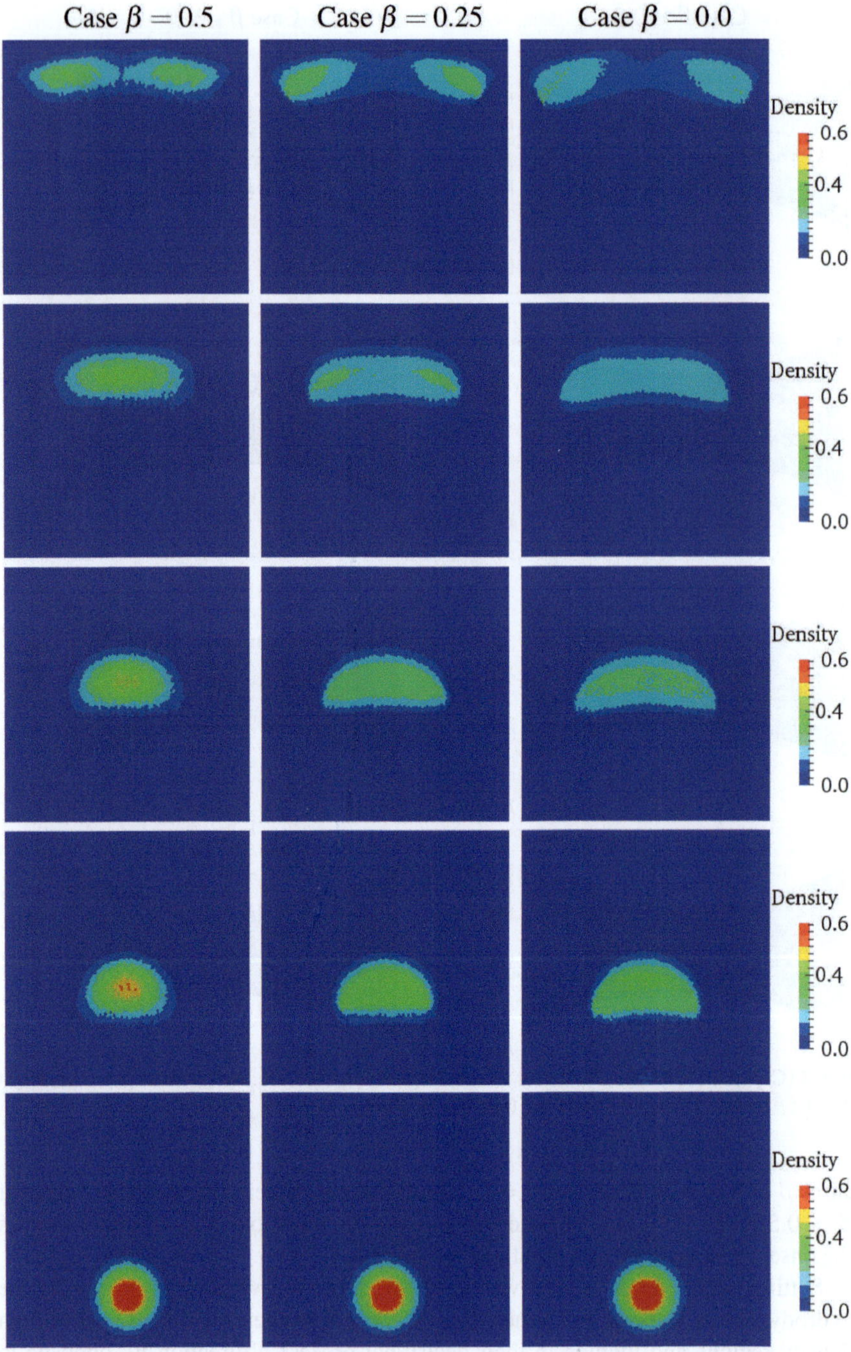

Fig. 3 Crowd density distribution for the two-exits room case. Left panels: $\beta = 0.5$. Center panels: $\beta = 0.25$. Right panels: $\beta = 0.0$. From bottom to top: t=0; t=1; t=2; t=3; and t=4.

starts dispersing which traduces into a larger crowd patch when compared to the case $\beta = 0.5$, see both Figure 2 and Figure 3. Besides, Figure 3 shows how the initial compact crowd patch separates faster while decreasing β.

5 Looking Ahead to Further Developments

This paper initiated a new approach to modeling crowd dynamics by conservation equations at the macroscopic scale closed by phenomenological relations modeling a decisional process by which walkers select their velocity direction and speed. This process includes the attempt of walkers to avoid the impact with walls. Simulations have been carried out for a homogeneous population only. Therefore a natural development consists in a detailed computational analysis of the heterogeneous case. However, various ideas toward research plans can be given in addition to the aforementioned obvious one. Let us discuss three of them selected according to the author bias.

1. The modeling of the parameter β can be refined by assuming that it depends on the local density. A natural way to achieve this improvement consists in weighting the parameter β so as to get a more realistic representation of the dynamics and avoid densities above the threshold $\rho = 1$.
2. Heterogeneity of the velocity distribution can be modeled by two scale models describing the dynamics of the velocity at the microscopic scale coupled with the mass conservation equations. The micro-scale model should be obtained by theoretical tools of stochastic games recently reviewed in [6].
3. Models with social dynamics can be obtained as natural development of the heterogeneous population model by inserting on the right-hand side of the conservation equations a source term modeling transition from one population to the other.

Moreover, some of the reasonings presented in this paper can be transferred to second order models involving mass and momentum equations. The key point is now the modeling of the decision process leading to the acceleration term in the linear momentum equation.

Acknowledgements The research leading to these results has received funding from the European Union's Seventh Framework Programme (FP7/2007-2013) under Grant Agreement Number 313161 (eVACUATE). Project title: "A holistic, scenario independent, situation-awareness and guidance system for sustaining the Active Evacuation Route for large crowds." This publication reflects the views only of the authors, and the Commission cannot be held responsible for any use which may be made of the information contained therein.

References

1. Agnelli, J.-P., Colasuonno, F., Knopoff, D.: A kinetic theory approach to the dynamics of crowd evacuation from bounded domains. Math. Models Methods Appl. Sci. **25**, 109–129 (2015)
2. Bellomo, N., Gibelli, L.: Toward a behavioral-social dynamics of pedestrian crowds. Math. Models Methods Appl. Sci. **25**, 2417 (2015)
3. Bellomo, N., Soler, J.: On the mathematical theory of the dynamics of swarms viewed as complex systems. Math. Models Methods Appl. Sci. **22**, paper No. 1140006 (2012)
4. Bellomo, N., Gibelli, L.: Behavioral crowds: Modeling and Monte Carlo simulations toward validation, Comput. Fluids, in press doi:10.1016/j.compfluid.2016.04.022
5. Bellomo, N., Bellouquid, A., Knopoff, D.: From the micro-scale to collective crowd dynamics. Multiscale Model. Simul. **11**, 943–963 (2013)
6. Bellomo, N., Knopoff, D., Soler, J.: On the difficult interplay between life, "complexity", and mathematical sciences Math. Models Methods Appl. Sci. **23**, 1861–1913 (2013)
7. Bellomo, N., Herrero, M.A., Tosin, A.: On the dynamics of social conflicts looking for the black swan. Kinet. Relat. Models **6**(3), 459–479 (2013)
8. Bellomo, N., Bellouquid, A., Nieto, J., Soler, J.: On the multiscale modeling of vehicular traffic: from kinetic to hydrodynamics. Discr. Contin. Dyn. Syst. Series B **19**, 1869–1888 (2014)
9. Bellouquid, A., De Angelis, E., Fermo, L.: Towards the modeling of vehicular traffic as a complex system: a kinetic theory approach. Math. Models Methods Appl. Sci. **22**, paper No. 1140003 (2012)
10. Borsche, R., Klar, A., Köhn, S., Meurer, A.: Coupling traffic flow networks to pedestrian motion. Math. Models Methods Appl. Sci. **24**, 359–380 (2014)
11. Cristiani, E., Piccoli, B., Tosin, A.: Multiscale Modeling of Pedestrian Dynamics. Springer, Berlin (2014)
12. Daamen, W., Hoogedorn, S.P.: Experimental research of pedestrian walking behavior. TRB Annual Meeting CD-ROM (2006)
13. Degond, P., Appert-Rolland, C., Moussaïd, M., Pettré, J., Theraulaz, G.: A hierarchy of heuristic-based models of crowd dynamics. J. Stat. Phys. **152**(6), 1033–1068 (2013)
14. Ferziger, J.H., Peric, M.: Computational Methods for Fluid Dynamics. Springer Science & Business Media, Berlin (2012)
15. Helbing, D.: Traffic and related self-driven many-particle systems. Rev. Modern Phys. **73**, 1067–1141 (2001)
16. Helbing, D., Johansson, A.: Pedestrian crowd and evacuation dynamics. In: Encyclopedia of Complexity and System Science, pp. 6476–6495. Springer, Berlin (2009)
17. Helbing, D., Johansson, A., Al-Abideen, H.Z.: Dynamics of crowd disasters: an empirical study. Phys. Rev. E **75**, 046109 (2007)
18. Hughes, R.L.: The flow of human crowds. Annu. Rev. Fluid Mech. **35**, 169–182 (2003)
19. Moussaïd, M., Theraulaz, G.: Comment les piétons marchent dans la foule. La Recherche **450**, 56–59 (2011)
20. Moussaïd, M., Helbing, D., Garnier, S., Johansson, A., Combe, M., Theraulaz, G.: Experimental study of the behavioural mechanisms underlying self-organization in human crowds. Proc. R. Soc. B **276**, 2755–2762 (2009)
21. Nowak, M.A.: Evolutionary Dynamics: Exploring the Equations of Life. Belknap Press of Harvard University Press, Cambridge, MA (2006)
22. Schadschneider, A., Seyfried, A.: Empirical results for pedestrian dynamics and their implications for modeling. Netw. Heterog. Media **6**, 545–560 (2011)
23. Seyfried, A., Steffen, B., Klingsch, W., Boltes, M.: The fundamental diagram of pedestrian movement revisited. J. Stat. Mech.: Theory Exp. **360**, 232–238 (2006)
24. Talischi, C., Paulino, G.H., Pereira, A., Menezes, I.F.M.: PolyMesher: a general-purpose mesh generator for polygonal elements written in Matlab. Struct. Multidiscip. Optim. **45**, 309–328 (2012)

Energy Inequalities and Outflow Boundary Conditions for the Navier–Stokes Equations

Norikazu Saito, Yoshiki Sugitani, and Guanyu Zhou

Abstract Artificial boundary conditions play important roles in numerical simulation of real-world flow problems. A typical example is a class of outflow boundary conditions for blood flow simulations in large arteries. The common outflow boundary conditions are a prescribed constant pressure or traction, a prescribed velocity profiles, and a free-traction (do-nothing) conditions. However, the flow distribution and pressure field are unknown and cannot be prescribed at the outflow boundary in many simulations. Moreover, with those boundary conditions, we are unable to obtain energy inequalities. This disadvantage may cause numerical instability in unstationary 3D simulations. In this paper, we examine some outflow boundary conditions for the Navier–Stokes equations from the viewpoint of energy inequalities. Further, we propose an energy-preserving unilateral condition and review mathematical results including the well-posedness, variational inequality formulations, and finite element approximations.

1 Introduction

In numerical simulation of real-world flow problems, we often encounter some issues concerning artificial boundary conditions. A typical and important example is the blood flow problem in large arteries, where the blood is assumed to be a viscous incompressible fluid (cf. [6, 14]). The blood vessel is modeled by a branched pipe as illustrated, for example, in Fig. 1. We are able to give a velocity profile at the *inflow boundary* S and the flow is supposed to be a perfect non-slip on the wall C. Then, the blood flow simulation is highly dependent on the choice of artificial boundary conditions posed on the *outflow boundary* Γ.

To be more specific, let $\Omega \subset \mathbb{R}^d$, $d = 2, 3$, be a bounded domain and let the boundary $\partial\Omega$ be composed of three parts S, C, and Γ. Those S, C, and Γ are assumed to be smooth surfaces, although the whole boundary $\partial\Omega$ itself is not

N. Saito (✉) • Y. Sugitani • G. Zhou
Graduate School of Mathematical Sciences, The University of Tokyo,
Komaba 3-8-1, Meguro, Tokyo 153-8914, Japan
e-mail: norikazu@ms.u-tokyo.ac.jp; sugitani@ms.u-tokyo.ac.jp; zhoug@ms.u-tokyo.ac.jp

Y. Bazilevs, K. Takizawa (eds.), *Advances in Computational Fluid-Structure Interaction and Flow Simulation*, Modeling and Simulation in Science, Engineering and Technology, DOI 10.1007/978-3-319-40827-9_24

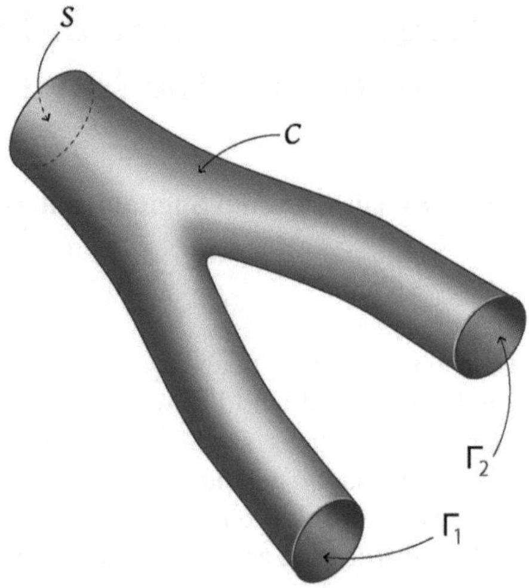

Fig. 1 Inflow boundary S, physical boundary (vessel wall) C and outflow boundary $\Gamma = \Gamma_1 \cup \Gamma_2$.

smooth; see Fig. 1. Then, for $T > 0$, we consider the Navier–Stokes equations

$$u_t + (u \cdot \nabla)u = \nabla \cdot \sigma(u,p) + f, \quad \nabla \cdot u = 0 \qquad \text{in} \Omega \times (0,T), \tag{1a}$$

$$u = b \qquad \text{on} S \times (0,T), \tag{1b}$$

$$u = 0 \qquad \text{on} C \times (0,T), \tag{1c}$$

$$u|_{t=0} = u_0 \qquad \text{on} \Omega \tag{1d}$$

for the velocity $u = (u_1, \ldots, u_d)$ and the pressure p with the density $\rho = 1$ and the kinematic viscosity ν of the viscous incompressible fluid under consideration. Therein, $\sigma(u,p) = (\sigma_{i,j}(u,p)) = -pI + 2\nu D(u)$ denotes the stress tensor, where $D(u) = (D_{i,j}(u)) = \left(\frac{1}{2}\left(\nabla u + \nabla u^{\mathrm{T}}\right)\right)$ the deformation-rate tensor and I the identity matrix. The prescribed functions $f = f(x,t)$ and $u_0 = u_0(x)$ denote the external force and initial velocity, respectively. Moreover, $b = b(x,t)$ denotes the prescribed inflow velocity with $b|_{\partial S} = 0$. Below, we set $\tau(u,p) = \sigma(u,p)n$, where n denotes the outward normal vector to $\partial\Omega$, which is called the traction vector on $\partial\Omega$. For a vector-valued function v defined on $\partial\Omega$, its normal and tangential components are denoted, respectively, as $v_n = v \cdot n$ and $v_T = v - v_n n$. Thus, $\tau_n(u,p) = \tau(u,p) \cdot n$ and $\tau_T(u) = \tau(u,p) - \tau_n(u,p)n$ are normal and tangential traction vectors, respectively. We write as $\tau = \tau(u,p)$, $\tau_n = \tau(u,p)$, and $\tau_T = \tau_T(u)$ if there is no fear of confusion.

As a common outflow boundary condition, the free-traction condition or the so-called do-nothing condition

$$\tau = 0 \quad \text{on } \Gamma \tag{2}$$

is still frequently used so far (cf. [7, 8]).

However, it is difficult to obtain reasonable flow rates at outflow boundaries Γ_j with (2). Formaggia et al. [5] proposed the so-called defective boundary condition in order to ensure prescribed flow rate,

$$\int_{\Gamma_j} u_n \, d\Gamma_j = Q_j \quad (j = 1, 2)$$

by Lagrange multiplier technique. Vignon-Clementel et al. [15] introduced the resistance boundary condition

$$\tau_n + p_0 + R_j \int_{\Gamma_j} u_n \, d\Gamma_j = 0, \quad \tau_T = 0 \quad \text{on } \Gamma_j \quad (j = 1, 2) \tag{3}$$

by setting the resistance R_j at Γ_j. Herein, p_0 denotes the pressure at the reference position. For example, R_j is chosen as $R_j = (4\nu L)/(a^2|\Gamma_j|)$, where L is the distance from the reference position to Γ_j, a the radius of tube, and $|\Gamma_j|$ the area of Γ_j to obtain the flow which is consistent with the Hagen–Poiseuille flow. On the other hand, Bruneau and Fabrie [3] proposed a certain nonlinear boundary condition from the viewpoint of the energy inequality. To describe it, we introduce a function satisfying

$$\nabla \cdot g = 0 \quad \text{in} \Omega, \quad g = b \text{on} S, \quad g = 0 \text{on} C. \tag{4}$$

For example, we could take the solution (g, π) of the Stokes equations $\nabla \cdot \sigma(g, \pi) = 0$ in Ω with (4). Then, we find (u, p) as the perturbation from (g, π), that is,

$$u = v + g, \quad p = q + \pi. \tag{5}$$

Then, if there exists a positive constant M_T, which depends only on the prescribed data, satisfying

$$\sup_{0 \le t \le T} \int_\Omega |v|^2 \, dx + 2\nu \int_0^T \int_\Omega D_{i,j}(v) D_{i,j}(v) \, dx ds \le M_T, \tag{6}$$

we say that the energy inequality is satisfied. This inequality is of use. It plays a crucial role in the construction of a solution of the Navier–Stokes equations. Furthermore, it is connected with the stability of numerical schemes from the viewpoint of numerical computation. That is, it is preferred that the energy inequality does not spoiled after discretizations. However, it is not certain the energy inequality to hold under (2). Bruneau and Fabrie [3] proposed

$$\tau(u, p) = -\frac{1}{2}[u_n]_- v + \tau(g, \pi) \text{ on } \Gamma \tag{7}$$

to realize (6), where $[a]_- = \max\{0, -a\}$. See also [2, 4]. Bazilevs et al. [1] introduce the regularized traction vector $\tilde{\tau}(u, p) = \tau(u, p) + [u_n]_- u$ and applied the resistance condition (3) to obtain

$$\tilde{\tau}(u,p)\cdot n + R\int_\Gamma u_n\, d\Gamma + p_0 = 0, \quad \tilde{\tau}(u,p) - [\tilde{\tau}(u,p)\cdot n]n = 0 \quad \text{on } \Gamma.$$

This condition is equivalently written as

$$\tau = -[u_n]_- u - \left(R\int_\Gamma u_n\, d\Gamma + p_0\right) n. \tag{8}$$

If $b = 0$ (then we can take $g = 0$ and $\pi = 0$), we derive the energy inequality under this condition. They offered several numerical results for medical problems and did not give any mathematical considerations. As a matter of fact, they described that (cf. [1, Rem. 4.1]) "*We found that the addition of this term is important for the overall stability of the computations. In its absence we sometimes experience rapid divergence, the outlet of the descending branch of the aorta being the most vulnerable location for initiation of outflow instabilities....* ". On the other hand, Labeur and Wells [11] considered essentially the same condition as (8) with $R = p_0 = 0$, where they studied energy stable hybrid discontinuous finite element method.

However, if applying (7), we need to get (g, π) before computation. We offered another boundary condition that ensures (6) to hold in [16]. That is, we propose a unilateral boundary condition of Signorini's type which is given as

$$u_n \geq 0, \quad \tau_n \geq 0, \quad u_n\tau_n = 0, \quad \tau_T = 0 \quad \text{on } \Gamma. \tag{9}$$

This is a generalization of (2) in the sense that

$$\begin{aligned} u_n > 0 \text{ on } \omega \subset \Gamma \quad &\Rightarrow \quad \tau_n = 0 \text{ on } \omega, \\ u_n = 0 \text{ on } \omega \subset \Gamma \quad &\Rightarrow \quad \tau_n \geq 0 \text{ on } \omega. \end{aligned}$$

In the actual computation, (9) should be combined with (3). Condition (9) is described by inequalities so that it is not suitable for computation directly. Instead, its penalty approximation

$$\tau_n = \frac{1}{\varepsilon}[u_n]_-, \ \tau_T = 0 \quad \text{on} \Gamma \tag{10}$$

is of use, where $\varepsilon > 0$ denotes the penalty parameter. As a matter of fact, Bazilevs' condition (8) is corresponding to a special choice of ε (cf. [16, §1]).

In the present paper, we review mathematical results for the Navier–Stokes and Stokes equations under (9) that are recently established in [13] and [16]. In Section 2, we briefly state the well-posedness of the problems. Our analysis is an application of the theory of variational inequalities. However, because of the limitation of the page number, we mention only the summary of results and skip the detail of analysis. Instead, we present variational formulations for a model Stokes problem in Section 3. Finite element approximation to the Stokes problem is introduced, and the stability and convergence results are stated in Section 4. All results will be stated without the proofs; See Saito et al. [13] and Zhou and Saito [16] for the complete proofs.

2 Well-Posedness

As is clearly stated in Introduction of Kashiwabara [9], weak solutions of Leray–Hopf's class are not suitable for the purpose of application to numerical analysis. We are strongly motivated by [9] and interested in constructing of strong solutions of Ladyzhenskaya's class (cf. Ladyzhenskaya [12]). That is, we want to find functions

$$u \in L^\infty(0,T;H^1(\Omega)^d), \quad u_t \in L^2(0,T;H^1(\Omega)^d) \cap L^\infty(0,T;L^2(\Omega)^d),$$
$$p \in L^\infty(0,T;L^2(\Omega))$$

that satisfy the Navier–Stokes equation (1) with the unilateral boundary condition (9) in the sense of distributions.

To this end, it suffices to find (v,q) satisfying the following perturbed Navier–Stokes problem (recall (5)):

(NS) For $t \in (0,T)$, find (v,q) such that

$$\begin{aligned}
&v_t + ((v+g)\cdot\nabla)v + (v\cdot\nabla)g - \nabla\cdot\sigma(v,q) = F, \ \ \nabla\cdot v = 0 && \text{in}\,\Omega,\\
&v = 0 && \text{on}\,S\cup C,\\
&v_n + g_n \geq 0, \quad \tau_n(v+g,q+\pi) \geq 0 && \text{on}\,\Gamma,\\
&(v_n+g_n)\tau_n(v+g,q+\pi) = 0, \quad \tau_T(v) = -\tau_T(g) && \text{on}\,\Gamma,\\
&v|_{t=0} = v_0 && \text{on}\,\Omega,
\end{aligned}$$

where $F = f - g_t - (g\cdot\nabla)g$ and $v_0 = u_0 - g(0)$.

Actually, under some appropriate assumptions on F, U_0, and (g,π), we succeeded in proving that (cf. [16, Theorem 2]) there exists a unique solution of (NS) satisfying

$$v \in L^\infty(0,T;H^1(\Omega)^d), \quad v_t \in L^2(0,T;H^1(\Omega)^d) \cap L^\infty(0,T;L^2(\Omega)^d),$$
$$q \in L^\infty(0,T;L^2(\Omega)).$$

For the penalty problem (1) with (10), we consider the following perturbed problem:

(NS$_\varepsilon$) Let $\varepsilon > 0$. For all $t \in (0, T)$, find $(v_\varepsilon, q_\varepsilon)$ such that

$$
\begin{aligned}
&v_{\varepsilon,t} + (v_\varepsilon + g \cdot \nabla)v_\varepsilon + (v_\varepsilon \cdot \nabla)g - \nabla \cdot \sigma(v_\varepsilon, q_\varepsilon) = F, \ \ \nabla \cdot v_\varepsilon = 0 && \text{in}\,\Omega,\\
&v_\varepsilon = 0 && \text{on}\,S \cup C,\\
&\tau_n(v_\varepsilon + g, q_\varepsilon + \pi) = \frac{1}{\varepsilon}[v_{\varepsilon n} + g_n]_-, \quad \tau_T(v_\varepsilon) = -\tau_T(g) && \text{on}\,\Gamma,\\
&v_\varepsilon|_{t=0} = v_0 && \text{on}\,\Omega.
\end{aligned}
$$

Then, there exists a unique solution of (NS$_\varepsilon$) satisfying

$$
v_\varepsilon \in L^\infty(0, T; H^1(\Omega)^d), \quad v_{\varepsilon,t} \in L^2(0, T; H^1(\Omega)^d) \cap L^\infty(0, T; L^2(\Omega)^d),
$$
$$
q_\varepsilon \in L^\infty(0, T; L^2(\Omega))
$$

for a sufficiently small ε; See [16, Theorem 3].

We proved those results by converting (NS) and (NS$_\varepsilon$) into suitable variational formulations after having re-defined the traction vectors as functionals. In the next section, we will illustrate those variational formulation by using a model Stokes problem.

3 Model Stokes Problem

In order to exemplify the variational formulation of (NS), we consider a (non-stationary) model Stokes problem which is given as

$$
\begin{aligned}
&-\nu\Delta u + \nabla p = f, \quad \nabla \cdot u = 0 && \text{in } \Omega, && \text{(11a)}\\
&u = 0 && \text{on } S \cup C, && \text{(11b)}\\
&u_n + g_n \geq 0, && \text{on } \Gamma, && \text{(11c)}\\
&\tau_n + \alpha_n \geq 0, \ (u_n + g_n)(\tau_n + \alpha_n) = 0, \ \tau_T + \alpha_T = 0 && \text{on } \Gamma, && \text{(11d)}
\end{aligned}
$$

where f and α are prescribed vector-valued functions.

For the sake of simplicity, we restrict our attention to a polyhedral domain $\Omega \subset \mathbb{R}^d$, $d = 2, 3$, with a polygon (line segment) Γ. We use the standard Lebesgue and Sobolev spaces, for example, $L^2(\Omega)$, $H^1(\Omega)$, $L^2(\Gamma)$, and $H^{1/2}(\Gamma)$. The basic function spaces of our consideration are

$$
V = \{v \in H^1(\Omega)^d \mid v = 0 \text{ on } S \cup C\} \quad \text{and} \quad Q = L^2(\Omega).
$$

They are Hilbert spaces equipped with the norms $\|v\|_1 = \|v\|_{H^1(\Omega)^d}$ and $\|q\| = \|q\|_{L^2(\Omega)}$, respectively. The space Q_0 denotes the set of all $L^2(\Omega)$ functions with the zero mean value. A convex subset $K = \{v \in V \mid v_n + g_n \geq 0 \text{ on } \Gamma\}$ of V plays an important role. We also use the so-called *Lions–Magenes space* $H_{00}^{1/2}(\Gamma)$. Below we set $M = H_{00}^{1/2}(\Gamma)$ and $\|\mu\|_{1/2,\Gamma} = \|\mu\|_{H_{00}^{1/2}(\Gamma)}$.

Let

$$(\cdot,\cdot) = \text{the standard inner product of } L^2(\Omega)^d,$$
$$[[\cdot,\cdot]] = [[\cdot,\cdot]]_{(M^d)',M^d} = \text{ the duality pairing between } (M^d)' \text{ and } M^d.$$

We use the following forms:

$$a(u,v) = 2\nu \int_\Omega D_{i,j}(u)D_{i,j}(v)\,dx \qquad (u,v \in H^1(\Omega)^d);$$
$$b(p,u) = -\int_\Omega p(\nabla\cdot u)\,dx \qquad (p \in Q,\ u \in H^1(\Omega)^d).$$

Below we assume

$$f \in L^2(\Omega)^d, \qquad \alpha \in (M^d)'.$$

Then, the variational formulation of (11) is described as (cf. [13, Theorem 10]):

(VI) Find $(u,p) \in K \times Q$ such that

$$a(u,v-u) + b(p,v-u) \geq (f,v-u) - [[\alpha, v-u]] \qquad (\forall v \in K),$$
$$b(q,u) = 0 \qquad (\forall q \in Q).$$

As a matter of fact, (VI) is not a new problem. In a classical monograph, Kikuchi and Oden [10], Chapter 7 is devoted to similar problems. However, their problem contains the traction condition $\tau = h$. More precisely, they suppose that C is divided into two parts C_1, C_2 and consider

$$u = 0 \quad \text{on } S \cup C_1, \qquad \tau(u,p) = h \quad \text{on } C_2$$

instead of (11b). Then, supposing

$$\overline{\Gamma} \cap \overline{(S \cup C_1)} = \emptyset, \tag{12}$$

we can prove that there exists a domain constant $\gamma > 0$ satisfying

$$\gamma\left[\|q\| + \|\tau\|_{H^{-1/2}(\Gamma)}\right] \leq \sup_{v \in H^1(\Omega)^d, v|_{S\cup C_1}=0} \frac{\int_\Omega q(\nabla\cdot v)\,dx + \int_\Gamma \tau v_n\,d\Gamma}{\|v\|_1} \tag{13}$$

for any $(q,\tau) \in L^2(\Omega) \times H^{-1/2}(\Gamma)$ (cf. [10, Theorem 7.2]). This inequality is usually referred to as the coupled Babuška–Brezzi condition. The well-posedness and error estimates of the corresponding penalty problem are direct consequences of this result from the general theory. In contrast, we are interested in establishing a formulation without the traction boundary condition. Unfortunately, if $C_2 = \emptyset$, (13) is not available and it makes analysis somewhat difficult. Moreover, we do not prefer assuming (12). Consequently, we have to develop a totally new method of analysis in this paper. In particular, we offer a new device to treat the pressure part; see [13].

4 Finite Element Approximation

As a regularization of $[s]_-(s \in \mathbb{R})$, we introduce a function $\phi_\delta : \mathbb{R} \to \mathbb{R}$ that satisfies

$$\phi_\delta \text{ is a non-increasing } C^1 \text{ function;}$$

$$|\phi_\delta(s) - [s]_-| \leq C\delta \quad (s \in \mathbb{R}); \qquad \left|\frac{d}{ds}\phi_\delta(s)\right| \leq C \qquad (s \in \mathbb{R});$$

$$\phi_\delta(s) = 0 \quad (s \geq 0), \qquad 0 \leq \phi_\delta(s) \leq -s \quad (s < 0).$$

where $\delta \in (0,1]$ is regularized parameter and C's are independent of δ. For example, we can take

$$\phi_\delta(s) = \begin{cases} 0 & (s \geq 0) \\ (\sqrt{s^2+\delta^2} - \delta) & (s < 0). \end{cases}$$

For penalty parameter $\varepsilon \in (0,1]$, we consider the following penalty problem:

(PE$_{\varepsilon,\delta}$) Find $(u,p) \in V \times Q$ such that

$$a(u,v) + b(p,v) - \frac{1}{\varepsilon}\int_\Gamma \phi_\delta(u_n + g_n)v_n\, d\Gamma = (f,v) - [[\alpha, v]] \quad (\forall v \in V),$$

$$b(q,u) = 0 \qquad (\forall q \in Q).$$

We use the so-called P1 bubble/P1 (P1b/P1) elements for discretization. Let $\{\mathscr{T}_h\}_h$ be a *regular* family of triangulations of Ω. As the granularity parameter, we have employed $h = \max\{h_T \mid T \in \mathscr{T}_h\}$, where h_T denotes the diameter of T. We introduce the following function spaces:

$$V_h = \{v_h \in C^0(\overline{\Omega}) \mid v_h = 0 \text{on} S \cup C,\ v_h|_T \in [\mathscr{P}_1^{(d)} \oplus \mathrm{span}\{\varphi_T\}]^d\ (\forall T \in \mathscr{T}_h)\},$$

$$Q_h = \{q_h \in C^0(\overline{\Omega}) \mid q_h|_T \in \mathscr{P}_1^{(d)}\ (\forall T \in \mathscr{T}_h)\}, \quad M_h = \{\mu_h = v_{hn}|_\Gamma \mid v_h \in V_h\},$$

where $\mathscr{P}_k^{(d)}$ denotes the set of all polynomials in $x_1, \ldots, x_d$ of degree $\leq k$, and $\varphi_T = \prod_{i=1}^{d+1} \lambda_{T,i}$ with $\lambda_{T,1}, \ldots, \lambda_{T,d+1}$ the barycentric co-ordinates of T.

Let us denote by $\mathscr{S}_h$ the $d-1$ dimensional triangulation of Γ inherited from $\mathscr{T}_h$. We obviously have

$$M_h = \{\mu_h \in C(\overline{\Gamma}) \mid \mu_h|_S \in \mathscr{P}_1^{(d-1)} \ (\forall S \in \mathscr{S}_h),\ \mu_h|_{\partial\Gamma} = 0\}.$$

Moreover, we introduce a projection operator $\Lambda : Q \to Q_0$ by setting

$$\Lambda q = q - m(q) \quad \text{with} \quad m(q) = \frac{1}{|\Omega|}\int_\Omega q\, dx \qquad (q \in Q).$$

Then, the finite element approximation for $(\mathrm{PE}_{\varepsilon,\delta,h})$ reads as follows:

$(\mathrm{PE}_{\varepsilon,\delta,h})$ Find $(u_h, p_h) \in V_h \times Q_h$ such that

$$a(u_h, v_h) + b(p_h, v_h) - \frac{1}{\varepsilon}\int_\Gamma \phi_\delta(u_n + g_n) v_{hn}\, d\Gamma = (f, v_h) - [[\alpha, v_h]] \quad (\forall v_h \in V_h),$$

$$b(q_h, u_h) = 0 \qquad (\forall q_h \in Q_h).$$

Then, we proved the following results (cf. [13, Theorems 14, 15, and 23]):

Theorem 1. *There exists a unique solution $(u_h, p_h) \in V_h \times Q_h$ of $(PE_{\varepsilon,\delta,h})$, and we have*

$$\|u_h\|_1 + \|\hat{p}_h\| + \left\| \frac{1}{\varepsilon}\phi_\delta(u_{hn} + g_n) + k_h \right\|_{M_h'} \leq C_*,$$

where $\hat{p}_h = \Lambda p_h$ and $k_h = m(p_h)$.

Hereinafter, C_* denotes a positive constant depending only on Ω, $\|f\|$, $\|g\|_1$, and $\|\alpha\|_{(M^d)'}$. If making some additional assumptions, we will obtain some fine stability estimates.

Theorem 2. *Assume*

(A1) the family $\{\mathscr{S}_h\}_h$ is of quasi-uniform;
(A2) there exists $\Gamma_1 \subset \Gamma$ with $|\Gamma_1| > 0$ which is independent of h, ε, δ, and Ω such that $u_{hn} + g_n > 0$ on Γ_1.

Then, the solution $(u_h, p_h) \in V_h \times Q_h$ of $(PE_{\varepsilon,\delta,h})$ admits the following estimates:

$$\|u_h\|_1 + \|p_h\| + \left\| \frac{1}{\varepsilon}\phi_\delta(u_{hn} + g_n) \right\|_{M_h'} \leq C_*;$$

$$\left\| \frac{1}{\varepsilon}\phi_\delta(u_{hn} + g_n) \right\|_{M'} + \frac{1}{\sqrt{\varepsilon}}\|[u_{hn} + g_n]_-\|_{L^2(\Gamma)} \leq C_*\left(1 + \frac{\delta}{\varepsilon}\right).$$

Remark 1. Condition (A2) is not restrictive; if $|b|$ is sufficiently large and h, ε, δ are suitably small, it is natural to suppose this condition.

Remark 2. If $\delta \le c_0\varepsilon$ with some $c_0 > 0$, we have $\|[u_{hn} + g_n]_-\|_\Gamma \to 0$ as $\varepsilon \to 0$.

Theorem 3. *Assume that (A1) and (A2) are satisfied. Let (u, p) and (u_h, p_h) be solutions of (PDE) and (PE$_{\varepsilon,\delta,h}$), respectively, and suppose that $(u, p) \in H^2(\Omega)^d \times H^1(\Omega)$ and $\tau_n(u, p) + \alpha_n \in M$. Moreover, assume that $h \le c_1\varepsilon$ with a constant $c_1 > 0$. Then, we have*

$$\|u - u_h\|_1 + \|\hat{p} - \hat{p}_h\| \le C_{**}(\sqrt{\varepsilon} + \sqrt{\delta} + \sqrt{h}),$$

where $\hat{p} = \Lambda p$, $\hat{p}_h = \Lambda p_h$, C_{**} denotes a positive constant depending only on Ω, $|u|_{H^2}$, $|p|_{H^1}$, $\|\tau_n(u, p) + \alpha_n\|_M$, $\|f\|$, $\|g\|_1$ and $\|\alpha\|_{(M^d)'}$. If, furthermore,

(A3) there exists $\Gamma_0 \subset \Gamma$ with $|\Gamma_0 \cap \Gamma_1| > 0$ such that $u_n + g_n > 0$ on Γ_0,

then we have

$$\|u - u_h\|_1 + \|p - p_h\| \le C_{**}(\sqrt{\varepsilon} + \sqrt{\delta} + \sqrt{h}).$$

Acknowledgements We thank Professors K. Takizazawa and H. Suito who brought the subject to our attention. We also thank Dr. T. Kashiwabara for his valuable suggestions. This work is supported by JST, CREST, and JSPS KAKENHI (23340023, 15H03635, 15K13454).

References

1. Bazilevs, Y., Gohean, J.R., Hughes, T.J.R., Moser, R.D., Zhang, Y.: Patient-specific isogeometric fluid-structure interaction analysis of thoracic aortic blood flow due to implantation of the Jarvik 2000 left ventricular assist device. Comput. Methods Appl. Mech. Eng. **198**, 3534–3550 (2009)
2. Boyer, F., Fabrie, P.: Mathematical Tools for the Study of the Incompressible Navier-Stokes Equations on Related Models. Springer, Berlin (2012)
3. Bruneau, C.H., Fabrie, P.: Effective downstream boundary conditions for incompressible Navier-Stokes equations. Int. J. Numer. Methods Fluids **19**, 693–705 (1994)
4. Bruneau, C.H., Fabrie, P.: New efficient boundary conditions for incompressible Navier-Stokes equations: a well-posedness result. RAIRO Modél. Math. Anal. Numér. **30**, 815–840 (1996)
5. Formaggia, L., Gerbeau, J.F., Nobile, F., Quarteroni, A.: Numerical treatment of defective boundary conditions for the Navier-Stokes equations. SIAM J. Numer. Anal. **40** 376–401 (2002)
6. Formaggia, L., Quarteroni A., Veneziani, A. (eds.): Cardiovascular Mathematics. Springer, Berlin (2009)
7. Gresho, P.M., Sani, R.L.: Incompressible Flow and the Finite Element Method. Isothermal Laminar Flow, vol. 2. Wiley, New York (2000)
8. Heywood, J.G., Rannacher, R., Turek, S.: Artificial boundaries and flux and pressure conditions for the incompressible Navier-Stokes equations. Int. J. Numer. Methods Fluids **22**, 325–352 (1996)
9. Kashiwabara, T.: On a strong solution of the non-stationary Navier-Stokes equations under slip or leak boundary conditions of friction type. J. Differ. Equ. **254**, 756–778 (2013)

10. Kikuchi, N., Oden, J.T.: Contact Problems in Elasticity. SIAM, Philadelphia (1988)
11. Labeur, R.J., Wells, G.N.: Energy stable and momentum conserving hybrid finite element method for the incompressible Navier-Stokes equations. SIAM J. Sci. Comput. **34**, 889–913 (2012)
12. Ladyzhenskaya, O.A.: The Mathematical Theory of Viscous Incompressible Flow. Gordon and Breach, London (1969)
13. Saito, N., Sugitani, Y., Zhou, G.: Unilateral problem for the Stokes equations: the well-posedness and finite element approximation. Appl. Numer. Math. **105**, 124–147 (2016). doi: 10.1016/j.apnum.2016.03.002
14. Taylor, C.A., Hughes, T.J., Zarins, C.K.: Finite element modeling of blood flow in arteries. Comput. Methods Appl. Mech. Eng. **158**, 155–196 (1998)
15. Vignon-Clementel, I.E., Figueroa, C.A., Jansenc, K.E., Taylor, C.A.: Outflow boundary conditions for three-dimensional finite element modeling of blood flow and pressure in arteries. Comput. Methods Appl. Mech. Eng. **195**, 3776–3796 (2006)
16. Zhou, G., Saito, N.: The Navier-Stokes equations under a unilateral boundary condition of Signorini's type. J. Math. Fluid Mech. (to appear). doi:10.1007/s00021-016-0248-7

Numerical Studies on the Stability of Mixed Finite Elements Over Anisotropic Meshes Arising from Immersed Boundary Stokes Problems

Ferdinando Auricchio, Franco Brezzi, Adrien Lefieux, and Alessandro Reali

Abstract Motivated by a recently proposed local refinement strategy for immersed interface problems, in this work we aim at dealing with the behavior of mixed finite elements for the Stokes problem in (strongly) anisotropic mesh situations, leading to severely distorted elements. In fact, the majority of the theoretical results present in the finite element literature has been carried out under the assumption of well-shaped elements. In the case such a condition is not satisfied, the inf-sup constant may degenerate, thus leading to the instability of the system.

To this aim, we herein use a generalized eigenvalue test problem that allows to conveniently investigate the behavior of different mixed finite elements over anisotropic mesh patterns, arising, e.g., in immersed interface Stokes problems. We then test and study the numerical stability of two 2D finite element pairs, namely the Hood-Taylor ($\mathbf{P}_2/P_1$) and the Hood-Taylor ($\mathbf{P}_2^+/P_1$) with a velocity field enhanced by a cubic bubble. On the contrary to the results presented in [3], we herein provide additional results on the potential presence of spurious modes and their locations. The present results corroborate those obtained in more complex cases described in [3] and [4].

F. Auricchio • A. Lefieux
Università degli Studi di Pavia, Pavia, Italy
e-mail: auricchio@unipv.it; adrien.lefieux@unipv.it

F. Brezzi
Istituto di Matematica Applicata e Tecnologie Informatiche "Enrico Magenes" – CNR, Pavia, Italy
e-mail: brezzi@imati.cnr.it

A. Reali (✉)
Università degli Studi di Pavia, Pavia, Italy

Institute of Advanced Study – Technical University of Munich, Garching, Germany
e-mail: alessandro.reali@unipv.it

Y. Bazilevs, K. Takizawa (eds.), *Advances in Computational Fluid-Structure Interaction and Flow Simulation*, Modeling and Simulation in Science, Engineering and Technology, DOI 10.1007/978-3-319-40827-9_25

1 Introduction

In a previous work (see [3]), the present authors discussed the inf-sup stability of mixed finite elements for the incompressible Stokes problem (namely, $\mathbf{P}_2/P_1$ and $\mathbf{P}_2^+/P_1$, i.e., the Hood-Taylor and the Hood-Taylor with an additional bubble in the velocity field) on anisotropic elements (i.e., elements very small angles).

More precisely, in [3], we presented a local remeshing strategy in the context of an immersed boundary approach for 2D problems with triangles. The simple idea is to keep the element geometry fixed and to proceed to a local (element-wise) remeshing, ergo highly anisotropic triangles are introduced in the remeshed ones. The use of anisotropic triangles is known to be allowed within the finite element method, as long as the largest angle of a triangle is bounded away from π (see, e.g., [5] and [1] for an extended discussion). However, their use within the mixed finite element largely remains an open problem, in particular from a theoretical point of view (see Appendix C in [10]).

In [3] our results are in accordance with those in [2]. More precisely, we showed that the $\mathbf{P}_2/P_1$ element fails to be inf-sup stable in the context of our locally anisotropic remeshing, but that $\mathbf{P}_2^+/P_1$ passes all our numerical tests. Additionally to [2], we showed that the instability of the Hood-Taylor element occurs in some very specific cases and, as a consequence, the Hood-Taylor element maybe used within the locally anisotropic remeshing approach for a very large set of problems without showing any sign of inf-sup stability issues. The present work deals with these findings in more details.

In the present work, we provide additional results to those published in [3], emphasizing on the set of problems which can be used with the Hood-Taylor element. Starting from the smallest generalized eigenvalue (SGE) test presented in [3], we discuss in detail the spurious modes associated with $\mathbf{P}_2/P_1$, not only by looking at the numerical inf-sup constant, but also by looking at the first 8 pressure modes (i.e., those associated with the smallest eigenvalues of the SGE test). We also analyze their associated eigenvectors, to discuss the locations of the spurious modes. On the contrary to the work performed in [3], we present the SGE tests on three simple meshes, instead of only one. The results presented here are coherent with those obtained with more complex problems in [3, 4] and [10].

In Section 2, we shortly present the anisotropic local refinement strategy and the context in which the method has been developed. In Section 3, we present the incompressible Stokes problem both in weak and discrete forms, as well as the inf-sup condition and the associated error estimates. We further stress the importance of having inf-sup stable elements, and how anisotropic elements may adverse the inf-sup stability. In Section 11 we present the numerical tests and results. In Appendix A.1, we provide the complete set of results.

2 An Anisotropic Local Remeshing Strategy

In this section we provide a very short introduction to the motivations for the anisotropic local remeshing strategy presented in [3]. Firstly, in Section 2.1, we discuss the geometry of the problem, which includes an immersed boundary and how a mesh can be associated with such a problem. Secondly, in Section 2.2, we discuss the construction of a discrete immersed boundary and a mesh the locally anisotropic remeshing strategy.

2.1 Geometry

In this section we recall some geometric aspects of the problem, i.e., the problem of the construction of a mesh conveniently discretizing the considered physical domain. Two approaches are available: "fitted" or "unfitted" (cf. Fig. 1).

In the fitted strategy the discretized domain fits the boundary of the problem, while in the unfitted one the physical domain is a subset of the discretization. More precisely, in the unfitted case, we consider a problem defined on $\Omega \subset \mathbb{R}^2$ such that a part of the boundary of $\partial\Omega$, denoted by Γ (named *immersed boundary*), is not fitted *a priori* by the triangulation of $\hat{\Omega}$, with $\Omega \subset \hat{\Omega}$. The part of the boundary $\partial\Omega$ that is fitted by the triangulation of $\hat{\Omega}$ is denoted by Σ.

The "unfitted" strategy presented in Figure 1(c) avoids the difficulties and the costs connected with the generation during the pre-processing phase of fitted meshes in complicated situations.

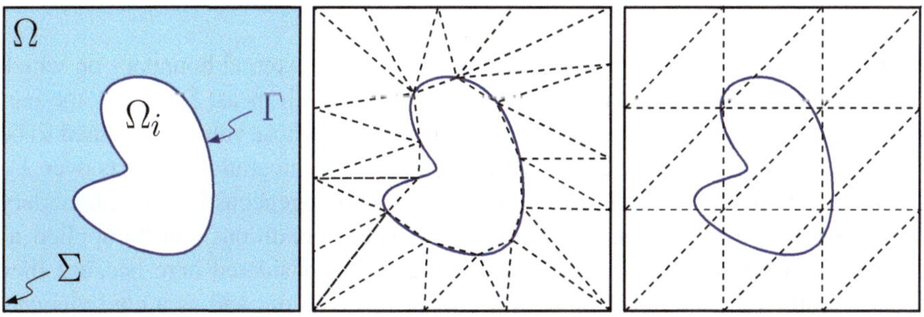

(a) Physical domain (b) Fitted grid (c) Unfitted grid

Fig. 1 Fitted and unfitted discretizations of the physical region Ω: Ω_i is the interior (non physical) domain, Γ is the immersed boundary, $\Sigma = \partial\hat{\Omega}$ is the external boundary, and $\hat{\Omega} := \Omega \cup \Omega_i \cup \Gamma$ is the discretized domain.

2.2 Discrete Immersed Boundary and Local Remeshing

We assume that a regular triangulation $\hat{\mathcal{T}}$ of $\hat{\Omega}$ (named *background* mesh) and the interface Γ satisfy the conditions presented in [9], that is, the boundary Γ crosses once two triangle edges. We note that there always exists a sufficiently fine triangulation of $\hat{\Omega}$ such that the conditions are fulfilled for any smooth immersed boundary. The reconstructed boundary of Γ is denoted Γ_h and it is the linear interpolation of all intersections with the background mesh edges. It follows that the reconstructed interface is a segment in each intersected element, and it defines a new domain Ω_h such that $\partial\Omega_h = \Sigma \cup \Gamma_h$. Domain Ω_h is referred to as the computational domain. The linear reconstruction of Γ is not a limitation of the method we propose and, in the case with a curved immersed boundary, isoparametric elements may be used, as well as more complex algorithms, to describe the boundary.

For triangles cut by the immersed boundary we consider the two cases, depending if the subelement belonging to Ω_h is: a) a triangle, or b) a quadrilateral. Since we wish to work only with triangles we consider subdividing the quadrilateral into two triangles. This subdivision of the quadrilateral is not unique, but we point out that the Delaunay triangulation leads to the best subdivision, i.e., it minimizes the distortion of the triangles by ensuring the best choice of element ratios (with the element ratio defined as $\sigma = h/r$, with h the diameter of the element and r the radius of its inscribed circle, see, e.g., [6]). The union of uncut elements belonging of $\hat{\mathcal{T}}$ to Ω_h and the remeshed ones defines a partition denoted hereafter as $\mathcal{T}_r$. Notice that $\mathcal{T}_r$ is a partition of Ω_h. More details are given in [3].

3 Model Problem: The Incompressible Stokes Problem

Let $\Sigma = \Sigma_D \cup \Sigma_N$ where Σ_D denotes the part of the external boundary on which we impose a Dirichlet boundary condition and Σ_N (with $\text{meas}(\Sigma_N) > 0$) the part on which we impose a Neumann boundary condition, whose value is assumed to be zero without loss of generality such that the term after integration by parts over Σ_N vanishes. On the other hand, on Γ, we consider a homogeneous Dirichlet boundary condition but nonhomogeneous Dirichlet boundary conditions can be applied as well. Neumann boundary conditions on Γ are not considered here because they can be enforced "naturally" in the variational formulation, and as a consequence, they are easier to tackle and a different framework might be more suitable than the present one.

3.1 The Incompressible Stokes Problem

3.1.1 Continuous Weak Formulation

The model problem we consider in this chapter is given by a standard weak form of the incompressible Stokes equations.

The weak formulation of the incompressible Stokes problem reads:

Problem 1. Find $(\mathbf{u}, p) \in V(\Omega) \times Q(\Omega)$ such that

$$\int_{\Omega} \nabla \mathbf{u} : \nabla \mathbf{v} - \int_{\Omega} p \,\mathrm{div}\,(\mathbf{v}) = \int_{\Omega} \mathbf{f} \cdot \mathbf{v} \qquad \forall \mathbf{v} \in V_0(\Omega), \tag{1a}$$

$$\int_{\Omega} q \,\mathrm{div}\,(\mathbf{u}) = 0 \qquad \forall q \in Q(\Omega). \tag{1b}$$

As functional spaces, with a standard notation for the Sobolev spaces, we employ

$$V(\Omega) := \{\mathbf{v} \in [H^1(\Omega)]^2;\ \mathbf{v}_{|\Sigma_D} = \mathbf{u}_D \text{ and } \mathbf{v}_{|\Gamma} = \mathbf{0}\}, \tag{2a}$$

$$V_0(\Omega) := \{\mathbf{v} \in [H^1(\Omega)]^2;\ \mathbf{v}_{|\Sigma_D} = \mathbf{0} \text{ and } \mathbf{v}_{|\Gamma} = \mathbf{0}\}, \tag{2b}$$

$$Q(\Omega) := L^2(\Omega). \tag{2c}$$

For the incompressible Stokes problem, $\mathbf{u}$ is the velocity of the flow and p is the hydrostatic pressure, more precisely a Lagrange multiplier associated with the incompressibility constraint (see (1b)).

3.1.2 Discrete Formulation

We provide an example of the discretized Stokes problem using the locally anisotropic strategy ($\mathcal{T}_r$ contains anisotropic triangles) with $\mathbf{P}_2/P_1$.

The discrete counter part of Problem 1 reads

Problem 2. Find $(\mathbf{u}_h, p_h) \in V^h \times Q^h$ such that

$$\int_{\Omega_h} \nabla \mathbf{u}_h : \nabla \mathbf{v}_h - \int_{\Omega_h} p_h \,\mathrm{div}\,(\mathbf{v}_h) = \int_{\Omega_h} \mathbf{f} \cdot \mathbf{v}_h \qquad \forall \mathbf{v}_h \in V_0^h, \tag{3a}$$

$$\int_{\Omega_h} q_h \,\mathrm{div}\,(\mathbf{u}_h) = 0 \qquad \forall q_h \in Q^h. \tag{3b}$$

The finite element spaces (Hood-Taylor) are given by

$$\mathbf{V}^h := \{\mathbf{v} \in C^0(\Omega_h); \mathbf{v}_{|T} \in [\mathcal{P}_2]^2, \mathbf{v}_{|\Sigma_D^h} = \mathbf{u}_D \ \&\ \mathbf{v}_{|\Gamma_h} = \mathbf{0}, \forall T \in \mathcal{T}_r\}, \tag{4a}$$

$$\mathbf{V}_0^h := \{\mathbf{v} \in C^0(\Omega_h); \mathbf{v}_{|T} \in [\mathcal{P}_2]^2, \mathbf{v}_{|\Sigma_D^h} = \mathbf{0} \ \&\ \mathbf{v}_{|\Gamma_h} = \mathbf{0}, \forall T \in \mathcal{T}_r\}, \tag{4b}$$

$$Q^h := \{q \in C^0(\Omega_h); q_{|T} \in [\mathcal{P}_1], \forall T \in \mathcal{T}_r\}. \tag{4c}$$

In this work, we also consider the so-called $\mathbf{P}_2^+/P_1$ element defined by

$$\mathbf{V}^h := \{\mathbf{v} \in C^0(\Omega_h); \mathbf{v}_{|T} \in [\mathcal{P}_2 \oplus B_3]^2, \mathbf{v}_{|\Sigma_D^h} = \mathbf{u}_D, \mathbf{v}_{|\Gamma_h} = \mathbf{0}, \forall T \in \mathcal{T}_r\} \tag{5a}$$

$$\mathbf{V}_0^h := \{\mathbf{v} \in C^0(\Omega_h); \mathbf{v}_{|T} \in [\mathcal{P}_2 \oplus B_3]^2, \mathbf{v}_{|\Sigma_D^h} = \mathbf{0}, \mathbf{v}_{|\Gamma_h} = \mathbf{0}, \forall T \in \mathcal{T}_r\} \tag{5b}$$

where B_3 denotes the space of cubic bubble functions (see [7] for more details) and the discrete pressure space is given in Equation (4c).

The interest for $\mathbf{P}_2^+/P_1$ over $\mathbf{P}_2/P_1$ will be cleared in subsequent sections.

3.2 The Inf-Sup Condition and Error Estimates

Given the approximations $\mathbf{u}_h = \sum_{i=1}^n \mathbf{N}_i\hat{\mathbf{u}}_i$ and $p_h = \sum_{i=1}^m M_i\hat{p}_i$, where $\mathbf{N}_i$ and M_i are the finite element bases for $\mathbf{V}_0^h$ and Q^h (with n and m the number of degrees of freedom, respectively) the discrete incompressible Stokes problem in matrix form reads

$$\begin{bmatrix} \mathbf{A} & \mathbf{B}^T \\ \mathbf{B} & \mathbf{0} \end{bmatrix} \begin{Bmatrix} \hat{\mathbf{u}} \\ \hat{\mathbf{p}} \end{Bmatrix} = \begin{Bmatrix} \hat{\mathbf{f}} \\ \hat{\mathbf{g}} \end{Bmatrix}, \tag{6}$$

where $\mathbf{A}$, $\mathbf{B}$ are the stiffness and divergence matrices, respectively. Explicit definitions are provided in [3].

The Euclidean norm is given by $||\hat{\mathbf{v}}||_0^2 = \hat{\mathbf{v}}^T\hat{\mathbf{v}}$ with $\hat{\mathbf{v}} \in \mathbb{R}^n$. We also consider the norm defined by the stiffness matrix $\mathbf{A}$, that is $||\hat{\mathbf{v}}||_A^2 = \hat{\mathbf{v}}^T\mathbf{A}^T\hat{\mathbf{v}}$ and its associated dual norm given by $||\hat{\mathbf{v}}||_{A'}^2 = \hat{\mathbf{v}}^T\mathbf{A}^{-T}\hat{\mathbf{v}}$. Let $\hat{\mathbf{q}} \in \mathbb{R}^m$, then the norm used for the pressure field is given by $||\hat{\mathbf{q}}||_M^2 = \hat{\mathbf{q}}^T\mathbf{M}^T\hat{\mathbf{q}}$ and its associated dual norm by $||\hat{\mathbf{q}}||_{M'}^2 = \hat{\mathbf{q}}^T\mathbf{M}^{-T}\hat{\mathbf{q}}$, where $\mathbf{M}$ is the pressure mass matrix.

It is well known that a key component for (6) to have a unique solution is the satisfaction of the following condition (see [7]):

Inf-sup: $\exists\beta_h > 0$ (independent of h) such that

$$\max_{\hat{\mathbf{v}} \in \mathbb{R}^n \setminus \{\mathbf{0}\}} \frac{\hat{\mathbf{v}}^T\mathbf{B}^T\hat{\mathbf{q}}}{||\hat{\mathbf{v}}||_A} \geq \beta_h ||\hat{\mathbf{q}}||_M \qquad \forall \hat{\mathbf{q}} \in \mathbb{R}^m. \tag{7}$$

Being $\hat{\mathbf{u}}^I$ and $\hat{\mathbf{p}}^I$ the vectors of analytical solutions, an error estimate is given by (see [7]):

$$||\hat{\mathbf{u}}^I - \hat{\mathbf{u}}||_A \leq C\left(||\hat{\mathbf{f}}||_{A'} + \beta_h^{-1}||\hat{\mathbf{g}}||_{M'}\right), \tag{8}$$

$$||\hat{\mathbf{p}}^I - \hat{\mathbf{p}}||_M \leq C\left(\beta_h^{-1}||\hat{\mathbf{f}}||_{A'} + \beta_h^{-2}||\hat{\mathbf{g}}||_{M'}\right), \tag{9}$$

where C denotes a general constant independent of h and β_h.

We clearly can see from (8) and (9) that if $\beta_h \to 0$ as the element ratio $\sigma \to \infty$ (i.e., when the triangle gets highly distorted), then the error for the pressure may not be bounded and it depends on $1/\beta_h^2$, while the velocity field may also not be bounded but it depends only on $1/\beta_h$. Notice that $\hat{\mathbf{g}}$ is empty only if $\mathbf{u}_D = \mathbf{0}$, which does not occur in practice. In that specific case, the velocity error is bounded and the pressure error depends only on $1/\beta_h$.

4 Numerical Tests

In the following numerical tests, we wish to evaluate the numerical inf-sup constant β_h (see Equation (10)). The constant β_h can easily be evaluated by solving the following generalized eigenvalue problem (see, e.g., [11] and [8]):

$$\mathbf{B}\mathbf{A}^{-1}\mathbf{B}^T\mathbf{q} = \lambda \mathbf{M}\mathbf{q}. \tag{10}$$

The numerical inf-sup constant is given by the square root of the smallest λ if $\text{meas}(\Sigma_N) > 0$ or by the square root of second smallest λ if $\text{meas}(\Sigma_N) = 0$. Indeed, in the latter case, the smallest λ is always zero since $\text{Ker}(\mathbf{B}^T) \supset \{\mathbf{1}\}$.

4.1 Smallest Generalized Eigenvalue Test Problems

We propose a constant flow problem with two sets of boundary conditions (see Fig. 2). The analytical solution for Problem D and Problem N is given by Equation (11). The test consists in moving an "immersed boundary" (depicted in red in Fig. 2) as $a \to 0$ or $b \to 0$, applying the locally anisotropic remeshing (examples are shown in Fig. 3), and by solving the eigenproblem (10). The tests are performed on three meshes, while in [3] only one mesh is considered. The mesh patterns used in these tests are depicted in Fig. 3.

$$\begin{cases} u_x(x,y) &= 1, \\ u_y(x,y) &= 0, \\ p(x,y) &= 0. \end{cases} \tag{11}$$

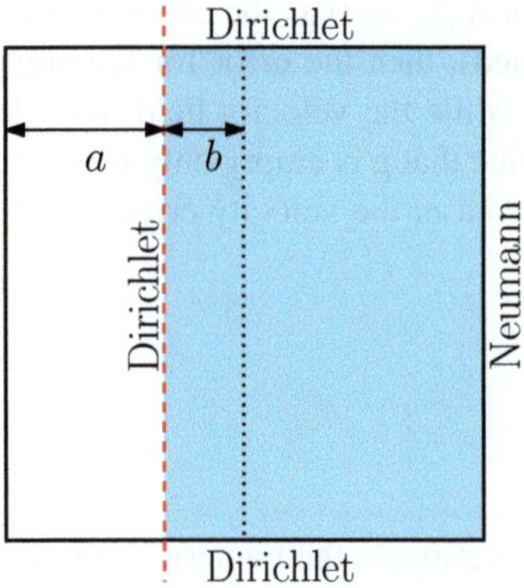

(a) Problem D: with Dirichlet boundary conditions on $y = \pm 1$.

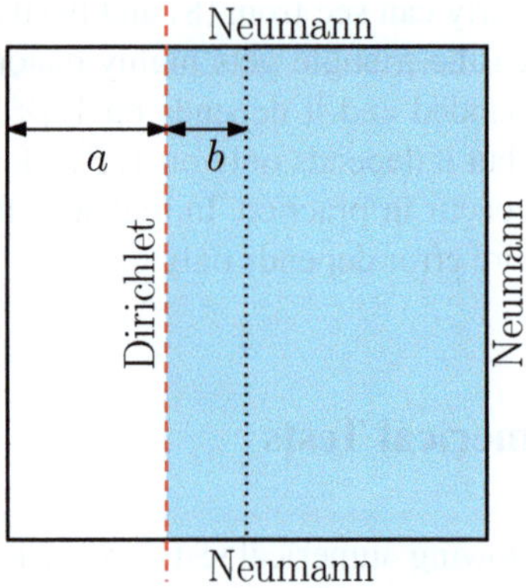

(b) Problem N: with Neumann boundary conditions on $y = \pm 1$.

Fig. 2 Boundary value problems under consideration for the eigenproblem (10). The Neumann condition here is intended as the free-stress condition.

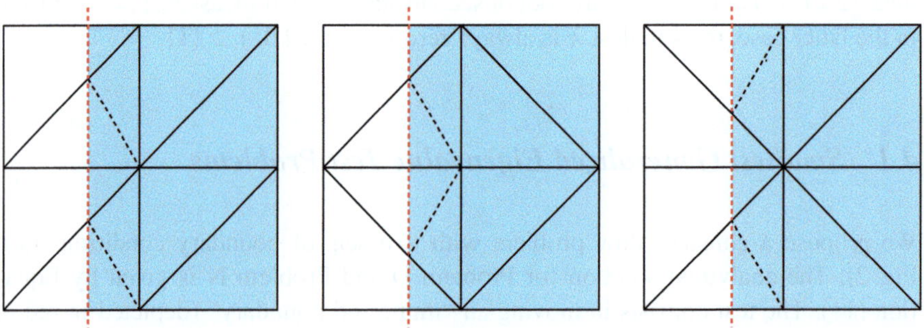

(a) Mesh 1 (b) Mesh 2 (c) Mesh 3

Fig. 3 The three background meshes (i.e., the partitions $\hat{\mathcal{T}}$) used for the SGE tests. The immersed boundary is depicted in dashed red. The computational domain Ω_h is depicted in blue. The remeshed partition $\mathcal{T}_r$ is also represented. The quadrilateral subdivision into triangles is represented in dashed black.

4.2 Eigenvalue and Eigenvector Analysis

We perform the tests with two mixed finite elements: $\mathbf{P}_2/P_1$ (see (4)) and $\mathbf{P}_2^+/P_1$ (see (5) and (4c) for the pressure space). It is known that both finite element schemes are inf-sup stable with isotropic meshes (see, e.g., [7] for a detailed presentation and stability proofs). We first provide a summary of the result regarding stability of the finite elements in Table 1. We then go into the details by analyzing the first smallest square rooted λ and its associated eigenvectors in case λ depends on a or b (Fig. 4).

Table 1 Summary of the results: if an element passes the test it is denoted by P (i.e., λ does not depend on a nor b). On the contrary, if an element fails the test, the table shows the number of spurious modes (i.e., how many λ depends on a or b).

Mesh:		1	2	3
$a \to 0$	$\mathbf{P}_2/P_1$	P	P	P
	$\mathbf{P}_2^+/P_1$	P	P	P
$b \to 0$	$\mathbf{P}_2/P_1$	1	2	P
	$\mathbf{P}_2^+/P_1$	P	P	P

(a) Problem D

Mesh		1	2	3
$a \to 0$	$\mathbf{P}_2/P_1$	P	P	P
	$\mathbf{P}_2^+/P_1$	P	P	P
$b \to 0$	$\mathbf{P}_2/P_1$	P	P	P
	$\mathbf{P}_2^+/P_1$	P	P	P

(b) Problem N

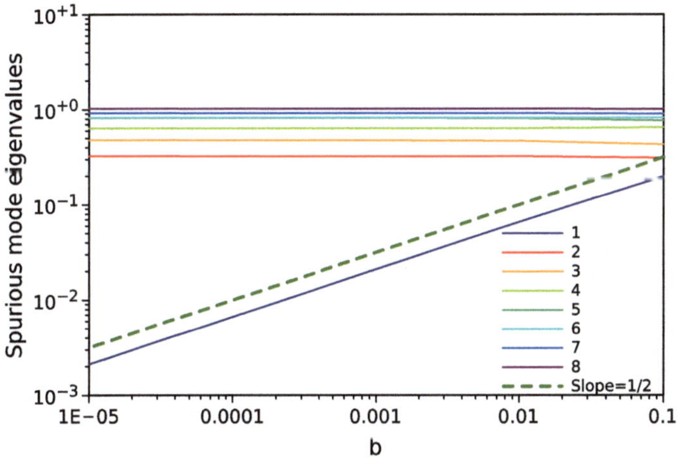

Fig. 4 Problem D for $\mathbf{P}_2/P_1$ with $b \to 0$ and Mesh 1. There is 1 $\mathcal{O}(\sqrt{b})$ spurious mode.

From Table 1, it may be observed that the instability of $\mathbf{P}_2/P_1$ comes from a single spurious pressure mode on corners with Dirichlet boundary conditions (see Figs. 4, 5, 6, and 7). For Problem N and for Problem D with the Mesh 3, the pair is stable for all tests, even when the smallest element area is $\mathcal{O}(b^2)$. These results are in accordance with the observations made in [3]. Indeed, they provide an explanation for the stability of the Hood-Taylor element in the test "Flow around a disk" in [3] and with the fluid–structure interaction of a rigid leaflet in [4]. The $\mathbf{P}_2^+/P_1$ element passes all tests. Detailed values are provided in Table A.1.

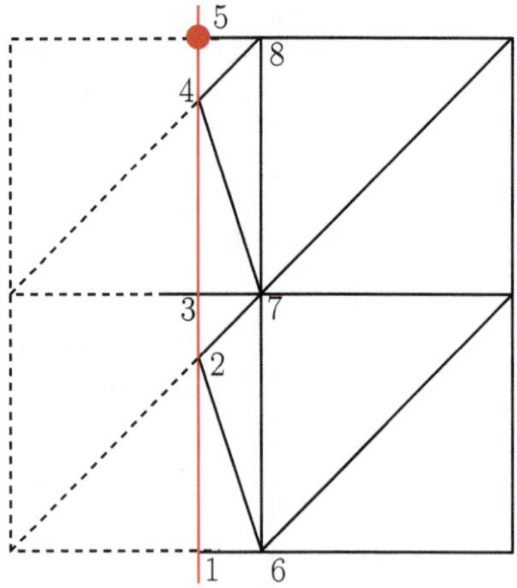

Fig. 5 Problem D for $\mathbf{P}_2/P_1$ with $b \to 0$ and Mesh 1. The spurious mode is localized only on the node that is on the smallest element area and not connected to other elements. This result is in accordance with the results of more general test performed in [3]. This test provides a clear location for the spurious mode.

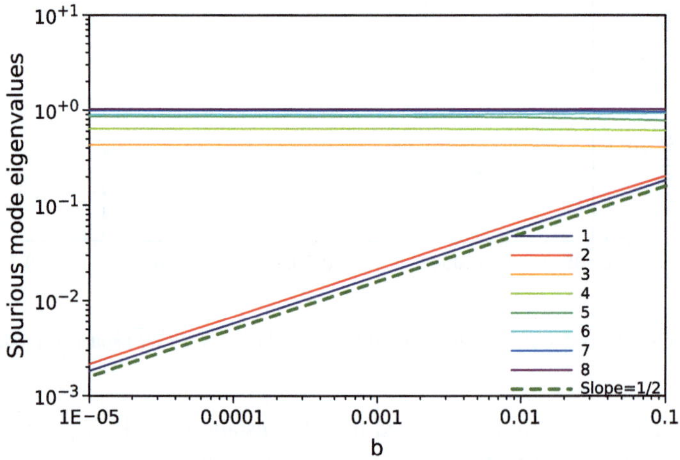

Fig. 6 Problem D for $\mathbf{P}_2/P_1$ with $b \to 0$ and Mesh 2. There are 2 $\mathcal{O}(\sqrt{b})$ spurious modes.

A.1 Spurious Mode Eigenvectors

Acknowledgements This work is funded by: the Cariplo Foundation through the Project no. 2009.2822; the European Research Council through the ERC Starting Grant Project no. 259229; Ministero dell'Istruzione, dell'Università e della Ricerca through the PRIN Project no. 2010BFXRHS.

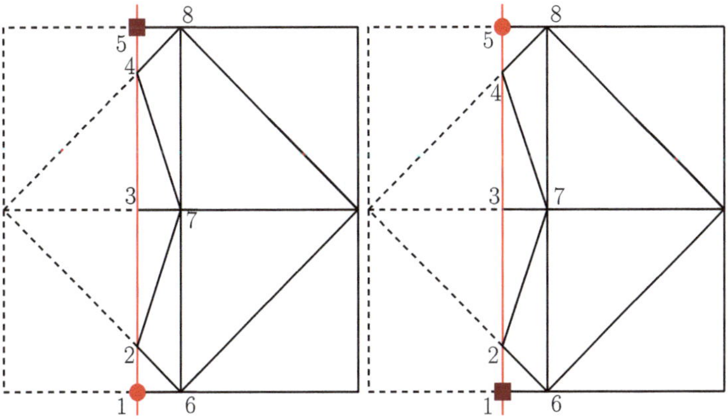

(a) Mode 1: $\mathcal{O}(\sqrt{b})$. (b) Mode 2: $\mathcal{O}(\sqrt{b})$.

Fig. 7 Problem D for $\mathbf{P}_2/P_1$ with $b \to 0$ and Mesh 2. This result is consistent with the results of Problem D with Mesh 1. Indeed, we have two elements in corners and thus two spurious modes. For Problem D with Mesh 3 there are no elements in a corner and as consequence the element is stable.

Table A.1 First 8 squared rooted values of the eigenvalues and associated eigenvectors of (10) with Problem D for $\mathbf{P}_2/P_1$ and $b \to 0$ on Mesh 1 and 2. The eigenvectors are orthogonalized with the Gram–Schmidt process.

β_h	**2.11e-03**	3.27e-01	4.80e-01	6.38e-01
1	-1.11e-06	9.41e-01	-1.71e-01	-1.92e-01
2	5.56e-07	-2.98e-01	-5.97e-02	-4.35e-01
3	-1.67e-06	1.53e-01	7.55e-01	3.26e-01
4	2.22e-06	-5.33e-02	-5.04e-01	-8.58e-03
5	**-1.00e+00**	-1.58e-06	-2.22e-06	-5.92e-07
6	1.69e-12	1.15e-06	-6.95e-02	1.50e-01
7	2.83e-12	-7.36e-08	1.87e-01	-4.04e-01
8	-4.47e-11	2.02e-08	-3.17e-01	6.83e-01

(a) Mesh 1

β_h	**1.81e-03**	**2.13e-03**	4.33e-01	6.35e-01
1	**9.99e-01**	**-3.33e-02**	-6.84e-06	2.57e-07
2	-6.84e-06	6.08e-08	-1.00e+00	-1.81e-02
3	5.16e-07	4.83e-07	2.32e-02	-5.28e-01
4	-2.28e-07	-1.83e-06	-8.38e-03	7.04e-01
5	**3.33e-02**	**9.99e-01**	-1.94e-07	1.55e-06
6	3.80e-14	-5.39e-12	-3.87e-06	2.10e-01
7	-4.15e-12	1.93e-11	-1.46e-06	-3.50e-01
8	3.39e-13	3.63e-12	5.20e-07	2.10e-01

(b) Mesh 2

References

1. Apel, T.: Anisotropic Finite Elements: Local Estimates and Applications. Teubner, Stuttgart (1999)
2. Apel, T., Randrianarivony, H.: Stability of discretizations of the stokes problem on anisotropic meshes. J. Math. Comput. Simul. **61**, 437–447 (2003)
3. Auricchio, F., Brezzi, F., Lefieux, A., Reali, A.: An "immersed" finite element method based on a locally anisotropic remeshing for the incompressible stokes problem. Comput. Methods Appl. Mech. Eng. **294**, 428–448 (2015)
4. Auricchio, F., Lefieux, A., Reali, A., Veneziani, A.: A locally anisotropic fluid–structure interaction remeshing strategy for thin structures with application to a hinged rigid leaflet. Int. J. Numer. Meth. Engng., **107**, 155–180 (2016)

5. Babuška, I., Aziz, A.: On the angle condition in the finite element method. SIAM J. Numer. Anal. **13**, 214–226 (1976)
6. Bern, M., Eppstein, D.: Mesh generation and optimal triangulation. In: Computing in Euclidian Geometry. Lecture Notes Series on Computing, vol. 1, pp. 23–90. World Scientific, Singapore (1992)
7. Boffi, D., Brezzi, F., Fortin, M.: Mixed Finite Element Methods. Springer, Berlin (2013)
8. Elman, H., Silvester, D., Wathen, A.: Finite Elements and Fast Iterative Solvers with Applications in Incompressible Fluid Dynamics. Oxford Press, Oxford (2005)
9. Hansbo, A., Hansbo, P.: An unfitted finite element method, based on Nitsche's method, for elliptic interface problems. Comput. Methods Appl. Mech. Eng. **191**, 5537–5552 (2002)
10. Lefieux, A.: On the use of anisotropic triangles in an immersed finite element approach with application to fluid-structure interaction problems. Ph.D. thesis, Istituto Universitario degli Studi Superiori di Pavia (2014)
11. Malkus, D.: Eigenproblems associated with the discrete LBB condition for incompressible finite elements. Int. J. Eng. Sci. **19**, 1299–1310 (1981)

Stabilized Lagrange–Galerkin Schemes of First- and Second-Order in Time for the Navier–Stokes Equations

Hirofumi Notsu and Masahisa Tabata

Abstract Two stabilized Lagrange–Galerkin schemes for the Navier–Stokes equations are reviewed. The schemes are based on a combination of the Lagrange–Galerkin method and Brezzi–Pitkäranta's stabilization method. They maintain the advantages of both methods: (i) They are robust for convection-dominated problems and the systems of linear equations to be solved are symmetric; and (ii) Since the P1 finite element is employed for both velocity and pressure, the numbers of degrees of freedom are much smaller than that of other typical elements for the equations, e.g., P2/P1. Therefore, the schemes are efficient especially for three-dimensional problems. The one of the schemes is of first-order in time by Euler's method and the other is of second-order by Adams–Bashforth's method. In the second-order scheme an additional initial velocity is required. A convergence analysis is done for the choice of the velocity obtained by the first-order scheme, whose theoretical result is also recognized numerically.

1 Introduction

In this chapter, we review two stabilized Lagrange–Galerkin schemes for the Navier–Stokes equations. The one is of first-order in time and the other is of second-order. We also discuss the choice of the initial velocities for the (two-step) second-order scheme.

The Lagrange–Galerkin (LG) method is a finite element method combined with the method of characteristics and is known to be powerful for flow problems. The idea is based on the discretization of the material derivative along the trajectory of

---

H. Notsu (✉)
Faculty of Mathematics and Physics, Kanazawa University, Kakuma, Kanazawa 920-1192, Japan
e-mail: notsu@se.kanazawa-u.ac.jp

M. Tabata
Department of Mathematics, Waseda University, 3-4-1 Ohkubo, Shinjuku, Tokyo 169-8555, Japan
e-mail: tabata@waseda.jp

Y. Bazilevs, K. Takizawa (eds.), *Advances in Computational Fluid-Structure Interaction and Flow Simulation*, Modeling and Simulation in Science, Engineering and Technology, DOI 10.1007/978-3-319-40827-9_26

the fluid particle and its approximation is natural from a physical point of view. The LG method has common advantages, robustness for convection-dominated problems and symmetry of the resultant matrix to be solved.

Many authors have proposed and analyzed *stable* LG schemes for flow problems, cf. [1, 3, 14, 18] and the references therein. The *stable* means that the scheme requires a pair of finite elements for velocity and pressure satisfying the conventional inf-sup condition [7], e.g., P2/P1. In order to perform efficiently three-dimensional computation by the LG method, we have originally proposed a stabilized LG scheme of first-order in time for the Navier–Stokes equations in [9, 10], which employs a cheap P1/P1 finite element thanks to Brezzi–Pitkäranta's pressure-stabilization [4]. Since the additional term of the stabilization method is also symmetric, the stabilized LG scheme maintains the advantages of the LG method.

Optimal error estimates of the stabilized LG scheme of first-order in time have been established in [11, 12]. Combining the idea of [3, 6], we have proposed a stabilized LG scheme of second-order in time to prove optimal error estimates in [13]. After reviewing these results for the Navier–Stokes equations, we discuss the choice of an initial velocity needed at the first time step of the two-step scheme, the stabilized LG scheme of second-order in time. We present a simple choice of the velocity to show the best possible error estimate for the velocity in the $L^2(\Omega)$-norm.

The chapter is organized as follows. In Section 2 stabilized LG schemes of first- and second-order in time for the Navier–Stokes equations are introduced. In Section 3 the convergence results for the schemes are presented. In Section 4 the choice of the initial velocities for the second-order scheme is discussed. In Section 5 convergence results obtained in the previous section are recognized numerically. In Section 6 the conclusions are stated.

2 Stabilized LG Schemes of First- and Second-Order in Time

We prepare the function spaces and the notation to be used throughout the chapter. Let Ω be a bounded domain in $\mathbb{R}^d (d = 2, 3)$, $\Gamma \equiv \partial\Omega$ the boundary of Ω, and T a positive constant. For a non-negative integer m we use the Sobolev spaces $H^m(\Omega)$, $H^1_0(\Omega)$, and $W^{1,\infty}_0(\Omega)$. We use the same notation $(\cdot, \cdot)$ to represent the $L^2(\Omega)$ inner product for scalar-, vector-, and matrix-valued functions. The norms on $L^2(\Omega)^d (= H^0(\Omega)^d)$, $H^1(\Omega)^d$ and $W^{1,\infty}_0(\Omega)^d$ are simply denoted by $\|\cdot\|_0 = \|\cdot\|_{L^2(\Omega)^d}$, $\|\cdot\|_1 = \|\cdot\|_{H^1(\Omega)^d}$, $\|\cdot\|_{1,\infty} = \|\cdot\|_{W^{1,\infty}(\Omega)^d}$, respectively, and the notations $\|\cdot\|_0$ and $\|\cdot\|_1$ are employed not only for vector-valued functions but also for scalar-valued ones. Let $L^2_0(\Omega)$ be a subspace of $L^2(\Omega)$ defined by $L^2_0(\Omega) \equiv \{q \in L^2(\Omega);\ (q, 1) = 0\}$. We often omit Ω and/or d if there is no confusion. The superscript "T" stands for the transposition. We employ c to represent a positive constant independent of discretization parameters in space and time, h and Δt, and it may take different values at different places.

We consider the Navier–Stokes problem; find $(u,p) : \Omega \times (0,T) \to \mathbb{R}^d \times \mathbb{R}$ such that

$$\frac{Du}{Dt} - \nabla \cdot \big[2\nu D(u)\big] + \nabla p = f \qquad \text{in } \Omega \times (0,T), \tag{1a}$$

$$\nabla \cdot u = 0 \qquad \text{in } \Omega \times (0,T), \tag{1b}$$

$$u = 0 \qquad \text{on } \Gamma \times (0,T), \tag{1c}$$

$$u = u^0 \qquad \text{in } \Omega, \text{ at } t = 0, \tag{1d}$$

where u is the velocity, p is the pressure, $f : \Omega \times (0,T) \to \mathbb{R}^d$ is a given external force, $u^0 : \Omega \to \mathbb{R}^d$ is a given initial velocity, $\nu > 0$ is a viscosity, $D(u) \equiv [\nabla u + (\nabla u)^T]/2$ is the strain-rate tensor, and $D/Dt \equiv \partial/\partial t + u \cdot \nabla$ is the material derivative.

The weak formulation of (1) can be written as follows; find $(u,p) : (0,T) \to V \times Q$ such that, for $t \in (0,T)$,

$$\Big(\frac{Du}{Dt}(t), v\Big) + \mathscr{A}\big((u,p)(t),(v,q)\big) = (f(t), v), \quad \forall (v,q) \in V \times Q, \tag{2}$$

with $u(0) = u^0$, where $V \equiv H_0^1(\Omega)^d$ and $Q \equiv L_0^2(\Omega)$ are the function spaces for velocity and pressure, and a, b, and $\mathscr{A}$ are bilinear forms defined by

$$a(u,v) \equiv 2\nu\big(D(u), D(v)\big), \quad b(v,q) \equiv -(\nabla \cdot v, q),$$
$$\mathscr{A}\big((u,p),(v,q)\big) \equiv a(u,v) + b(v,p) + b(u,q).$$

Let Δt be a time increment and $t^n \equiv n\Delta t$ for $n \in \mathbb{N} \cup \{0\}$. For a function g defined in $\Omega \times (0,T)$ we denote generally $g(\cdot, t^n)$ by g^n. Let $g^{(n-1)*}$ be a second-order approximate function of g^n defined by

$$g^{(n-1)*} \equiv 2g^{n-1} - g^{n-2}.$$

Let $X : (0,T) \to \mathbb{R}^d$ be a solution of the system of ordinary differential equations,

$$\frac{dX}{dt} = u(X,t). \tag{3}$$

Then, it holds that for smooth u

$$\frac{Du}{Dt}(X(t),t) = \frac{d}{dt}u\big(X(t),t\big).$$

Let $X(\cdot; x, t^n)$ be the solution of (3) subject to an initial condition $X(t^n) = x$. For a velocity $w : \Omega \to \mathbb{R}^d$ let $X_1(w,\Delta t) : \Omega \to \mathbb{R}^d$ be a mapping defined by

$$X_1(w,\Delta t)(x) \equiv x - w(x)\Delta t.$$

Based on Euler's and (two-step) Adams–Bashforth's methods along the trajectory of the characteristic-curve $X(t; x, t^n)$, we can consider first- and second-order approximations of $(Du/Dt)(x, t^n)$ as follows:

$$\text{first-order:} \qquad \frac{u^n - u^{n-1} \circ X_1(u^{n-1}, \Delta t)}{\Delta t}(x),$$

$$\text{second-order:} \qquad \frac{3u^n - 4u^{n-1} \circ X_1(u^{(n-1)*}, \Delta t) + u^{n-2} \circ X_1(u^{(n-1)*}, 2\Delta t)}{2\Delta t}(x),$$

where the symbol "$\circ$" stands for the composition of functions, $(v \circ w)(x) \equiv v(w(x))$, for $v : \Omega \to \mathbb{R}^d$ and $w : \Omega \to \Omega$.

$X_1(w, \Delta t)(x)$ is called an upwind point of x with respect to the velocity w and the time increment Δt. The next proposition gives sufficient conditions to guarantee that all upwind points by $X_1(w, \Delta t)$ are in Ω and that its Jacobian is around 1.

Proposition 1 ([16, 20]).

(i) Let $w \in W_0^{1,\infty}(\Omega)^d$ be a given velocity. Then, under the condition $\Delta t|w|_{1,\infty} < 1$, $X_1(w, \Delta t) : \Omega \to \Omega$ is bijective.

(ii) Furthermore, under the condition $\Delta t|w|_{1,\infty} \le 1/4$, the estimate $1/2 \le J \equiv \det(\partial X_1(w, \Delta t)/\partial x) \le 3/2$ holds.

For the sake of simplicity we assume that Ω is a polygonal ($d = 2$) or polyhedral ($d = 3$) domain. Let $\mathscr{T}_h = \{K\}$ be a triangulation of $\bar{\Omega}$ $(= \bigcup_{K \in \mathscr{T}_h} K)$, h_K a diameter of $K \in \mathscr{T}_h$, and $h \equiv \max_{K \in \mathscr{T}_h} h_K$ the maximum element size. We consider a regular family of triangulations $\{\mathscr{T}_h\}_{h \downarrow 0}$ satisfying the inverse assumption [5], which leads to the next inverse inequalities:

$$\|v_h\|_{0,\infty} \le ch^{-d/6}\|v_h\|_1, \quad \|v_h\|_{1,\infty} \le ch^{-d/2}\|v_h\|_1, \quad \forall v_h \in V_h. \tag{4}$$

Let X_h, M_h, V_h, and Q_h be function spaces defined by $X_h \equiv \{v_h \in C(\bar{\Omega})^d;\ v_{h|K} \in P_1(K)^d,\ \forall K \in \mathscr{T}_h\}$, $M_h \equiv \{q_h \in C(\bar{\Omega});\ q_{h|K} \in P_1(K),\ \forall K \in \mathscr{T}_h\}$, $V_h \equiv X_h \cap V$, and $Q_h \equiv M_h \cap Q$, respectively, where $P_1(K)$ is the space of linear functions on $K \in \mathscr{T}_h$. Let $N_T \equiv \lfloor T/\Delta t \rfloor$ be a total number of time steps, δ_0 a positive constant, and $(\cdot, \cdot)_K$ the $L^2(K)^d$ inner product. We define the bilinear forms $\mathscr{C}_h$ and $\mathscr{A}_h$ by

$$\mathscr{C}_h(p, q) \equiv \delta_0 \sum_{K \in \mathscr{T}_h} h_K^2 (\nabla p, \nabla q)_K,$$

$$\mathscr{A}_h\big((u, p), (v, q)\big) \equiv a(u, v) + b(v, p) + b(u, q) - \mathscr{C}_h(p, q).$$

The bilinear form $\mathscr{C}_h$ is known as Brezzi–Pitkäranta's pressure-stabilization [4].

Assume $f \in C([0, T]; L^2(\Omega)^d)$. Suppose that an approximate function $u_h^0 \in V_h$ of u^0 is given. Our stabilized LG scheme of first-order in time for (1) [9, 10, 12] is to find $\{(u_h^n, p_h^n)\}_{n=1}^{N_T} \subset V_h \times Q_h$ such that, for $n = 1, \cdots, N_T$,

$$\Big(\frac{u_h^n - u_h^{n-1} \circ X_1(u_h^{n-1}, \Delta t)}{\Delta t}, v_h\Big) + \mathscr{A}_h\big((u_h^n, p_h^n), (v_h, q_h)\big) = (f^n, v_h),$$

$$\forall (v_h, q_h) \in V_h \times Q_h. \qquad \text{(S1)}$$

Suppose that approximate functions $u_h^i \in V_h$ of $u^i \in V$, $i = 0, 1$, are given. Our stabilized LG scheme of second-order in time for (1) [13] is to find $\{(u_h^n, p_h^n)\}_{n=2}^{N_T} \subset V_h \times Q_h$ such that, for $n = 2, \cdots, N_T$,

$$\frac{1}{2\Delta t}\big(3u_h^n - 4u_h^{n-1} \circ X_1(u_h^{(n-1)*}, \Delta t) + u_h^{n-2} \circ X_1(u_h^{(n-1)*}, 2\Delta t), v_h\big)$$

$$+\mathscr{A}_h\big((u_h^n, p_h^n), (v_h, q_h)\big) = (f^n, v_h), \quad \forall (v_h, q_h) \in V_h \times Q_h. \qquad \text{(S2)}$$

3 Convergence Results

In this section we state the convergence results for the schemes (S1) and (S2).

We use the next norms and seminorm, $\|u\|_{\ell^\infty(X)} \equiv \max\{\|u^n\|_X;\ n = 0, \cdots, N_T\}$, $\|u\|_{\ell^2(X)} \equiv \{\Delta t \sum_{n=1}^{N_T} \|u^n\|_X^2\}^{1/2}$, and $|p|_h \equiv \{\sum_{K\in\mathscr{T}_h} h_K^2 (\nabla p, \nabla p)_K\}^{1/2}$ for $X = L^2(\Omega)$ and $H^1(\Omega)$. Let $\overline{D}_{\Delta t}$ and $\overline{D}_{\Delta t}^{(2)}$ be the backward difference operators of first- and second-order in Δt defined by $\overline{D}_{\Delta t} u^n \equiv (u^n - u^{n-1})/\Delta t$ and $\overline{D}_{\Delta t}^{(2)} u^n \equiv (3u^n - 4u^{n-1} + u^{n-2})/(2\Delta t)$, respectively.

Hereafter, for the sake of simplicity, we assume that the solution of problem (1) is smooth enough, and that the Stokes problem is regular, i.e., for any $g \in L^2(\Omega)^d$ the solution $(w, r) \in V \times Q$ of the Stokes problem,

$$\mathscr{A}\big((w, r), (v, q)\big) = (g, v), \quad \forall (v, q) \in V \times Q,$$

belongs to $H^2(\Omega)^d \times H^1(\Omega)$ and the estimate $\|w\|_2 + \|r\|_1 \le c_R \|g\|_0$ holds, where c_R is a positive constant independent of g, w, and r. The latter assumption holds, e.g., if Ω is convex in $\mathbb{R}^2$, cf. [7]. For the initial velocity $u_h^0 \in V_h$, we always set the first component of a Stokes projection of $(u^0, 0) \in V \times Q$, where, in general, the Stokes projection $(\hat{w}_h, \hat{r}_h) \in V_h \times Q_h$ of $(w, r) \in V \times Q$ is defined by the solution of the variational problem:

$$\mathscr{A}_h\big((\hat{w}_h, \hat{r}_h), (v_h, q_h)\big) = \mathscr{A}\big((w, r), (v_h, q_h)\big), \quad \forall (v_h, q_h) \in V_h \times Q_h. \qquad (5)$$

We note that $\hat{w}_h$ satisfies for $(w, r) \in H^2(\Omega)^d \times H^1(\Omega)$

$$\|\hat{w}_h - w\|_i = O(h^{2-i}), \quad i = 0, 1. \qquad (6)$$

The next theorem holds for scheme (S1).

Theorem 1 ([12]). *There exist positive constants h_1 and c_1 independent of h and Δt such that, for any pair $(h, \Delta t)$ satisfying $h \in (0, h_1]$ and $\Delta t \le c_1 h^{d/4}$, the solution $(u_h, p_h) = \{(u_h^n, p_h^n)\}_{n=1}^{N_T} \subset V_h \times Q_h$ of scheme (S1) satisfies*

$$\begin{aligned} \|u_h - u\|_{\ell^\infty(H^1)},\ \|p_h - p\|_{\ell^2(L^2)} &= O(\Delta t + h), \\ \|u_h - u\|_{\ell^\infty(L^2)} &= O(\Delta t + h^2). \end{aligned} \tag{7}$$

For scheme (S2) we need initial velocities u_h^0 and u_h^1. As mentioned above u_h^0 is taken as the first component of the Stokes projection (5) of $(u^0, 0)$. Supposing the following hypothesis on u_h^1 and p_h^1, we have the convergence results for scheme (S2).

Hypothesis 1. *$(u_h^1, p_h^1) \in V_h \times Q_h$ satisfies*

$$b(u_h^1, q_h) - \mathscr{C}_h(p_h^1, q_h) = 0, \quad \forall q_h \in Q_h,$$

$$\|u_h^1 - u^1\|_1,\ |p_h^1 - p^1|_h,\ \sqrt{\Delta t}\|\overline{D}_{\Delta t} u_h^1 - \overline{D}_{\Delta t} u^1\|_0,\ \sqrt{\Delta t}\|p_h^1 - p^1\|_0 = O(\Delta t^2 + h),$$

$$\|u_h^1 - u^1\|_0 = O(\Delta t^2 + h^2).$$

Theorem 2 ([13]). *Suppose that Hypothesis 1 holds. Then, there exist positive constants h_2 and c_2 independent of h and Δt such that, for any pair $(h, \Delta t)$ satisfying $h \in (0, h_2]$ and $\Delta t \le c_2 h^{d/6}$, the solution $(u_h, p_h) = \{(u_h^n, p_h^n)\}_{n=2}^{N_T} \subset V_h \times Q_h$ of scheme (S2) satisfies*

$$\begin{aligned} \|u_h - u\|_{\ell^\infty(H^1)},\ \|p_h - p\|_{\ell^2(L^2)} &= O(\Delta t^2 + h), \\ \|u_h - u\|_{\ell^\infty(L^2)} &= O(\Delta t^2 + h^2). \end{aligned} \tag{8}$$

4 Choice of u_h^1 for Scheme (S2)

In order to supply an initial velocity $u_h^1 \in V_h$ of scheme (S2) we use scheme (S1). At first we set $(u_h^1, p_h^1) \equiv (\tilde{u}_h^{\tilde{N}_{\Delta t}}, \tilde{p}_h^{\tilde{N}_{\Delta t}})$ for the solution $\{(\tilde{u}_h^m, \tilde{p}_h^m)\}_{m=1}^{\tilde{N}_{\Delta t}} \subset V_h \times Q_h$ such that, for $m = 1, \cdots, \tilde{N}_{\Delta t}$,

$$\Big(\frac{\tilde{u}_h^m - \tilde{u}_h^{m-1} \circ X_1(\tilde{u}_h^{m-1}, \tau)}{\tau}, v_h\Big) + \mathscr{A}_h((\tilde{u}_h^m, \tilde{p}_h^m), (v_h, q_h)) = (\tilde{f}^m, v_h),$$

$$\forall (v_h, q_h) \in V_h \times Q_h, \tag{9}$$

where $\tilde{u}_h^0 \equiv u_h^0$, $\tau \in (0, \Delta t]$, $\tilde{N}_{\Delta t} \equiv \lfloor \Delta t / \tau \rfloor$, and $\tilde{f}^m \equiv f(\cdot, m\tau)$. When we employ scheme (9) with $\tau = \Delta t^2$, Hypothesis 1 is satisfied from Theorem 1 and its proof, cf. [12]. Then, Theorem 2 holds. This choice of (u_h^1, p_h^1) is not so efficient since the number of iteration $\tilde{N}_{\Delta t}$ is proportional to N_T.

Next, we consider a simple choice of (u_h^1, p_h^1) by scheme (9) with $\tau = \Delta t$ $(\tilde{N}_{\Delta t} = 1)$, i.e., the solution (u_h^1, p_h^1) of (S1), which leads to the following result:

Proposition 2. *There exist positive constants h_3 and c_3 independent of h and Δt such that, for any pair $(h, \Delta t)$,*

$$h \in (0, h_3], \quad \Delta t \le c_3 h^{d/5}, \tag{10}$$

the solution $(u_h, p_h) = \{(u_h^n, p_h^n)\}_{n=2}^{N_T} \subset V_h \times Q_h$ of scheme (S2) with (u_h^1, p_h^1), the solution of (S1), satisfies

$$\|u_h - u\|_{\ell^\infty(H^1)}, \ \|p_h - p\|_{\ell^2(L^2)} = O(\Delta t^{3/2} + h), \tag{11}$$

$$\|u_h - u\|_{\ell^\infty(L^2)} = O(\Delta t^2 + h^2). \tag{12}$$

Remark 1. The above proposition implies that there is no loss of convergence order for the velocity in the $L^2(\Omega)$-norm, cf. (8).

Before the proof we prepare equations for the error. Let $(\hat{u}_h, \hat{p}_h)(t) \in V_h \times Q_h$ be the Stokes projection of $(u, p)(t)$ by (5) for $t \in [0, T]$, and let $e_h^n \equiv u_h^n - \hat{u}_h^n$ and $\epsilon_h^n \equiv p_h^n - \hat{p}_h^n$. We have that

$$(\overline{D}_{\Delta t} e_h^1, v_h) + \mathscr{A}_h\big((e_h^1, \epsilon_h^1), (v_h, q_h)\big) = \langle R_h^1, v_h\rangle, \quad \forall (v_h, q_h) \in V_h \times Q_h, \tag{13a}$$

$$(\overline{D}_{\Delta t}^{(2)} e_h^n, v_h) + \mathscr{A}_h\big((e_h^n, \epsilon_h^n), (v_h, q_h)\big) = \langle R_h^n, v_h\rangle, \quad \forall (v_h, q_h) \in V_h \times Q_h, \tag{13b}$$

for $n \ge 2$, where

$$R_h^n = \begin{cases} \dfrac{Du^1}{Dt} - \overline{D}_{\Delta t}\hat{u}_h^1 - \dfrac{u_h^0 - u_h^0 \circ X_1(u_h^0, \Delta t)}{\Delta t} & (n = 1), \\[2ex] \dfrac{Du^n}{Dt} - \overline{D}_{\Delta t}^{(2)}\hat{u}_h^n - \dfrac{2}{\Delta t}\{u_h^{n-1} \quad u_h^{n-1} \circ X_1(u_h^{(n-1)*}, \Delta t)\} & \\[2ex] \qquad + \dfrac{1}{2\Delta t}\{u_h^{n-2} - u_h^{n-2} \circ X_1(u_h^{(n-1)*}, 2\Delta t)\} & (n \ge 2). \end{cases}$$

Proof of Proposition 2. Since the difference between the schemes in Theorem 2 and Proposition 2 is the choice of u_h^1 only, we outline the proof by focusing on the estimates of u_h^1, i.e., $\|u_h^1 - u^1\|_i$, $i = 0, 1$, which determine the final convergence orders. For the detail of the rest of the proof, we refer to [13].

We prove (11). There exists a positive constant h_0 independent of h and Δt such that for any $h \in (0, h_0]$ and $\Delta t \le h^{d/5}$

$$\|u_h^0\|_{0,\infty} \le \|u\|_{C(L^\infty)} + 1, \tag{14a}$$

$$\Delta t \|u_h^0\|_{1,\infty} \le 1/4, \tag{14b}$$

which are proved in a similar way to [12] by using (4) and (6). We need $h_3 \le h_0$ and $c_3 \le 1$. From Proposition 1 and (14b) there exists a unique solution (u_h^1, p_h^1) of (S1). (14a) and (6) lead to

$$\|R_h^1\|_0 \le c(\Delta t + h). \tag{15}$$

Substituting $(\overline{D}_{\Delta t} e_h^1, 0)$ into (v_h, q_h) in (13a), we have

$$\|\overline{D}_{\Delta t} e_h^1\|_0^2 + a(e_h^1, \overline{D}_{\Delta t} e_h^1) + b(\overline{D}_{\Delta t} e_h^1, \epsilon_h^1) \le \|R_h^1\|_0 \|\overline{D}_{\Delta t} e_h^1\|_0. \tag{16}$$

From (13a) with $q_h = 0$ and the definitions of (u_h^0, p_h^0) and $(\hat{u}_h^0, \hat{p}_h^0)$, we also have

$$b(\overline{D}_{\Delta t} e_h^1, q_h) = \mathscr{C}_h(\overline{D}_{\Delta t} \epsilon_h^1, q_h), \quad \forall q_h \in Q_h.$$

Combining the above relation with (16) and using (15), we get

$$\overline{D}_{\Delta t}\Big(\nu \|D(e_h^1)\|_0^2 + \frac{\delta_0}{2}|\epsilon_h^1|_h^2\Big) + \frac{1}{2}\|\overline{D}_{\Delta t} e_h^1\|_0^2 \le c(\Delta t^2 + h^2),$$

which yields

$$\|u_h^1 - u^1\|_1, \ |p_h^1 - p^1|_h, \ \sqrt{\Delta t}\|\overline{D}_{\Delta t} u_h^1 - \overline{D}_{\Delta t} u^1\|_0 \le c(\Delta t^{3/2} + h). \tag{17}$$

Let $\Pi_h : C(\bar{\Omega})^d \to X_h$ be the Lagrange interpolation operator, which has a property $\|\Pi_h w - w\|_i = O(h^{2-i})$, $i = 0, 1$, for $w \in H^2(\Omega)^d$. From the first inequality of (17) we have

$$\|u_h^{1*}\|_{0,\infty} \le \|u\|_{C(L^\infty)} + 1, \tag{18a}$$

$$2\Delta t \|u_h^{1*}\|_{1,\infty} \le 1/4, \tag{18b}$$

by the estimates,

$$\begin{aligned}
\|u_h^{1*}\|_{0,\infty} &\le \|\Pi_h u^{1*}\|_{0,\infty} + \|u_h^{1*} - \Pi_h u^{1*}\|_{0,\infty} \\
&\le \|\Pi_h u^{1*}\|_{0,\infty} + 2\|u_h^1 - \Pi_h u^1\|_{0,\infty} + \|u_h^0 - \Pi_h u^0\|_{0,\infty} \\
&\le \|\Pi_h u^{1*}\|_{0,\infty} + ch^{-d/6}(\|u_h^1 - \Pi_h u^1\|_1 + \|u_h^0 - \Pi_h u^0\|_1) \quad \text{(by (4))} \\
&\le \|u^2\|_{0,\infty} + c\big\{\Delta t^2 + h^{-d/6}(\Delta t^{3/2} + h)\big\} \quad (\text{by } u^{1*} = u^2 + O(\Delta t^2)) \\
&\le \|u\|_{C(L^\infty)} + c\big\{c_3^2 h^{2d/5} + c_3^{3/2} h^{2d/15} + h^{1-d/6}\big\} \quad \text{(by (10))} \\
&\le \|u\|_{C(L^\infty)} + 1 \quad (\text{if } h_3 \text{ is small enough}),
\end{aligned}$$

$$\begin{aligned}
2\Delta t \|u_h^{1*}\|_{1,\infty} &\le 2\Delta t (2\|u_h^1 - \Pi_h u^1\|_{1,\infty} + \|u_h^0 - \Pi_h u^0\|_{1,\infty} + \|\Pi_h u^{1*}\|_{1,\infty}) \\
&\le 2\Delta t \big\{ch^{-d/2}(\|u_h^1 - \Pi_h u^1\|_1 + \|u_h^0 - \Pi_h u^0\|_1) + \|\Pi_h u^{1*}\|_{1,\infty}\big\} \quad \text{(by (4))}
\end{aligned}$$

$$
\begin{aligned}
&\leq c\Delta t\{h^{-d/2}(\Delta t^{3/2} + h) + 1\} \\
&\leq c\{c_3^{5/2} + c_3 h^{1-3d/10} + c_3 h^{d/5}\} \quad \text{(by (10))} \\
&\leq 1/4 \quad \text{(if } c_3 \text{ and } h \text{ are small enough)}.
\end{aligned}
$$

From (18b) scheme (S2) works for $n = 2$. Thanks to (18a), we can get the estimate $\|R_h^2\|_0 = O(\Delta t^{3/2} + h)$. We omit the rest of the proof of (11), cf. [13].

We prove (12). From (18b) the existence of the solution $(u_h, p_h) = \{(u_h^n, p_h^n)\}_{n=2}^{N_T}$ of scheme (S2) with (u_h^1, p_h^1), the solution of (S1), and its stability in $\ell^\infty(L^\infty)$-norm are ensured under (10). Substituting $(e_h^1, -\epsilon_h^1)$ into (v_h, q_h) in (13a), we have

$$
(\overline{D}_{\Delta t} e_h^1, e_h^1) + 2\nu \|D(e_h^1)\|_0^2 + \delta_0 |\epsilon_h^1|_h^2 \leq \|R_h^1\|_0 \|e_h^1\|_0,
$$

which leads to

$$
\|e_h^1\|_0 \leq \|e_h^0\|_0 + \Delta t \|R_h^1\|_0 \leq c(\Delta t^2 + h^2) \tag{19}
$$

by dropping the last two terms in the left-hand side and using (6) and (15). Thus, we have

$$
\|u_h^1 - u^1\|_0 = O(\Delta t^2 + h^2). \tag{20}
$$

For $n \geq 2$, substituting $(e_h^n, -\epsilon_h^n)$ into (v_h, q_h) in (13b) and using the next identity of $(\overline{D}_{\Delta t}^{(2)} e_h^n, e_h^n)$ [15] and estimate of $\|R_h^n\|_0$ [12],

$$
\begin{aligned}
(\overline{D}_{\Delta t}^{(2)} e_h^n, e_h^n) &= \frac{1}{\Delta t}\Big\{\frac{3}{4}\|e_h^n\|_0^2 - \|e_h^{n-1}\|_0^2 + \frac{1}{4}\|e_h^{n-2}\|_0^2 + \frac{1}{4}\|e_h^n - 2e_h^{n-1} + e_h^{n-2}\|_0^2 \\
&\quad + \frac{1}{2}\big(\|e_h^n - e_h^{n-1}\|_0^2 - \|e_h^{n-1} - e_h^{n-2}\|_0^2\big)\Big\}, \\
\|R_h^n\|_0 &\leq c(\|e_h^{n-1}\|_0 + \|e_h^{n-2}\|_0 + \Delta t^2 + h^2),
\end{aligned}
$$

we have

$$
\begin{aligned}
\frac{1}{\Delta t}\Big\{\frac{3}{4}\|e_h^n\|_0^2 - \|e_h^{n-1}\|_0^2 + \frac{1}{4}\|e_h^{n-2}\|_0^2 + \frac{1}{2}\big(\|e_h^n - e_h^{n-1}\|_0^2 - \|e_h^{n-1} - e_h^{n-2}\|_0^2\big)\Big\} \\
\leq c(\|e_h^{n-1}\|_0^2 + \|e_h^{n-2}\|_0^2 + \Delta t^4 + h^4),
\end{aligned}
$$

which implies (12) by Gronwall's inequality and (20). □

5 Numerical Results

In this section two- and three-dimensional test problems are computed by scheme (S2) with (u_h^1, p_h^1), the solution of (S1), in order to recognize the convergence results of Proposition 2 numerically.

Quadrature formulae [17] of degree five for $d = 2$ and 3 are employed for the computation of the integrals of composite functions in (S2) and (9). We set $\delta_0 = 1$. The system of linear equations is solved by MINRES [2].

Example 1. In problem (1) we set $\Omega = (0,1)^d, T = 1$, and $\nu = 10^{-k}, k = 1, \cdots, 4$. The functions f and u^0 are given so that the exact solution is as follows:

for $d = 2$:

$$u(x,t) = \Big(\frac{\partial \psi}{\partial x_2}, -\frac{\partial \psi}{\partial x_1}\Big)(x,t),\ \ p(x,t) = \sin\{\pi(x_1 + 2x_2 + t)\},\ \ \alpha_0 = \sqrt{3}/(2\pi),$$

$$\psi(x,t) \equiv \alpha_0 \sin^2(\pi x_1)\sin^2(\pi x_2)\sin\{\pi(x_1 + x_2 + t)\},$$

for $d = 3$:

$$u(x,t) = \mathrm{rot}\,\Psi(x,t),\ \ p(x,t) = \sin\{\pi(x_1 + 2x_2 + x_3 + t)\},\ \ \alpha_1 = 8\sqrt{3}/(27\pi),$$

$$\Psi_1(x,t) \equiv \alpha_1 \sin(\pi x_1)\sin^2(\pi x_2)\sin^2(\pi x_3)\sin\{\pi(x_2 + x_3 + t)\},$$

$$\Psi_2(x,t) \equiv \alpha_1 \sin^2(\pi x_1)\sin(\pi x_2)\sin^2(\pi x_3)\sin\{\pi(x_3 + x_1 + t)\},$$

$$\Psi_3(x,t) \equiv \alpha_1 \sin^2(\pi x_1)\sin^2(\pi x_2)\sin(\pi x_3)\sin\{\pi(x_1 + x_2 + t)\}.$$

These solutions are normalized so that $\|u\|_{C(L^\infty)} = \|p\|_{C(L^\infty)} = 1$.

Let N be the division number of each side of the domain. We set $N = 16, 32, 64, 128$, and 256 for $d = 2$ and $N = 16, 32$, and 64 for $d = 3$, and (re)define $h \equiv 1/N$. Example 1 is solved by scheme (S2) with (u_h^1, p_h^1), the solution of (S1). For the solution (u_h, p_h) of the scheme we define the relative errors $Er1$ and $Er2$ by

$$Er1 \equiv \frac{\|u_h - \Pi_h u\|_{\ell^2(H^1)} + \|p_h - \Pi_h p\|_{\ell^2(L^2)}}{\|\Pi_h u\|_{\ell^2(H^1)} + \|\Pi_h p\|_{\ell^2(L^2)}}, \qquad Er2 \equiv \frac{\|u_h - \Pi_h u\|_{\ell^\infty(L^2)}}{\|\Pi_h u\|_{\ell^\infty(L^2)}}.$$

We set $\Delta t = h$ in order to observe the convergence order of (12). The right-hand sides of (11) and (12) become of order $O(\Delta t^{3/2} + h) = O(\Delta t)$ and $O(\Delta t^2 + h^2) = O(\Delta t^2)$, respectively. Figure 1 shows the graphs of $Er1$ versus Δt (the left two, left: $d = 2$, right: $d = 3$) and $Er2$ versus Δt (the right two, left: $d = 2$, right: $d = 3$) in logarithmic scale, where the symbols are summarized in Table 1. We can see that the slopes of $Er1$ are between 1 and 2, and that $Er2$ is almost of second-order in Δt. These results are consistent with Proposition 2.

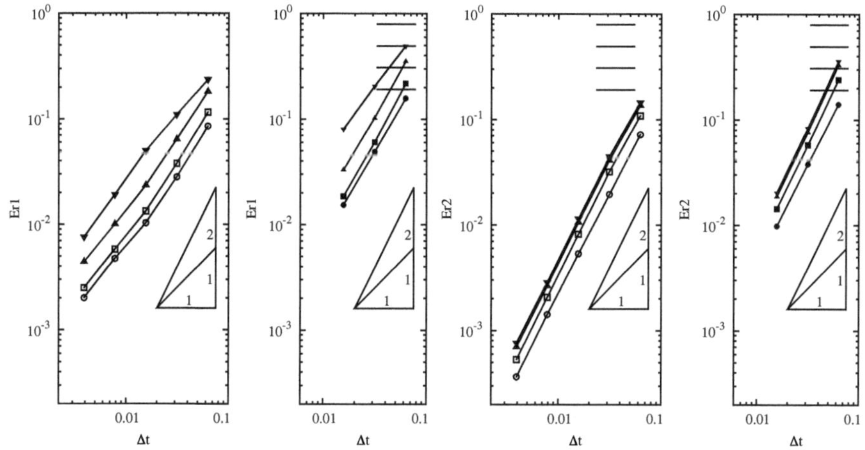

Fig. 1 Graphs of $Er1$ vs. Δt (the left two, left: $d = 2$, right: $d = 3$) and $Er2$ vs. Δt (the right two, left: $d = 2$, right: $d = 3$) in logarithmic scale for $\Delta t = h$.

Table 1 Symbols used in Figure 1.

	ν			
d	10^{-1}	10^{-2}	10^{-3}	10^{-4}
2	○	□	△	▽
3	●	■	▲	▼

6 Conclusions

We have reviewed stabilized Lagrange–Galerkin schemes of first- and second-order in time for the Navier–Stokes equations and their convergence results. Since the schemes maintain the advantages of the *stable* LG method and the pressure-stabilization method, i.e., robustness for convection-dominated problems, symmetry of the resulting matrix to be solved, and small number of degrees of freedom, they are efficient especially in three-dimensional problems.

The first-order scheme can be used to supply an initial velocity at the first step of the second-order scheme. When we employ the first-order scheme with a small time increment of order Δt^2, the optimal error estimates of the second-order scheme are obtained in the consumption of computation time. A more efficient way is to set the initial velocity from the first-step solution of the first-order scheme with the time increment Δt. Then, the best possible convergence order for the velocity in the $L^2(\Omega)$-norm is kept, which is recognized numerically.

It is known that a rough numerical quadrature may cause instability of LG schemes, cf. [8, 19], since the composite functions of LG schemes are not smooth. Recently, the instability has been overcome by a locally linearized velocity technique in [20], which can also be applied to the stabilized LG schemes in this chapter.

Acknowledgements This work was supported by JSPS (the Japan Society for the Promotion of Science) under the Japanese–German Graduate Externship (Mathematical Fluid Dynamics) and Grant-in-Aid for Scientific Research (S), No. 24224004. The authors are indebted to JSPS also for Grant-in-Aid for Young Scientists (B), No. 26800091 to the first author and for Grant-in-Aid for Scientific Research (C), No. 25400212 to the second author.

References

1. Achdou, Y., Guermond, J.-L.: Convergence analysis of a finite element projection/Lagrange–Galerkin method for the incompressible Navier–Stokes equations. SIAM J. Numer. Anal. **37**, 799–826 (2000)
2. Barrett, R., Berry, M., Chan, T.F., Demmel, J., Donato, J., Dongarra, J., Eijkhout, V., Pozo, R., Romine, C., van der Vorst, H.: Templates for the Solution of Linear Systems: Building Blocks for Iterative Methods. SIAM, Philadelphia (1994)
3. Boukir, K., Maday, Y., Métivet, B., Razafindrakoto, E.: A high-order characteristics/finite element method for the incompressible Navier–Stokes equations. Int. J. Numer. Methods Fluids **25**, 1421–1454 (1997)
4. Brezzi, F., Pitkäranta, J.: On the stabilization of finite element approximations of the Stokes equations. In: Hackbusch, W. (ed.) Efficient Solutions of Elliptic Systems. Vieweg, Wiesbaden (1984)
5. Ciarlet, P.G.: The Finite Element Method for Elliptic Problems. North-Holland, Amsterdam (1978)
6. Ewing, R.E., Russell, T.F.: Multistep Galerkin methods along characteristics for convection-diffusion problems. In: Vichnevetsky, R., Stepleman, R.S. (eds.) Advances in Computer Methods for Partial Differential Equations. IMACS, vol. IV, pp. 28–36 (1981)
7. Girault, V., Raviart, P.-A.: Finite Element Methods for Navier–Stokes Equations, Theory and Algorithms. Springer, Berlin (1986)
8. Morton, K.W., Priestley, A., Süli, E.: Stability of the Lagrange–Galerkin method with non-exact integration. ESAIM: Math. Model. Numer. Anal. **22**, 625–653 (1988)
9. Notsu, H.: Numerical computations of cavity flow problems by a pressure stabilized characteristic-curve finite element scheme. Trans. Jpn. Soc. Comput. Eng. Sci. **2008**, 20080032 (2008)
10. Notsu, H., Tabata, M.: A combined finite element scheme with a pressure stabilization and a characteristic-curve method for the Navier–Stokes equations. Trans. Jpn. Soc. Ind. Appl. Math. **18**, 427–445 (2008) (in Japanese)
11. Notsu, H., Tabata, M.: Error estimates of a pressure-stabilized characteristics finite element scheme for the Oseen equations. J. Sci. Comput. **65**, 940–955 (2015)
12. Notsu, H., Tabata, M.: Error estimates of a stabilized Lagrange–Galerkin scheme for the Navier–Stokes equations. ESAIM: Math. Model. Numer. Anal. **50**, 361–380 (2016)
13. Notsu, H., Tabata, M.: Error estimates of a stabilized Lagrange–Galerkin scheme of second-order in time for the Navier–Stokes equations. In: Shibata, Y., Suzuki, Y. (eds.) Mathematical Fluid Dynamics, Present and Future. Springer, Berlin (to appear)
14. Pironneau, O.: On the transport-diffusion algorithm and its applications to the Navier–Stokes equations. Numer. Math. **38**, 309–332 (1982)
15. Ravindran, S.S.: Convergence of extrapolated BDF2 finite element schemes for unsteady penetrative convection model. Numer. Funct. Anal. Optim. **33**, 48–79 (2012)
16. Rui, H., Tabata, M.: A second order characteristic finite element scheme for convection-diffusion problems. Numer. Math. **92**, 161–177 (2002)
17. Stroud, A.H.: Approximate Calculation of Multiple Integrals. Prentice-Hall, Englewood Cliffs, NJ (1971)

18. Süli, E.: Convergence and nonlinear stability of the Lagrange–Galerkin method for the Navier–Stokes equations. Numer. Math. **53**, 459–483 (1988)
19. Tabata, M.: Discrepancy between theory and real computation on the stability of some finite element schemes. J. Comput. Appl. Math. **199**, 424–431 (2007)
20. Tabata, M., Uchiumi, S.: A Lagrange–Galerkin scheme with a locally linearized velocity for the Navier–Stokes equations. arXiv:1505.06681 [math.NA] (2015)

Part VI
Biomedical Applications

On Three-Dimensional ALE Finite Element Model For Simulating Interstitial Medium Deformation in the Presence of a Moving Needle

Yannick Deleuze, Marc Thiriet, and Tony W.H. Sheu

Abstract The effects of inserted needle on the subcutaneous interstitial flow are studied. A goal is to describe the physical stress affecting cells during acupuncture needling. The convective Brinkman equations are considered to describe the flow through a fibrous medium. Three-dimensional simulations are carried out employing an ALE finite element model. Numerical studies illustrate the acute physical stress developed by the implantation of a needle.

1 Introduction

In recent years, computational techniques have been widely used by researchers to investigate and simulate biological flow within three-dimensional context. Applications include blood flow models, air flow models in the respiratory tract, interstitial

Y. Deleuze
Sorbonne Universités, UPMC Univ Paris 06, UMR 7598, Laboratoire Jacques-Louis Lions, F-75005 Paris, France

Department of Engineering Science and Ocean Engineering, National Taiwan University, No. 1, Sec. 4, Roosevelt Road, Taipei, Taiwan
e-mail: yannick.deleuze@ljll.math.upmc.fr

M. Thiriet
CNRS, UMR 7598, Laboratoire Jacques-Louis Lions, F-75005 Paris, France

Sorbonne Universités, UPMC Univ Paris 06, UMR 7598, Laboratoire Jacques-Louis Lions, F-75005 Paris, France

INRIA-Paris-Rocquencourt, EPC REO, Domaine de Voluceau, BP105, 78153 Le Chesnay Cedex, France
e-mail: marc.thiriet@upmc.fr

T.W.H. Sheu (✉)
Department of Engineering Science and Ocean Engineering, National Taiwan University, No. 1, Sec. 4, Roosevelt Road, Taipei, Taiwan

Center for Advanced Study in Theoretical Sciences, National Taiwan University, No. 1, Sec. 4, Roosevelt Road, Taipei, Taiwan
e-mail: twhsheu@ntu.edu.tw

Y. Bazilevs, K. Takizawa (eds.), *Advances in Computational Fluid-Structure Interaction and Flow Simulation*, Modeling and Simulation in Science, Engineering and Technology, DOI 10.1007/978-3-319-40827-9_27

flow models, and chemical mediators transport. Most of the structure and fluid interactions have been considered with simplified rigid wall or deformable wall models.

Methods to predict flows that account for moving domains or domain deformability using the finite element method are based on fixed mesh methods or moving mesh methods. On the one hand, fixed mesh methods include the immersed boundary formulation and the fictitious domain formulation. These methods are well adapted to moving bodies in the fluid or fluid–structure computation with interface of a highly geometric complexity. On the other hand, moving mesh methods include the Lagrangian method, the moving finite element method, the deformation map method, the geometric conservation law method, the space/time finite element method, and the arbitrarily Lagrangian–Eulerian (ALE) method for the solution of fluid dynamic problems.

The mathematically rigorous ALE framework has been well accepted to be applicable to simulate transport phenomena in time and allows some freedom in the description of mesh motion. A theoretical analysis of the ALE method can be found in [5, 8]. However, ALE equations are computationally expensive when considering a large domain because of the necessity of continuously updating the geometry of the fluid and structural mesh. Interface tracking with time discretization also raises some implementation questions. The implementation of the ALE method can be done in FreeFem++ [2].

Study of biological flows plays a central role in acupuncture research. For a description of the underlying acupuncture mechanism, one can refer to [3, 14, 15]. Interstitial flow models take into account interstitial fluid, cell membrane interaction, and fiber interactions [10]. Mastocytes, among other cells, are able to respond to fluidic stimuli via mechanotransduction pathways leading to the degranulation and liberation of chemical mediators [7]. Degranulation mechanisms include interaction of the cell membrane with interstitial and cytosolic flow [16]. Ion transport in narrow ion channels is another challenging task to model. Indeed, degranulation of chemical mediators upon stimulation can be triggered by a rapid Ca^{2+} entry in the cytosol [12].

Modeling the three-dimensional interstitial flow in tissues is extremely challenging for a large number of reasons: a complex geometry of the tissue, an accurate constitutive description of the behavior of the tissue, and flow rheology are only a few examples. Macroscopic models for incorporating complex microscopic structure are essential for applications [1, 10, 12, 19, 20]. In the context of acupuncture, the interstitial flow has been modeled by the Brinkman equations [3, 19, 20].

In this paper, a porous medium formulation of the interstitial fluid is presented for modeling mastocyte-needle interaction in deformable connective tissues. This formulation is based on a conventional ALE characteristic/Galerkin finite element model for an unsteady flow thought a porous medium modeled by the incompressible Brinkman's equations in a three-dimensional moving domain. The motion of the needle in the fluid is taken into account. The main features of the model can be summarized as follows:

1. The loose connective tissue of the hypodermis is constituted of scattered cells immersed in extracellular matrix. The extracellular matrix contains relatively sparse fibers and abundant interstitial fluid. The interstitial fluid contains water, ions, and other small molecules. Such a fluid corresponds to plasma without macromolecules and interacts with the ground substance, thereby forming a viscous hydrated gel that can stabilize fiber network [11, 13].
2. The Darcys law is used to approximate fibers of the media as a continuum and allows us to compute the actual microscopic flow phenomena that occur in the fibrous media.
3. Brinkman's law then allows us to describe the flow field around solid bodies such as the embedded cells in extracellular matrix.
4. Transient convective Brinkman's equations are applied to simulate the interstitial flow in a fibrous medium driven by a moving needle.

Although the previously stated approach cannot give information on microscopic events, it can describe macroscale flow patterns in porous media. Focus is given to the effects of interstitial fluid flow during implantation of an acupuncture needle until the tip has reached the desired location within the hypodermis. The objective of this work is to give a description of the physical stress (shear stress and pressure) influencing tissue and cells.

2 Methods

Due to biological complexity, the interstitium is considered as a fluid-filled porous material. The interstitial flow is simulated using the incompressible convective Brinkman equation.

2.1 *Flow Equations*

The governing equations of the unsteady flow of an incompressible fluid through a porous medium (with mass density ρ, dynamic viscosity μ, and kinematic viscosity $\nu = \mu/\rho$) can be derived as [17]:

$$\frac{\rho}{\alpha_f}\left(\frac{\partial \bar{\mathbf{u}}}{\partial t} + \bar{\mathbf{u}} \cdot \nabla\left(\frac{\bar{\mathbf{u}}}{\alpha_f}\right)\right) - \mu \nabla^2 \bar{\mathbf{u}} + \frac{1}{\alpha_f} \nabla(\alpha_f p_f) = -\frac{\mu}{P} \bar{\mathbf{u}} \quad \text{in} \Omega(t), \tag{1}$$

$$\nabla \cdot \bar{\mathbf{u}} = 0, \tag{2}$$

$$\bar{\mathbf{u}}(\mathbf{x}, 0) = \bar{\mathbf{u}}_0(\mathbf{x}), \tag{3}$$

where $-\frac{\mu}{P}\bar{\mathbf{u}}$ denotes the Darcy drag, P the Darcy permeability, $\bar{\mathbf{u}}$ the averaged velocity vector, and p_f the pressure. The averaged velocity is defined as

$$\bar{\mathbf{u}} = \alpha_f \mathbf{u}_f, \tag{4}$$

where $\mathbf{u}_f$ is the fluid velocity and

$$\alpha_f = \frac{\text{fluid volume}}{\text{total volume}} \tag{5}$$

is the fluid volume fraction. This volume fraction corresponds to the effective porosity of the medium. The fluid fractional volume α_f is taken as a space-dependent parameter to model the distinguished properties of an acupoint.

The system of equations (1–2) is applied to the case of a flow driven by the motion of a needle in the deformable domain $\Omega(t)$ [3]. The domain boundary can be decomposed into three surfaces: the needle boundary denoted by Γ_1, an impervious boundary (wall) denoted by Γ_2, the mastocyte membrane denoted by Γ_3, and the open boundary on the sides denoted by Γ_4. The classical no-slip condition is applied to the needle surface Γ_1, the rigid wall Γ_2, and the cell surface Γ_3. At the outer boundary Γ_4 a traction-free boundary condition is prescribed. Thus, the entire set of boundary conditions reads as

$$\bar{\mathbf{u}} = \mathbf{v}_{\text{needle}}, \text{on } \Gamma_1, \tag{6}$$

$$\bar{\mathbf{u}} = 0, \text{on } \Gamma_2, \tag{7}$$

$$\bar{\mathbf{u}} = 0, \text{on } \Gamma_3, \tag{8}$$

$$-\mu \nabla \bar{\mathbf{u}} \cdot \mathbf{n} + p_f \mathbf{n} = 0, \text{on } \Gamma_4. \tag{9}$$

2.2 *Finite Element Model*

The governing equations in section 2.2.1 are solved using the finite element software FreeFem++ [6]. This code programs the discrete equations derived from the finite element weak formulation of the problem presented in section 2.2.3 using a characteristic/Galerkin model to stabilize convection terms.

2.2.1 Scaling and Setting for Numerical Simulations

L denotes the characteristic length, that is, the needle width and V is the characteristic velocity set to be the needle maximum velocity. Rescaling the variables leads to

$$\mathbf{x}' = \frac{\mathbf{x}}{\mathsf{L}}, t' = \frac{t}{(\mathsf{L}/\mathsf{V})}, p' = \frac{p_f}{(\rho \mathsf{V}^2)}, \mathbf{u}' = \frac{\bar{\mathbf{u}}}{\mathsf{V}}. \tag{10}$$

In the resulting dimensionless form, after removing the prime in the rescaled variables, the dimensionless incompressible convective Brinkman equations read as

$$\frac{1}{\alpha_f}\frac{\partial \mathbf{u}}{\partial t} + \frac{1}{\alpha_f}\mathbf{u} \cdot \nabla\left(\frac{\mathbf{u}}{\alpha_f}\right) - \frac{1}{\text{Re}}\nabla^2 \mathbf{u} + \frac{1}{\alpha_f}\nabla(\alpha_f p) = -\frac{1}{\text{Da Re}}\mathbf{u}, \tag{11}$$

$$\nabla \cdot \mathbf{u} = 0. \tag{12}$$

where Re is the Reynolds number and Da is the Darcy number. The previous dimensionless parameters are defined as

$$\mathrm{Re} = \frac{\rho \mathsf{LV}}{\mu}, \ \mathrm{Da} = \frac{\mathscr{P}}{\mathsf{L}^2}. \tag{13}$$

In considering the above dimensionless governing equations, the normalized boundary conditions on the domain boundary are prescribed as

$$\mathbf{u} = \mathbf{v} \text{ on } \Gamma_1, \tag{14}$$

$$\mathbf{u} = 0 \text{ on } \Gamma_2, \tag{15}$$

$$\mathbf{u} = 0 \text{ on } \Gamma_3, \tag{16}$$

$$-\frac{1}{\mathrm{Re}}\nabla\mathbf{u}\cdot\mathbf{n} + p\mathbf{n} = 0 \text{ on } \Gamma_{sides}. \tag{17}$$

2.2.2 ALE Implementation on Moving Meshes

In the present paper, the ALE framework built in FreeFem++ is employed to compute the flow in the moving domain. In the current problem setting, the motion of needle is prescribed with respect to time. The boundary of the domain is thus exactly known at each time so that an area preserving mesh can be precisely generated.

The framework of the ALE approach employed is briefly described below. Let $\Omega(t)$ be the domain at each time t with regular boundary $\partial\Omega(t)$. In the Eulerian description, the fluid is described by

$$\mathbf{u}(\mathbf{x}, t) \text{ and } p(\mathbf{x}, t), \forall \mathbf{x} \in \Omega(t). \tag{18}$$

To follow a moving domain, one can define the ALE map as

$$\tilde{\mathscr{A}} : \tilde{\omega} \times \mathbb{R}^+ \to \mathbb{R}^2 \quad (\tilde{\mathbf{x}}, t) \to \tilde{\mathscr{A}}(\tilde{\mathbf{x}}, t) := \tilde{\mathscr{A}}_t, \tag{19}$$

such that $\omega(t) = \tilde{\mathscr{A}}(\tilde{\omega}, t)$, where $\tilde{\omega}$ is the reference computational domain. Given an ALE field $\tilde{q} : \tilde{\omega} \times \mathbb{R}^+ \to \mathbb{R}$, its Eulerian description is given by

$$\forall \mathbf{x} \in \Omega(t), q(\mathbf{x}, t) = \tilde{q}(\tilde{\mathscr{A}}_t^{-1}(\mathbf{x}), t) \tag{20}$$

In ALE framework, the computational domain velocity (or ALE velocity or grid velocity) is defined as

$$\tilde{\mathbf{a}}(\tilde{\mathbf{x}}, t) = \frac{\partial \tilde{\mathscr{A}}}{\partial t}(\tilde{\mathbf{x}}, t), \quad \forall \tilde{\mathbf{x}} \in \tilde{\omega}, \tag{21}$$

so that we can get

$$\mathbf{a}(\mathbf{x},t) = \tilde{\mathbf{a}}(\tilde{\mathscr{A}}_t^{-1},t). \tag{22}$$

The ALE time-derivative is defined as

$$\left.\frac{\partial q}{\partial t}\right|_{\tilde{\mathscr{A}}} = \frac{\mathrm{d}}{\mathrm{d}t} q(\tilde{\mathscr{A}}(\tilde{\mathbf{x}},t),t), \tag{23}$$

and the following identity holds:

$$\left.\frac{\partial q}{\partial t}\right|_{\tilde{\mathscr{A}}} = (\mathbf{a}.\nabla)q + \frac{\partial q}{\partial t}. \tag{24}$$

A general method is used to construct the mapping or, equivalently, the domain velocity **a**. The domain velocity is computed by solving the following Laplace equation subjected to the Dirichlet boundary condition [4]:

$$-\nabla^2\mathbf{a} = 0, \qquad \mathbf{a}_{|\partial\Omega} = \mathbf{v}. \tag{25}$$

In the ALE framework, equations (11–12), subject to a prescribed needle motion, become

$$\left.\frac{\partial(\mathbf{u}/\alpha_f)}{\partial t}\right|_{\tilde{\mathscr{A}}} + \left(\left(\frac{\mathbf{u}}{\alpha_f} - \mathbf{a}\right)\cdot\nabla\right)\frac{\mathbf{u}}{\alpha_f} - \frac{1}{\mathrm{Re}}\nabla^2\mathbf{u} + \frac{1}{\alpha_f}\nabla(\alpha_f p) = -\frac{\mathbf{u}}{\mathrm{Da\,Re}}, \tag{26}$$

$$\nabla\cdot\mathbf{u} = 0. \tag{27}$$

The solutions **u** and p are sought subject to the initial condition (3) and the boundary conditions (7–9) described in section 2.2.1.

2.2.3 Finite Element Discretization

The convective incompressible Brinkman equations are approximated with the method of characteristics for the nonlinear convection term and a Galerkin method for the rest of the spatial derivative terms. The time discretization gives

$$\frac{1}{\Delta t}\left(\frac{\mathbf{u}^{n+1}}{\alpha_f} - \left(\frac{\mathbf{u}^n}{\alpha_f}\right)\circ\mathbf{X}^n\right) - \frac{1}{\mathrm{Re}}\nabla^2\mathbf{u}^{n+1} + \frac{1}{\alpha_f}\nabla(\alpha_f p^{n+1}) = -\frac{\mathbf{u}^{n+1}}{\mathrm{Da\,Re}}, \tag{28}$$

$$\nabla\cdot\mathbf{u}^{n+1} + \varepsilon p^{n+1} = 0, \tag{29}$$

in Ω^{n+1}. Note that $\mathbf{X}^n$ is approximated by $\mathbf{X}^n \approx \mathbf{x} - \left(\frac{\mathbf{u}^n}{\alpha_f} - \mathbf{a}^n\right)(\mathbf{x})\,\Delta t$. A small stabilization parameter ε is introduced following the so-called artificial compressibility method.

For all $\varphi \in H^{1/2}(\Gamma_1)$, let us introduce the product space

$$V_\varphi = \left\{(\mathbf{w}, q) \in [H^1(\Omega)]^2 \times L^2(\Omega), \mathbf{w} = \varphi \text{ on } \Gamma_{\text{needle}}, \mathbf{w} = 0 \text{ on } \Gamma_{\text{wall}}\right\}. \quad (30)$$

Let

$$(a, b) = \int_{\Omega^{n+1}} ab \, d\mathbf{x}. \quad (31)$$

The weak formulation becomes the following finite dimensional linear system: find $(\mathbf{u}^{n+1}, p^{n+1}) \in V_g$ such that

$$\begin{aligned} \frac{1}{\Delta t}\left(\frac{\mathbf{u}^{n+1}}{\alpha_f} - \left(\frac{\mathbf{u}^n}{\alpha_f}\right) \circ \mathbf{X}^n, \mathbf{w}\right) + \frac{1}{\text{Re}}\left(\frac{1}{\alpha_f}\nabla \mathbf{u}^{n+1}, \nabla \mathbf{w}\right) & \\ - \left(\alpha_f p^{n+1}, \nabla \cdot \left(\frac{\mathbf{w}}{\alpha_f}\right)\right) + \frac{1}{\text{Da Re}}\left(\mathbf{u}^{n+1}, \mathbf{w}\right) &= 0, \\ \left(\nabla \cdot \mathbf{u}^{n+1}, q\right) + \varepsilon\left(p^{n+1}, q\right) &= 0, \end{aligned} \quad (32)$$

for all $(\mathbf{w}, q) \in V_0$.

The Taylor–Hood $\mathbb{P}_2$–$\mathbb{P}_1$ elements are adopted to ensure satisfaction of the LBB stability condition [9]. Note that temporal accuracy order of the presented characteristic/Galerkin method is one. Meshes are generated within FreeFem++ and mesh adaptation is performed prior to simulations so as to enhance mesh quality around the needle and the cell.

3 Results

In the present work, the needling direction is perpendicular to the skin surface. In practice, it is possible that the needling direction is oblique to the skin surface. The simulation results show that the insertion of an acupuncture needle can influence interstitial fluid flow. The computed velocity field shows that at a location away from the needle, the effect of the stress field vanishes (Fig. 1). Furthermore, when the needle reaches its maximum speed, the interstitial pressure gradient becomes higher at a location close to the needle tip (Fig. 2). The changes in the interstitial fluid flow and the high pressure gradient can affect the activities of the mastocyte pools in the stimulated area.

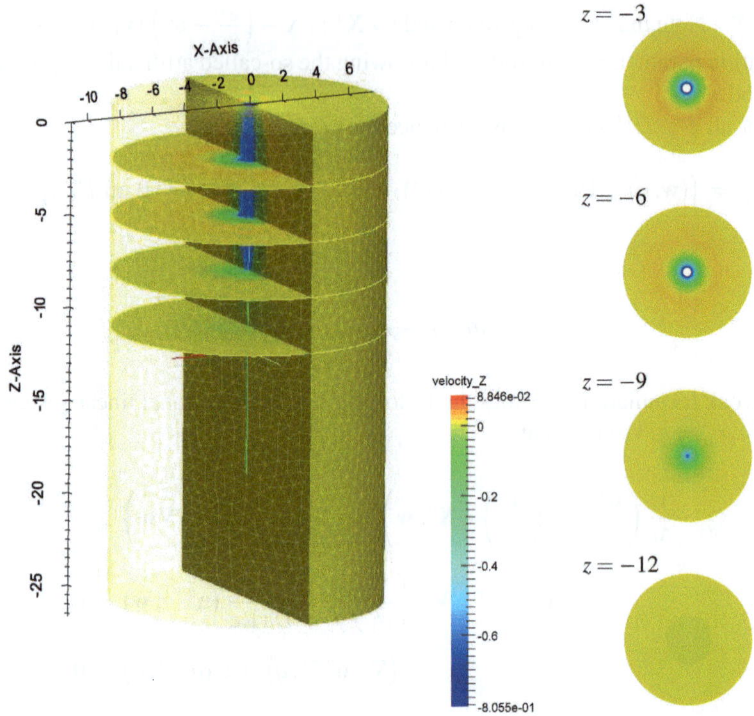

Fig. 1 The predicted contours of velocity along the z-direction resulting from the needle (blue) motion in interstitial fluid.

Another subject of interest is the effects of the fluidic stimuli on an interstitial cell. Local mechanical forces can trigger the activation of mechanoresponsive proteins on the cell surface [14, 18] so that Ca^{++} is allowed to enter the cytosol via pressure and shear stress gated ion channels. Simulations are carried out by considering fixed cells and no-slip boundary condition prescribed at the cell surface. Figure 3 shows the pressure contours on the surface of a cell added closely to the needle. The pressure on the cell surface is higher in the region closest to the needle tip.

4 Conclusions

The proposed three-dimensional ALE finite element model is able to describe the interstitial flow and pressure from the macroscopic point of view when a needle is inserted and moving within the hypodermis. High local fluid pressure and shear

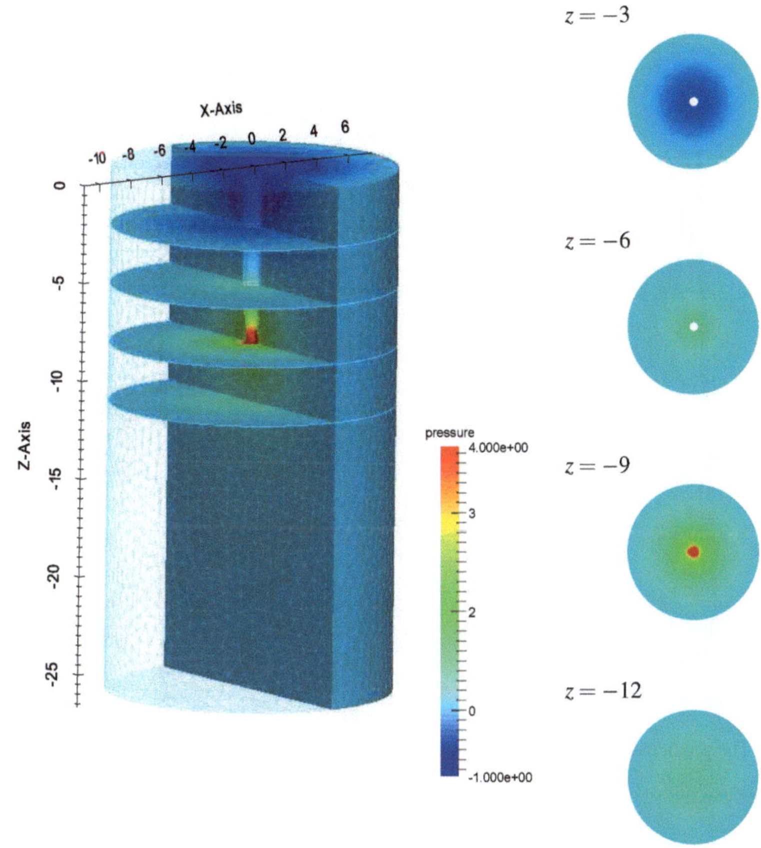

Fig. 2 The predicted contours of velocity along the z-direction resulting from the needle (blue) motion in interstitial fluid.

stress on cells are most likely to appear near the needle tip region. However, the proposed method does not allow the rotation of the needle to be taken into account. When considering the rotation of the needle, a large deformation of the tissue is observed with the twisting of the fibers around the needle, that in turn makes the corresponding change in interstitial flow. A fluid/structure model taking into account the mechanics of the fibers should then be considered. This study has shown that the numerical prediction of the interstitial pressure and shear stress is an essential tool to gain a better understanding of the mechanism involved in acupuncture needling.

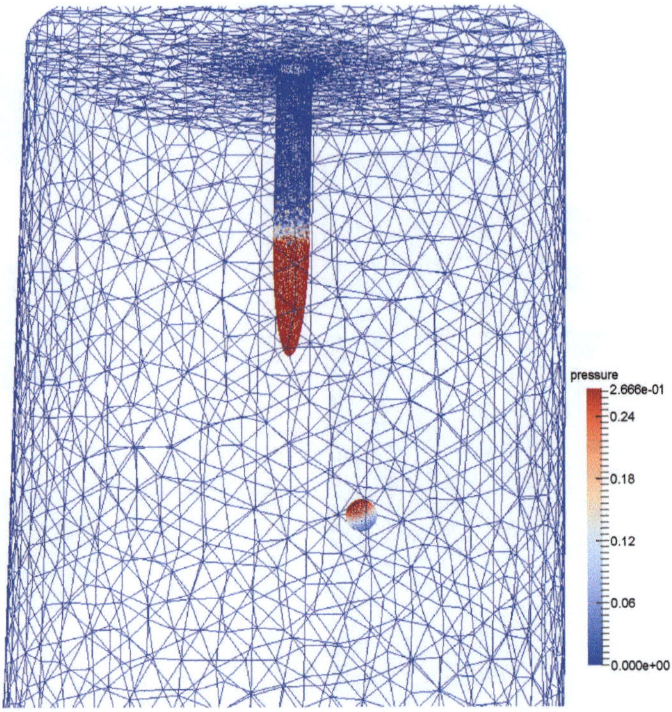

Fig. 3 The predicted pressure contours on the needle and cell surfaces as the needle moves toward the cell.

References

1. Blasselle, A.: Modélisation mathématique de la peau. Thèse de doctorat, Université Pierre et Marie Curie, Paris (2011)
2. Decoene, A., Maury, B.: Moving meshes with freefem++. J. Numer. Math. **20**, 195 (2013)
3. Deleuze, Y., Thiriet, M., Sheu, T.W.H.: Modeling and simulation of local physical stress on the mastocytes created by the needle manipulation during acupuncture. Commun. Comput. Phys. **18**, 850–867 (2015)
4. Fernández, M.A., Formaggia, L., Gerbeau, J.F., Quarteroni, A.: The derivation of the equations for fluids and structure. In: L. Formaggia, A. Quarteroni, A. Veneziani (eds.) Cardiovascular Mathematics, vol. 1 in MS&A, pp. 77–121. Springer, Milan (2009)
5. Grandmont, C., Maday, Y.: Fluid-structure interaction: a theoretical point of view. Rev. Eur. Élém. **9**(6–7), 633–653 (2000)
6. Hecht, F.: New development in FreeFem++. J. Numer. Math. **20**(3–4), 251 (2013)
7. Park, J.Y., Yoo, S.J., Patel, L., Lee, S.H., Lee, S.H.: Cell morphological response to low shear stress in a two-dimensional culture microsystem with magnitudes comparable to interstitial shear stress. Biorheology **47**(3), 165–178 (2010)
8. Quarteroni, A., Tuveri, M., Veneziani, A.: Computational vascular fluid dynamics: problems, models and methods. Comput. Vis. Sci. **2**(4), 163–197 (2000)
9. Raviart, P.A., Thomas, J.M.: Introduction À L'analyse Numérique Des Équations Aux Dérivées Partielles. Masson, Paris (1983)

10. Swartz, M.A., Fleury, M.E.: Interstitial flow and its effects in soft tissues. Annu. Rev. Biomed. Eng. **9**(1), 229–256 (2007)
11. Thiriet, M.: Biology and Mechanics of Blood Flows: Part I: Biology. CRM Series in Mathematical Physics. Springer, New York (2008)
12. Thiriet, M.: Cell and Tissue Organization in the Circulatory and Ventilatory Systems. Biomathematical and Biomechanical Modeling of the Circulatory and Ventilatory Systems, vol. 1. Springer, New York (2011)
13. Thiriet, M.: Cells and tissues. In: Cell and Tissue Organization in the Circulatory and Ventilatory Systems. Biomathematical and Biomechanical Modeling of the Circulatory and Ventilatory Systems, vol. 1, pp. 11–67. Springer, New York (2011)
14. Thiriet, M.: Intracellular Signaling Mediators in the Circulatory and Ventilatory Systems. Biomathematical and Biomechanical Modeling of the Circulatory and Ventilatory Systems, vol. 4. Springer, New York (2013)
15. Thiriet, M., Deleuze, Y., Sheu, T.W.: A biological model of acupuncture and its derived mathematical modeling and simulations. Commun. Comput. Phys. **18**, 831–849 (2015)
16. Tseng, Y.H., Huang, H.: An immersed boundary method for endocytosis. J. Comput. Phys. **273**, 143–159 (2014)
17. Vafai, K., Tien, C.L.: Boundary and inertia effects on flow and heat transfer in porous media. Int. J. Heat Mass Transf. **24**(2), 195–203 (1981)
18. Wei, F., Shi, X., Chen, J., Zhou, L.: Fluid shear stress-induced cytosolic calcium signalling and degranulation dynamics in mast cells. Cell Biol. Int. Rep. **19**(2), 45–51 (2012)
19. Yao, W., Ding, G.H.: Interstitial fluid flow: simulation of mechanical environment of cells in the interosseous membrane. Acta Mech. Sin. **27**(4), 602–610 (2011)
20. Yao, W., Li, Y., Ding, G.: Interstitial fluid flow: the mechanical environment of cells and foundation of meridians. Evid. Based Complement. Alternat. Med. **2012**, 1–9 (2012)

Time-Dependent Outflow Boundary Conditions for Blood Flow in the Arterial System

JaeHyuk Kwack, Soonpil Kang, Geetha Bhat, and Arif Masud

Abstract We present time-dependent outflow boundary conditions for blood flow simulations in the arterial system. The new method allows for embedding clinically obtained patient-specific data into the patient-specific geometric models of the circulation system. Blood rheology is accounted for by shear-rate dependent models for blood. Our recently developed stabilized finite element method for non-Newtonian fluid models is extended to include downstream effects by incorporating clinically measured downstream resistance via a novel functional form for the outflow boundary conditions. Patient-specific flow-rate and pressure profiles measured clinically (e.g., ultrasound device, CT, or MRI) are used to determine time-dependent resistance functions. For verification of the new method, we compare the clinically measured time-dependent resistance outflow boundary conditions to the constant pressure, constant resistance, and the impedance outflow boundary conditions. Numerical tests verify that the time-dependent outflow boundary conditions proposed in this work impose the most accurate downstream effects that are caused by the non-Newtonian behavior of blood as well as the geometrical complexity of the branching arteries. Our numerical tests show that the reduced geometry with the proposed outflow boundary conditions results in an order of magnitude reduction in computational cost as compared to that of the full arterial geometry model.

J. Kwack • S. Kang • A. Masud (✉)
University of Illinois at Urbana-Champaign, Urbana, IL, USA
e-mail: amasud@illinois.edu

G. Bhat
Advocate Christ Medical Center, Oak Lawn, IL, USA

Y. Bazilevs, K. Takizawa (eds.), *Advances in Computational Fluid-Structure Interaction and Flow Simulation*, Modeling and Simulation in Science, Engineering and Technology, DOI 10.1007/978-3-319-40827-9_28

1 Introduction

Blood is composed of various cell types (i.e., red blood cells, white blood cells, and platelets) that are suspended in blood plasma (i.e., water, dissipated proteins, glucose, mineral ions, hormones, and carbon dioxide). The complex rheological response of blood is a function of a variety of interactions between cells and plasma, that manifest themselves via shear-rate dependent viscous stress and viscoelastic stress [1–5]. A literature review shows that several mathematical models have been developed for the non-Newtonian behavior of fluids [6–11]. Masud and Kwack have developed variational multi-scale (VMS)-based stabilized methods for viscoelastic [12, 13] and shear-rate dependent [14–16] non-Newtonian fluids wherein a variety of convergent finite elements are presented. Blood flow simulations in the human cardiovascular system that account for the complexity of the patient-specific geometries, transition to turbulence in arteries, and blood–artery interactions have also been reported in the literature [17–23].

The cardiovascular system is comprised of the heart and blood vessels that form a closed network. Developing a computational model for the entire system is not only difficult due to its geometric complexity, but it is also computationally expensive. Consequently, developing appropriate outflow boundary conditions that can take into account the effects of the circulation system downstream of the zone of interest is an essential ingredient of blood flow modeling. With appropriate boundary conditions applied on the outflow surfaces, blood flow simulations can be carried out only in the zones of interest, it can substantially reduce the cost of computation while preserving the accuracy of the physics involved. To account for the downstream effects, Olufsen [24] introduced dynamic boundary conditions in the impedance spectra for large systemic arteries, while Esmaily-Moghadam et al. [25] presented outflow boundary conditions that prevent backflow in blood flow simulations. A sophisticated method for the constant resistance and impedance outflow boundary conditions is proposed by Vignon-Clementel et al. [26].

The present paper is inspired by [26] wherein we develop a systematic procedure to embed clinically measured flow parameters via a transient resistance function that describes the outflow boundary condition. An outline of the paper is as follow: Section 2 presents a VMS-based stabilized method for incompressible shear-rate dependent fluids. It includes strong form of the governing equations along with the shear-rate dependent non-Newtonian constitutive model. The nonlinear stabilized form is presented next. The outflow boundary conditions for the constant pressure, the constant resistance, the impedance, and the time-dependent resistance function are introduced in Section 3. Section 4 presents numerical tests in simple channel geometry, as well as in idealized arterial tree geometry. Conclusions are drawn in Section 5.

2 A Variational Multi-Scale (VMS) Framework for Non-Newtonian Fluids

Masud and Kwack have developed a VMS framework for shear-rate dependent non-Newtonian fluids [14–16] and non-Newtonian viscoelastic fluids [12, 13]. In the present work the VMS-based formulation for shear-rate dependent fluids [15] is employed. The governing equations for shear-rate dependent fluids are given by momentum balance equations:

$$\rho \boldsymbol{v}_{,t} + \rho \boldsymbol{v} \cdot \nabla \boldsymbol{v} - \nabla \cdot \boldsymbol{\sigma}_{v}(\boldsymbol{v}) + \nabla p = \boldsymbol{f} \qquad \text{in } \Omega \times]0, T[\tag{1}$$

$$\nabla \cdot \boldsymbol{v} = 0 \qquad \text{in } \Omega \times]0, T[\tag{2}$$

where $\boldsymbol{v}_{,t}$ is the time derivative of the velocity field $\boldsymbol{v}$, p is the thermodynamic pressure, ρ is the density, $\boldsymbol{f}$ is the body force vector, $\boldsymbol{\sigma}_v = 2\eta(\gamma)\,\boldsymbol{\varepsilon}(\boldsymbol{v})$ is shear-rate dependent stress tensor, $\boldsymbol{\varepsilon}(\boldsymbol{v})$ is the rate-of-deformation tensor, γ is the shear-rate defined as $\gamma := \sqrt{2\boldsymbol{\varepsilon}(\boldsymbol{v}) : \boldsymbol{\varepsilon}(\boldsymbol{v})}$, and $\eta(\gamma)$ is shear-rate dependent viscosity. Although different forms of $\eta(\gamma)$ are admissible for the shear-rate dependent stress tensor $\boldsymbol{\sigma}_v$, we have employed the Carreau-Yasuda model [6] for the viscosity function.

$$\eta(\gamma) \quad := \quad \mu_\infty + (\mu_0 - \mu_\infty)\left(1 + (\lambda\gamma)^a\right)^{(n-1)/a} \tag{3}$$

The VMS-stabilized formulation [15] is based on an overlapping sum decomposition of the trial solution (i.e., $\boldsymbol{v}$) and weighting function (i.e., $\boldsymbol{w}$) into coarse- or resolvable-scales and fine- or subgrid-scales.

$$\boldsymbol{v}(\boldsymbol{x},t) = \underbrace{\bar{\boldsymbol{v}}(\boldsymbol{x},t)}_{\text{coarse scale}} + \underbrace{\boldsymbol{v}'(\boldsymbol{x},t)}_{\text{fine scale}}, \qquad \boldsymbol{w}(\boldsymbol{x}) = \underbrace{\bar{\boldsymbol{w}}(\boldsymbol{x})}_{\text{coarse scale}} + \underbrace{\boldsymbol{w}'(\boldsymbol{x})}_{\text{fine scale}} \tag{4}$$

Substituting (4) in the standard Galerkin forms for (1) and (2) leads to splitting of the weak forms into coarse-scale and fine-scale sub-problems. The derivation for the VMS-based stabilized formulation is comprised of two steps: First, the fine-scale sub-problem is solved locally to extract a model for the fine scales, and then the fine-scale model is embedded in the coarse-scale variational equation. The details of the derivation are given in [15] along with the specification of the appropriate spaces of functions. Herein we present the final nonlinear stabilized form for the shear-rate dependent fluids where $\boldsymbol{V} = \{\boldsymbol{v},\ p\}$ is the vector of unknown velocity and pressure fields and $\boldsymbol{W} = \{\boldsymbol{w},\ q\}$ is the vector of the corresponding weighting functions.

$$\mathcal{B}_{\text{stab}}(\boldsymbol{W}, \boldsymbol{V}) = \mathcal{L}_{\text{stab}}(\boldsymbol{W}) \tag{5}$$

where $\mathcal{B}(\cdot,\cdot)$ is linear with respect to the first slot and is nonlinear with respect to the second slot.

$$\mathcal{B}_{\text{stab}}\left(\boldsymbol{W},\boldsymbol{V}\right) = \mathcal{B}\left(\boldsymbol{W},\boldsymbol{V}\right) + \left(\boldsymbol{\chi},\boldsymbol{\tau}\left(\rho\boldsymbol{v}_{,t} + \rho\boldsymbol{v}\cdot\nabla\boldsymbol{v} - 2\eta\nabla\cdot\boldsymbol{\varepsilon}\left(\boldsymbol{v}\right) - 2\nabla\eta\cdot\boldsymbol{\varepsilon}\left(\boldsymbol{v}\right) + \nabla p\right)\right) \tag{6a}$$

$$\mathcal{L}_{\text{stab}}\left(\boldsymbol{W}\right) = \mathcal{L}\left(\boldsymbol{W}\right) + \left(\boldsymbol{\chi},\boldsymbol{\tau}\boldsymbol{f}\right) \tag{6b}$$

$$\mathcal{B}\left(\boldsymbol{W},\boldsymbol{V}\right) = \rho\left(\boldsymbol{w},\boldsymbol{v}_{,t}\right) + \rho\left(\boldsymbol{w},\boldsymbol{v}\cdot\nabla\boldsymbol{v}\right) + \left(\nabla\boldsymbol{w},2\eta\boldsymbol{\varepsilon}\left(\boldsymbol{v}\right)\right) - \left(\nabla\cdot\boldsymbol{w},p\right) + \left(q,\nabla\cdot\boldsymbol{v}\right) \tag{6c}$$

$$\mathcal{L}\left(\boldsymbol{W}\right) = \left(\boldsymbol{w},\boldsymbol{f}\right) + \left(\boldsymbol{w},\boldsymbol{h}\right)_{\Gamma_h} \tag{6d}$$

where $(\cdot,\cdot) = \int_{\Omega}(\cdot)\,d\Omega$ is the $L_2(\Omega)$—inner product, and $\mathcal{B}(\mathbf{W},\mathbf{V})$ and $\mathcal{L}(\mathbf{W})$ are the variational forms emanating from the underlying standard Galerkin method. The stabilization terms in (6a) and (6b) are defined as follows:

$$\boldsymbol{\tau} = b^e\left[\begin{array}{l}\rho\int(b^e)^2(\nabla\boldsymbol{v})^T\,d\Omega + \rho\int b^e\boldsymbol{v}\cdot\nabla b^e\,d\Omega\,\boldsymbol{I}\\ \quad + \int\eta\left(\nabla b^e\otimes\nabla b^e\right)\,d\Omega + \int\eta|\nabla b^e|^2\,d\Omega\,\boldsymbol{I}\end{array}\right]^{-1}\int_{\Omega^e}b^e d\Omega \tag{7a}$$

$$\begin{aligned}\boldsymbol{\chi} = \rho\left(-\nabla\boldsymbol{v}\cdot\boldsymbol{w} + \left(\nabla\cdot\boldsymbol{v}\right)\boldsymbol{w} + \boldsymbol{v}\cdot\nabla\boldsymbol{w}\right) + \eta\left(\nabla\left(\nabla\cdot\boldsymbol{w}\right) + \Delta\boldsymbol{w}\right)\\ + \nabla\eta\cdot\left(\left(\nabla\cdot\boldsymbol{w}\right)\boldsymbol{I} + \nabla\boldsymbol{w}\right) + \nabla q\end{aligned} \tag{7b}$$

where b^e represents bubble functions that are employed to expand the fine-scale velocity field.

The VMS-stabilized form in (5) inherits features of Streamline Upwind/Petrov-Galerkin (SUPG) method [27], and Galerkin/Least-Square (GLS) method [28], as discussed in [29]. It is implemented in the 3D context using hexahedral and tetrahedral elements with equal-order linear and quadratic Lagrange shape functions for the velocity and pressure fields. The method involves consistent mass and tangent tensors for nonlinear solution procedures as presented in [15].

3 Outflow Boundary Conditions (Outflow BCs)

The narrower downstream branches in the arterial system result in an increased resistance to the flow of blood. Appropriate boundary conditions (BCs) can account for these downstream effects on the zone of interest that constitutes our computational domain. In this section, we discuss four types of outflow BCs along with their weak forms. The first three conditions (i.e., constant pressure, constant resistance, and impedance boundary conditions) have been presented in the literature and we have employed them to verify the accuracy of our proposed method that embeds the clinically measured data as time-dependent resistance BC in our computational scheme. In the numerical section, we compare these different BCs in terms of how accurately they impose the downstream effects on to the computational domain.

3.1 Constant Pressure BC

A common procedure for applying outflow boundary condition is to apply Neumann boundary condition with constant pressure over the outlet surfaces. Accordingly, the traction term in (6d) is expressed as follows:

$$(\boldsymbol{w}, \boldsymbol{h})_{\Gamma_{\text{out}}} = (\boldsymbol{w}, P_{\text{mean}}\boldsymbol{n})_{\Gamma_{\text{out}}} \tag{8}$$

where Γ_{out} is the outlet surface, P_{mean} is the mean pressure on Γ_{out}, $\boldsymbol{n}$ is the normal vector to the outflow surface. Even though this BC satisfies the mean values of pressure at each outlet, the pressure is not a function of time; therefore, this BC cannot represent pressure variations from the downstream network.

3.2 Constant Resistance BC

The downstream resistance can be assumed to have a linear relation between the flow-rate and the pressure. The mean pressures and flow-rates at outlets determine the constant resistance values at these outlets.

$$\overline{R} = P_{\text{mean}}/Q_{\text{mean}} \tag{9}$$

where P_{mean} and Q_{mean} are the mean pressure and flow-rate for one cardiac cycle at the outlet, respectively. Taylor and colleagues [26] presented a weak form for the constant resistance BC as follows:

$$(\boldsymbol{w}, \boldsymbol{h})_{\Gamma_{\text{out}}} = \left(\boldsymbol{w}, \boldsymbol{n}\left(\overline{R}\int_{\Gamma_{\text{out}}} \boldsymbol{v}\cdot\boldsymbol{n}\, ds + \boldsymbol{n}\cdot\boldsymbol{\sigma}_v\cdot\boldsymbol{n}\right) - \boldsymbol{\sigma}_v\cdot\boldsymbol{n}\right)_{\Gamma_{\text{out}}} \tag{10}$$

Although the constant resistance BC produces transient pressure profiles at outlets in [26], it turns out that the variation in the pressure between systole and diastole is much larger than that of the actual pressure profile in the downstream network. In addition, the constant resistance BC does not produce time lag between flow-rate and pressure profiles that is usually observed in the downstream branches.

3.3 Impedance BC

The lack of time lag and the excessive pressure variation due to the constant resistance BC leads us to impedance BC. The impedance is determined in frequency domain as follows:

$$\widehat{Z} = \widehat{P}/\widehat{Q} \tag{11}$$

where $\widehat{P}$ and $\widehat{Q}$ are the Fourier transforms of pressure and flow-rate for one cardiac cycle at the outlet, respectively. $Z(t)$ is the inverse Fourier transform of $\widehat{Z}$, and it is embedded in the outflow BC as presented in [26].

$$(\boldsymbol{w}, \boldsymbol{h})_{\Gamma_{\text{out}}} = \left(\boldsymbol{w}, \boldsymbol{n}\left(\frac{1}{T}\int_{t-T}^{t} Z\left(t-t_1\right)\int_{\Gamma_{\text{out}}} \boldsymbol{v}\left(t_1\right)\cdot\boldsymbol{n}\; dsdt_1 + \boldsymbol{n}\cdot\boldsymbol{\sigma}_v\cdot\boldsymbol{n}\right) - \boldsymbol{\sigma}_v\cdot\boldsymbol{n}\right)_{\Gamma_{\text{out}}} \tag{12}$$

The impedance BC produces time lag between pressure and flow-rate profiles as well as lowers the pressure variation between systole and diastole as compared to the constant resistance BC as described in Section 3.2. However, round-off errors from the Fourier transforms of pressure and flow-rate as well as from the inverse Fourier transform of $\widehat{Z}$ may cause spurious oscillations in the numerical results. Round-off errors depend on complexity of the profiles, as discussed in Section 4.

3.4 Clinically Calibrated Time-Dependent Resistance BC

In order to accurately represent the observed time lags and pressure variations at outflow, we assume that downstream resistance varies during a cardiac cycle. Downstream blood rheology invariably manifests its non-Newtonian behavior due to relatively smaller diameters; downstream resistance dynamically varies with changes in flow-rate and pressure conditions. To incorporate patient-specific data in the simulation, patient-specific pressure and flow-rate profiles can be measured in the clinical setting. If the measured profiles do not vary abruptly between multiple cardiac cycles, then measured data can be used to construct reliable resistance profiles for a typical cardiac cycle. From these measured pressure and flow-rate values obtained as function of time, one can determine the time plot of resistance at that cross-section, as follows:

$$R(t) = P(t)/Q(t) \tag{13}$$

where $P(t)$ and $Q(t)$ are the pressure and flow-rate profiles for a generic cardiac cycle at the outlet, respectively. The corresponding BC is updated from (10), as follows:

$$(\boldsymbol{w},\boldsymbol{h})_{\Gamma_{\text{out}}} = \left(\boldsymbol{w},\boldsymbol{n}\left(R(t)\int_{\Gamma_{\text{out}}} \boldsymbol{v}\cdot\boldsymbol{n}\,ds + \boldsymbol{n}\cdot\boldsymbol{\sigma}_v\cdot\boldsymbol{n}\right) - \boldsymbol{\sigma}_v\cdot\boldsymbol{n}\right)_{\Gamma_{\text{out}}} \tag{14}$$

4 Numerical Results

The VMS-based stabilized method for shear-rate dependent fluids (i.e., equations (5)–(7)) with four different types of outflow BCs presented in (8), (10), (12), and (14) is implemented for equal-order, linear and quadratic, hexahedral and tetrahedral elements. We present two numerical test cases to verify our method and provide a comparative study between the four types of outflow BCs.

4.1 Unidirectional Flow in a Rectangular Channel

The first test case is flow in a rectangular channel of dimension 10 cm × 1 cm × 1 cm as presented in Figure 1. We have adjusted the height of the channel to keep it in the range of the diameter of human carotid artery. At the inflow we apply mean profiles for flow-rate and pressure that are typically encountered in the human carotid artery during a typical cardiac cycle. A parabolic velocity profile with flow-rate shown in Figure 2 is prescribed as the inflow boundary condition at $x = 0$, and the four types of outflow BCs are applied at $x = 10$ cm (i.e., on the blue surface in Figure 1). Non-slip BC is imposed at y = 0 and y = 1.0 cm while slip BC is applied at $z = 0$ and $z = 1.0$ cm. The material parameters used in the Carreau-Yasuda model employed here are as follows: $\mu_\infty = 0.005$ Pa · sec, $\mu_0 = 0.056$ Pa · sec, $\lambda = 1.902$ sec, $n = 0.22$, and $a = 1.25$. Density ρ of the fluid is set equal to 1050.0 kg/m^3. The time duration for one cardiac cycle is equal to 1.0 sec and Δt for time integration is set equal to 0.01 sec. In order to minimize the effects

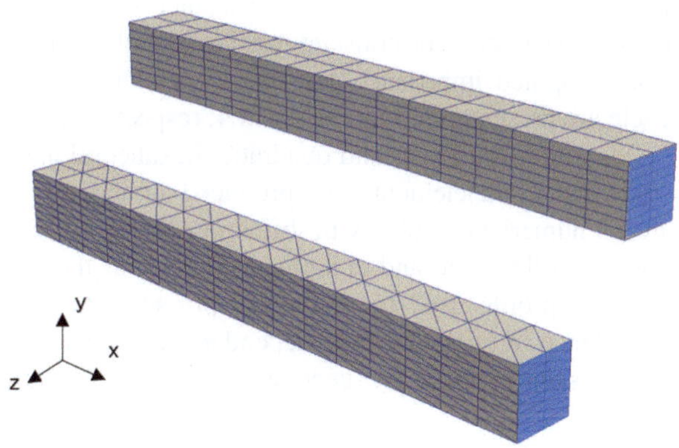

Fig. 1 A channel with hexahedral and tetrahedral element

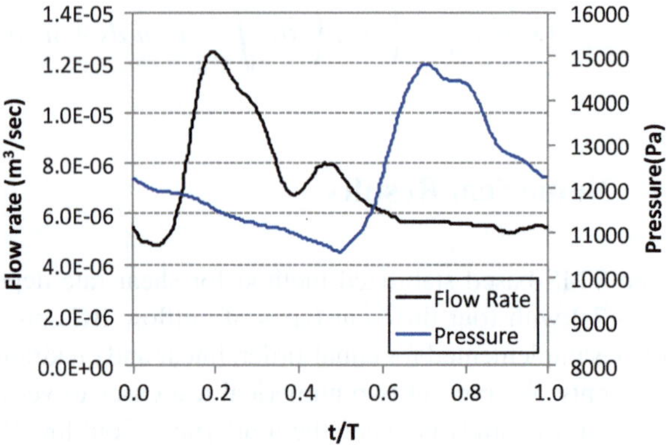

Fig. 2 Flow-rate and pressure profiles

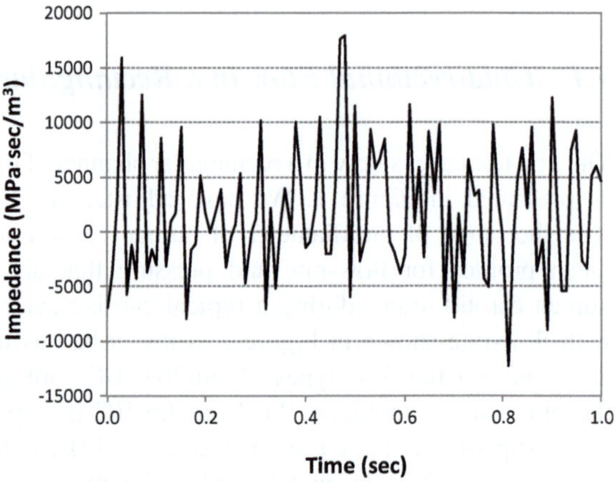

Fig. 3 Impedance for one cardiac cycle

from the stationary initial conditions, three cardiac cycles are simulated and the data from the last cycle is used for post-processing.

Using the flow-rate and pressure profiles given in Figure 2, constant resistance, impedance, and time-dependent resistance values are computed via (9), (11), and (13), respectively. The constant resistance applied in this test is 1736.8 $\text{MPa} \cdot \text{sec/m}^3$. The computed impedance and time-dependent resistance profiles for one cardiac cycle are presented in Figures 3 and 4, respectively.

We employed linear and quadratic, hexahedral and tetrahedral elements for this study and all the element types produced numerically converged solutions. Figure 5 shows numerical results with different outflow BCs: the black and red lines show computed flow-rate and computed pressure at the outlet, and the blue line is the pressure profile used for the mean pressure in (8), constant resistance in (10), impedance in (12), and time-dependent resistance in (14). Figure 5(a) shows that the constant pressure BC generates stationary pressure at the outlet. The constant

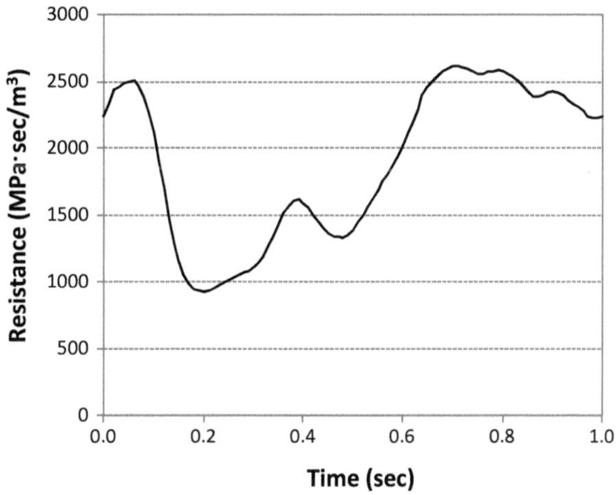

Fig. 4 Time-dependent resistance for one cardiac cycle

resistance BC generates much larger variation in the computed pressure (i.e., red line in Figure 5(b)) as compared to the actual pressure variation (i.e., blue line in Figure 5(b)) without any time lag from the computed flow-rate. The pressure variation for impedance BC (i.e., red line in Figure 5(c)) virtually lies on top of the actual pressure profile (i.e., blue line in Figure 5(c)). In Figure 5(d) for the time-dependent resistance BC, the computed pressure at the outlet (i.e., red line in Figure 5(d)) overlies the observed pressure for (14) (i.e., blue line in Figure 5(d)). In summary, the impedance and time-dependent resistance BCs produce the most accurate downstream effects for shear-rate dependent flows with one outlet in this test case.

4.2 *Idealized Arterial Geometry*

This numerical section investigates the effects of outflow BCs (e.g., constant resistance, impedance, and time-dependent resistance BCs) for multiple outlets. First, we construct and simulate idealized arterial tree geometry (i.e., full geometry) including all downstream branches (i.e., the vertical length from the inlet to the outlet at the bottom is equal to 116.8 cm) in Figure 6. Second, we reduce the full geometry to only the upstream main branches (see the right-hand side in Figure 6), and then at the location of outlets of the reduced geometry (i.e., the vertical length from the inlet to the outlets is equal to 43.7 cm), we measure flow-rates and pressure profiles from the computations performed on the full idealized arterial tree geometry. These computed values are then used to determine constant resistance, impedance, and time-dependent resistance data for outflow BCs. The simulated results from the full geometry also serve as reference solutions for the numerical results from the reduced geometry with various outflow BCs prescribed. Two types

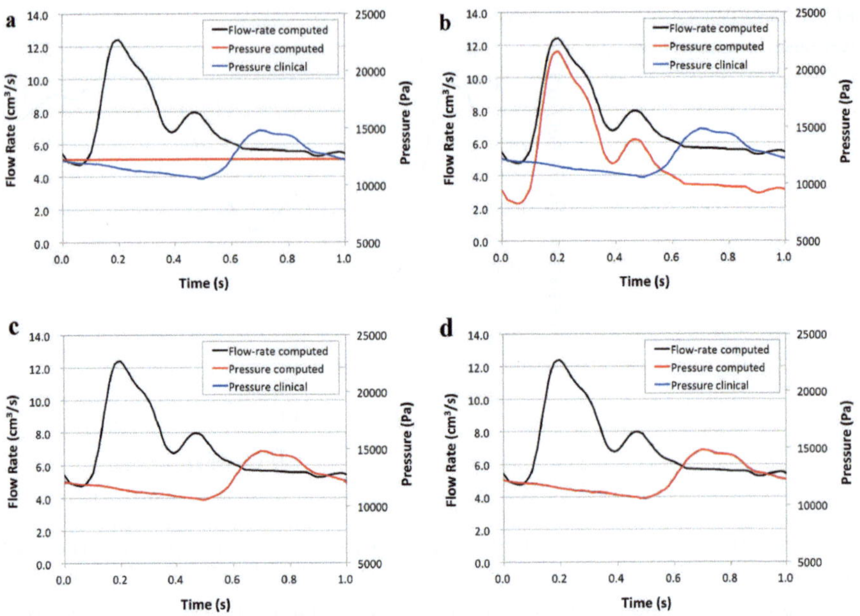

Fig. 5 Flow-rate and pressure profiles at the outlets. (a) Constant pressure BC. (b) Constant resistance BC. (c) Impedance BC. (d) Time-dependent resistance BC

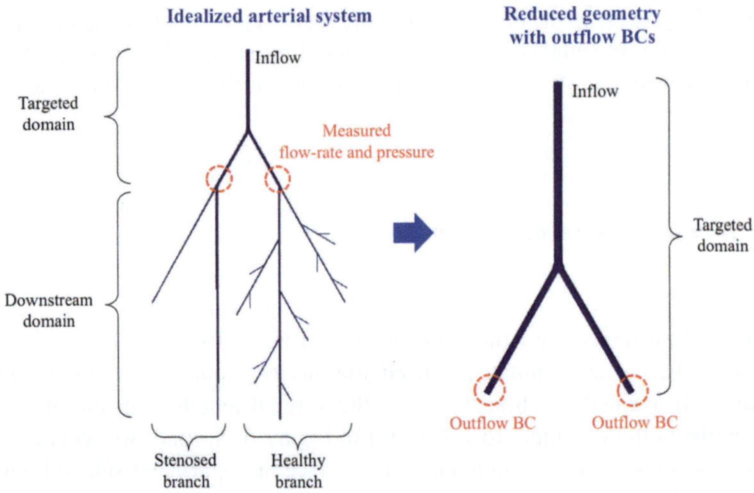

Fig. 6 Idealized arterial geometry and the reduced model

of downstream branches (i.e., stenosed branch and healthy branch) are designed in the full geometry in order to create unsymmetrical downstream effects. Table 1 shows mesh characteristics for the full geometry as well as the reduced geometry. Both geometries employ quadratic tetrahedral (T10) element, and are subject to the

Table 1 Mesh characteristics for the idealized arterial geometry and the reduced geometry

	Full geometry	Reduced geometry
No. of elements	127,144	9,569
No. of nodes	217,365	16,687
No. of unknowns	869,460	66,748
No. of CPUs	24	12
Wall time (sec)	11,018	1,599 ~ 1,616
CPU time (sec)	264,432	19,128 ~ 19,392

same cardiac cycle (i.e., time period $T = 0.9$ sec) with the same time increment (i.e., $\Delta t = 0.01$ sec). In order to minimize the effects of the stationary initial flow conditions, three cardiac cycles are simulated and the data from the last cycle is used for post-processing.

4.2.1 Idealized Arterial Tree Geometry

The diameter at the inlet of the most upstream branch is 1.26 cm and it symmetrically bifurcates into two daughter branches with diameter 0.98 cm each. The bifurcation angle is 60.0 degree. After the first bifurcation, the two daughter branches with a diameter D_p unsymmetrically bifurcate into a large branch with diameter $0.911D_p$, and a small branch with diameter $0.584D_p$, as presented in Olufsen [24]. From the second bifurcation onwards, the bifurcation angle is set equal to 30.0 degree. Each branch keeps the same diameter (i.e., D_i) along a length, $20D_i$. The healthy branch (i.e., the right branch in Figure 7) is constructed by following the above geometric ratios. For the stenosed branch (i.e., the left branch in Figure 7), we assume that the smaller downstream arteries are blocked at the bifurcations. Consequently, only large branches are constructed, while maintaining the same change in the diameters at the corresponding bifurcation locations as in the healthy branch. Table 2 presents the diameters as well as lengths of all the branches of the full geometry shown in Figure 7.

A parabolic velocity profile is prescribed at the inflow as Dirichlet BC wherein we have employed patient-specific flow-rate that is typically observed in the human femoral artery. The constant pressure BC is applied at each outlet surface with $P_{mean} = 1{,}500$ Pa. Figure 8 shows the velocity magnitude in the upstream branches. Since the stenosed branch has larger downstream resistance as compared to the healthy branch due to its blocked smaller arteries, the healthy branch (i.e., right bottom portion in Figure 8) shows larger velocity magnitude than in the stenosed branch (i.e., left bottom portion in Figure 8). Figure 9(a) shows the prescribed flow-rate (i.e., black line) and computed pressure (i.e., blue line) profiles at the inlet, and Figure 9(b) and (c) shows computed flow-rate (i.e., black lines) and pressure (i.e., blue lines) profiles at the beginning parts of the stenosed branch and the healthy branch, respectively. The stenosed branch in Figure 9(b) shows higher pressure and

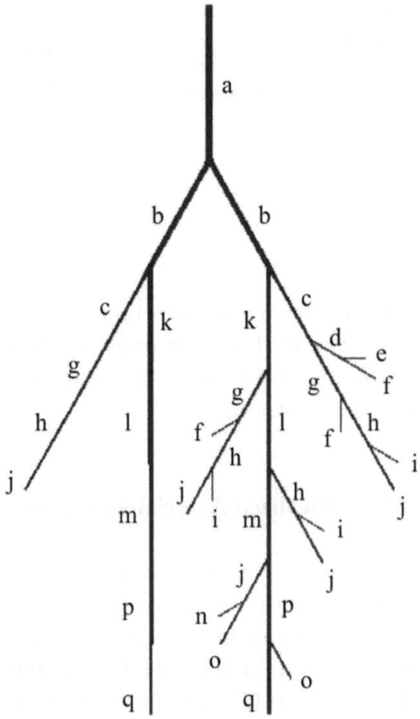

Fig. 7 Full geometry and labels for diameters and lengths of branches

Table 2 Diameters and lengths of full geometry

Label	a	b	c	d	e	f	g	h	i
Diameter (mm)	12.6	9.8	5.72	3.34	1.95	3.04	5.21	4.75	2.77
Length (mm)	252	196	114.4	66.7	38.9	60.8	104.2	95	55.4
Label	j	k	l	m	n	o	p	q	
Diameter (mm)	4.33	8.93	8.14	7.42	2.53	3.94	6.76	6.16	
Length (mm)	86.6	178.6	162.8	148.4	50.5	78.9	135.2	123.2	

lower flow-rates than the healthy branch in Figure 9(c). It shows that the downstream resistance in the stenosed branch is much larger than that in the healthy branch. The measurements in Figure 9(b) and (c) are used to determine the parameters to be employed in the outflow BCs for the reduced geometric model presented below. They also serve as the reference solutions because they include all nonlinear downstream effects including the non-Newtonian effects of blood.

4.2.2 Reduced Artery Geometry with Constant Resistance Outflow BC

Using equation (9) with the mean flow-rates and pressure shown in Figure 9(b) and (c) yields the constant resistance values at outlets. The constant resistance $\overline{R}_s$ at the

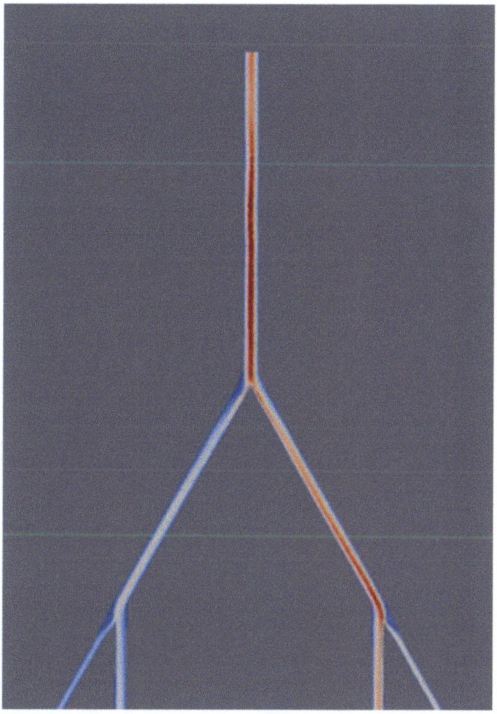

Fig. 8 Velocity magnitude contour in upstream branches

stenosed outlet is equal to 276.5 $\text{MPa} \cdot \text{sec/m}^3$, while the constant resistance $\overline{R}_\text{h}$ at the healthy outlet is equal to 145.8 $\text{MPa} \cdot \text{sec/m}^3$. Figure 10 shows computed flow-rate and mean pressure profiles at the stenosed and the healthy outlets for one typical cardiac cycle. The constant resistance outflow BC generates the same shape of the profiles for the flow-rates (see the profiles in Figure 10(a) and (b)). The computed pressure profiles with the constant resistance outflow BC are substantially different from the reference solutions as presented in Figure 10(c) and (d).

4.2.3 Reduced Artery Geometry with Impedance Outflow BC

Employing the flow-rate and pressure profiles given in Figure 9(b) and (c) in equation (11) results in impedance functions at stenosed and healthy branches and is presented in Figure 11. Since the stenosed branch has larger resistance than the healthy branch, the amplitude of impedance function for the stenosed branch is larger than the amplitude for the healthy branch. Figure 12 shows the computed flow-rate and mean pressure profiles at the stenosed and healthy outlets for one cardiac cycle. As compared to the constant resistance outflow BC, the impedance outflow BC results are apparently better in the profile of flow-rate and pressure at the outlets. However, the computed surface mean pressure in Figure 12(c) and (d)

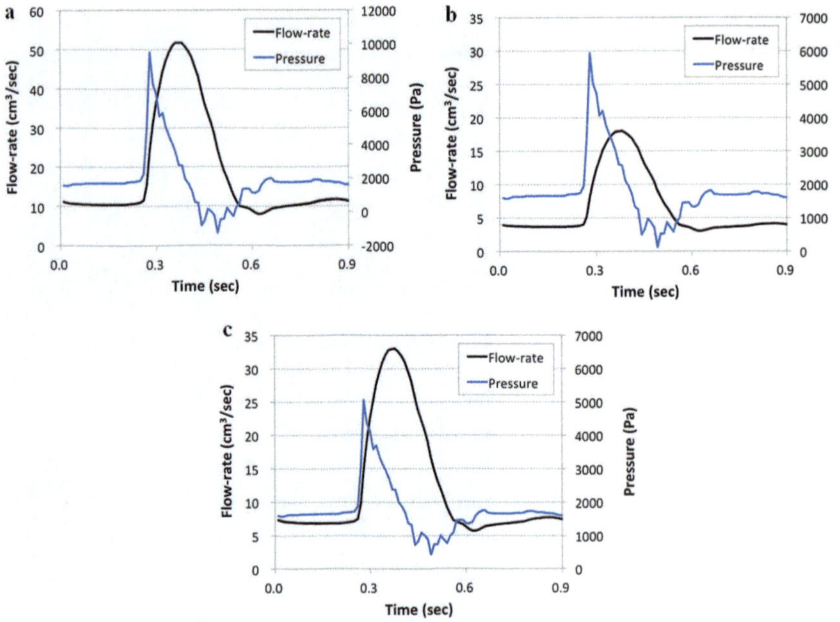

Fig. 9 Flow-rate and pressure profiles at the inlet and the locations of the outlets for the reduced geometry. (a) Inlet. (b) Stenosed branch. (c) Healthy branch

shows oscillations with a higher frequency than that of the heart rate. One possibility is that the round-off errors from the Fourier and the inverse Fourier transforms for the calculation of the impedance function are introducing the artificial oscillation in the numerical solution. Depending on the complexity of the patient-specific profiles for pressure and flow-rate, this numerical oscillation may become significant and therefore pollute the numerical simulations. This seems to be an obvious drawback of the impedance boundary condition for practical application to patient-specific cases.

4.2.4 Reduced Artery Geometry with Time-Dependent Resistance Outflow BC

Using equation (13) with the flow-rate and pressure profiles given in Figure 9(b) and (c) we develop time-dependent resistance functions at the stenosed and the healthy branches. These are presented in Figure 13. Due to the blockage of the smaller arteries, time-dependent resistance function for the stenosed branch (i.e., red line in Figure 13) is always higher than the resistance function for the healthy branch (i.e., blue line in Figure 13). Figure 14 shows the computed flow-rate and pressure profiles at each outlet for one cardiac cycle. In Figure 14(c) and (d), the surface mean pressure profile computed with the time-dependent resistance outflow BC matches well with the reference pressure profile that was computed with the full arterial tree

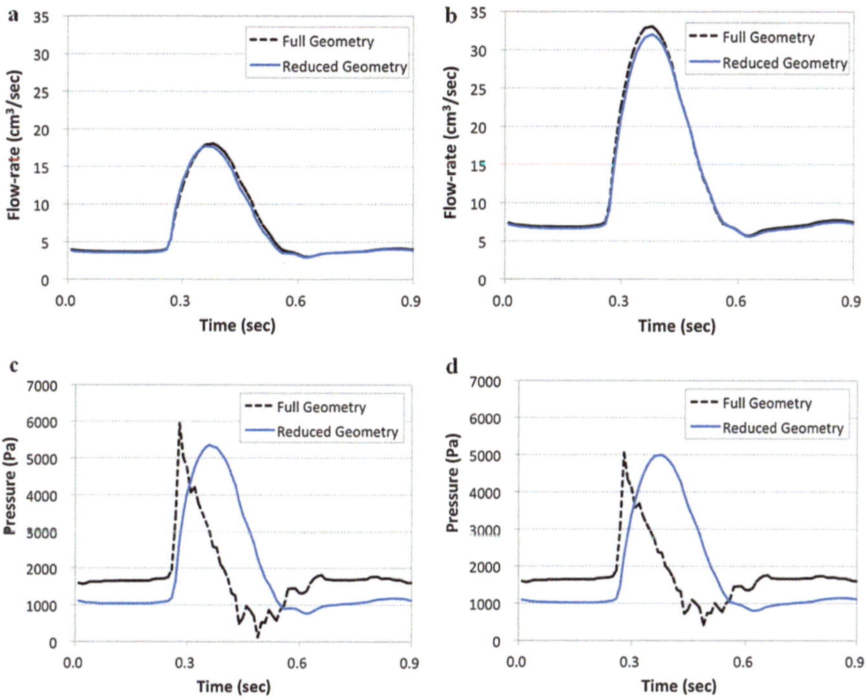

Fig. 10 Computed flow-rate and mean pressure profiles with the constant resistance outflow BC. (a) Flow-rate at the stenosed outlet. (b) Flow-rate at the healthy outlet. (c) Mean pressure at the stenosed outlet. (d) Mean pressure at the healthy outlet

geometry. If we compare these pressure profiles to the results for other three outflow BCs, the simulation with the time-dependent outflow BC is more accurate than the other BCs.

5 Conclusions

We have presented a stabilized finite element method for incompressible shear-rate dependent fluids that is augmented with patient-specific outflow BC that is developed based on clinically measured pressure and flow-rate profiles. We formulate time-dependent downstream resistance function at each outlet with the patient-specific data, and apply it at the outlet surface as a weakly imposed boundary condition. For verification of the new method, we have implemented the constant pressure outflow BC, the constant resistance outflow BC, and the impedance outflow BC. We provide a comparative study of the three types of BCs with the proposed outflow BC in simple rectangular channel geometry, as well as idealized arterial tree geometry.

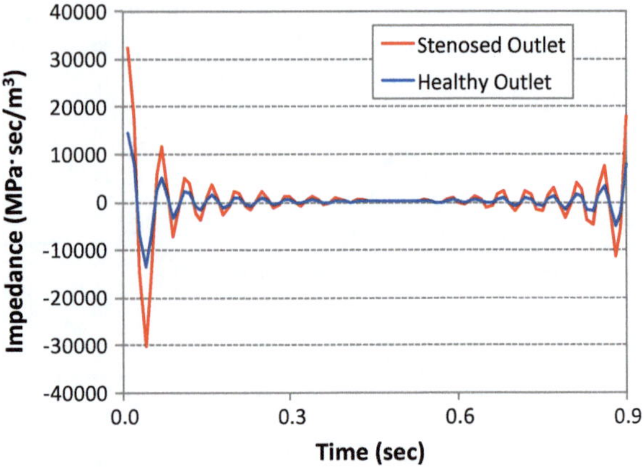

Fig. 11 Impedance functions for the outlets

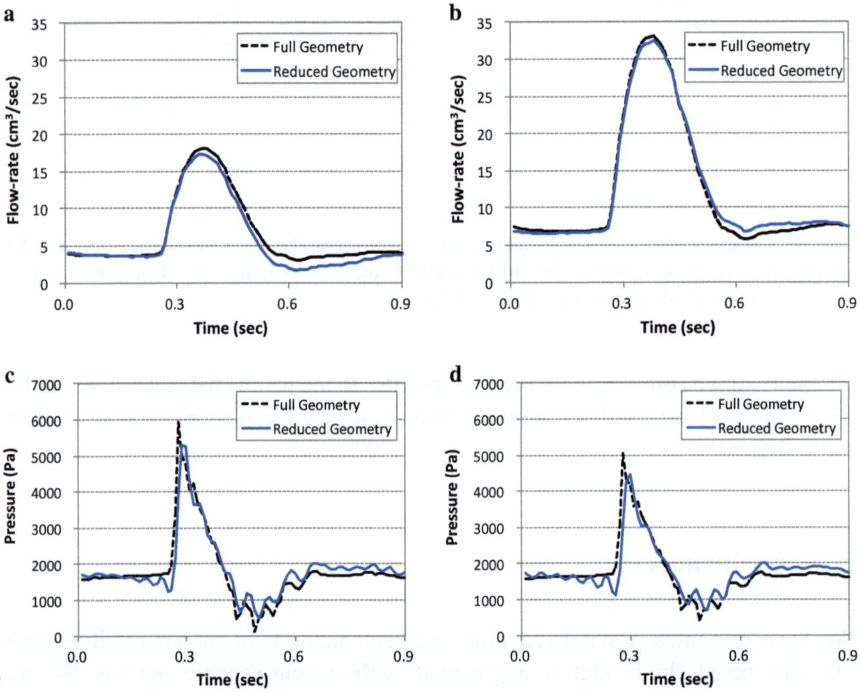

Fig. 12 Computed flow-rate and mean pressure profiles with the impedance outflow BC. (a) Flow-rate at the stenosed outlet. (b) Flow-rate at the healthy outlet. (c) Mean pressure at the stenosed outlet. (d) Mean pressure at the healthy outlet

Fig. 13 Time-dependent resistance functions for the outlets

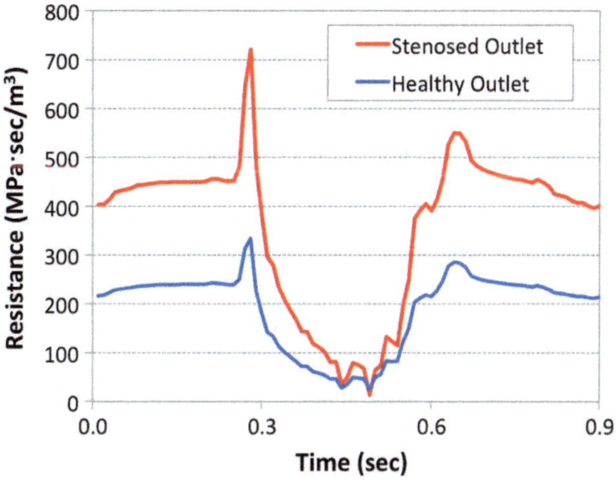

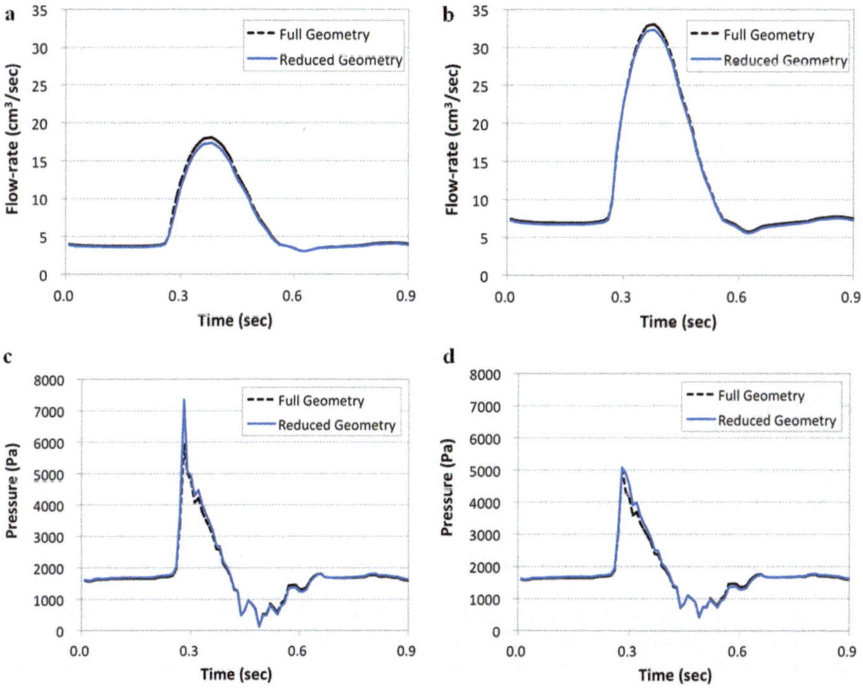

Fig. 14 Computed flow-rate and mean pressure profiles with the time-dependent resistance outflow BC. (a) Flow-rate at the stenosed outlet. (b) Flow-rate at the healthy outlet. (c) Mean pressure at the stenosed outlet. (d) Mean pressure at the healthy outlet

The constant pressure outflow BC produces a stationary pressure at the outlet that does not account for the transient effects from the downstream branches. The constant resistance outflow BC produces the same profiles for flow-rate and pressure; therefore, it cannot reproduce the time lag between flow-rate and pressure.

In addition the induced transient downstream effects are substantially different from that of the reference solution. The impedance outflow BC generates similar profiles for the flow-rate and the pressure as are in the reference data; however, the computed pressure profiles show high-frequency oscillations that never exist in the full geometry including all downstream branches. This is due to numerical error that is initiated by round-off errors that arise in the calculation of the impedance function. This is a significant weakness of the impedance BC for application to patient-specific cases. The proposed outflow BC (i.e., clinically measured time-dependent resistance outflow BC) produces most accurate results when compared to the reference solutions. In addition, it accurately predicts time lag between flow-rate and pressure as shown in Figure 14.

The proposed outflow BC is able to reproduce most accurate downstream effects when compared with the competing methods. With this outflow BC, we can considerably reduce the size as well as the complexity of the geometry for patient-specific models by ignoring the downstream branches, and this can substantially decreases the computational costs both in storage and in time. For the second test case involving idealized arterial tree geometry, the reduced geometry with the proposed outflow BC took 19,188 sec in CPU time, while the full geometry took 264,432 sec. This results in 13.8 times reduction in the cost of computation with the reduced geometry (see Table 1) as compared to the full arterial tree geometry.

Acknowledgements This research is part of the Blue Waters sustained-petascale computing project, which is supported by the National Science Foundation (awards OCI-0725070 and ACI-1238993) and the state of Illinois. Blue Waters is a joint effort of the University of Illinois at Urbana-Champaign and its National Center for Supercomputing Applications.

References

1. Charm, S., Kurland, G.: Viscometry of human blood for shear rates of 0-100,000 sec^{-1}. Nature **206**(4984), 617–618 (1965)
2. Chien, S., Usami, S., Taylor, H.M., Lundberg, J.L., Gregersen, M.I.: Effect of hematocrit and plasma proteins on human blood rheology at low shear rates. J. Appl. Physiol. **21**(1), 81–87 (1966)
3. Chien, S., Usami, S., Dellenback, R.J., Gregersen, M.I.: Blood viscosity: influence of erythrocyte aggregation. Science **157**(3790), 829–831 (1967)
4. Cross, M.M.: Rheology of non-Newtonian fluids: a new flow equation for pseudoplastic systems. J. Colloid Sci. **20**, 417–437 (1965)
5. Walburn, F.J., Schneck, D.J.: A constitutive equation for whole human blood. Biorheology **13**, 201–210 (1976)
6. Yasuda, K., Armstrong, R.C., Cohen, R.E.: Shear flow properties of concentrated solutions of linear and star branched polystyrenes. Rheol. Acta **20**, 163–178 (1981)
7. Anand, M., Rajagopal, K.R.: A shear-thinning viscoelastic fluid model for describing the flow of blood. Int J Cardiovasc Med Sci **4**, 59–68 (2004)
8. Rajagopal, K.R.: Mechanics of non-Newtonian fluids. In: Galdi, G.P., Necas, J. (eds.) Recent Developments in Theoretical Fluids Mechanics. Pitman Research Notes in Mathematics, vol. 291, pp. 129–162. Longman, New York (1993)

9. Anand, M., Rajagopal, K., Rajagopal, K.R.: A model incorporating some of the mechanical and biochemical factors underlying clot formation and dissolution in flowing blood. Comput. Math. Methods Med. **5**, 183–218 (2003)
10. Rajagopal, K.R., Srinivasa, A.R.: A thermodynamic frame work for rate type fluid models. J. Non-Newtonian Fluid Mech. **88**, 207–227 (2000)
11. Yeleswarapu, K.K., Kameneva, M.V., Rajagopal, K.R., Antaki, J.F.: The flow of blood in tubes: theory and experiment. Mech. Res. Commun. **25**, 257–262 (1998)
12. Kwack, J., Masud, A.: A three-field formulation for incompressible viscoelastic fluids. Int. J. Eng. Sci. **48**(11), 1413–1432 (2010)
13. Anand, M., Kwack, J., Masud, A.: A new generalized Oldroyd-B model for blood flow in complex geometries. Int. J. Eng. Sci. **72**, 78–88 (2013)
14. Masud, A., Kwack, J.: A stabilized mixed finite element method for the incompressible shear-rate dependent non-Newtonian fluids: Variational Multiscale framework and consistent linearization. Comput. Methods Appl. Mech. Eng. **200**(5-8), 577–596 (2011)
15. Kwack, J., Masud, A.: A stabilized mixed finite element method for shear-rate dependent non-Newtonian fluids: 3D benchmark problems and application to blood flow in bifurcating arteries. Comput. Mech. **53**(4), 751–776 (2014)
16. Weddell, J., Kwack, J., Imoukhuede, P.I., Masud, A.: Hemodynamic analysis in an idealized artery tree: differences in wall shear stress between Newtonian and non-Newtonian blood models. PLoS ONE **10**(4), e0124575 (2015). doi:10.1371/journal.pone.0124575
17. Taylor, C.A., Hughes, T.J.R., Zarins, C.K.: Finite element modeling of blood flow in arteries. Comput. Methods Appl. Mech. Eng. **158**, 155–196 (1998)
18. Taylor, C.A., Cheng, C.P., Espinosa, L.A., Tang, B.T., Parker, D., Herfkens, R.J.: In vivo quantification of blood flow and wall shear stress in the human abdominal aorta during lower limb exercise. Ann. Biomed. Eng. **30**, 402–408 (2002)
19. Bazilevs, Y., Calo, V.M., Zhang, Y., Hughes, T.J.R.: Isogeometric fluid-structure interaction analysis with applications to arterial blood flow. Comput. Mech. **38**, 310–322 (2006)
20. Zhang, Y., Bazilevs, Y., Goswami, S., Bajaj, C.L., Hughes, T.J.R.: Patient-specific vascular NURBS modeling for isogeometric analysis of blood flow. Comput. Methods Appl. Mech. Eng. **196**, 2943–2959 (2007)
21. Tezduyar, T.E., Sathe, S., Schwaab, M., Conklin, B.S.: Arterial fluid mechanics modeling with the stabilized space-time fluid-structure interaction technique. Int. J. Numer. Methods Fluids **57**, 601–629 (2008)
22. Bazilevs, Y., Hsu, M.C., Zhang, Y., Wang, W., Liang, X., Kvamsdal, T., Brekken, R., Isaksen, J.G.: A fully-coupled fluid-structure inter- action simulation of cerebral aneurysms. Comput. Mech. **46**, 3–16 (2010)
23. Takizawa, K., Schjodt, K., Puntel, A., Kostov, N., Tezduyar, T.E.: Patient-specific computational analysis of the influence of a stent on the unsteady flow in cerebral aneurysms. Comput. Mech. **51**, 1061–1073 (2013)
24. Olufsen, M.S.: Structured tree outflow condition for blood flow in larger systemic arteries. Am. Physiol. Soc. **276**(1), H257–H268 (1999)
25. Esmaily-Moghadam, M., Bazilevs, Y., Hsia, T.Y., Vignon-Clementel, I.E., Marsden, A.L.: A comparison of outlet boundary treatments for prevention of backflow divergence with relevance to bloodflow simulations. Comput. Mech. **48**, 277–291 (2011)
26. Vignon-Clementel, I.E., Alberto Figueroa, C., Jansen, K.E., Taylor, C.A.: Outflow boundary conditions for three-dimensional finite element modeling of blood flow and pressure in arteries. Comput. Methods Appl. Mech. Eng. **195**, 3776–3796 (2006)
27. Brooks, A.N., Hughes, T.J.R.: Streamline upwind/Petrov- Galerkin formulations for convection dominated flows with particular emphasis on the incompressible Navier-Stokes equations. Comput. Methods Appl. Mech. Eng. **32**, 199–259 (1982)
28. Hughes, T.J.R., Franca, L.P., Hulbert, G.M.: A new finite element formulation for computational fluid dynamics: VIII. The Galerkin-least-squares method for advective-diffusive equations. Comput. Methods Appl. Mech. Eng. **73**, 173–189 (1989)
29. Masud, A., Khurram, R.: A multiscale finite element method for the incompressible Navier-Stokes equations. Comput. Methods Appl. Mech. Eng. **195**, 1750–1777 (2006)

A Geometrical-Characteristics Study in Patient-Specific FSI Analysis of Blood Flow in the Thoracic Aorta

Hiroshi Suito, Kenji Takizawa, Viet Q.H. Huynh, Daniel Sze, Takuya Ueda, and Tayfun E. Tezduyar

Abstract This chapter is on fluid–structure interaction (FSI) analysis of blood flow in the thoracic aorta. The FSI is handled with the Sequentially Coupled Arterial FSI technique. We focus on the relationship between the aorta centerline geometry and the wall shear stress (WSS) distribution. The model centerlines are extracted from the CT scans, and we assume a constant diameter for the artery segment. Then, torsion-free model geometries are generated by projecting the original centerline to its averaged plane of curvature. The WSS distributions for the original and projected geometries are compared to examine the influence of the torsion.

1 Introduction

In this chapter, we examine the relationship between the geometrical characteristics of the thoracic aorta and the wall shear stress (WSS) distribution, which is thought to play an important role in development of cardiovascular diseases. Thoracic-aortic aneurysm is one of the life-threatening diseases, which is defined as a

H. Suito (✉) • V.H.Q. Huynh
Graduate School of Environmental and Life Sciences, Okayama University and JST, CREST, 3-1-1 Tsushima-naka, Okayama 700-8530, Japan
e-mail: suito@ems.okayama-u.ac.jp

K. Takizawa
Department of Modern Mechanical Engineering, Waseda University and JST, CREST, 1-6-1 Nishi-Waseda, Shinjuku-ku, Tokyo 169-8050, Japan

D. Sze
Stanford University School of Medicine, 300 Pasteur Dr S072, MC 5105, Stanford, CA 94305, USA

T. Ueda
Department of Radiology, Seikeikai Chiba Medical Center and JST, CREST, 1-7-1 Minami-cho, Chiba 260-0842, Japan

T.E. Tezduyar
Mechanical Engineering, Rice University, MS 321, 6100 Main Street, Houston, TX 77005, USA

Y. Bazilevs, K. Takizawa (eds.), *Advances in Computational Fluid-Structure Interaction and Flow Simulation*, Modeling and Simulation in Science, Engineering and Technology, DOI 10.1007/978-3-319-40827-9_29

condition of dilatation of the aorta to greater than 1.5 times its normal diameter [1]. The aneurysm slowly grows with the advancing age of the patient and increases its diameter, and may eventually rupture. An aortic aneurysm may be treated with surgery or endovascular repair, depending on the predicted rupture risk and condition of the patient [2]. An important factor in determining the treatment method is balancing the rupture risk against the complications of the operation itself. It has been reported in many articles that the rupture risk is proportional to the aneurysm size. However, the natural development history of these aneurysms has not been fully understood [3].

Another clinically important question is where the aneurysm would occur, as the aneurysm location is relevant to the treatment risks and prognostic implications. Our target in this chapter is the thoracic aorta, which consists of the ascending aorta, aortic arch, and descending aorta. The ascending aorta rises up from the heart and curves posteriorly (aortic arch). After the strong curve of the aortic arch, the aorta goes downward to the abdomen (descending aorta). The aortic arch has a wide variety of shapes, and the curvature and torsion distributions present from the aortic arch to the descending aorta vary from patient to patient.

The WSS is thought to play an important role in the natural history of the progress of atherosclerosis in aortic aneurysms. The WSS distribution is derived from the flow field, which is governed by the flow equations, initial and boundary conditions, physical parameters involved in the equations, and the geometry of the flow domain. Here, we concentrate on the domain geometry, which is different from patient to patient and is expected to be an important factor in distinguishing the high-risk patients from other patients and in predicting where the aneurysm would develop. Differences in the blood vessel geometry bring about differences in the flow characteristics, which generate different WSS distributions. To gain insight in this context, we provide a fluid–structure interaction (FSI) analysis of the blood flow and examine how the WSS distribution depends on the geometrical characteristics.

In our analysis, the centerlines of the aorta models are extracted from the CT scans, and we assume a constant diameter for the artery segment. We focus on the torsion, which is defined by the Frenet–Serret formulas. We generate torsion-free model geometries by projecting the original centerline to its averaged plane of curvature. We compare the WSS patterns corresponding to the original and projected geometries and discern the influence of the torsion.

The computational methods and problem setup are summarized in Section 2. The results are presented in Section 3, and the concluding remarks are given in Section 4.

2 Computational Methods and Problem Setup

The arterial FSI analysis is handled with the Sequentially-Coupled Arterial FSI (SCAFSI) technique [4–9], which is quite often suitable for classes of FSI problems with temporally-periodic FSI dynamics. First we carry out the structural mechanics computation to obtain the deformation of the artery under an observed blood pressure profile in a cardiac cycle. Then we do the fluid mechanics computation over a mesh that moves to follow the lumen as the artery deforms. The fluid mechanics computations with the moving meshes are carried out with the space–time variational multiscale (ST-VMS) method [9–13], which is the VMS version of the Deforming-Spatial-Domain/Stabilized ST (DSD/SST) method [14–18]. The VMS components of the ST-VMS method are from the residual-based VMS method given in [19, 20]. For mesh moving and remeshing methods, we refer the interested reader to [9, 13, 18, 21–29]. Earlier work [30] showed that a single pair of fluid and structural mechanics computations gives results that are very comparable to those obtained with full FSI computation, especially in terms of the WSS values.

Simplified geometries for the thoracic aorta are used in this study. Branches such as the brachiocephalic trunk, left common carotid artery, and left subclavian artery are neglected. Moreover, we assume a constant diameter for the aorta because our focus is on understanding the differences brought about by the shape of the aorta centerline. We consider three patient-specific cases, identified as Case A, Case B, and Case C. The lumen diameter is 3.0 cm in all three cases. The three cases have different curvature and torsion distributions. Figure 1 shows the original geometries for the three cases, indicated by the suffix WT. Figure 2 shows the torsion-free geometries, indicated by the suffix TF, obtained by projecting the original centerline to its averaged plane of curvature.

The fluid mechanics meshes are made of 4-node tetrahedral elements, and the structural mechanics meshes 8-node hexahedral elements. We build layers of refined fluid mechanics meshes near the arterial walls. The blood is assumed to be a Newtonian fluid. The density and kinematic viscosity are 1.0×10^3 kg/m^3 and 4.0×10^{-6} m^2/s. A velocity waveform pulsating over the cardiac cycle (with period $T = 1.0$ s) is applied as a plug flow at the proximal (inflow) boundary. Figure 3 shows the flow-rate waveform at the inflow boundary. The material model for the arterial wall structure is neo-Hookean. The density, Young's modulus, Poisson's ratio, and wall thickness are 1.0×10^3 kg/m^3, 400 kPa, 0.4 and 0.3 cm. Since no residual stresses are considered in the computations, the pressure is applied in the range 0 to 40 mm Hg.

Fig. 1 Original (with-torsion) models for the three cases.

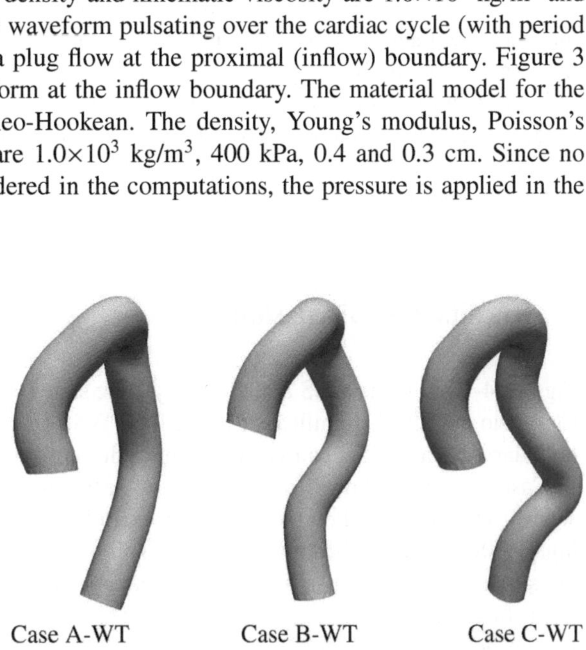

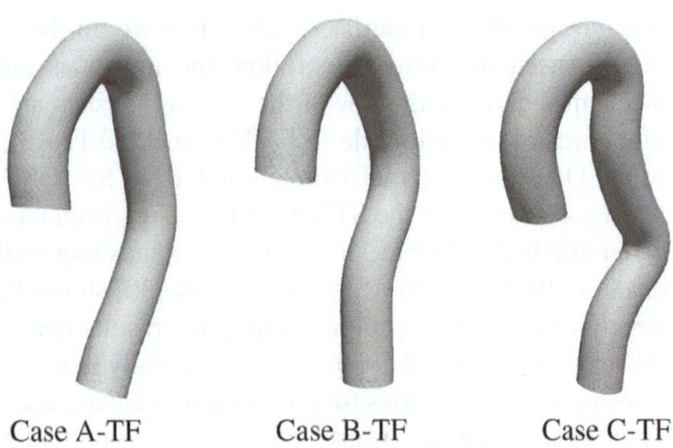

Fig. 2 Projected (torsion-free) models for the three cases.

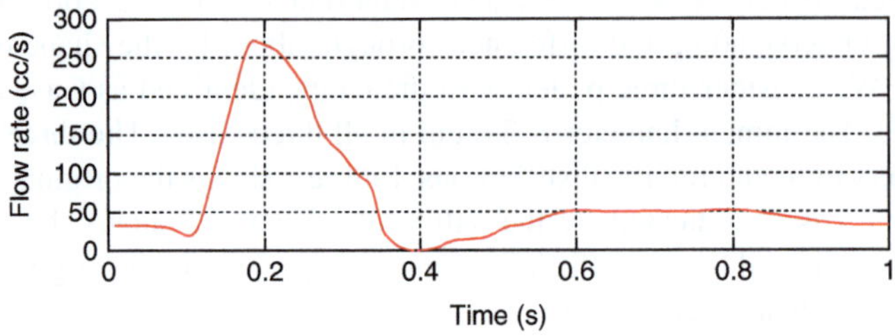

Fig. 3 Flow-rate waveform at the inflow boundary.

The mesh-moving method is based on solving the large-deformation solid mechanics equations, as in [25], and the constitutive model in those equations is St. Venant–Kirchhoff, also as in [25], with a Poisson's ratio of 0.0. The initial fluid mechanics mesh serves as the reference configuration in this large-deformation model. The mesh motion is driven by the motion of the fluid mechanics mesh at the interface, which is obtained by projection from the motion of the structural mechanics mesh there. For more on the computational conditions, see [31].

3 Results and Discussion

Figures 4–5 show the WSS at the peak systolic state ($t = 0.2$ s) for Cases A, B, and C. We do not attach significance to the high WSS seen in all cases near the inflow boundary, because it is caused by the plug flow profile specified at that boundary. In Cases A and B, shown in Figures 4 and 5, the high WSS around the aortic arch seems to be intensified by the torsion there. In Figure 4, it can also be seen that the high WSS area expands downstream in a spiral way. In Case C, shown in Figure 6, the situation is different. The high-WSS area seen around the aortic arch in Case C-TF loses intensity and is dispersed in Case C-WT. We note that the high WSS is seen not only at the aortic arch but also at the descending aorta because of the torsion effect.

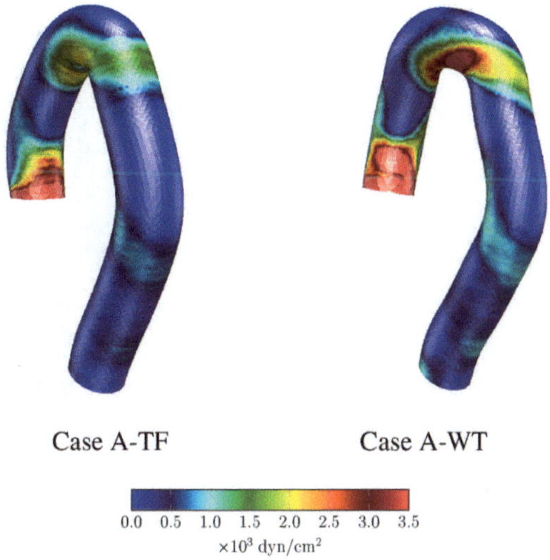

Fig. 4 WSS for Case A.

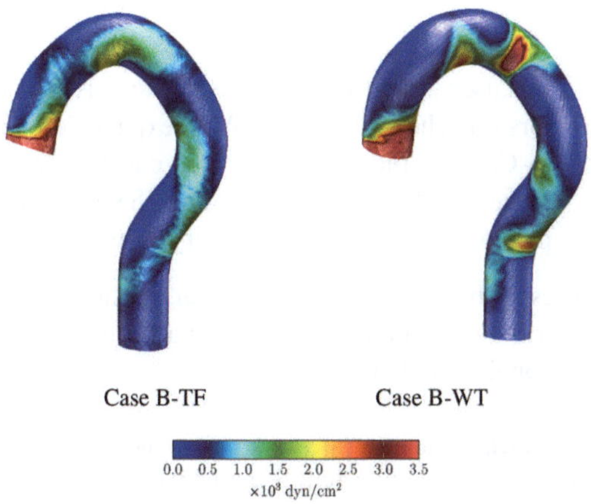

Fig. 5 WSS for Case B.

4 Concluding Remarks

We have presented a geometrical-characteristics study in patient-specific FSI analysis of blood flow in the thoracic aorta. The FSI was handled with the SCAFSI technique, and the fluid mechanics computations were based on the ST-VMS method. The centerlines of the aorta models were extracted from the CT scans, and we used a constant diameter for the artery segment. In our study, we focused on the effect of the torsion on the WSS. For this purpose, torsion-free model geometries were generated by projecting the original centerline to its averaged plane of curvature, where the centerline has almost the same curvature but without the torsion. We summarize our observations in two items.

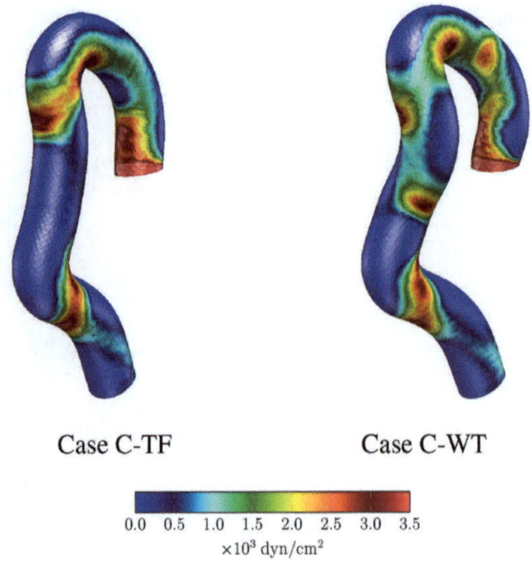

Fig. 6 WSS for Case C.

1. In Cases A and B, the WSS in the aortic arches is intensified in the presence of torsion. The area of high WSS expands downstream.
2. In Case C, the high-WSS area can be seen also in the descending aorta in the presence of torsion, where the WSS is not high when there is no torsion. In this patient's case, the aortic aneurysm was at the descending aorta.

These observations suggest that torsion has a significant effect on the WSS distribution in blood vessels, and a statistical identification of that effect would be a reasonable next step.

Acknowledgements This work was supported by the JST-CREST Mathematics program. The authors thank Dr. Ryo Torii (University College London) for providing the observed data for the pressure profile in the aorta.

References

1. Isselbacher, E.M.: Thoracic and abdominal aortic aneurysms. Circulation **111**, 816–828 (2005)
2. Elefteriades, J.A.: Natural history of thoracic aortic aneurysms: indications for surgery, and surgical versus nonsurgical risks. Ann. Thorac. Surg. **74**, S1877–S1880 (2002)
3. Davies, R.R., Goldstein, L.J., Coady, M.A., Tittle, S.L., Rizzo, J.A., Kopf, G.S., Elefteriades, J.A.: Yearly rupture or dissection rates for thoracic aortic aneurysms: simple prediction based on size. Ann. Thorac. Surg. **73**, 17–28 (2002)
4. Tezduyar, T.E., Sathe, S., Schwaab, M., Conklin, B.S.: Arterial fluid mechanics modeling with the stabilized space–time fluid–structure interaction technique. Int. J. Numer. Methods Fluids **57**, 601–629 (2008). doi:10.1002/fld.1633
5. Tezduyar, T.E., Schwaab, M., Sathe, S.: Sequentially-coupled arterial fluid–structure interaction (SCAFSI) technique. Comput. Methods Appl. Mech. Eng. **198**, 3524–3533 (2009). doi:10.1016/j.cma.2008.05.024

6. Tezduyar, T.E., Takizawa, K., Moorman, C., Wright, S., Christopher, J.: Multiscale sequentially-coupled arterial FSI technique. Comput. Mech. **46**, 17–29 (2010). doi:10.1007/s00466-009-0423-2
7. Tezduyar, T.E., Takizawa, K., Brummer, T., Chen, P.R.: Space–time fluid–structure interaction modeling of patient-specific cerebral aneurysms. Int. J. Numer. Methods Biomed. Eng. **27**, 1665–1710 (2011). doi:10.1002/cnm.1433
8. Takizawa, K., Bazilevs, Y., Tezduyar, T.E.: Space–time and ALE-VMS techniques for patient-specific cardiovascular fluid–structure interaction modeling. Arch. Comput. Methods Eng. **19**, 171–225 (2012). doi:10.1007/s11831-012-9071-3
9. Bazilevs, Y., Takizawa, K., Tezduyar, T.E.: Computational Fluid–Structure Interaction: Methods and Applications. Wiley, Chichester (2013). ISBN 978-0470978771
10. Takizawa, K., Tezduyar, T.E.: Multiscale space–time fluid–structure interaction techniques. Comput. Mech. **48**, 247–267 (2011). doi:10.1007/s00466-011-0571-z
11. Takizawa, K., Tezduyar, T.E.: Space–time fluid–structure interaction methods. Math. Models Methods Appl. Sci. **22**, 1230001 (2012). doi:10.1142/S0218202512300013
12. Takizawa, K., Bazilevs, Y., Tezduyar, T.E., Long, C.C., Marsden, A.L., Schjodt, K.: ST and ALE-VMS methods for patient-specific cardiovascular fluid mechanics modeling. Math. Models Methods Appl. Sci. **24**, 2437–2486 (2014). doi:10.1142/S0218202514500250
13. Takizawa, K.: Computational engineering analysis with the new-generation space–time methods. Comput. Mech. **54**, 193–211 (2014). doi:10.1007/s00466-014-0999-z
14. Tezduyar, T.E.: Stabilized finite element formulations for incompressible flow computations. Adv. Appl. Mech. **28**, 1–44 (1992). doi:10.1016/S0065-2156(08)70153-4
15. Tezduyar, T.E., Behr, M., Liou, J.: A new strategy for finite element computations involving moving boundaries and interfaces—the deforming-spatial-domain/space–time procedure: I. The concept and the preliminary numerical tests. Comput. Methods Appl. Mech. Eng. **94**, 339–351 (1992). doi:10.1016/0045-7825(92)90059-S
16. Tezduyar, T.E., Behr, M., Mittal, S., Liou, J.: A new strategy for finite element computations involving moving boundaries and interfaces—the deforming-spatial-domain/space–time procedure: II. Computation of free-surface flows, two-liquid flows, and flows with drifting cylinders. Comput. Methods Appl. Mech. Eng. **94**, 353–371 (1992). doi:10.1016/0045-7825(92)90060-W
17. Tezduyar, T.E.: Computation of moving boundaries and interfaces and stabilization parameters. Int. J. Numer. Methods Fluids **43**, 555–575 (2003). doi:10.1002/fld.505
18. Tezduyar, T.E., Sathe, S.: Modeling of fluid–structure interactions with the space–time finite elements: solution techniques. Int. J. Numer. Methods Fluids **54**, 855–900 (2007). doi:10.1002/fld.1430
19. Hughes, T.J.R.: Multiscale phenomena: green's functions, the Dirichlet-to-Neumann formulation, subgrid scale models, bubbles, and the origins of stabilized methods. Comput. Methods Appl. Mech. Eng. **127**, 387–401 (1995)
20. Bazilevs, Y., Calo, V.M., Cottrell, J.A., Hughes, T.J.R., Reali, A., Scovazzi, G.: Variational multiscale residual-based turbulence modeling for large eddy simulation of incompressible flows. Comput. Methods Appl. Mech. Eng. **197**, 173–201 (2007)
21. Tezduyar, T., Aliabadi, S., Behr, M., Johnson, A., Mittal, S.: Parallel finite-element computation of 3D flows. Computer **26**, 27–36 (1993). doi:10.1109/2.237441
22. Johnson, A.A., Tezduyar, T.E.: Mesh update strategies in parallel finite element computations of flow problems with moving boundaries and interfaces. Comput. Methods Appl. Mech. Eng. **119**, 73–94 (1994). doi:10.1016/0045-7825(94)00077-8
23. Tezduyar, T.E.: Finite element methods for flow problems with moving boundaries and interfaces. Arch. Comput. Methods Eng. **8**, 83–130 (2001). doi:10.1007/BF02897870
24. Takizawa, K., Henicke, B., Puntel, A., Kostov, N., Tezduyar, T.E.: Space–time techniques for computational aerodynamics modeling of flapping wings of an actual locust. Comput. Mech. **50**, 743–760 (2012). doi:10.1007/s00466-012-0759-x

25. Takizawa, K., Tezduyar, T.E., Boben, J., Kostov, N., Boswell, C., Buscher, A.: Fluid–structure interaction modeling of clusters of spacecraft parachutes with modified geometric porosity. Comput. Mech. **52**, 1351–1364 (2013). doi:10.1007/s00466-013-0880-5
26. Takizawa, K., Tezduyar, T.E., Buscher, A., Asada, S.: Space–time interface-tracking with topology change (ST-TC). Comput. Mech. **54**, 955–971 (2014). doi:10.1007/s00466-013-0935-7
27. Takizawa, K., Bazilevs, Y., Tezduyar, T.E., Hsu, M.-C., Øiseth, O., Mathisen, K.M., Kostov, N., McIntyre, S.: Engineering analysis and design with ALE-VMS and space–time methods. Arch. Comput. Methods Eng. **21**, 481–508 (2014). doi:10.1007/s11831-014-9113-0
28. Takizawa, K., Tezduyar, T.E., Buscher, A., Asada, S.: Space–time fluid mechanics computation of heart valve models. Comput. Mech. **54**, 973–986 (2014). doi:10.1007/s00466-014-1046-9
29. Takizawa, K., Tezduyar, T.E., Buscher, A.: Space–time computational analysis of MAV flapping-wing aerodynamics with wing clapping. Comput. Mech. **55**, 1131–1141 (2015). doi:10.1007/s00466-014-1095-0
30. Takizawa, K., Brummer, T., Tezduyar, T.E., Chen, P.R.: A comparative study based on patient-specific fluid–structure interaction modeling of cerebral aneurysms. J. Appl. Mech. **79**, 010908 (2012). doi:10.1115/1.4005071
31. Suito, H., Takizawa, K., Huynh, V.Q.H., Sze, D., Ueda, T.: FSI analysis of the blood flow and geometrical characteristics in the thoracic aorta. Comput. Mech. **54**, 1035–1045 (2014). doi:10.1007/s00466-014-1017-1

Particle Method Simulation of Thrombus Formation in Fontan Route

Ken-ichi Tsubota, Koichi Sughimoto, Kazuki Okauchi, and Hao Liu

Abstract A computer simulation was carried out for thrombus formation under the influence of blood flow after Fontan operation. Blood was modeled by computed particles assigned as normal blood or thrombus. Blood flow was calculated using a moving particle semi-implicit method. In a model of blood coagulation that causes thrombi, a normal blood particle changed to a thrombus when its shear rate was lower than a threshold. A spring force was employed to express the coagulation, and was substituted into the NS equations as the external force to couple the coagulation and the blood flow. In simulations, thrombus formation was affected by blood flow behaviors, such as stagnation and recirculation. The atrio-pulmonary connection (APC) square model showed the highest incidence for thrombus formation in the right atrium due to flow stagnation, followed by the APC round, whereas no thrombus was formed in the total cavopulmonary connection model. This result suggests that local hemodynamic behavior associated with the complex channel geometry plays a major role in thrombus formation in the Fontan route.

1 Introduction

It has long been suggested that thrombus formation is greatly affected by blood flow behavior [1, 2]. Flow stagnation causes venous thrombus, as referred to in Virchow's triad [3], in which the deposition of blood cells at low flow velocity is a key mechanical factor [4–8]. Conversely, a high fluid shear stress induces von Willebrand factor release followed by platelet activation and aggregation [9, 10], resulting in arterial thrombus formation. Thus, between flow stagnation and high flow velocity, blood flow plays different roles in thrombus formation [1], and both

K. Tsubota (✉) • K. Okauchi • H. Liu
Department of Mechanical Engineering, Chiba University, 1-33 Yayoi, Chiba 263-8622, Japan
e-mail: tsubota@faculty.chiba-u.jp

K. Sughimoto
Department of Cardiovascular Surgery, Kitasato University School of Medicine, Sagamihara, Japan

Y. Bazilevs, K. Takizawa (eds.), *Advances in Computational Fluid-Structure Interaction and Flow Simulation*, Modeling and Simulation in Science, Engineering and Technology, DOI 10.1007/978-3-319-40827-9_30

flow states should be considered to better understand the mechanism of thrombosis and aid the further development of improved treatments.

This study focuses on thrombosis of single ventricle patients, where a thrombus formation by flow stagnation likely to occur. Single ventricle is defined as a heart with only one ventricle connected with atria, or with two ventricles but unable to sustain separately pulmonary and systemic circulations [11]. Single ventricle generally undergoes "Fontan circulation" that is the reconstructed blood flow circulation by cardiac operation. Classical Fontan, namely atrio-pulmonary connection (APC), has been strongly associated with a dilatation of the right atrium that could occur over time, causing arrhythmia and thickening of the atrial wall, thereby resulting in thrombus formation [12–16]. Converting the venous return flow from the inferior vena cava (IVC) from the APC to a total cavopulmonary connection (TCPC) by placing a conduit has gained acceptable outcomes [17–19]. However, the reason why a TCPC is superior to an APC remains controversial in terms of hemodynamics [20] and coagulation [15, 16].

In this study, a two-dimensional (2D) computer simulation of thrombus formation in the Fontan route was carried out using a particle method [21–25]. In a model of blood coagulation that causes thrombi, a normal blood particle changed to a thrombus when its shear rate was lower than a threshold. A spring force was employed to express the coagulation, and was substituted into the NS equations as the external force to couple the coagulation and the blood flow [21, 25]. Simple hemodynamics models of an APC and a TCPC were created for simulations of thrombus formation under the influence of blood flows.

2 Methods

Blood coagulation was assumed to occur at local flow stagnation. Hemodynamics in the Fontan route was simulated using equations of continuity and Navier– Stokes. In a computer simulation of thrombus formation, blood coagulation and its associated hemodynamics was coupled using a particle method.

2.1 *Particle Method for Fluid Analysis*

A blood region was discretized by an assembly of computed particles. Incompressible viscous flow of the blood, which was based on the equation of continuity

$$\frac{\mathrm{D}\rho}{\mathrm{D}t} = 0 \tag{1}$$

and the NS equations

$$\frac{\mathrm{D}\mathbf{u}}{\mathrm{D}t} = -\frac{1}{\rho}\nabla P + \frac{\mu}{\rho}\nabla^2\mathbf{u} + \mathbf{f}, \tag{2}$$

was expressed by calculating particle motion using the moving particle semi-implicit (MPS) method [22]. In Eqs. (1) and (2), ρ is density, t is time, $\boldsymbol{u}$ is velocity, P is pressure, μ is viscosity, $\boldsymbol{f}$ is external force, and D/Dt is the material derivative. Gradient ∇ and Laplacian ∇^2 of physical quantities are discretized by a particle interaction model [22].

2.2 Biomechanics Model of Stagnation Thrombus Formation

Two types of blood component, normal blood and thrombus, were considered in a computer simulation of thrombus formation [25], and their properties were assigned to a computed blood particle (Fig. 1). To express thrombus formation at flow stagnation according to Virchow's triads [3], a spring force, $\boldsymbol{f}_{ij}{}^{\mathrm{C}}$, for coagulation, was applied between two thrombus particles i and j [21]:

$$f_{ij}^{C} = -f_{ji}^{C} = -k^{C}\left(\left|r_{ij}\right| - r_{0}^{C}\right)\frac{r_{ij}}{\left|r_{ij}\right|} \qquad \left(\left|r_{ij}\right| < r_{cut}^{C}\right), \tag{3}$$

In Eq. (3), $\boldsymbol{r}_{ij} = \boldsymbol{r}_j - \boldsymbol{r}_i$ and $\boldsymbol{r}_i$ are $\boldsymbol{r}_j$ the position vectors of particles i and j, respectively; $\boldsymbol{r}_0{}^{\mathrm{C}}$ is the natural length of the spring; k^{C} is the spring constant; and $\boldsymbol{r}_0{}^{\mathrm{C}}$ is the threshold distance for particle interaction. In the MPS method for solving blood flow, the blood coagulation was coupled with the blood flow by substituting $\boldsymbol{f}_{ij}{}^{\mathrm{C}}$ into the NS equation (2) as the external force $\boldsymbol{f}$ [21, 23, 24]. A normal blood particle was assumed to change to a thrombus when it neighbored another thrombus particle and its principal shear strain rate $\dot{\gamma}$ (a measure of stagnation, and hereafter called shear rate) was smaller than a threshold $\dot{\gamma}_{\mathrm{th}}^{\mathrm{C}}$.

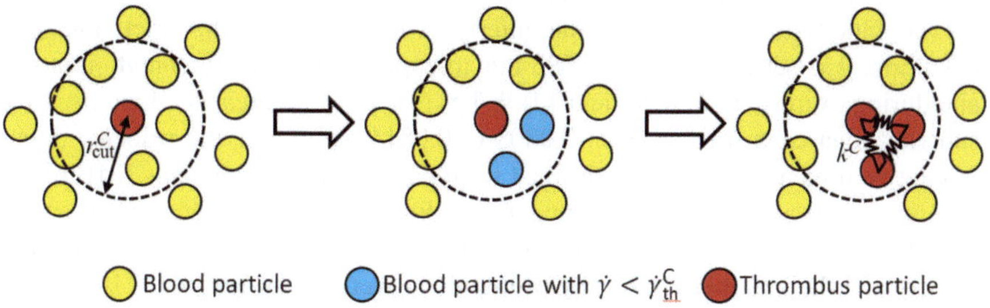

Fig. 1 Model of thrombus formation with particle method. Blood particle i changes to a thrombus if shear rate $\dot{\gamma}_i$ is smaller than a threshold $\dot{\gamma}_{\mathrm{th}}^{\mathrm{C}}$. Coagulation is expressed by connecting a thrombus particle to other thrombus particles by springs.

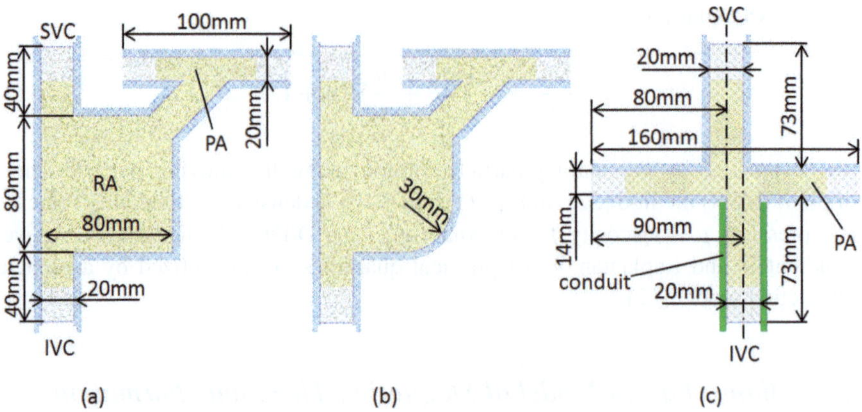

Fig. 2 Two-dimensional model of blood flow channel for atrio-pulmonary connection (APC) and total cavopulmonary connection (TCPC) of Fontan route. (a) APC square, (b) APC round, and (c) TCPC.

2.3 Hemodynamics Models of Fontan Route

Two-dimensional adult-sized Fontan geometries of a blood flow channel were modeled as an APC square, an APC round, and a TCPC (Fig. 2). “Square” or “round” in APC models denotes the shape of the right lower corner of the right atrium (RA). APC models were designed as an adult patient in failing Fontan with dilated right atrium. To focus on the difference mainly between the failing Fontan patient with dilated right atrium and the patient with TCPC route, the size of the APC square model was designed as follows: the square right atrium of 80 mm in length on each side, the vena cavae of 40 mm in length and 20 mm in width, and the pulmonary artery (PA) of 100 mm in length and 14 mm in width. The right atrium of the APC round model has the rounded lower right corner with a 30-mm curvature radius. With regard to the size of the TCPC model, the length and width of the extracardiac conduit was 73 and 20 mm, respectively.

The blood flow velocity in the inferior vena cavae, as the inlet boundary condition, was set at 0.1025 – 0.2083 m/s to express the velocity condition at rest and during exercise (0.5 W/kg and 1.0 W/kg) (Table 1), whereas the flow from the superior vena cava was set from 0.0808 to 0.0814 m/s. Here, the velocity values in the two dimensions were determined from published measurements [14]. At the pulmonary arteries as the outlet boundary condition, free-outflow and zero-pressure conditions were assumed. A non-slip condition was assumed at the wall. The density and the kinematic viscosity of the blood were set as 1.06×10^3 kg/m^3 and 4.43×10^{-6} m^2/s, respectively, giving the Reynolds number an order of a hundred. Thus, the 2D hemodynamics of the Fontan route was determined by the channel geometries of blood flow and velocity values at the vena cavae.

Table 1 Velocity from the SVC and IVC. 3D data is referenced from published measurements [14]. IVC inferior vena cava; SVC superior vena cava; 2D two-dimensional; 3D three-dimensional.

	SVC		IVC	
	3D	2D mean velocity	3D	2D mean velocity
	l/min/m^2	m/s	l/min/m^2	m/s
Rest	1.26	0.0808	1.6	0.1025
0.5 W/kg	1.27	0.0814	2.58	0.1654
1.0 W/kg	1.27	0.0814	3.25	0.2083

With respect to the set of simulation parameters, the spatial resolution was $d_0 = 1$ mm in terms of mean particle distance, and the total number of computed particles was 15,925 in the APC square, 15,224 in the APC round, and 11,142 in the TCPC. In Eq. (2) of the coagulation force, $k^C = 5.0 \times 10^{-2}$ N/m; $r_0 = d_0$; and $r_0{}^C = 1.5 d_0$. A threshold shear rate for blood coagulation was set as $\dot{\gamma}^C_{th} = 0$.

3 Results and Discussion

3.1 Effects of Fontan Route Geometry

Thrombus formation was simulated for the APC and TCPC models with the venous velocity at rest. In the APC square, the blood flow stagnated at the lower right corner of the atrium (Fig. 3a). Accordingly, thrombus was formed and grown at the corner of the atrium (Fig. 3b). A portion of the thrombus was collapsed by fluid flow, and flew out from the atrium into PA. In the APC round, the blood flow showed less stagnation in the atrium, and thus thrombus formation was moderate (Fig. 4a). The TCPC route showed a smooth flow from the vena cavae to the pulmonary arteries without significant stagnation, resulting in thrombus hardly formed (Fig. 4b).

3.2 Effects of Venous Flow Velocity

To examine the effects of blood flow velocity at the vena cavae, thrombus formation was compared among three velocity conditions at rest as well as in the presence of light and moderate exercise using the APC square model. The results showed that every condition resulted in thrombus formation at the right lower corner over time (Fig. 5). A lower flow rate caused more thrombus formation. A reduction in thrombus formation was observed during mild exercise (0.5 W/kg), whereas this was more evident during moderate exercise (1.0 W/kg).

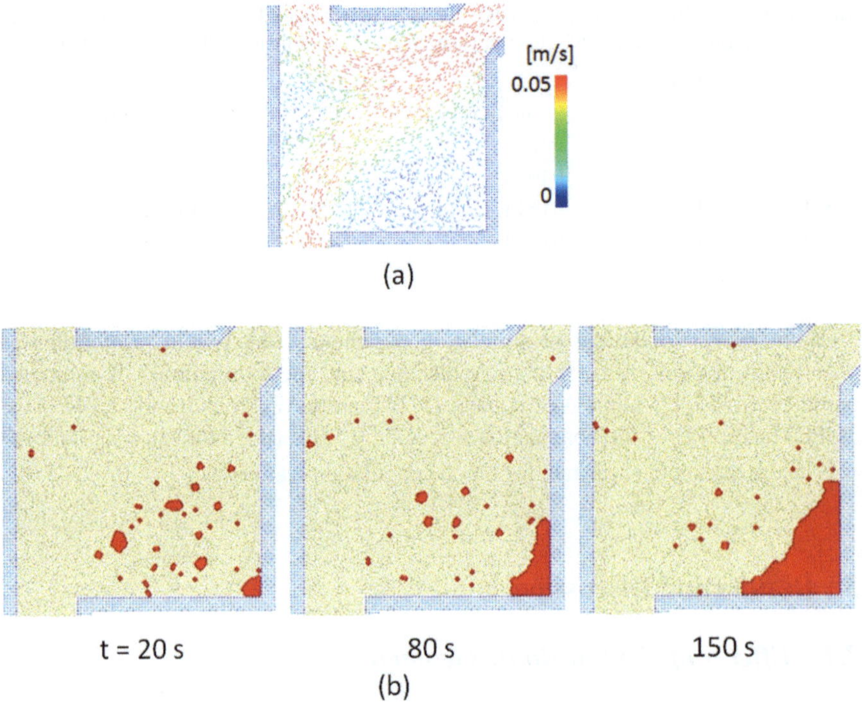

Fig. 3 Simulation result by APC square model with inflow velocity at rest. (a) Flow velocity and (b) thrombus formation according to blood flow.

4 Discussion and Conclusion

A 2D computer simulation method based on a simple coagulation rule and local complex hemodynamics demonstrated the mechanism of thrombus formation in the Fontan route. TCPC is superior to the classical APC in terms of preventing thrombus formation. The flow channel geometry of the APC route stagnates the blood flow, which in turn facilitates thrombus formation. A slower venous flow from the IVC is more likely to form a thrombus. The results of this study suggest that a complex local flow based on the geometry of the blood vessel is a major factor that influences thrombus formation in the Fontan route. The proposed simulation model can contribute to the assessment and prediction of the most efficient surgical approach that would simultaneously minimize the occurrence of local thrombus formation.

Conversion from the APC to the TCPC by using a conduit may ameliorate venous flow delivery to the lungs [26], ensuring a better quality of life. Although the indication and timing for TCPC conversion is mainly determined through experience-based decisions [18, 27, 28], the underlying mechanism of thrombus formation in the RA and the reason why the TCPC is superior to the classical

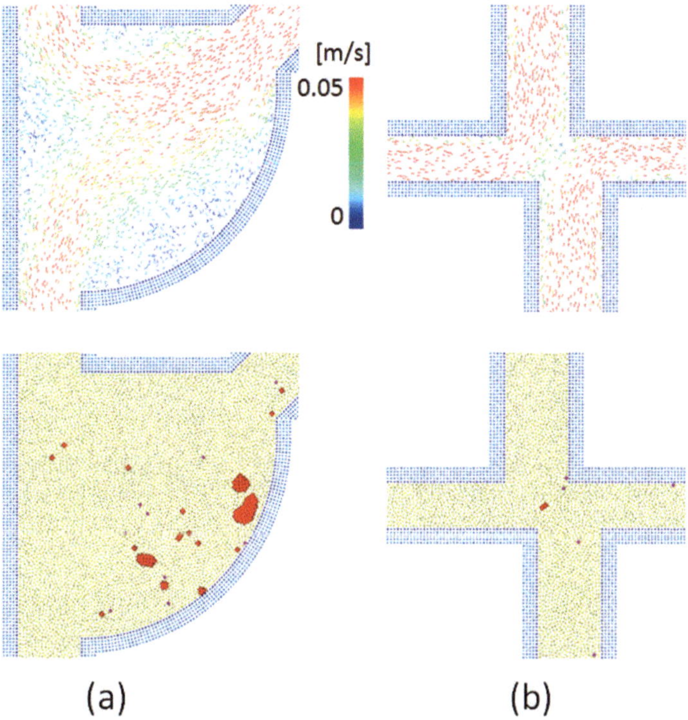

Fig. 4 Simulated flow velocity (top row) and thrombus formation (bottom row) for (a) APC round and (b) TCPC.

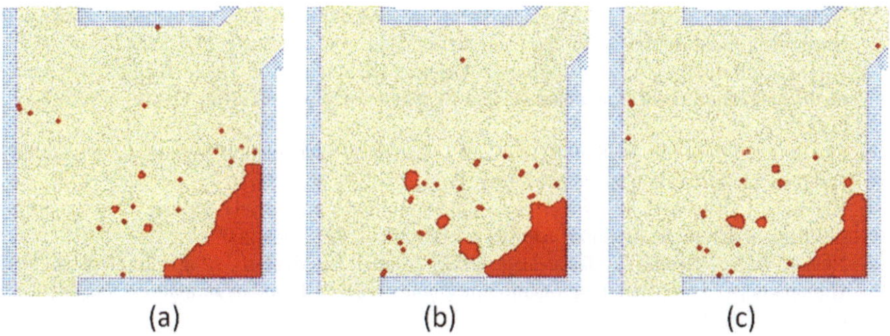

Fig. 5 Thrombus formation of APC square by different inflow velocities of vena cava at (a) rest, (b) light exercise (0.5 W/kg), and (c) moderate exercise (1.0 W/kg).

route in terms of hemodynamics [29, 30] and prevents thrombus formation are not conclusive yet. The simulation results demonstrated that thrombus formation is largely influenced by the following: 1) the configuration of the Fontan route from the vena cavae to the PA and 2) the speed of the flow in the Fontan pathway.

It was shown that once a thrombus is created, it grows in size, and a portion of the thrombus is disengaged and released into the outflow toward the PA. Thrombi embolization into the PA in a Fontan patient is life threatening; thus once a thrombus is detected, a TCPC conversion is required immediately [31]. In this respect, simulation results demonstrated that the shape of the Fontan route greatly influences thrombus formation process via local blood flow dynamics, as well as the flow rate from the IVC (i.e., cardiac output). This point is also related with treatment with anticoagulation therapies such as warfarin [32].

The simulation results shown here were based on rather simplified models; they should be improved to include detailed coagulation systems, three-dimensional fluid mechanics by patient-specific cardiovascular geometries, and endothelial injuries. These improvements will realize quantitative estimation of a risk of thrombosis for single ventricle patients.

Acknowledgements This research was partly funded by Grants-in-Aid for Scientific Research (15H03915 and 25630046), JSPS. KT was supported by MEXT SPIRE Supercomputational Life Science. KS was supported by the Inohana Alumni Association of the Chiba University of Medicine (12046).

References

1. Hathcock, J.J.: Flow effects on coagulation and thrombosis. Arterioscler. Thromb. Vasc. Biol. **26**, 1729–1737 (2006)
2. Slack, S.M., Cui, Y., Turitto, V.T.: The effects of flow on blood coagulation and thrombosis. Thromb. Haemost. **70**, 129–134 (1993)
3. Wolberg, A.S., Aleman, M.M., Leiderman, K., Machlus, K.R.: Procoagulant activity in hemostasis and thrombosis: Virchow's triad revisited. Anesth. Analg. **114**, 275–285 (2012)
4. Affeld, K., Reininger, A.J., Gadischke, J., Grunert, K., Schmidt, S., Thiele, F.: Fluid mechanics of the stagnation point flow chamber and its platelet deposition. Artif. Organs. **19**, 597–602 (1995)
5. David, T., Thomas, S., Walker, P.G.: Platelet deposition in stagnation point flow: an analytical and computational simulation. Med. Eng. Phys. **23**, 299–312 (2001)
6. Kaibara, M.: Rheological study on coagulation of blood with special reference to the triggering mechanism of venous thrombus formation. J Biorheol. **23**, 2–10 (2009)
7. Nesbitt, W.S., Westein, E., Tovar-Lopez, F.J., Tolouei, E., Mitchell, A., Fu, J., et al.: A shear gradient-dependent platelet aggregation mechanism drives thrombus formation. Nat. Med. **15**, 665–675 (2009)
8. Sorensen, E.N., Burgreen, G.W., Wagner, W.R., Antaki, J.F.: Computational simulation of platelet deposition and activation: II. Results for Poiseuille flow over collagen. Ann. Biomed. Eng. **27**, 449–458 (1999)
9. Casa, L.D., Deaton, D.H., Ku, D.N.: Role of high shear rate in thrombosis. J. Vasc. Surg. **61**, 1068–1080 (2015)
10. Ruggeri, Z.M., Ware, J.: von Willebrand factor. FASEB J. Off. Publ. Fed. Am. Soc. Exp. Biol. **7**, 308–316 (1993)
11. Frescura, C., Thiene, G.: The new concept of univentricular heart. Front. Pediatr. **2**, 62 (2014)
12. Balling, G., Vogt, M., Kaemmerer, H., Eicken, A., Meisner, H., Hess, J.: Intracardiac thrombus formation after the Fontan operation. J. Thorac. Cardiovasc. Surg. **119**, 745–752 (2000)

13. Coon, P.D., Rychik, J., Novello, R.T., Ro, P.S., Gaynor, J.W., Spray, T.L.: Thrombus formation after the Fontan operation. Ann. Thorac. Surg. **71**, 1990–1994 (2001)
14. Hjortdal, V.E., Emmertsen, K., Stenbog, E., Frund, T., Schmidt, M.R., Kromann, O., et al.: Effects of exercise and respiration on blood flow in total cavopulmonary connection: a real-time magnetic resonance flow study. Circulation **108**, 1227–1231 (2003)
15. McCrindle, B.W., Manlhiot, C., Cochrane, A., Roberts, R., Hughes, M., Szechtman, B., et al.: Factors associated with thrombotic complications after the Fontan procedure: a secondary analysis of a multicenter, randomized trial of primary thromboprophylaxis for 2 years after the Fontan procedure. J. Am. Coll. Cardiol. **61**, 346–353 (2013)
16. Monagle, P., Cochrane, A., Roberts, R., Manlhiot, C., Weintraub, R., Szechtman, B., et al.: A multicenter, randomized trial comparing heparin/warfarin and acetylsalicylic acid as primary thromboprophylaxis for 2 years after the Fontan procedure in children. J. Am. Coll. Cardiol. **58**, 645–651 (2011)
17. Mavroudis, C., Backer, C.L., Deal, B.J., Johnsrude, C., Strasburger, J.: Total cavopulmonary conversion and maze procedure for patients with failure of the Fontan operation. J. Thorac. Cardiovasc. Surg. **122**, 863–871 (2001)
18. Sheikh, A.M., Tang, A.T., Roman, K., Baig, K., Mehta, R., Morgan, J., et al.: The failing Fontan circulation: successful conversion of atriopulmonary connections. J. Thorac. Cardiovasc. Surg. **128**, 60–66 (2004)
19. van Son, J.A., Mohr, F.W., Hambsch, J., Schneider, P., Hess, H., Haas, G.S.: Conversion of atriopulmonary or lateral atrial tunnel cavopulmonary anastomosis to extracardiac conduit Fontan modification. Eur. J. Cardiothorac. Surg. **15**, 150–157 (1999). discussion 7–8
20. Jacobs, M.L., Pourmoghadam, K.K.: Thromboembolism and the role of anticoagulation in the Fontan patient. Pediatr. Cardiol. **28**, 457–464 (2007)
21. Kamada, H., Tsubota, K., Nakamura, M., Wada, S., Ishikawa, T., Yamaguchi, T.: A three-dimensional particle simulation of the formation and collapse of a primary thrombus. Int. J. Numer. Methods Biomed. Eng. **26**, 488–500 (2010)
22. Koshizuka, S., Oka, Y.: Moving-particle semi-implicit method for fragmentation of incompressible fluid. Nucl. Sci. Eng. **123**, 421–434 (1996)
23. Tsubota, K., Wada, S.: Effect of the natural state of an elastic cellular membrane on tank-treading and tumbling motions of a single red blood cell. Phys. Rev. E Stat. Nonlin. Soft Matter Phys. **81**, 011910 (2010)
24. Tsubota, K., Wada, S., Yamaguchi, T.: Simulation study on effects of hematocrit on blood flow properties using particle method. J. Biomech. Sci. Eng. **1**, 159–170 (2006)
25. Sughimoto, K., Okauchi, K., Zannino, D., Brizard, C., Liang, F., Sugawara, M., et al.: Cavopulmonary connection is superior to atriopulmonary connection Fontan in preventing thrombus formation: computer simulation of flow-related blood coagulation. Pediatr. Cardiol. **36**, 1436–1441 (2015)
26. Giardini, A., Pace Napoleone, C., Specchia, S., Donti, A., Formigari, R., Oppido, G., et al.: Conversion of atriopulmonary Fontan to extracardiac total cavopulmonary connection improves cardiopulmonary function. Int. J. Cardiol. **113**, 341–344 (2006)
27. Jang, W.S., Kim, W.H., Choi, K., Nam, J., Choi, E.S., Lee, J.R., et al.: The mid-term surgical results of Fontan conversion with antiarrhythmia surgery. Eur. J. Cardiothorac. Surg. **45**, 922–927 (2014)
28. Weinstein, S., Chan, D.: Extracardiac Fontan conversion, cryoablation, and pacemaker placement for patients with a failed Fontan. Semin. Thorac. Cardiovasc. Surg. **17**, 170–178 (2005)
29. Lardo, A.C., Webber, S.A., Friehs, I., del Nido, P.J., Cape, E.G.: Fluid dynamic comparison of intra-atrial and extracardiac total cavopulmonary connections. J. Thorac. Cardiovasc. Surg. **117**, 697–704 (1999)
30. Liang, F., Sughimoto, K., Matsuo, K., Liu, H., Takagi, S.: Patient-specific assessment of cardiovascular function by combination of clinical data and computational model with applications to patients undergoing Fontan operation. Int. J. Numer. Methods Biomed. Eng. **30**, 1000–1018 (2014)

31. Hedrick, M., Elkins, R.C., Knott-Craig, C.J., Razook, J.D.: Successful thrombectomy for thrombosis of the right side of the heart after the Fontan operation. Report of two cases and review of the literature. J. Thorac. Cardiovasc. Surg. **105**, 297–301 (1993)
32. Cheung, Y.F., Chay, G.W., Chiu, C.S., Cheng, L.C.: Long-term anticoagulation therapy and thromboembolic complications after the Fontan procedure. Int. J. Cardiol. **102**, 509–513 (2005)

Computational Study of Aortic Hemodynamics: From Simplified to Patient-Specific Geometries

A. Lefieux, F. Auricchio, M. Conti, S. Morganti, A. Reali, S. Trimarchi, and A. Veneziani

Abstract The investigation of aortic hemodynamics in physiological and pathological conditions by computational fluid dynamics is still one of the major topic of vascular biomechanics. In particular, thanks to the developments of new endovascular technologies such as Thoracic EndoVAscular Repair (TEVAR), a lot of attention is paid to the hemodynamics analysis of thoracic aorta. In this work, we aim at performing a sensitivity analysis of morphological aspects by comparing numerical results about three cases: (i) an idealized aortic arch with a *candy cane* shape; (ii) a patient-specific healthy arch; and (iii) a patient-specific dissected aorta. For the idealized aortic arch case we also compare the obtained results with respect to the theoretical and experimental literature dedicated to curved pipes.

A. Lefieux
Department of Civil Engineering and Architecture (DICAr), Pavia University, Pavia, Italy

Emory University. Cardiology Division. Atlanta, GA, USA
e-mail: adrien.lefieux@emory.edu

F. Auricchio • M. Conti (✉)
Department of Civil Engineering and Architecture (DICAr), Pavia University, Pavia, Italy
e-mail: auricchio@unipv.it; michele.conti@unipv.it

A. Reali
Department of Civil Engineering and Architecture (DICAr), Pavia University, Pavia, Italy

Technische Universität München – Institute for Advanced Study, Lichtenbergstrasse 2a, 85748, Garching, Germany

S. Morganti
Department of Electrical, Computer, and Biomedical Engineering, Pavia University, Pavia, Italy
e-mail: simone.morganti@unipv.it

S. Trimarchi
Thoracic Aortic Research Center, Policlinico San Donato IRCCS, University of Milan, Milan, Italy
e-mail: santi.trimarchi@unimi.it

A. Veneziani
MathCS, Emory University, Atlanta, GA, USA
e-mail: ale@mathcs.emory.edu

Y. Bazilevs, K. Takizawa (eds.), *Advances in Computational Fluid-Structure Interaction and Flow Simulation*, Modeling and Simulation in Science, Engineering and Technology, DOI 10.1007/978-3-319-40827-9_31

1 Introduction

Aorta is the main artery of our circulatory system; the investigation of its hemodynamics in physiological and pathological conditions is still one of the major topic of vascular biomechanics because it represents a driving factor in the evaluation of mechanisms underlying the aortic diseases, such as dissection or aneurysm, which are characterized by high mortality if untreated [6]. In particular, thanks to the developments of new endovascular technologies such as thoracic endovascular repair (TEVAR), which is rapidly replacing the classical open-surgery, a lot of attention is paid to the hemodynamics analysis of thoracic aorta, (see, e.g., [1, 11]), i.e., the aortic region from the sino-tubular junction to the celiac bifurcation. Such an analysis accounts also for aortic arch embedding the supra-aortic branches, which is considered the new frontier for further developments of endovascular approaches aimed at treating challenging clinical situations located in the region close to the heart (see, e.g., [18]). Furthermore, there are several aspects that make aortic hemodynamics very peculiar from patient to patient. First of all, the morphology of the arch triggers special features (see, e.g., [3]) that generally have a major impact on the fluid dynamics. Other aspects refer to the flow regime and the disturbances of the flow that in dependence on the patient's conditions require special numerical treatments. In this work we aim at performing a sensitivity analysis of morphological aspects by comparing numerical results about three cases: (i) an idealized aortic arch with a *candy cane* shape; (ii) a patient-specific healthy arch; (iii) a patient-specific dissected aorta. For the idealized aortic arch case we also compare the obtained results with respect to the theoretical and experimental literature dedicated to curved pipes.

2 Fluid Model and Numerical Method

The fluid problem is modeled by the incompressible Navier–Stokes equations (that is, a Newtonian isothermal fluid with a homogeneous density and viscosity distribution). Such an approximation is known to be acceptable in the aorta (see, e.g., [7]). In particular, we solve

$$\rho\left(\frac{\partial \mathbf{u}}{\partial t} + \mathbf{u}\cdot\nabla\mathbf{u}\right) + \nabla p - \nabla\cdot\mu\left(\nabla\mathbf{u} + \nabla\mathbf{u}^T\right) = \mathbf{f}, \tag{1}$$

$$\nabla\cdot\mathbf{u} = 0, \tag{2}$$

where ρ is the density (1.06 g/cm^3), μ is the dynamic viscosity (4 cP), $\mathbf{u}$ is the velocity, p is the pressure, and $\mathbf{f}$ a possible load.

For the geometries of interest, the analytical solutions of equations (1–2) are not available and we need to resort to numerical approximations. As for a consolidated practice, computational tools provide reliable quantitative results to investigate physiopathological conditions (see, e.g., [7]).

In particular, we consider a classical finite element approximation applied to the primitive variable formulation of the incompressible fluid equations given in (1–2). The special nature of the problem requires care in the selection of the piecewise polynomial approximation of the discrete velocity and pressure fields. In our case, we use piecewise linear pressures while velocities are approximated by piecewise linear functions added by a cubic correction (called "bubble"). We use unstructured tetrahedral meshes within the LifeV Library (https://www.lifev.org). Time advancing is obtained by a second order BDF scheme with an algebraic splitting of velocity and pressure computation—Yosida scheme, see [16]. Initial conditions are set to zero, their impact on the solution is negligible after simulating a judicious number of heart beats [17]. Boundary conditions are specified later on (see section 4.1).

3 Considered Geometries

In the present work, we consider three vascular geometries.

- A simplified *candy cane* geometry, depicted in Fig. 1, basically represented by four parameters: pipe diameter a (3.2 cm), curvature radius R (2.9 cm), proximal L_1 (5 cm), and distal extension L_2 (14 cm) of the pipe.
- A healthy patient-specific aorta: a dataset of medical images has been acquired at Ospedale Maggiore in Milano (Italy), using a Siemens SOMATOM Definition Flash Dual-Source CT Scanner, able to capture 10 time frames per cardiac cycle. The original computed tomography angiography set refers to a 72-year-old male patient and covers the entire length of the thoracic and abdominal aorta from

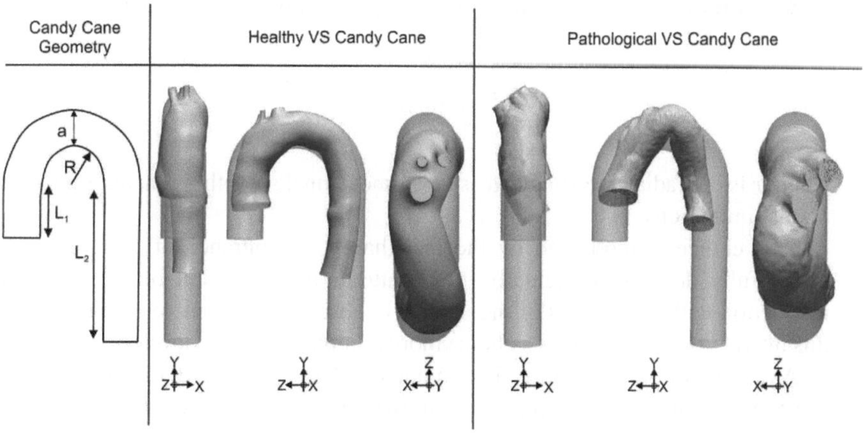

Fig. 1 Considered geometries, from left to right: the *candy cane*, the *healthy* and *pathological* aortas

the aortic root to the bifurcation into the iliac arteries. From this set, the region including the aortic arch has been selected and a level set segmentation has been performed.

- A pathological patient-specific aorta: this anatomy is obtained by the segmentation of pre-operative multislice computed tomography regarding a 51-year-old male patient suffering hypertension and having an asymptomatic post-dissecting thoracic aortic aneurysm. It is worth noting that in this case we do not include the distal tract of the descending aorta. As a matter of fact, the presence of the false lumen and of one entry tear complicates the simulation and are, however, out of our region of interest, i.e., ascending, arch, and proximal descending aorta.

  Both *healthy* and *pathological* aortas have an average diameter of 2.8cm.

4 Boundary Conditions

We assumed different boundary conditions when performing the different test cases as detailed in the following. The prescription of boundary conditions in computational hemodynamics is critical. In patient specific cases, we typically experience a lack of measures that prevents a completely data-driven analysis. Simplifying assumptions and surrogate models are necessary to have mathematically consistent problems to simulate. Here we present some possible choices aiming at being reliable for the aortic problems and at having significantly comparable cases.

4.1 Inflow Conditions

1. **Steady inflow**: we choose as an almost flat profile given by the equation:

$$f(r) = \begin{cases} 1 & r \leq \alpha a, \\ \frac{(r-a)}{(\alpha-1)a} & r > \alpha a, \end{cases} \tag{3}$$

   where r is the radial coordinate, α is an adimensional smoothing factor, set to 0.8 in our simulations.

   This choice is motivated by the fact that at the entrance of the arch flow is not fully developed. Actually, flow patterns are the result of the complex interaction with the leaflets of the valves. In absence of a complete modeling of this interaction, flat profile is a choice more reliable than fully developed flow [3].

   As a matter of fact, the inlet flow in our simulations is not yet fully developed. We believe that this is not an analysis drawback because, at the inlet of the arch, the flow is mainly driven by the heart during systolic phase with a profile close to a flat shape more than to a fully developed paraboloid (see [3]).

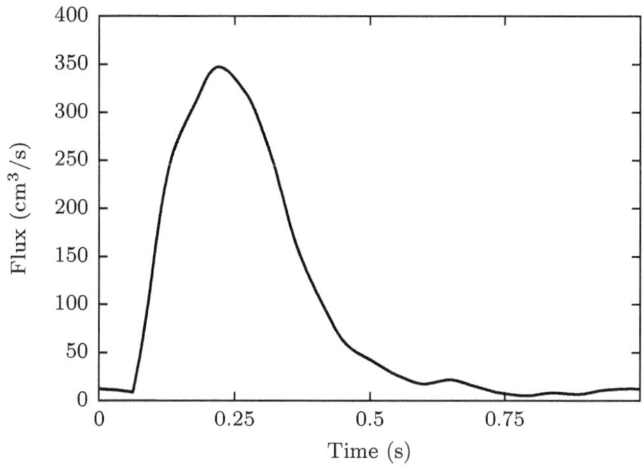

Fig. 2 Considered pulsatile physiological entry-flow rate (see [10])

2. **Unsteady inflow**: For the unsteady test cases, on the inflow section, we modulate the flat profile in eq. (3) by a time waveform retrieved from [10] and displayed in Fig. 2.

4.2 Outflow Conditions

Outlet conditions are critical for the reliability of the simulations. A popular approach is to incorporate the presence of distal districts by means of surrogate low fidelity and low dimensional models like lumped parameter systems. In simple idealized cases we can simply prescribe homogeneous Neumann conditions when no peripheral circulation needs to be included. This motivates the following choices.

- **3-elements Windkessel outflow**: For the patient-specific aortas, we prescribed a classical 3-elements Windkessel modeling of the distal circulation. This means that the peripheral impedance at each outflow section is represented by two resistances R_1 and R_2 and one compliance C (RCR model). The specific values of those parameters are taken from [8].
- **Traction free outflow**: For the *candy cane* geometries, we impose the free-stress boundary condition. As there is only one outflow boundary and this choice basically shifts the pressure only up to a hydrostatic constant.
- **Wall boundary condition**: we prescribe null velocity on the lumen walls. Despite such an approach is commonly adopted in literature, a more accurate model would include the interaction of fluid and structure, but the computational costs would be significantly higher and the accuracy advantage questionable

because the structural model for the arterial wall (differently from the blood model given by the Navier–Stokes equations) is affected by several uncertainties.

5 Numerical Tests

As mentioned before we consider three geometries and four numerical tests.

1. *Candy cane* geometry with *constant entry-flow rate*, whose value is the maximal flow rate at the systolic peak of the pulsatile physiological entry flow (347 cm^3/s), see Fig. 2, and free-stress outflow.
2. *Candy cane* geometry with a *pulsatile physiological entry-flow rate* (see Fig. 2) and free-stress outflow.
3. *Healthy aorta* geometry with a *pulsatile physiological entry-flow rate* (see Fig. 2) and 3-Windkessel outflows (see section 4).
4. *Pathological aorta* geometry with a *pulsatile physiological entry-flow rate* (see Fig. 2) and 3-Windkessel outflows (see section 4).

In what follows, we first present a discussion on the flow and wall shear stress in curved pipes for steady and pulsatile flows; however we do not review extensively previous contributions, since our main objective is to discuss trends observed in simplified geometries with respect to the two patient-specific geometries at hand. Nevertheless, we provide many important references on the subject for the interested reader. We conclude by comparing the flows and wall shear stresses in the patient specific aortas with respect to the results obtained with the *candy cane* geometry. Simulations are carried out with the C++ Object Oriented library LifeV (www.lifev.org), in particular with the solvers developed in the Emory-Pavia branch described in [16].

5.1 *Curved Pipes*

Steady flows in curved pipes with a constant curvature are characterized by the following adimensional group, known as the Dean number (see [12]):

$$D := \mathrm{Re}\sqrt{\frac{a}{R}} = \frac{2\rho Q}{\mu\pi}\sqrt{\frac{1}{aR}},$$

where Q is the flux, a is the radius of the pipe, R the radius of curvature, and Re is the Reynolds number, i.e., ($Re = 2\rho Ua/\mu$) with U the inflow mean velocity.

The maximum Dean number for our application is about 2717 with the maximal entry flux-rate being of 347 cm^3/s at the systolic peak (see Fig. 2) and, as a consequence, our application belongs to the category of large Dean numbers for which no analytical solutions exist (see [12]). It is worth pointing out that not all

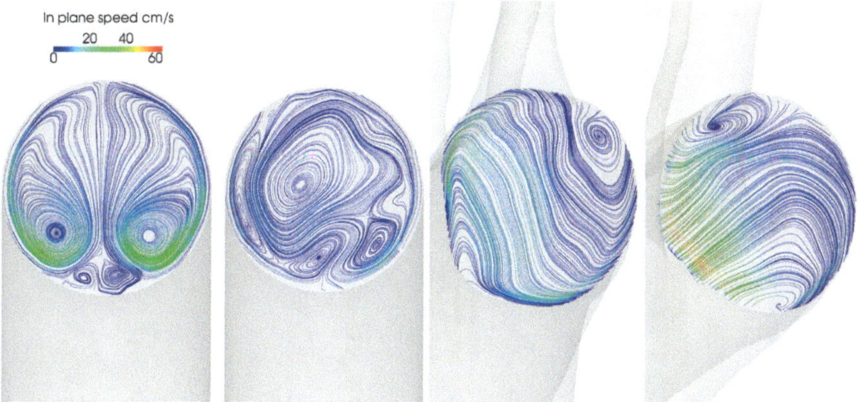

(a) In plane cross section velocity streamlines.

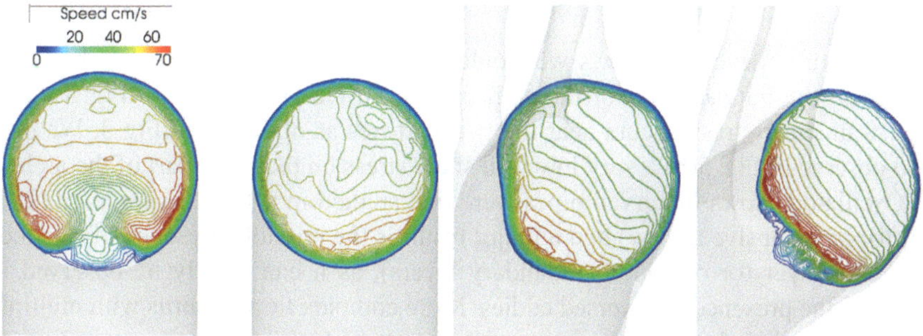

(b) Velocity contour plot.

Fig. 3 Streamlines (top) and velocity contour plots (bottom) for the four cases considered. From left to right: candy cane steady, candy cane pulsatile, healthy patient, pathological case. The cross sections displayed are taken in the center of the aortic arch. Bottom side corresponds to the internal bend. Results refer to the systolic peak for the three rightmost (unsteady) simulations

the definitions of the Dean numbers are consistent, which renders cumbersome the comparisons of the results with the literature (see [2] for a discussion on the subject).

Major results from the analysis of Dean's equations are, firstly, that symmetric secondary flows are introduced since the curvature of the pipe induces a gradient of pressure from the inner bend to the outer one and, secondly, that the fastest flow tends to be pushed toward the outer bend as the Dean number grows (see [12]). These analytical and numerical results assume that the flow is fully developed. For such a high Dean number used in our application, we would therefore expect the fastest flow to be much closer to the outward part of the bend than the one we obtain (see Figs. 3 and 4). The discrepancy comes from the fact that we do not employ a fully developed but a flat profile in inflow (see Sec. 4). Instead, we observe two vortexes on the inner part of the arch; a similar result is obtained in [13] with an experimental setting and a flat inflow profile (see also [3]). A major implication is

Fig. 4 Streamlines and speed at systolic peak. From left to right, we have: steady and unsteady flows in *candy cane*, unsteady flows in the healthy and pathological aorta

that with fully developed flows we would expect the wall shear stress to be much higher on the outward part of the arch. Instead, it clearly appears in Fig. 5 that a high wall shear stress occurs near the two vortexes as observed in Fig 3. We may also observe in Fig. 5(b) a very low wall shear stress on the inner part of the arch.

Oscillatory flows in curved pipes were first studied in [9] (see again, e.g., [2, 3, 12] for an extensive review) and a major result is the possible presence of a Stokes boundary layer (or oscillatory boundary layers), as it can clearly be observed in Fig. 3 by the presence of reversed eddies. More complex flow patterns with multiple vortexes are also observed since we employ a physiological pulsatile entry flow, as in [5]. As opposed to [5] in which symmetry of the secondary flows is assumed, we can see in Fig. 3 that the secondary flows are not symmetric, as shown in, e.g., [15], while the symmetry is always present in the steady entry flow case with finite curvature, as shown in [14].

Another major difference between steady and pulsatile flows in curved pipes is that, in the case of pulsatile flows, the maximal axial speed is pushed toward the inner bend, as clearly observed in Fig. 4. As a consequence, the wall shear stress is higher in the inner part of the pipe, as it can be seen in Fig. 5, and as opposed to the steady flow case, in particular with fully developed entry flows, as shown experimentally in [4] and theoretically in [9].

5.2 Patient-Specific Aortas

Two important shared trends are observed from the pulsatile entry flow analysis between the simplified geometry and the patient-specific ones:

1. the highest axial speed is bended toward the inner part of the bend (see Fig. 4);
2. high wall shear stress remains on the lower part of the aorta (see Fig. 5).

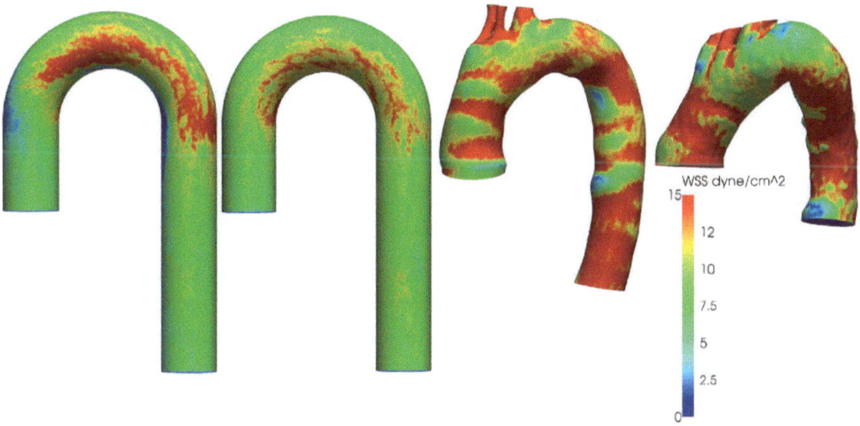

(a) Anterior view.

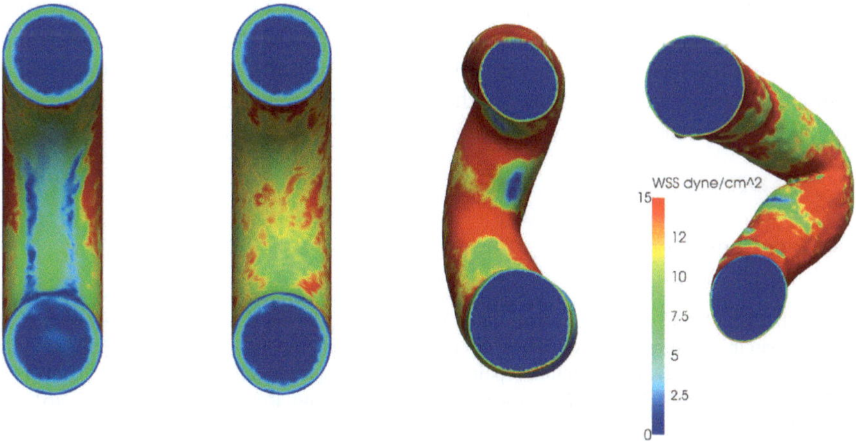

(b) Interior view.

Fig. 5 Wall shear stresses for the different cases under investigation

Furthermore, for the healthy aorta, the flow boundary layer appears to be of the same size as the *candy cane* with a pulsatile entry flow and the speed is of the same order of magnitude. On the contrary, due to a strong bending in the pathological aorta much faster flow is observed at the inner part of the arch, as well as a much thinner boundary layer. Also, results in Fig.3 suggest completely different and irregular flow patterns from the idealized and patient-specific aortas; meaning that irregular diseased conditions may even accelerate and drive the evolution of the pathology.

However, it clearly appears that only a patient-specific analysis can provide a precise distribution of the wall shear stress which is severely impacted from the changes of topology of the aorta wall. Also, the pathological aorta bluff induces a boundary-layer flow separations, resulting in a disturbed flow in the upper part of the descending aorta, leading thus to a much greater Reynolds number in this part of the aorta.

6 Conclusion

As well known, numerical results in computational hemodynamics are majorly influenced by the vascular geometry and the prescription of the boundary conditions. Image and geometrical processing techniques enable an accurate reconstruction of patient-specific cases, while for the boundary conditions the gap between available measures and needed data for the mathematical consistency of the problems to solve is still challenging. In this paper, we tested our numerical solver on nontrivial cases for the aortic flow. Aorta has a complex fully 3D morphology that triggers specific flow patterns and simplifying assumptions in the numerical process may prevent a correct capturing of the relevant features. In cross-checking our results and comparing available experimental data, we assess the reliability of our specific numerical solver while pointing out the importance of an accurate reconstruction of the patient-specific geometry and of the selection of boundary data.

References

1. Auricchio, F., Conti, M., Lefieux, A., Morganti, S., Reali, A., Sardanelli, F., Secchi, F., Trimarchi, S., Veneziani, A.: Patient-specific analysis of post-operative aortic hemodynamics: a focus on thoracic endovascular repair (TEVAR). Comput. Mech. **54**(4), 943–953 (2014)
2. Berger, S., Talbot, L., Yao, L.: Flow in curved pipes. Annu. Rev. Fluid Mech. **15**(1), 461–512 (1983)
3. Chandran, K.B.: Flow dynamics in the human aorta: Techniques and applications. Cardiovascular Techniques, vol. II. CRC Press (2001)
4. Chandran, K., Swanson, W., Ghista, D., Vayo, H.: Oscillatory flow in thin-walled curved elastic tubes. Ann. Biomed. Eng. **2**(4), 392–412 (1974)
5. Chang, L.J., Tarbell, J.: Numerical simulation of fully developed sinusoidal and pulsatile (physiological) flow in curved tubes. J. Fluid Mech. **161**, 175–198 (1985)
6. Erbel, R., Aboyans, V., Boileau, C., Bossone, E., Di Bartolomeo, R., Eggebrecht, H., Evangelista, A., Falk, V., Frank, H., Gaemperli, O., et al.: 2014 ESC guidelines on the diagnosis and treatment of aortic diseases. Eur. Heart J. **35**(41), 2873–2926 (2014)
7. Formaggia, L., Quarteroni, A., Veneziani, A.: Cardiovascular Mathematics: Modeling and Simulation of the Circulatory System, vol. 1. Springer Science and Business Media, Milan (2010)
8. Kim, H.J., Vignon-Clementel, I.E., Figueroa, C.A., LaDisa, J.F., Jansen, K.E., Feinstein, J.A., Taylor, C.A.: On coupling a lumped parameter heart model and a three-dimensional finite element aorta model. Ann. Biomed. Eng. **37**(11), 2153–2169 (2009)
9. Lyne, W.: Unsteady viscous flow in curved pipe. J. Fluid Mech. **45**, 13–31 (1971)

10. Morbiducci, U., Ponzini, R., Rizzo, G., Cadioli, M., Esposito, A., De Cobelli, F., Del Maschio, A., Montevecchi, F.M., Redaelli, A.: In vivo quantification of helical blood flow in human aorta by time-resolved three-dimensional cine phase contrast magnetic resonance imaging. Ann. Biomed. Eng. **37**(3), 516–531 (2009)
11. Pasta, S., Cho, J.S., Dur, O., Pekkan, K., Vorp, D.A.: Computer modeling for the prediction of thoracic aortic stent graft collapse. J. Vasc. Surg. **57**(5), 1353–1361 (2013)
12. Pedley, T.J.: The Fluid Mechanics of Large Blood Vessels. Cambridge University Press, London (1980)
13. Scarton, H.A., Shah, P.M., Tsapogas, M.J.: Relationship of the spatial evolution of secondary flow in curved tubes to the aortic arch. In: Dubey, R.N., Lind, N.C. (eds.) Mechanics in Engineering, p. 111. U of Waterloo Press, Waterloo (1977)
14. Siggers, J., Waters, S.: Steady flows in pipes with finite curvature. Phys. Fluids (1994-present) **17**(7), 077102 (2005)
15. Siggers, J.H., Waters, S.L.: Unsteady flows in pipes with finite curvature. J. Fluid Mech. **600**, 133–165 (2008)
16. Veneziani, A., Villa, U.: Aladins: An {ALgebraic} splitting time {ADaptive} solver for the incompressible Navier-Stokes equations. J. Comput. Phys. **238**, 359–375 (2013)
17. Vignon-Clementel, I.E., Figueroa, C.A., Jansen, K.E., Taylor, C.A.: Outflow boundary conditions for three-dimensional finite element modeling of blood flow and pressure in arteries. Comput. Methods Appl. Mech. Eng. **195**(29), 3776–3796 (2006)
18. Weidman, J.M., Desai, M., Iftekhar, A., Boyle, K., Greengard, J.S., Fisher, L.M., Thomas, R.L., Zannetti, S.: Engineering goals for future thoracic endografts—how can we make them more effective? Prog. Cardiovasc. Dis. **56**(1), 92–102 (2013)

An Image-Based Computational Framework for Analyzing Disease Occurrence and Treatment Outcome in Patients with Peripheral Arterial Disease

Shaolie S. Hossain

Abstract Peripheral arterial disease (PAD) is generally attributed to the progressive vascular accumulation of lipoproteins that lead to the formation of atherosclerotic plaques in the lower extremities. The disease process is largely regulated by the local hemodynamics and biophysical conditions. In this work, an isogeometric analysis framework is presented to analyze blood flow and vascular deposition of circulating nanoparticles (NPs) in the superficial femoral artery (SFA) of a PAD patient followed up over 24 months. The patient-specific geometry of the blood vessel and the hemodynamic conditions are derived from magnetic resonance imaging (MRI), performed at baseline and 24-months after stent-implantation. A dramatic improvement in blood flow dynamics is observed post-intervention resulting in a 500 % increase in peak flow rate as a consequence of luminal enlargement. Furthermore, flow simulations reveal a 32 % drop in mean oscillatory shear index, indicating reduced disturbed flow post-intervention. The same patient-specific data is then used to predict in silico the vascular deposition of systemically injected nanomedicine. These NPs are targeted to inflammatory cell adhesion molecules (CAMs) such as vascular cell adhesion molecules (VCAM-1) that are known to overexpress at the diseased site. Results show that VCAM-1 directed NPs preferentially accumulate near the stenosis in the baseline configuration. Such selective deposition of NPs within the stenosis could be effectively used for the detection and treatment of plaques forming in the SFA. The proposed MRI based computational protocol can be used to analyze data from clinical trials to explore possible correlations between hemodynamics and disease progression in PAD patients, and potentially predict disease occurrence as well as the outcome of an intervention.

S.S. Hossain (✉)
Department of Molecular Cardiology, Texas Heart Institute, 6770 Bertner Avenue, Houston, TX 77030, USA
e-mail: sshossain@texasheart.org

Y. Bazilevs, K. Takizawa (eds.), *Advances in Computational Fluid-Structure Interaction and Flow Simulation*, Modeling and Simulation in Science, Engineering and Technology, DOI 10.1007/978-3-319-40827-9_32

1 Introduction

Peripheral arterial disease (PAD) is a vascular disease inducing the progressive formation of plaques (atherogenesis), specifically in the arteries that carry blood to the limbs, head, kidneys, and stomach [1–4]. These blockages can lead to impaired leg function, reduced quality of life, and possibly loss of limb. PAD management can be categorized into pharmacological (lipid lowering treatments), surgical (deployment of stents, revascularization, and vascular surgery), and a combination of the two [1, 5]. The first step in atherogenesis generally is the accumulation of circulating low density lipoprotein particles into the sub-intimal space of the arterial wall [6]. This accumulation is generally higher at sites of disturbed flow [7–12], where typically larger rates of endothelial cell (EC) proliferation and apoptosis, higher vascular permeability, over-expression of EC adhesion molecules (ICAM1, VCAM1, E-selectin), and the release of chemo-attractants from ECs are observed [13, 14]. This clearly emphasizes the importance of an accurate determination of the local flow field to potentially predict the occurrence of the disease and the outcome of any intervention. The main objective of this work is to bring together a number of available technologies to build a patient-specific computational framework that can be used to analyze data from clinical trials to provide insights into PAD management.

To that end, the geometry and hemodynamic conditions for an SFA of a PAD patient are extracted from MR imaging at baseline and 24-months post-stent-implantation. Then finite element-based isogeometric analysis [15, 16] is used to simulate the blood flow dynamics within the SFA over a full cardiac cycle. Near wall quantities like time-averaged wall shear stress (TAWSS) and oscillatory shear index (OSI) are also investigated. Finally, the vascular transport of systemically injected nanoparticles (NPs), targeted to different inflammatory vascular molecules such as VCAM-1 that are known to overexpress at the diseased site, is analyzed. These CAMs are regarded as early markers of atherosclerosis due to their upregulation at the site of inflammation. Possible relationship between local hemodynamics and particle distribution pattern is also studied.

2 MRI Data Acquisition and Mesh Generation

The MRI data utilized in this analysis were obtained from the Effect of Lipid Modification on PAD after Endovascular Intervention Trial (ELIMIT) [17]. The patients underwent MR imaging of the distal SFA at baseline, 6-months, 12-months, and 24-months with a 3.0 T system (Signa Excite, GE Healthcare, Milwaukee, Wisconsin) using a unilateral phased array coil with a field of view (FOV) of 8 cm (along z-axis) and 12 cm (in plane x and y axes; Pathway Biomedical, Inc.). The geometries were extracted from proton-density-weighted (PDW) fast spin-echo scans. ELIMIT participants were also imaged using gated 2D phase-contrast (2D-PC) sequences, acquired at select locations within the corresponding PDW volumes—typically proximally and distally to SFA lesions. Gated 2D-PC were acquired during the same exam with slice-thickness = 4 mm, repetition time (TR) = 10.6 ms, echo time

(TE) = 4.97 ms, echo train length (ETL) = 1, bandwidth = 244 Hz/pixel, 20 frames per cardiac cycle, and phase-contrast encoding velocity (VENC) of 120 cm/s. Serial MRI scans were carefully co-registered across follow-up visits using anatomical landmarks (artery, vein, and muscle) [17]. Velocity profiles over the cardiac cycle were obtained at each location using the luminal cross-sectional areas (region of interest [ROI]), the mean signal intensity of the ROIs, and the VENC factor.

An image-based modeling pipeline, as detailed in [18], is adopted to construct hexahedral solid NURBS meshes from the MR images of the SFA. Briefly, the image quality is first improved by passing the raw imaging data through a preprocessing pipeline for enhancing the contrast, filtering noise, classifying and segmenting regions of various materials. The target surfaces (the blood vessel wall) are then extracted by isocontouring the pre-segmented imaging data, followed by the extraction of the vascular skeleton (vessel centerlines) [19, 20]. Finally, an optimized skeleton-based sweeping method is used to construct hexahedral control meshes employing an in-house numerical code [18], from which hexahedral solid NURBS are constructed (Fig. 1).

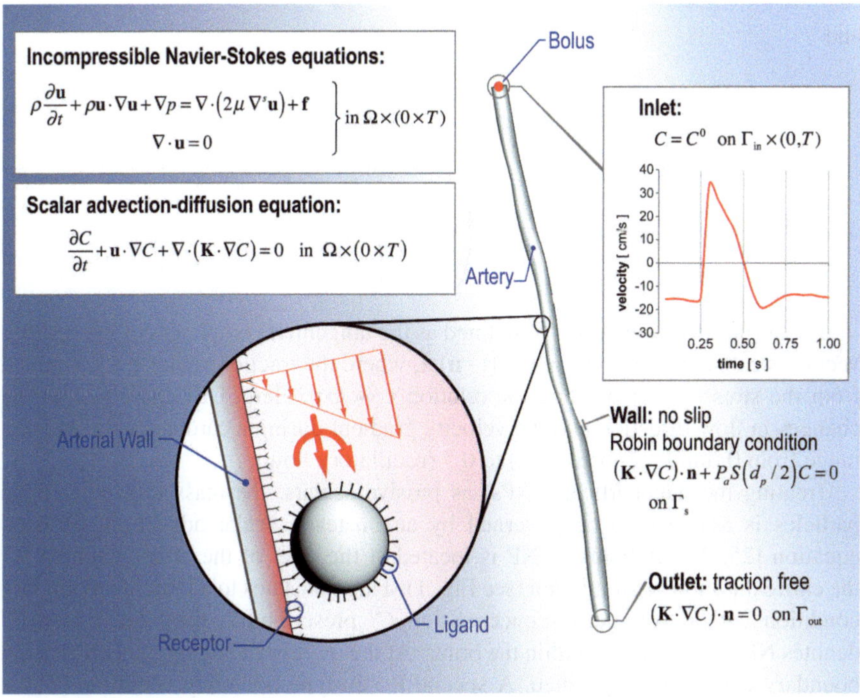

Fig. 1 MRI-based computational model—the problem setup. The geometry of the superficial femoral artery (SFA) and the inflow conditions are extracted from the patient-specific MRI data. The adhesion of spherical NPs to the vessel walls of the SFA is mediated by ligand–receptor interactions. The main governing equations and boundary conditions for the problem are included.

3 Governing Equations and Solution Approach

A continuum-based approach was adopted to simulate blood flow and particle transport within the SFA. The details of the governing equations and the methodology are reported elsewhere [21–24]. Briefly, considering blood as an incompressible Newtonian fluid with a dynamic viscosity (μ) of 0.003 Pa-s and a density (ρ) of 1060 kg/m^3, blood flow was assumed to be governed by the unsteady Navier–Stokes equations for incompressible flow subjected to appropriate boundary conditions (see Fig. 1). An inflow velocity profile was specified at the inlet, a no-slip boundary condition was prescribed at the rigid and impermeable wall, and a traction-free outflow boundary condition was set at the branch outlet. In order to analyze the flow field characteristics and its near wall behavior, two wall shear-based quantities were calculated over a cardiac cycle with a time period T:

$$\mathrm{TAWSS} = \left| \frac{1}{T} \int_0^T \mathbf{t}_s \, dt \right| \tag{1}$$

and

$$\mathrm{OSI} = \frac{1}{2} \left(1 - \frac{\left| \frac{1}{T} \int_0^T \mathbf{t}_s \, dt \right|}{\frac{1}{T} \int_0^T |\mathbf{t}_s| dt} \right) \tag{2}$$

The surface traction vector $\mathbf{t}_s$, defined as the tangential component of the traction vector, is calculated from $\mathbf{t}_s = \mathbf{t} - (\mathbf{t} \cdot \mathbf{n})\, \mathbf{n}$, where the traction vector $\mathbf{t}$ is computed from the stress tensor σ using the relation $\mathbf{t} = \sigma \mathbf{n}$. Here, OSI characterizes the changes in flow direction and the velocity gradient during a cardiac cycle, and can range from 0 (unidirectional flow) to 0.5 (oscillatory flow).

Treating the nanoparticles (NPs) as passive scalars, the mass transport of the particles is assumed to be governed by an unsteady scalar advection–diffusion equation [25, 26]. A bolus of NP is located at the inlet of the artery segment for the entire duration of simulation (see Fig. 1). This translates to a Dirichlet boundary condition with a volumetric concentration C^0 prescribed at the inlet, where C^0 denotes NP concentration within the bolus. At the outflow, a homogenous Neumann boundary condition is specified. A special Robin-type boundary condition [21] is prescribed at the lumen-wall interface to account for NP deposition to the vessel wall (Fig. 1). The mass flux of particles diffusing through the lumen-wall interface and adhering to the vessel wall is assumed to be a function of particle size d_p, local wall shear rate S, and probability of adhesion P_a, defined as the probability of having at least one ligand–receptor bond (see Fig. 1). The propensity of a particle to adhere

Table 1 Particle adhesion parameters used in the simulation

Parameters	Values
Surface density of ligand molecules	$m_l = 10^{15}$ #/m^2
Surface density of receptor molecules	$m_r = 10^{13}$ #/m^2
Ligand-receptor affinity constant at zero load	$K_a{}^0 = 2.3 \times 10^{-17}$ m^2
Characteristic length of ligand-receptor bond	$\lambda = 1 \times 10^{-10}$m
Dynamic viscosity of water	$\mu = 0.001$ N-s/m^2
NP-substrate separation distance, at equilibrium	$\Delta = 0.68$ nm
Boltzmann's constant	$k_B T = 4.142 \times 10^{-21}$ J
Drag coefficient on the spherical particle	$F^s = 1.668$

firmly to the vessel wall withstanding dislodging hydrodynamic forces is strongly influenced by the physico-chemical properties of the NPs. Accordingly, Decuzzi et al. [27] modeled P_a as a function of physiological parameters like target receptor density and local WSS, as well as certain design parameters including particle shape, size, ligand type, and density:

$$P_a \simeq \alpha_1 \alpha_2 \pi \left(d_p/2\right)^2 \exp\left[-\beta \frac{\mu S}{\alpha_2}\right] \quad (3)$$

where, $\alpha_1 = m_l K_A^0$, $\alpha_2 = m_r[1 - (1 - \frac{\Delta}{d_p/2})^2]$, and $\beta = \frac{\lambda 6 F^S}{k_B T}$. A summary of the parameters and their values used is provided in Table 1. This particle firm-adhesion model, which has been validated in vitro in a previous work [21] for spherical particles in point-contact with the vessel wall, is utilized herein to account for NP wall deposition. For a more comprehensive description of the modeling approach and parameter selection, see [21, 28] and the references therein.

The governing equations are solved by applying finite element-based isogeometric analysis [23, 29] using quadratic NURBS to describe the exact geometry as well as the solution space [16, 18]. A residual-based variational multiscale method [30] is implemented to solve the system of equations, utilizing a Newton–Raphson procedure with a multi-stage predictor–corrector algorithm applied at each time step. The generalized $-\alpha$ method [31, 32], an implicit second-order time-accurate method that is also unconditionally stable, is used for time advancement. The readers are referred to the numerical procedures described in [16, 24, 29, 33, 34] for further details.

4 Results and Discussion

In Fig. 2, the velocity magnitude is presented for baseline (top) and 24-months post-intervention case. Significant luminal enlargement is observed post-intervention because of stent-implantation, resulting in a 453 %, 313 %, and 264 % increase

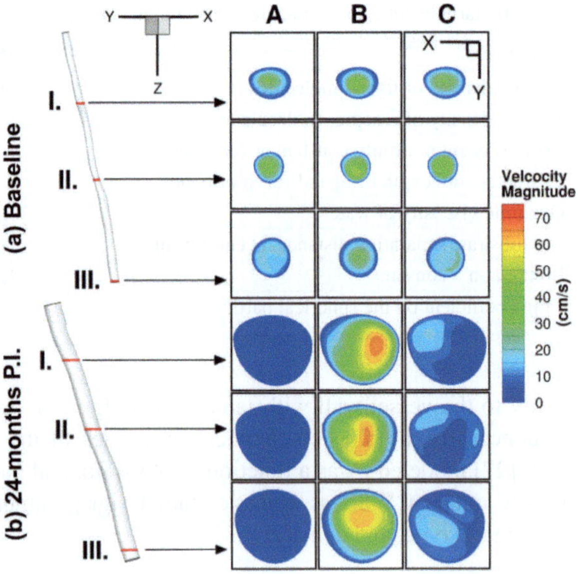

Fig. 2 Velocity magnitude is shown for baseline (top) and 24-months post-intervention (P.I.) at three different cross-sections: (I.) proximal, (II.) mean, and (III.) distal. In addition, three different times in the cardiac cycle are presented: end-diastole (column A), post-peak systole (column B), and post-systole (column C).

in cross-sectional area at the stenosis ("mean"), proximal and distal locations, respectively. At baseline, the velocity magnitude levels seem to drop noticeably from the proximal to the distal sections consistently during the cardiac cycle because of flow restriction at the stenosis. However, this phenomenon appears to be less pronounced around peak systole. Mean flow velocity magnitude measured via PC MRI also shows markedly decreased values distal to the SFA narrowing (data not shown here, see [35]). On the other hand, because of the implanted stent in the 24-months post-intervention SFA geometry blood flow is no longer restricted, resulting in a 500 % increase in peak flow rate [35]. Consequently, there is less of a variation in velocity magnitude between proximal and distal slices in contrast with the baseline case.

Fig. 3 reports the spatial distribution of TAWSS and OSI. In the baseline case, alternate areas of lower (<1 Pa) and higher (>5 Pa) TAWSS regions appear near the stenosis. 24-months after stent-implantation, relatively lower levels of TAWSS magnitude (<5 Pa) are observed that also appear to be more uniformly distributed throughout the SFA segment when compared with the baseline case. At 24-months post-intervention, OSI levels appear to diminish considerably (32 % drop in mean OSI), especially near the original constriction, in some areas by a factor of two, when compared with the baseline case, signifying reduced disturbed flow post-intervention. It appears that regions of relatively higher OSI generally correspond to lower WSS values.

Next, the computational tool is used to predict the wall deposition patterns in a simulation of systemically injected NPs in the pre- and post-intervention SFA to explore possible use of nanomedicine in the non-invasive imaging and treatment of PAD patients. To that end, the vessel wall deposition of vascular cell adhesion

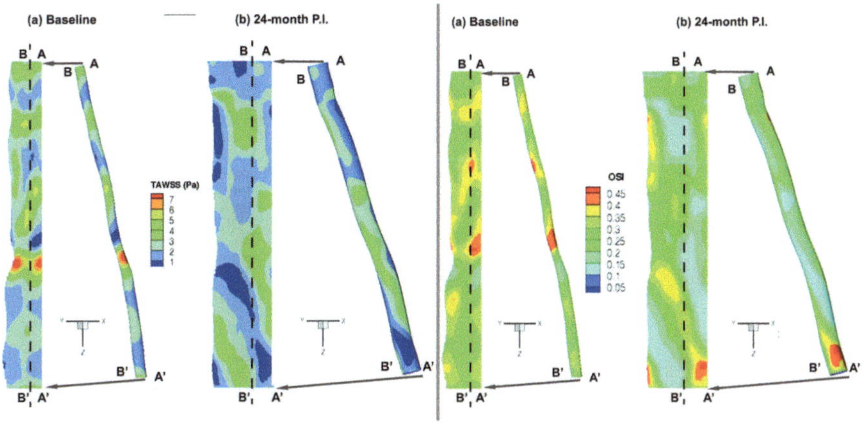

Fig. 3 Left panel shows the vascular distribution patterns for time-averaged wall shear stress (TAWSS) in Pa at (a) baseline and (b) 24-months post-intervention, while the right panel depicts the same for the oscillatory shear index (OSI)

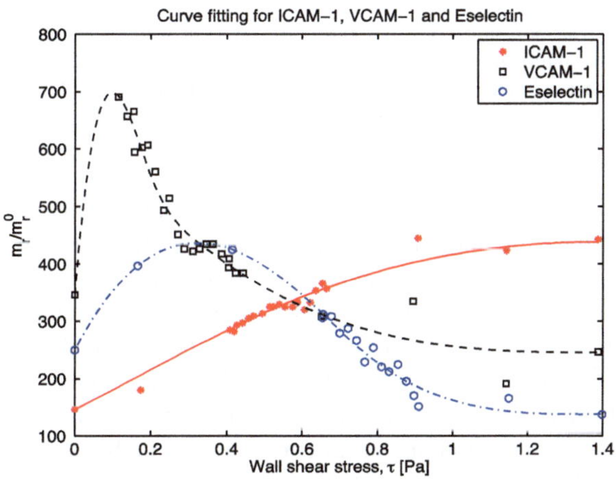

Fig. 4 Receptor surface density vs. wall shear stress relationship determined by curve fitting to in-vitro data for TNF-α stimulated CAM expression, obtained from [36]. The respective CAM (receptor) surface density m_r is reported as percent (%) of unstimulated CAM expression under static conditions $m_r{}^0$.

molecule (VCAM-1) directed particles is investigated. First, a phenomenological model (see Fig. 4) correlating CAM expression to local WSS as detailed in [21, 28, 36] is utilized to estimate the surface density of the target receptor with respect to their unstimulated expression under static conditions (see Fig. 5A).

Spherical NPs ($d_p = 100$ nm) uniformly coated with anti-VCAM-1 antibodies (aVCAM1-NPs) are introduced into the bloodstream by placing a bolus of NPs at the SFA inlet. Carried by the blood flow, these NPs can recognize VCAM-1 expression (receptors) and firmly adhere to the vessel wall through ligand-receptor bond formation. Flow-coupled particle transport simulations are then carried out for ten cardiac cycles. The resulting spatial distribution of NPs in terms of particle surface density (cm^{-2}) is presented in Fig. 5B for the baseline and 24-months

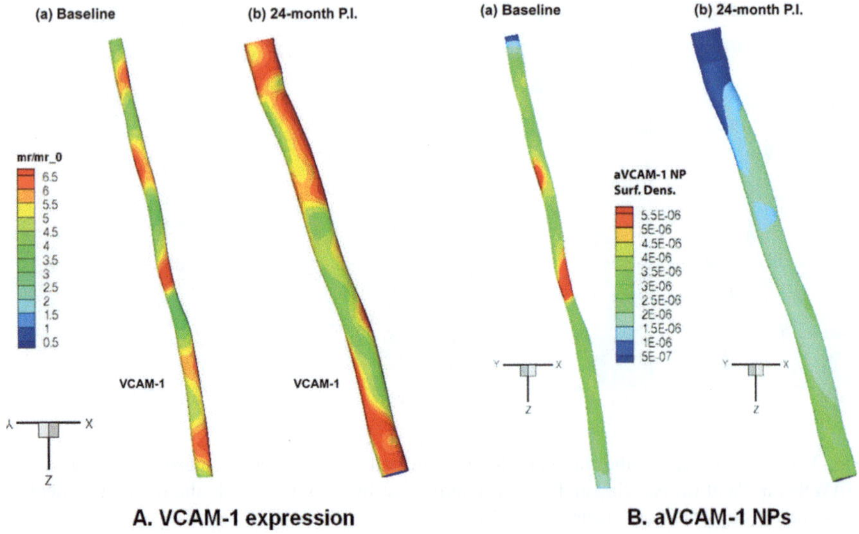

Fig. 5 A) The panel on the left shows the normalized surface density of VCAM-1 molecules (m_r/m_r^0), estimated as a function of the local wall shear rates at (a) baseline and (b) 24-months post-intervention (P.I.). B) The panel on the right depicts the spatial distribution patterns for 100 nm spherical nanoparticles adhering on the superficial femoral artery (SFA) in terms of #/cm^2 for (a) baseline and (b) 24-months post-intervention (P.I.) configurations.

post-intervention cases. In addition to local hemodynamics, receptor surface density appears to modulate particle adhesion, resulting in a spatially inhomogeneous particle distribution pattern in both cases. That is, the higher the receptor density and higher the OSI, the greater the particle concentration at a given site. aVCAM-1 NPs appear to preferentially accumulate near the diseased region with a surface density up to 1.9 times higher than the mean in the baseline case. This is encouraging from a therapeutic (and diagnostic) perspective as this may potentially translate to a higher drug/imaging agent concentration in the targeted area (i.e., the stenosis).

In order to investigate possible spatial correlation between local hemodynamics and NP adhesion, the OSI and aVCAM-1 NP distribution color maps were merged for the baseline and 24-months post-intervention cases (see Fig. 6 for the baseline case). Qualitatively there appears to be a strong spatial correlation between the two quantities throughout the computational domain, except for small areas very close to the inlet and outlet boundaries. Generally regions of high OSI are associated with regions of greater aVCAM-1 deposition. In a similar qualitative comparison made between TAWSS and aVCAM-1, the spatial correlations are not as evident, especially in the baseline case (data not shown here, see [35]). In regions downstream of the stenosis, the aVCAM-1 NP deposition seems to follow the pattern of high OSI values rather than that of low TAWSS values.

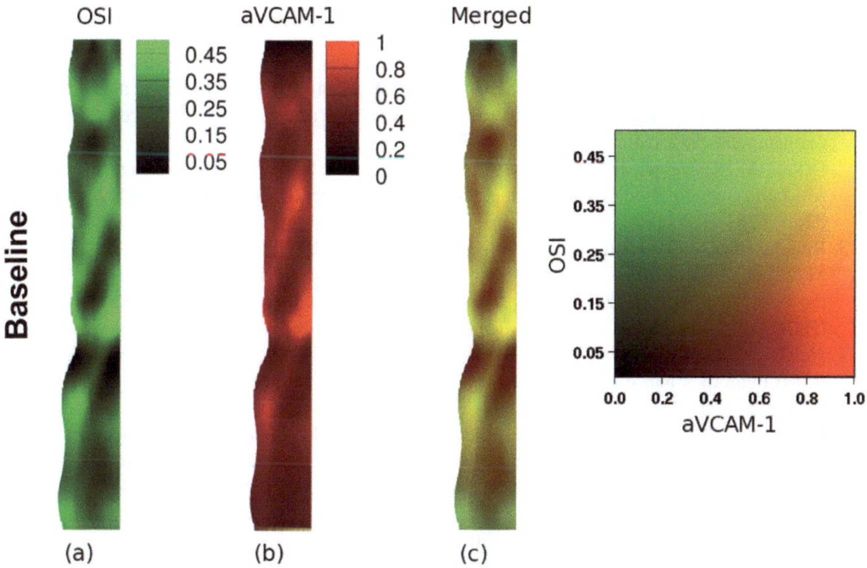

Fig. 6 Two different contours, (a) OSI and (b) aVCAM-1 NP concentration (normalized), are merged (c) for baseline configuration. Here the unrolled geometries are presented. Bright yellow regions in (c) indicate high OSI along with higher concentrations of aVCAM-1 particles.

5 Conclusions

An isogeometric analysis framework has been used to predict the blood flow and vascular deposition in a simulation of systemically injected nanomedicines into the SFA of a PAD patient. Upon intervention, a dramatic improvement in blood flow dynamics has been documented by the in-silico simulations showing a 32 % reduction in OSI and an approximate 500 % increase in overall peak blood flow rate. Also, it has been observed that systemically injected nanomedicines, targeted to inflammatory vascular molecules such as VCAM-1, would preferentially accumulate near the stenosis in the baseline configuration. In particular, for the VCAM-1 targeted 100 nm nanoparticles, the vascular accumulation has shown a maximum at the stenosis with a surface density up to 1.9 times higher than the mean. This high, specific vascular accumulation could be effectively used for the non-invasive detection and treatment of plaques forming in the SFA. The development of accurate image-guided patient-specific predictive tools, such as the isogeometric analysis framework presented here, can help in assessing disease burden, and monitor progression/regression of atherosclerosis in response to therapeutic and interventional treatment.

References

1. Lumsden, A.B., Rice, T.W., Chen, C., Zhou, W., Lin, P.H., Bray, P., et al.: Peripheral arterial occlusive disease: magnetic resonance imaging and the role of aggressive medical management. World J. Surg. **31**(4), 695–704 (2007)
2. Criqui, M.H., Langer, R.D., Fronek, A., Feigelson, H.S., Klauber, M.R., McCann, T.J., et al.: Mortality over a period of 10 years in patients with peripheral arterial disease. N. Engl. J. Med. **326**(6), 381–386 (1992)
3. Regensteiner, J.G., Hiatt, W.R., Coll, J.R., Criqui, M.H., Treat-Jacobson, D., McDermott, M.M., et al.: The impact of peripheral arterial disease on health-related quality of life in the Peripheral Arterial Disease Awareness, Risk, and Treatment: New Resources for Survival (PARTNERS) Program. Vasc. Med. **13**(1), 15–24 (2008)
4. Nehler, M.R., McDermott, M.M., Treat-Jacobson, D., Chetter, I., Regensteiner, J.G.: Functional outcomes and quality of life in peripheral arterial disease: current status. Vasc. Med. **8**(2), 115–126 (2003)
5. DeRubertis, B.G., Faries, P.L., McKinsey, J.F., Chaer, R.A., Pierce, M., Karwowski, J., et al.: Shifting paradigms in the treatment of lower extremity vascular disease: a report of 1000 percutaneous interventions. Ann. Surg. **246**(3), 415–422 (2007). discussion 22-4
6. Libby, P., Aikawa, M.: Stabilization of atherosclerotic plaques: new mechanisms and clinical targets. Nat. Med. **8**(11), 1257–1262 (2002)
7. Libby, P., Aikawa, M., Jain, M.K.: Vascular endothelium and atherosclerosis. Handb. Exp. Pharmacol. **176 Pt 2**, 285–306 (2006)
8. Hahn, C., Schwartz, M.A.: Mechanotransduction in vascular physiology and atherogenesis. Nat. Rev. Mol. Cell Biol. **10**(1), 53–62 (2009)
9. Prosi, M., Perktold, K.: Computer simulation of coupled luminal and transmural mass transport processes in a carotid bifurcation model. J. Biomech. **39**, S376 (2006)
10. Formaggia, L., Quarteroni, A.M., Veneziani, A.: Cardiovascular Mathematics. Springer, Milan (2009)
11. Steinman, D.A., Vorp, D.A., Ethier, C.R.: Computational modeling of arterial biomechanics: insights into pathogenesis and treatment of vascular disease. J. Vasc. Surg. **37**(5), 1118–1128 (2003)
12. Wada, S., Karino, T.: Computational study on LDL transfer from flowing blood to arterial walls. In: Yamaguchi, T. (ed.) Clinical Application of Computational Mechanics to the Cardiovascular System, pp. 157–173. Springer, Japan (2000)
13. Caro, C.G.: Discovery of the role of wall shear in atherosclerosis. Arterioscler. Thromb. Vasc. Biol. **29**(2), 158–161 (2009)
14. Peiffer, V., Sherwin, S.J., Weinberg, P.D.: Does low and oscillatory wall shear stress correlate spatially with early atherosclerosis? A systematic review. Cardiovasc. Res. **99**(2), 242–250 (2013)
15. Bazilevs, Y., Michler, C., Calo, V.M., Hughes, T.J.R.: Isogeometric variational multiscale modeling of wall-bounded turbulent flows with weakly enforced boundary conditions on unstretched meshes. Comput. Methods Appl. Mech. Eng. **199**(13–16), 780–790 (2010)
16. Hughes, T.J.R., Cottrell, J.A., Bazilevs, Y.: Isogeometric analysis: CAD, finite elements, NURBS, exact geometry and mesh refinement. Comput. Methods Appl. Mech. Eng. **194**(39–41), 4135–4195 (2005)
17. Brunner, G., Yang, E.Y., Kumar, A., Sun, W., Virani, S.S., Negi, S.I., et al.: The effect of lipid modification on peripheral artery disease after endovascular intervention trial (ELIMIT). Atherosclerosis **231**(2), 371–377 (2013)
18. Zhang, Y., Bazilevs, Y., Goswami, S., Bajaj, C.L., Hughes, T.J.R.: Patient-specific vascular NURBS modeling for isogeometric analysis of blood flow. Comput. Methods Appl. Mech. Eng. **196**(29–30), 2943–2959 (2007)
19. Schneider, C.A., Rasband, W.S., Eliceiri, K.W., Schindelin, J., Arganda-Carreras, I., Frise, E., et al.: NIH image to imageJ: 25 years of image analysis. Nat. Methods. **9**(7), 671–675 (2012)

20. Tagliasacchi, A., Alhashim, I., Olson, M., Zhang, H. (eds.): Mean curvature skeletons. Comput. Graph. Forum (2012) Wiley Online Library
21. Hossain, S.S., Zhang, Y., Liang, X., Hussain, F., Ferrari, M., Hughes, T.J.R., et al.: In silico vascular modeling for personalized nanoparticle delivery. Nanomedicine **8**(3), 343–357 (2012)
22. Hossain, S.S., Kopacz, A.M., Zhang, Y., Lee, S.-Y., Lee, T.-R., Ferrari, M., et al.: Multiscale modeling for the vascular transport of nanoparticles. In: Espinosa, H.D., Bao, G. (eds.) Nano and Cell Mechanics, pp. 437–459. Wiley, Oxford (2013)
23. Hossain, S., Hossainy, S., Bazilevs, Y., Calo, V., Hughes, T.: Mathematical modeling of coupled drug and drug-encapsulated nanoparticle transport in patient-specific coronary artery walls. Comput. Mech. **49**(2), 213–242 (2012)
24. Hossain, S.S.: Mathematical modeling of coupled drug and drug-encapsulated nanoparticle transport in patient-specific coronary artery walls. Dissertation, University of Texas at Austin, Austin (2009)
25. Olgac, U., Poulikakos, D., Saur, S.C., Alkadhi, H., Kurtcuoglu, V.: Patient-specific three-dimensional simulation of LDL accumulation in a human left coronary artery in its healthy and atherosclerotic states. Am. J. Physiol. Heart Circ. Physiol. **296**(6), H1969–H1982 (2009)
26. Stangeby, D.K., Ethier, C.R.: Computational analysis of coupled blood-wall arterial LDL transport. J. Biomech. Eng. **124**(1), 1–8 (2002)
27. Decuzzi, P., Ferrari, M.: The adhesive strength of non-spherical particles mediated by specific interactions. Biomaterials **27**(30), 5307–5314 (2006)
28. Hossain, S.S., Hughes, T.J.R., Decuzzi, P.: Vascular deposition patterns for catheter-injected nanoparticles in an inflamed patient-specific arterial tree. Biomech. Model. Mechanobiol. **13**(3), 585–597 (2013)
29. Bazilevs, Y., Calo, V.M., Zhang, Y., Hughes, T.J.R.: Isogeometric fluid–structure interaction analysis with applications to arterial blood flow. Comput. Mech. **38**(4), 310–322 (2006)
30. Bazilevs, Y., Calo, V.M., Cottrell, J.A., Hughes, T.J.R., Reali, A., Scovazzi, G.: Variational multiscale residual-based turbulence modeling for large eddy simulation of incompressible flows. Comput. Methods Appl. Mech. Eng. **197**(1–4), 173–201 (2007)
31. Chung, J., Hulbert, G.M.: A time integration algorithm for structural dynamics with improved numerical dissipation: the generalized-method. J. Appl. Mech. **60**, 371–375 (1993)
32. Jansen, K.E., Whiting, C.H., Hulbert, G.M.: A generalized-[alpha] method for integrating the filtered Navier-Stokes equations with a stabilized finite element method. Comput. Methods Appl. Mech. Eng. **190**(3–4), 305–319 (2000)
33. Bazilevs, Y., da Veiga, L.B., Cottrell, J.A., Hughes, T.J.R., Sangalli, G.: Isogeometric analysis: approximation, stability and error estimates for h-refined meshes. Math. Models Methods Appl. Sci. **16**(7), 1031–1090 (2006)
34. Cottrell, J.A., Hughes, T.J.R., Reali, A.: Studies of refinement and continuity in isogeometric structural analysis. Comput. Methods Appl. Mech. Eng. **196**(41–44), 4160–4183 (2007)
35. Hossain, S.S., Zhang, Y., Fu, X., Brunner, G., Singh, J., Hughes, T.J., et al.: Magnetic resonance imaging-based computational modelling of blood flow and nanomedicine deposition in patients with peripheral arterial disease. J. R. Soc. Interface **12**(106), 20150001 (2015)
36. Tsou, J.K., Gower, R.M., Ting, H.J., Schaff, U.Y., Insana, M.F., Passerini, A.G., et al.: Spatial regulation of inflammation by human aortic endothelial cells in a linear gradient of shear stress. Microcirculation **15**(4), 311–323 (2008)

Part VII
Fluid–Structure Interaction

Modal Analysis of Liquid–Structure Interaction

Roger Ohayon and Jean-Sébastien Schotté

Abstract The computation of the linear vibrations of structures partially filled with liquids is of prime importance in various industries such as aerospace, naval, civil and nuclear engineering. Here, a formulation is proposed for the modal analysis of incompressible liquids in elastic tanks taking into account sloshing, pressurization and elastogravity effects obtained through an adapted procedure of linearization. A corresponding finite element formulation is then presented.

1 Introduction

The case of structures partially filled with liquids (propellants, cooling liquids, liquid natural gas, etc.) is very common and many industrial domains are interested in this issue especially in the transport and aerospace industries (for instance, for liquid propelled launchers and satellites). Computational aspects of fluid–structure interaction and sloshing (with or without surface tension effects) may be found in [1, 4–7, 9, 13, 15, 16, 22]. We are here interested in the linear vibrational response of elastic structures partially filled with a liquid. As we are primarily concerned with the low frequency domain, the liquid may be considered as inviscid and incompressible (for compressibility effects, we refer to [13–15]). As a result, the liquid small motions are irrotational. This constraint can be taken into account through various procedures [2, 8]. Using a scalar description for the fluid, alternative symmetric variational formulations have been derived [10, 13].

R. Ohayon (✉)
Structural Mechanics and Coupled System Laboratory, Conservatoire National des Arts et Métiers, Paris, France
e-mail: roger.ohayon@cnam.fr

J.-S. Schotté
Aeroelasticity and Structural Dynamics Department, ONERA the French Aerospace Lab, Chatillon, France
e-mail: schotte@onera.fr

Y. Bazilevs, K. Takizawa (eds.), *Advances in Computational Fluid-Structure Interaction and Flow Simulation*, Modeling and Simulation in Science, Engineering and Technology, DOI 10.1007/978-3-319-40827-9_33

In the present work, through a specific derivation of the linearized equations of the coupled system, particular phenomena, such as tank ullage gas pressure and elastogravity stiffness effects, are analyzed [17, 18, 20]. Firstly, the nonlinear equations are posed, followed by an adapted linearization procedure. Secondly a finite element discretization, leading to a symmetric matrix system, is presented through an appropriate variational formulation.

2 Boundary Value Equations in the Deformed Configuration

We suppose that the referential is a Galilean frame. Thermo-mechanical coupling is not considered here, therefore all mechanical parameters of the system are supposed given at a nominal temperature and are considered constant during the observation time. The dynamic state of this coupled problem is characterized by the structure displacement field $\mathbf{U}^S(\mathbf{x}, t)$, the fluid velocity field $\mathbf{v}^F(\mathbf{x}, t)$, the fluid pressure field $P(\mathbf{x}, t)$, and the tank ullage gas pressure $P^G(\mathbf{x}, t)$, which are the unknown fields of the problem. The description of the system is given in Fig. 1. Each domain will be designated with a ' when it refers to its time-dependent position.

2.1 *Fluid Boundary Value Problem*

We consider an inviscid heavy fluid with a free surface Γ inside a deformable structure. The tank ullage is filled by an inert pressurized gas (no phase or chemical transformations are taken into account). The local equations are written on the deformed configuration, denoted by'.

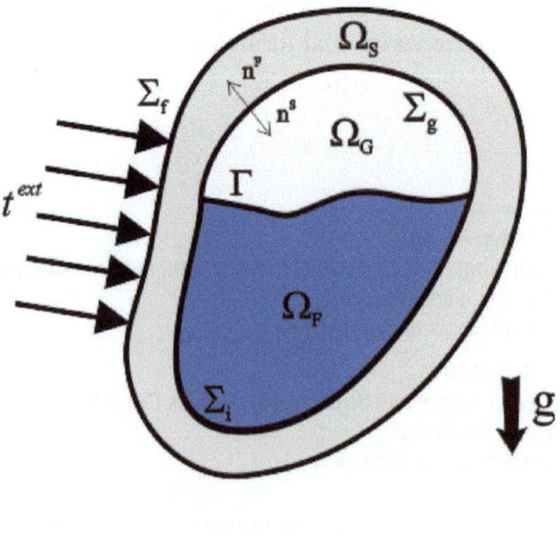

Fig. 1 Structure partially filled with liquid

$$\rho^F \frac{d\mathbf{v}^F}{dt} = -\nabla_{\mathbf{x}} P + \rho^F \mathbf{g} \qquad \text{in } \Omega'_F \tag{1a}$$

$$\frac{d\rho^F}{dt} + \rho^F \nabla_{\mathbf{x}}.\mathbf{v}^F = 0 \qquad \text{in } \Omega'_F \tag{1b}$$

$$P = P^G \qquad \text{on } \Gamma' \tag{1c}$$

$$\mathbf{v}^F.\mathbf{n}'_F + \mathbf{v}^S.\mathbf{n}'_S = 0 \qquad \text{on } \Sigma'_i \tag{1d}$$

Equation (1a) is the momentum balance equation of the inviscid heavy fluid and equation (1b) expresses the mass conservation of the fluid (continuity equation). Equation (1c) is the equilibrium equation of the fluid free surface when capillarity forces are neglected. Equation (1d) yields the sliding condition on the structure wall for an inviscid fluid. This system is completed with Cauchy initial conditions.

2.2 Structure Boundary Value Problem

The structure is supposed elastic (**H** denotes the tensor of elastic coefficients) and submitted to external surface forces, including the external and internal pressures.

$$\text{Div}_{\mathbf{x}}\boldsymbol{\sigma} + \rho^S \mathbf{g} = \rho^S \mathbf{a} \qquad \text{in } \Omega'_S \tag{2a}$$

$$\boldsymbol{\sigma} = \mathbf{H}\,\boldsymbol{\varepsilon} \qquad \text{in } \Omega'_S \tag{2b}$$

$$\boldsymbol{\varepsilon} = \frac{1}{2}\left(\mathbf{D}_{\mathbf{x}}(\mathbf{U}^S) + {}^t\mathbf{D}_{\mathbf{x}}(\mathbf{U}^S)\right) \qquad \text{in } \Omega'_S \tag{2c}$$

$$\boldsymbol{\sigma}\,\mathbf{n}'_S = \mathbf{t}^{\text{ext}} - P^{\text{ext}}\,\mathbf{n}'_S \qquad \text{on } \Sigma'_f \tag{2d}$$

$$\boldsymbol{\sigma}\,\mathbf{n}'_S = -P^G\,\mathbf{n}'_S \qquad \text{on } \Sigma'_g \tag{2e}$$

$$\boldsymbol{\sigma}\,\mathbf{n}'_S = -P\,\mathbf{n}'_S \qquad \text{on } \Sigma'_i \tag{2f}$$

Equation (2a) is the local balance of momentum (in a fixed frame of reference) where **a** is the structure acceleration. Equations (2b) and (2c) yield the classical definition of the Cauchy's stress tensor ff and strain tensor " ($\mathbf{D}_{\mathbf{x}}$ denotes the gradient operator of a vector), and their relation through Hooke's law. In equation (2d), P^{ext} is the external pressure (the atmospheric pressure, for instance) and $\mathbf{t}^{\text{ext}}$ are additional local surface loads (thrust or aerodynamic forces, for instance).

2.3 Local Equations in the Tank Ullage Gas

The tank ullage being supposed to be pressurized by a gas, a simple gas behavior model will be used to represent its effect on the dynamics of the fluid–structure system. As the acoustic phenomena in the gas enclosed in the tank ullage are generally decoupled from the low frequency fluid–structure vibrations which are the frequency range of interest of the present paper, the acoustic field in the gas will be indeed represented by a uniform value $P^G(t)$ in Ω_G and the isentropic law of transformation of a perfect gas will be used [12, 18] :

$$P^G|\Omega'_G|^\gamma = P^G_0|\Omega_G|^\gamma \qquad \text{in}\,\Omega'_G \tag{3}$$

where γ is the heat capacity ratio of the gas.

3 Equilibrium Configuration for the Coupled System

The geometry of the reference state will be described by the variable $\mathbf{X}$. In the reference configuration, the structure and fluid are supposed at rest. The local equations defining this reference state are then :

$$\mathbf{v}_0^F = \mathbf{0} \qquad \text{in } \Omega_F \tag{4a}$$

$$\nabla_{\mathbf{X}} P_0 = \rho_0^F\,\mathbf{g} \qquad \text{in } \Omega_F \tag{4b}$$

$$P_0 = P_0^G \qquad \text{on } \Gamma \tag{4c}$$

$$\mathbf{v}_0^S = \mathbf{0} \qquad \text{in } \Omega_S \tag{4d}$$

$$\text{Div}_{\mathbf{X}}\boldsymbol{\sigma}_0 + \rho_0^S\,\mathbf{g} = \mathbf{0} \qquad \text{in } \Omega_S \tag{4e}$$

$$\boldsymbol{\sigma}_0 = \mathbf{H}\,\boldsymbol{\varepsilon}_{\mathbf{X}}(\mathbf{u}_0^S) \qquad \text{in } \Omega_S \tag{4f}$$

$$\boldsymbol{\varepsilon}_{\mathbf{X}}(\mathbf{u}_0^S) = \frac{1}{2}\left(\mathbf{D}_{\mathbf{X}}(\mathbf{u}_0^S) + {}^t\mathbf{D}_{\mathbf{X}}(\mathbf{u}_0^S)\right) \qquad \text{in } \Omega_S \tag{4g}$$

$$\boldsymbol{\sigma}_0\,\mathbf{n}_S = \mathbf{t}_0^{\text{ext}} - P^{\text{ext}}\mathbf{n}_S \qquad \text{on } \Sigma_f \tag{4h}$$

$$\boldsymbol{\sigma}_0\,\mathbf{n}_S = -P_0^G\mathbf{n}_S \qquad \text{on } \Sigma_g \tag{4i}$$

$$\boldsymbol{\sigma}_0\,\mathbf{n}_S = -P_0\mathbf{n}_S \qquad \text{on } \Sigma_i \tag{4j}$$

where $\mathbf{t}_0^{\text{ext}}$ is the stationary part of the external loads. The external pressure P^{ext} is supposed to be a constant field. To solve this problem, a nonlinear iterative procedure is generally needed since these equations are written on domains whose position depends on the unknowns $(\mathbf{u}_0^S, \mathbf{u}_0^F)$ [21].

4 Linearized Local Equations

The solution of the static problem (4) is supposed known and we are interested here in the small amplitude response of the coupled system around this static reference state (described by the variable $\mathbf{X}$). Using the procedure developed in [20], a first order Taylor expansion of the nonlinear dynamic equations (1,2,3) around the equilibrium position yields the expression of the linearized equations. In the following, $\delta[h](\mathbf{u})$ will denote the derivative of a function h with respect to the displacement $\mathbf{U}$, at the static position $\mathbf{U} = \mathbf{0}$ and in direction $\mathbf{u}$.

4.1 Linearization of the Fluid Equations

To linearize equation (1a), we first need to differentiate the gradient term $\nabla_{\mathbf{x}} P$ which is a spatial derivative with respect to the coordinates $\mathbf{x}$ of the deformed fluid domain, depending, by definition, on the displacement $\mathbf{U}$ since $\mathbf{x} = \mathbf{X} + \mathbf{U}$. Using the following relation :

$$\delta[\nabla_{\mathbf{x}} P](\mathbf{u}^F) = \nabla_{\mathbf{X}} \delta[P](\mathbf{u}^F) - \nabla_{\mathbf{X}}\left(\nabla_{\mathbf{X}} P_0 . \mathbf{u}^F\right) \tag{5}$$

and the value of $\nabla_{\mathbf{X}} P_0$ given by (4b), we finally obtain

$$\delta[\nabla_{\mathbf{x}} P](\mathbf{u}^F) = \nabla_{\mathbf{X}} p_{\mathscr{L}}(\mathbf{u}^F) - \nabla_{\mathbf{X}}(\rho_0^F \, \mathbf{g}.\mathbf{u}^F) \tag{6}$$

where $p_{\mathscr{L}}(\mathbf{u}^F) = \delta[P](\mathbf{u}^F)$ is the Lagrangian pressure fluctuation in the liquid, i.e., the variation of the pressure observed when following a fluid particle.

Using the classical linearization of the acceleration term (the convection term disappears) and the fact that the Lagrangian density fluctuation $\delta[\rho^F](\mathbf{u}^F)$ in an incompressible liquid is null, the linearized version of Euler's equation (1a) can be written:

$$\rho_0^F \frac{\partial^2 \mathbf{u}^F}{\partial t^2} = \nabla_{\mathbf{X}}\left(\rho_0^F \, \mathbf{g}.\mathbf{u}^F - p_{\mathscr{L}}(\mathbf{u}^F)\right) \tag{7}$$

Using the classical expression of the Lagrangian derivative, the mass conservation equation (1b) can be written

$$\frac{\partial \rho^F}{\partial t} + \mathbf{v}^F . \nabla_{\mathbf{x}} \rho^F + \rho^F \, \nabla_{\mathbf{x}} . \mathbf{v}^F = 0 \tag{8}$$

The linearization of this equation can be achieved with the same approach as previously. Considering that the liquid is at rest in its equilibrium position ($\mathbf{v}_0^F = \mathbf{0}$), that it is supposed incompressible ($\delta[\rho^F](\mathbf{u}^F) = 0$) and that its density in the reference state is homogeneous ($\nabla_{\mathbf{X}} \rho_0^F = 0$), then the linearized form of the mass conservation equation (8) can be written

$$\frac{\partial}{\partial t}\left(\rho_0^F \, \nabla_{\mathbf{X}} . \mathbf{u}^F\right) = 0 \tag{9}$$

The same approach applied to equation (1d) yields the following linearized equation:

$$\frac{\partial}{\partial t}\left(\mathbf{u}^F.\mathbf{n}_F + \mathbf{u}^S.\mathbf{n}_S\right) = 0 \tag{10}$$

We deduce therefore that the continuity of the normal displacement is satisfied at any time t as soon as it is satisfied by the initial conditions. The linearized equations of the fluid around the reference state are finally ($\forall t \in \mathbb{R}^+$):

$$\rho_0^F \frac{\partial^2 \mathbf{u}^F}{\partial t^2} = \nabla_{\mathbf{X}}\left(\rho_0^F \, \mathbf{g}.\mathbf{u}^F - p_{\mathscr{L}}(\mathbf{u}^F)\right) \quad \text{in } \Omega_F \tag{11a}$$

$$\nabla_{\mathbf{X}}.\mathbf{u}^F = 0 \quad \text{in } \Omega_F \tag{11b}$$

$$p_{\mathscr{L}} = p^G \quad \text{on } \Gamma \tag{11c}$$

$$\mathbf{u}^F.\mathbf{n}_F + \mathbf{u}^S.\mathbf{n}_S = 0 \quad \text{on } \Sigma_i \tag{11d}$$

where $p^G = \delta[P^G](\mathbf{u}^F, \mathbf{u}^S)$ is the pressure fluctuation in the gas when the fluid and structure are deformed. To these equations, we add Cauchy initial conditions. Equation (9) has been simplified by supposing that the initial conditions satisfy the mass conservation equation.

4.2 *Linearization of the Structural Equations*

The local equations of the structure (2) are now linearized using the same approach. However, to simplify the derivation of the equations, before the differentiation, we use a Lagrangian transformation to write those equations in the reference configuration, as follows:

$$\mathrm{Div}_{\mathbf{X}}(\mathbf{FS}) = \rho_0^S \, \frac{\mathrm{d}^2 \mathbf{U}^S}{\mathrm{d}t^2} - \rho_0^S \, \mathbf{g} \quad \text{in } \Omega_S \tag{12a}$$

$$\mathbf{S}(\mathbf{U}^S) = \mathbf{HE}(\mathbf{u}_0^S + \mathbf{U}^S) \tag{12b}$$

$$\mathbf{FS}\,\mathbf{n}_S = \mathbf{t}^{\mathrm{ext}} \det(\mathbf{F}) \|{}^{\mathrm{t}}\mathbf{F}^{-1}\mathbf{n}_S\| - P^{\mathrm{ext}} \det(\mathbf{F})\, {}^{\mathrm{t}}\mathbf{F}^{-1}\mathbf{n}_S \quad \text{on } \Sigma_f \tag{12c}$$

$$\mathbf{FS}\,\mathbf{n}_S = -P^G \det(\mathbf{F})\, {}^{\mathrm{t}}\mathbf{F}^{-1}\mathbf{n}_S \quad \text{on } \Sigma_g \tag{12d}$$

$$\mathbf{FS}\,\mathbf{n}_S = -P \det(\mathbf{F})\, {}^{\mathrm{t}}\mathbf{F}^{-1}\mathbf{n}_S \quad \text{on } \Sigma_i \tag{12e}$$

In these equations, $\mathbf{E}$ is the (nonlinear) Green-Lagrange strain tensor and $\mathbf{F}$ is the transformation gradient tensor, defined by $\mathbf{F} = \mathbf{D}_{\mathbf{X}}(\mathbf{x})$. The linearization of (12a) is classical and gives

$$\mathrm{Div}_{\mathbf{X}}\left(\mathbf{D}_{\mathbf{X}}(\mathbf{u}^S)\boldsymbol{\sigma}_0 + \mathbf{H}\boldsymbol{\varepsilon}_0(\mathbf{u}^S)\right) = \rho_0^S \, \frac{\partial^2 \mathbf{u}^S}{\partial t^2} \tag{13}$$

where

$$\boldsymbol{\sigma}_0 = \mathbf{H}\,\mathbf{E}(\mathbf{u}_0^S) \tag{14}$$

$$\boldsymbol{\varepsilon}_0(\mathbf{u}^S) = \delta[\mathbf{E}](\mathbf{u}^S) \tag{15}$$

For the differentiation of equation (12e), the following term has to be computed:

$$\delta[P\det(\mathbf{F})\,{}^{\mathrm{t}}\mathbf{F}^{-1}](\mathbf{u}^S) = \delta[P](\mathbf{u}^S)\,\mathbf{I} + P_0\,\delta[\det(\mathbf{F})\,{}^{\mathrm{t}}\mathbf{F}^{-1}](\mathbf{u}^S) \tag{16}$$

In the first term of this expression, we recognize a Lagrangian fluctuation of the pressure on the fluid–structure interface which is denoted by $p^S_{\mathscr{L}}(\mathbf{u}^S)$. Let us remark that this term is defined as a pressure fluctuation related to a structural point and then is different from the Lagrangian pressure fluctuation $p_{\mathscr{L}}(\mathbf{u}^F)$ defined previously in equation (6) for a fluid particle. This difficulty is a consequence of the sliding of the inviscid liquid on the structure wall.

To explicit the second term of (16), we use the following expressions:

$$\delta[\det(\mathbf{F})](\mathbf{u}) = \mathrm{Tr}\left[\mathbf{D}_{\mathbf{X}}\mathbf{u}\right]\mathbf{I} = (\nabla_{\mathbf{X}}.\mathbf{u}^S)\,\mathbf{I} \tag{17}$$

$$\delta[{}^{\mathrm{t}}\mathbf{F}^{-1}](\mathbf{u}) = -{}^{\mathrm{t}}\mathbf{D}_{\mathbf{X}}\mathbf{u} \tag{18}$$

Then, we obtain

$$\delta[\det(\mathbf{F})\,{}^{\mathrm{t}}\mathbf{F}^{-1}](\mathbf{u}^S) = (\nabla_{\mathbf{X}}.\mathbf{u}^S)\,\mathbf{I} - {}^{\mathrm{t}}\mathbf{D}_{\mathbf{X}}(\mathbf{u}^S) \tag{19}$$

Using this relation in (16) yields the linearized expression of (12e):

$$\left(\mathbf{D}_{\mathbf{X}}(\mathbf{u}^S)\boldsymbol{\sigma}_0 + \mathbf{H}\,\boldsymbol{\varepsilon}_0(\mathbf{u}^S)\right)\mathbf{n}_S = -p^S_{\mathscr{L}}(\mathbf{u}^S)\,\mathbf{n}_S - P_0\,\boldsymbol{\tau}(\mathbf{u}^S) \tag{20}$$

where the vector $\boldsymbol{\tau}$ is defined as

$$\boldsymbol{\tau}(\mathbf{u}^S) = (\nabla_{\mathbf{X}}.\mathbf{u}^S)\mathbf{n}_S - {}^{\mathrm{t}}\mathbf{D}_{\mathbf{X}}(\mathbf{u}^S)\,\mathbf{n}_S \tag{21}$$

The comparison of this expression with the following one, that we would have obtained by a direct differentiation of the right-hand term of equation (2f):

$$\delta[-P\mathbf{n}'_S\,\mathrm{d}\Sigma'](\mathbf{u}^S) = -p^S_{\mathscr{L}}(\mathbf{u}^S)\mathbf{n}_S\,\mathrm{d}\Sigma - P_0\,\delta[\mathbf{n}'_S\,\mathrm{d}\Sigma'](\mathbf{u}^S) \tag{22}$$

shows that $\boldsymbol{\tau}$ can be interpreted as the linearization of the normal vector rotation:

$$\boldsymbol{\tau}(\mathbf{u}^S)\,\mathrm{d}\Sigma = \delta[\mathbf{n}'_S\,\mathrm{d}\Sigma'](\mathbf{u}^S) \tag{23}$$

The same treatment can be applied to equation (12d). However, for equation (12c), we have to deal with the external force term $\mathbf{t}^{\mathrm{ext}}$. The differentiation of this term yields

$$\delta\Big[\mathbf{t}^{\mathrm{ext}}\det(\mathbf{F})\|{}^{\mathrm{t}}\mathbf{F}^{-1}\mathbf{n}_S\|\Big](\mathbf{u}^S) = \delta[\mathbf{t}^{\mathrm{ext}}](\mathbf{u}^S) + \mathbf{t}_0^{\mathrm{ext}}\delta\Big[\det(\mathbf{F})\|{}^{\mathrm{t}}\mathbf{F}^{-1}\mathbf{n}_S\|\Big](\mathbf{u}^S) \tag{24}$$

A direct computation of the second term can be avoided if we remember that

$$\delta\Big[\det(\mathbf{F})\|{}^{\mathrm{t}}\mathbf{F}^{-1}\mathbf{n}_S\|\Big](\mathbf{u}^S)\,\mathrm{d}S = \delta[\mathrm{d}S'](\mathbf{u}^S) \tag{25}$$

where $\mathrm{d}S$ and $\mathrm{d}S'$ are, respectively, the elementary surfaces on the reference and actual geometries. Since, from (23), we have $\boldsymbol{\tau}(\mathbf{u}^S)\mathrm{d}S = \delta[\mathbf{n}'\,\mathrm{d}S'](\mathbf{u}^S) = \delta[\mathbf{n}'](\mathbf{u}^S)\,\mathrm{d}S + \mathbf{n}\,\delta[\mathrm{d}S'](\mathbf{u}^S)$ and $\mathbf{n}.\delta[\mathbf{n}'](\mathbf{u}^S) = 0$, we can deduce from (21) that

$$\delta\Big[\det(\mathbf{F})\|{}^{\mathrm{t}}\mathbf{F}^{-1}\mathbf{n}_S\|\Big](\mathbf{u}^S) = \mathbf{n}_S.\boldsymbol{\tau}(\mathbf{u}^S) = \nabla_{\mathbf{X}}.\mathbf{u}^S - \mathbf{n}_S.\mathbf{D}_{\mathbf{X}}(\mathbf{u}^S)\mathbf{n}_S \tag{26}$$

Finally, the linearized expression of (12c) can be written

$$\Big(\mathbf{D}_{\mathbf{X}}(\mathbf{u}^S)\boldsymbol{\sigma}_0 + \mathbf{H}\boldsymbol{\varepsilon}_0(\mathbf{u}^S)\Big)\mathbf{n}_S = \mathbf{t}^d + \tilde{\mathbf{t}}(\mathbf{u}^S) + \mathbf{t}_0^{\mathrm{ext}}\mathbf{n}_S.\boldsymbol{\tau}(\mathbf{u}^S) - P^{\mathrm{ext}}\,\boldsymbol{\tau}(\mathbf{u}^S) \tag{27}$$

where we have distinguished two terms in the variation of the external surface loads $\delta[\mathbf{t}^{\mathrm{ext}}](\mathbf{u}^S)$: a dynamic time-dependent part $\mathbf{t}^d$ and a position-dependent part $\tilde{\mathbf{t}}(\mathbf{u}^S)$. Let us verify that if the external surface load $\mathbf{t}^{\mathrm{ext}}$ is a pressure, i.e., $\mathbf{t}^{\mathrm{ext}} = -P_{\mathrm{t}}\mathbf{n}_S$, then equations (24) and (25) lead to coherent expressions of (27) and (20):

$$\begin{aligned}\delta[\mathbf{t}^{\mathrm{ext}}]\,\mathrm{d}\Sigma + \mathbf{t}_0^{\mathrm{ext}}\,\mathbf{n}_S.\boldsymbol{\tau}\,\mathrm{d}\Sigma &= \delta[-P_{\mathrm{t}}\mathbf{n}_S']\,\mathrm{d}\Sigma - P_{\mathrm{t}0}\,\mathbf{n}_S\,\delta[\mathrm{d}\Sigma'] \\ &= -\delta[P_{\mathrm{t}}]\,\mathbf{n}_S\,\mathrm{d}\Sigma - P_{\mathrm{t}0}\,\delta[\mathbf{n}_S'\,\mathrm{d}\Sigma']\end{aligned} \tag{28}$$

By using (23), we denote then that expressions (20) and (27) are equivalent.

The linearized equations of the structure established in this section are summarized hereafter:

$$\mathrm{Div}_{\mathbf{X}}\mathbf{s}(\mathbf{u}^S) = \rho_0^S\,\frac{\partial^2\mathbf{u}^S}{\partial t^2} \qquad \text{in } \Omega_S \tag{29a}$$

$$\mathbf{s}(\mathbf{u}^S) = \mathbf{D}_{\mathbf{X}}(\mathbf{u}^S)\,\boldsymbol{\sigma}_0 + \mathbf{H}\,\boldsymbol{\varepsilon}_0(\mathbf{u}^S) \qquad \text{in } \Omega_S \tag{29b}$$

$$\mathbf{s}\,\mathbf{n}_S = \mathbf{t}^d + \tilde{\mathbf{t}}(\mathbf{u}^S) + \mathbf{t}_0^{\mathrm{ext}}\mathbf{n}_S.\boldsymbol{\tau}(\mathbf{u}^S) - P^{\mathrm{ext}}\,\boldsymbol{\tau}(\mathbf{u}^S) \qquad \text{on } \Sigma_f \tag{29c}$$

$$\mathbf{s}\,\mathbf{n}_S = -p^G(\mathbf{u}^S, \mathbf{u}^F)\,\mathbf{n}_S - P_0^G\,\boldsymbol{\tau}(\mathbf{u}^S) \qquad \text{on } \Sigma_g \tag{29d}$$

$$\mathbf{s}\,\mathbf{n}_S = -p_{\mathscr{L}}^S(\mathbf{u}^S)\,\mathbf{n}_S - P_0\,\boldsymbol{\tau}(\mathbf{u}^S) \qquad \text{on } \Sigma_i \tag{29e}$$

To these equations, Cauchy initial conditions have to be added.

4.3 Linearization of the Gas Equation

The differentiation of the gas equation (3) yields

$$\delta[P^G](\mathbf{u}^S, \mathbf{u}^F)\,|\Omega_G|^\gamma + P_0^G\gamma\,|\Omega_G|^{\gamma-1}\,\delta\Big[|\Omega_G'|\Big](\mathbf{u}^S, \mathbf{u}^F) = 0 \tag{30}$$

Since the gas volume $|\Omega'_G|$ is defined by

$$|\Omega'_G| = \int_{\Omega'_G} \mathrm{d}\Omega' = \int_{\Omega_G} \det(\mathbf{F})\, \mathrm{d}\Omega \tag{31}$$

Using the relation (17) and Stokes theorem, we obtain

$$\delta\Big[|\Omega'_G|\Big](\mathbf{u}^S, \mathbf{u}^F) = \int_{\Omega_G} \nabla_{\mathbf{X}}.\mathbf{u}\, \mathrm{d}\Omega = -\int_{\Sigma_g} \mathbf{u}^S.\mathbf{n}_S\, \mathrm{d}\Sigma - \int_{\Gamma} \mathbf{u}^F.\mathbf{n}_F\, \mathrm{d}\Gamma \tag{32}$$

Finally the linearized equation of the gas can be written

$$p^G(\mathbf{u}^S, \mathbf{u}^F) = \frac{p_0^G \gamma}{|\Omega_G|} \left(\int_{\Sigma_g} \mathbf{u}^S.\mathbf{n}_S\, \mathrm{d}\Sigma + \int_{\Gamma} \mathbf{u}^F.\mathbf{n}_F\, \mathrm{d}\Gamma \right) \quad \text{in } \Omega_G \tag{33}$$

Equations (11), (29), and (33) form the coupled problem we want to solve. In these equations, all the spatial derivatives are carried out with respect to the coordinate vector $\mathbf{X}$ and then, for sake of convenience, this subscript $_\mathbf{X}$ will be omitted in the following.

5 Fluid–Structure Problem Solving

5.1 Potential Fluid Variational Formulation

Taking the curl of equation (11a) shows that the small motions $\mathbf{u}^F$ of an incompressible fluid are irrotational and then can be represented by the gradient of a displacement potential φ defined by

$$\nabla \wedge \mathbf{u}^F = 0 \quad \Longrightarrow \quad \mathbf{u}^F = \nabla\varphi \tag{34}$$

However, this relation does not ensure the uniqueness of φ. Therefore, it is necessary to specify a unicity condition on φ, generically written $\ell(\varphi) = 0$, where ℓ is a linear form such that $\ell(1) \neq 0$. This unicity condition will be chosen explicitly in the following.

This new variable φ can be substituted to $\mathbf{u}^F$ in the fluid equations, in particular in (11a) which yields, after integration with respect to $\mathbf{X}$:

$$p_{\mathscr{L}} = \rho_0^F \mathbf{g}.\nabla\varphi - \rho_0^F \frac{\partial^2 \varphi}{\partial t^2} + \pi \tag{35}$$

where π is an integration constant which represents a uniform pressure in the fluid whose value depends on the choice of the unicity condition $\ell(\varphi)$. The expression (35) of $p_{\mathscr{L}}$ as a function of φ allows to eliminate $p_{\mathscr{L}}$ from the equations. However, before establishing the variational formulation, we choose to introduce a scalar variable η to describe the normal component of the fluid free surface displacement:

$$\eta = \mathbf{u}^F.\mathbf{n}_F = \nabla\varphi.\mathbf{n}_F \quad \text{on } \Gamma \tag{36}$$

Since $\mathbf{n}_F = \mathbf{i}_z$ on Γ, the local equations of the boundary problem (11) are now written for an incompressible fluid in terms of $(\varphi, \eta, \pi, p^G)$ as follows:

$$\Delta\varphi = 0 \quad \text{in } \Omega_F \tag{37a}$$

$$\mathbf{n}_F.\nabla\varphi = \mathbf{u}^S.\mathbf{n}_F \quad \text{on } \Sigma_i \tag{37b}$$

$$\mathbf{n}_F.\nabla\varphi = \eta \quad \text{on } \Gamma \tag{37c}$$

$$\rho_0^F g\, \eta = -\rho_0^F \frac{\partial^2\varphi}{\partial t^2} + \pi - p^G \quad \text{on } \Gamma \tag{37d}$$

$$p^G = \frac{P_0^G \gamma}{|\Omega_G|} \int_{\Sigma_i \cup \Sigma_g} \mathbf{u}^S.\mathbf{n}_S \, \mathrm{d}\Sigma \quad \text{in } \Omega_G \tag{37e}$$

$$\ell(\varphi) = 0 \tag{37f}$$

where equation (37e) has been obtained from (33) by considering the following relation given by the integration of property (11b) on the fluid domain:

$$0 = \int_{\Sigma_i} \mathbf{u}^F.\mathbf{n}_F \, \mathrm{d}\Sigma + \int_{\Gamma} \mathbf{u}^F.\mathbf{n}_F \, \mathrm{d}\Gamma \tag{38}$$

The equation (37e) shows that, due to the incompressibility of the fluid, p^G is now a function depending only of the tank wall deformation $\mathbf{u}^S$.

From the local equations (37a), (37b), and (37c), we can write the following variational formulation for the liquid, where C_φ^ℓ is the admissible space of solutions φ defined by (considering t as a parameter)

$$C_\varphi^\ell = \{\varphi \in \mathcal{H}^1(\Omega_F) / \ell(\varphi) = 0\} \tag{39}$$

For $\varphi \in C_\varphi^\ell$, wehave$\forall\ \delta\varphi \in C_\varphi^\ell$

$$\mathcal{F}(\varphi, \delta\varphi) - \mathcal{B}(\eta, \delta\varphi) = \mathcal{C}(\delta\varphi, \mathbf{u}^S) \tag{40}$$

$$\text{with}\mathcal{F}(\varphi, \delta\varphi) = \rho_0^F \int_{\Omega_F} \nabla\varphi.\nabla\delta\varphi \, \mathrm{d}\Omega \tag{41}$$

$$\mathcal{B}(\eta, \delta\varphi) = \rho_0^F \int_{\Gamma} \eta\, \delta\varphi \, \mathrm{d}\Gamma \tag{42}$$

$$\mathcal{C}(\delta\varphi, \mathbf{u}^S) = \rho_0^F \int_{\Sigma_i} \mathbf{u}^S.\mathbf{n}_F \, \delta\varphi \, \mathrm{d}\Sigma \tag{43}$$

In this expression, $\mathcal{F}$ is the symmetric positive semi-definite bilinear form associated with the fluid kinetic energy and $\mathcal{C}$ is the classical fluid–structure coupling operator.

We also consider the following relation equivalent to (38), which expresses the incompressibility of the fluid:

$$-\mathscr{B}(\eta, 1) = \mathscr{C}(1, \mathbf{u}^S) \tag{44}$$

The variational formulation corresponding to the local equation (37d) is written

$$\text{For}\eta \in \mathrm{C}_\eta, \text{wehave}\forall\ \delta\eta \in \mathrm{C}_\eta,$$

$$\mathscr{S}_g(\eta, \delta\eta) + \mathscr{B}\left(\delta\eta, \frac{\partial^2\varphi}{\partial t^2}\right) - \frac{\pi}{\rho_0^F}\,\mathscr{B}(\delta\eta, 1) + \frac{p^G}{\rho_0^F}\,\mathscr{B}(\delta\eta, 1) = 0 \tag{45}$$

$$\text{with } \mathscr{S}_g(\eta, \delta\eta) = \rho_0^F\, g \int_\Gamma \eta\, \delta\eta\, \mathrm{d}\Gamma \tag{46}$$

where C_η is the admissible space for η, defined as $\mathrm{C}_\eta = \left\{\eta \in \mathscr{L}^2(\Gamma)\right\}$ and $\mathscr{S}_g$ is the symmetric positive definite bilinear form associated with the fluid gravity potential energy.

A final equation is obtained from the expression (37e) of p^G:

$$\chi^G\, \frac{p^G}{\rho_0^F} - \mathscr{B}(\eta, 1) = \mathscr{C}_g(1, \mathbf{u}^S) \tag{47}$$

$$\text{with } \mathscr{C}_g(\delta\varphi, \mathbf{u}^S) = \rho_0^F \int_{\Sigma_g} \mathbf{u}^S.\mathbf{n}_S\, \delta\varphi\, \mathrm{d}\Sigma \tag{48}$$

$$\text{and}\chi^G = \frac{\rho_0^{F2}\, |\Omega_G|}{P_0^G \gamma} \tag{49}$$

5.2 *Structure Variational Formulation*

From the local equations of the structure given in (29), a variational formulation for the structure can be written, where $\mathrm{C}_u = \left\{\mathbf{u}^S \in \mathscr{H}^1(\Omega_S)^3\right\}$ denotes the admissible space (considering t as a parameter). In the following, the fluid pressure P_0 in the reference configuration will be replaced by its expression obtained from (4b) and (4c):

$$P_0(\mathbf{X}) = P_0^G + \rho_0^F\, \mathbf{g}.\mathbf{X} \tag{50}$$

where O (the origin of the position vector $\mathbf{X}$) is a point of the liquid free surface.

Then the variational principle for the structure is written

Given f^{ext}, find $\mathbf{u}^S \in C_u$ such that $\forall\ \boldsymbol{\delta}\mathbf{u} \in C_u$,

$$(\mathscr{K}_E + \mathscr{K}_G + \mathscr{K}_P + \mathscr{K}_t)(\mathbf{u}^S, \boldsymbol{\delta}\mathbf{u}) + \mathscr{M}\left(\frac{\partial^2 \mathbf{u}^S}{\partial t^2}, \boldsymbol{\delta}\mathbf{u}\right) + \frac{p^G}{\rho_0^F}\mathscr{C}_g(1, \boldsymbol{\delta}\mathbf{u}) \cdots$$

$$+ \int_{\Sigma_i} p^S_{\mathscr{L}}(\mathbf{u}^S)\, \mathbf{n}_S . \boldsymbol{\delta}\mathbf{u} + (\rho_0^F\, \mathbf{g}.\mathbf{X})\boldsymbol{\tau}(\mathbf{u}^S).\boldsymbol{\delta}\mathbf{u}\, \mathrm{d}\Sigma = f^{\text{ext}}(\boldsymbol{\delta}\mathbf{u}) \quad (51)$$

$$\text{with}\ \mathscr{K}_E(\mathbf{u}^S, \boldsymbol{\delta}\mathbf{u}) = \int_{\Omega_S} \mathbf{H}\, \boldsymbol{\varepsilon}_0(\mathbf{u}^S) : \mathbf{D}_{\mathbf{x}}(\boldsymbol{\delta}\mathbf{u})\, \mathrm{d}\Omega \quad (52)$$

$$\mathscr{K}_G(\mathbf{u}^S, \boldsymbol{\delta}\mathbf{u}) = \int_{\Omega_S} \mathbf{D}_{\mathbf{x}}(\mathbf{u}^S)\, \boldsymbol{\sigma}_0 : \mathbf{D}_{\mathbf{x}}(\boldsymbol{\delta}\mathbf{u})\, \mathrm{d}\Omega \quad (53)$$

$$\mathscr{K}_P(\mathbf{u}^S, \boldsymbol{\delta}\mathbf{u}) = \int_{\Sigma_f} P^{\text{ext}}\boldsymbol{\tau}(\mathbf{u}^S).\boldsymbol{\delta}\mathbf{u}\, \mathrm{d}\Sigma + \int_{\Sigma_g \cup \Sigma_i} P_0^G \boldsymbol{\tau}(\mathbf{u}^S).\boldsymbol{\delta}\mathbf{u}\, \mathrm{d}\Sigma \quad (54)$$

$$\mathscr{K}_t(\mathbf{u}^S, \boldsymbol{\delta}\mathbf{u}) = -\int_{\Sigma_f} \tilde{\mathbf{t}}(\mathbf{u}^S).\boldsymbol{\delta}\mathbf{u} + \mathbf{n}_S.\boldsymbol{\tau}(\mathbf{u}^S)\, \mathbf{t}_0^{\text{ext}}.\boldsymbol{\delta}\mathbf{u}\, \mathrm{d}\Sigma \quad (55)$$

$$\mathscr{M}(\mathbf{u}^S, \boldsymbol{\delta}\mathbf{u}) = \int_{\Omega_S} \rho_0^S\, \mathbf{u}^S.\boldsymbol{\delta}\mathbf{u}\, \mathrm{d}\Omega \quad (56)$$

$$f^{\text{ext}}(\boldsymbol{\delta}\mathbf{u}) = \int_{\Sigma_f} \mathbf{t}^d.\boldsymbol{\delta}\mathbf{u}\, \mathrm{d}\Sigma \quad (57)$$

where $\mathscr{K}_E$ is the elastic stiffness operator of the structure deformed in its static configuration. If these initial deformations $\mathbf{u}_0^S$ are small, $\boldsymbol{\varepsilon}_0$ tends to $\boldsymbol{\varepsilon}$ and $\mathscr{K}_E$ is then the standard symmetric positive stiffness operator. $\mathscr{M}$ is the classical symmetric positive definite operator associated with the mass of the structure and $\mathscr{K}_G$ is the symmetric operator, due to the initial prestress $\boldsymbol{\sigma}_0$ in the structure, classically known as the geometric stiffness. $\mathscr{K}_P$ is the stiffness operator arising from the follower force effect of the initial pressures applied on the structure external and internal walls (the external pressure P^{ext} and the pressurization P_0^G). $\mathscr{K}_t$ is a similar term due to the presence of a static external surface load $\mathbf{t}^{\text{ext}}$ which can also be a follower force. f^{ext} is a linear form representing the dynamic part of this load.

We need now an expression of $p^S_{\mathscr{L}}(\mathbf{u}^S)$ as a function of the unknowns $\mathbf{u}^S$ and $\mathbf{u}^F$. This expression is obtained by considering a fluid particle in contact with a structural point of the fluid interface in the reference configuration. At time t, the pressure at this structural point is $P(\mathbf{U}^S)$ and the pressure of the same fluid particle is $P(\mathbf{U}^F)$. The difference can be written as follows:

$$P(\mathbf{U}^S) - P(\mathbf{U}^F) = \nabla_{\mathbf{x}} P.(\mathbf{U}^S - \mathbf{U}^F) \quad (58)$$

By differentiating this expression at the reference configuration and in the direction $(\mathbf{u}^S, \mathbf{u}^F)$, we obtain

$$p^S_{\mathscr{L}}(\mathbf{u}^S) - p_{\mathscr{L}}(\mathbf{u}^F) = \nabla_{\mathbf{X}} P_0.(\mathbf{u}^S - \mathbf{u}^F) = \rho^F_0\, \mathbf{g}.(\mathbf{u}^S - \mathbf{u}^F) \tag{59}$$

Since the expression of $p_{\mathscr{L}}$ is given by (35), we can write the following expression:

$$p^S_{\mathscr{L}}(\mathbf{u}^S) = \rho^F_0\, \mathbf{g}.\mathbf{u}^S - \rho^F_0 \frac{\partial^2 \varphi}{\partial t^2} + \pi \tag{60}$$

By introducing this relation into the previously given variational formulation of the structure, we obtain

$$\exists\, \mathbf{u}^S \in \mathrm{C}_u \text{such that} \forall\, \boldsymbol{\delta}\mathbf{u} \in \mathrm{C}_u, \tag{61}$$

$$\mathscr{K}_T(\mathbf{u}^S, \boldsymbol{\delta}\mathbf{u}) + \frac{p^G}{\rho^F_0}\mathscr{C}_g(1, \boldsymbol{\delta}\mathbf{u}) - \frac{\pi}{\rho^F_0}\mathscr{C}(1, \boldsymbol{\delta}\mathbf{u}) + \mathscr{M}\left(\frac{\partial^2 \mathbf{u}^S}{\partial t^2}, \boldsymbol{\delta}\mathbf{u}\right) + \mathscr{C}\left(\frac{\partial^2 \varphi}{\partial t^2}, \boldsymbol{\delta}\mathbf{u}\right) = f^{\mathrm{ext}}(\boldsymbol{\delta}\mathbf{u})$$

$$\text{with}\mathscr{K}_T(\mathbf{u}^S, \boldsymbol{\delta}\mathbf{u}) = (\mathscr{K}_E + \mathscr{K}_G + \mathscr{K}_P + \mathscr{K}_t + \mathscr{K}_g)(\mathbf{u}^S, \boldsymbol{\delta}\mathbf{u}) \tag{62}$$

$$\mathscr{K}_g(\mathbf{u}^S, \boldsymbol{\delta}\mathbf{u}) = \int_{\Sigma_i} (\rho^F_0\, \mathbf{g}.\mathbf{u}^S)\, \mathbf{n}_S.\boldsymbol{\delta}\mathbf{u} + (\rho^F_0\, \mathbf{g}.\mathbf{X})\, \boldsymbol{\tau}(\mathbf{u}^S).\boldsymbol{\delta}\mathbf{u}\ \mathrm{d}\Sigma \tag{63}$$

where $\mathscr{K}_g$ is the stiffness operator related to the follower force effect of the fluid hydrostatic pressure on the structure wall (it can be shown that this operator is symmetric [18]).

5.3 Finite Element Discretization

The classical finite element method is applied to discretize the space part of the unknown fields. We consider here compatible meshes for the structure and the fluid. Since φ belongs to $\mathrm{C}^{\ell}_{\varphi}$, the condition $\ell(\varphi) = 0$ has to be taken into account in strong form. One way to do is to choose $\ell(\varphi)$ such that one component i of φ is equal to 0 ($\ell(\varphi) = \boldsymbol{\varphi}_i = 0$). Then, all the row and columns associated with these zero values are removed from the matrix system. To simplify the notations, vectors of nodal values will be denoted by the same letter as the associated continuous variable. The matrix operators associated with the bilinear and linear forms are denoted as follows:

$$\mathscr{B}(\delta\eta, \varphi) \implies {}^{\mathrm{t}}\boldsymbol{\delta\eta}\, \mathbf{B}\, \boldsymbol{\varphi} \qquad \text{and} \qquad \mathscr{B}(\delta\eta, 1) \implies {}^{\mathrm{t}}\boldsymbol{\delta\eta}\, \mathbf{b} \tag{64}$$

The fluid equation (40) shows that φ can be expressed as a function of η and $\mathbf{u}^S$. After a finite element discretization of this equation, since $\mathbf{F}$ is invertible [13], we obtain the following expression:

$$\varphi = \mathbf{F}^{-1}\,{}^{t}\mathbf{C}\,\mathbf{u}^S + \mathbf{F}^{-1}\mathbf{B}\,\eta \tag{65}$$

In the frequency domain (after a Fourier transform), denoting by ω the circular frequency, the discretization of equations (61), (45), (44), and (47), using the elimination relation (65), leads to the following symmetric matrix system:

$$\left(\begin{bmatrix} \mathbf{K}_\mathcal{T} & \mathbf{0} & -\mathbf{c} & \mathbf{c}_g \\ \mathbf{0} & \mathbf{S}_g & -\mathbf{b} & \mathbf{b} \\ -{}^{t}\mathbf{c} & -{}^{t}\mathbf{b} & \mathbf{0} & \mathbf{0} \\ {}^{t}\mathbf{c}_g & {}^{t}\mathbf{b} & \mathbf{0} & -\chi^G \end{bmatrix} - \omega^2 \begin{bmatrix} \mathbf{M}+\mathbf{C}\mathbf{F}^{-1}\,{}^{t}\mathbf{C} & \mathbf{C}\mathbf{F}^{-1}\mathbf{B} & \mathbf{0} & \mathbf{0} \\ {}^{t}\mathbf{B}\mathbf{F}^{-1}\,{}^{t}\mathbf{C} & {}^{t}\mathbf{B}\mathbf{F}^{-1}\mathbf{B} & \mathbf{0} & \mathbf{0} \\ \mathbf{0} & \mathbf{0} & \mathbf{0} & \mathbf{0} \\ \mathbf{0} & \mathbf{0} & \mathbf{0} & \mathbf{0} \end{bmatrix} \right) \begin{pmatrix} \mathbf{u}^S \\ \eta \\ \pi/\rho_0^F \\ p^G/\rho_0^F \end{pmatrix} = \begin{pmatrix} \mathbf{f}^{ext} \\ \mathbf{0} \\ 0 \\ 0 \end{pmatrix} \tag{66}$$

6 Applications

In order to compute the vibrations of a pressurized elastic structure containing an inviscid incompressible liquid with a free surface, one has to solve equation (66) in the frequency domain (and then obtain the time response by an inverse Fourier transform). To reduce the CPU-time, a modal synthesis method could be used [20]. It is based on the projection of the liquid equation onto its sloshing eigenmodes which are the liquid eigensolutions of equations (37) for a rigid motionless cavity ($\mathbf{u}^S = \mathbf{0}$). Equation (37e) shows that, if the structure is undeformed, $p^G = 0$. Then, the sloshing modes $(\eta_\alpha, \pi_\alpha)\cos(\omega_\alpha t)$ are solutions of:

$$\left(\begin{bmatrix} \mathbf{S}_g & -\mathbf{b} \\ -{}^{t}\mathbf{b} & \mathbf{0} \end{bmatrix} - \omega_\alpha^2 \begin{bmatrix} {}^{t}\mathbf{B}\mathbf{F}^{-1}\mathbf{B} & \mathbf{0} \\ \mathbf{0} & \mathbf{0} \end{bmatrix} \right) \begin{pmatrix} \eta_\alpha \\ \pi_\alpha/\rho_0^F \end{pmatrix} = \begin{pmatrix} \mathbf{0} \\ 0 \end{pmatrix} \tag{67}$$

If the matrix ${}^{t}\mathbf{B}\mathbf{F}^{-1}\mathbf{B}$ is denoted by $\mathbf{M}_\Gamma$, we deduce that the sloshing eigenmodes of the liquid are only described by their free surface normal displacement $\boldsymbol{\eta}_\alpha$ which are the solutions of the eigenvalue problem $\mathbf{S}_g\,\boldsymbol{\eta}_\alpha = \omega_\alpha^2\,\mathbf{M}_\Gamma\,\boldsymbol{\eta}_\alpha$ with ${}^{t}\mathbf{b}\,\boldsymbol{\eta}_\alpha = 0$, where $\mathbf{S}_g$ is the gravity potential energy operator, $\mathbf{M}_\Gamma$ is the mass operator of the liquid in a rigid motionless cavity, condensed on its free surface, and π has the role of a Lagrange multiplier associated with the incompressibility condition ${}^{t}\mathbf{b}\,\boldsymbol{\eta}_\alpha = 0$ (which represents the fluid volume invariance). Since $\mathbf{S}_g$ and $\mathbf{M}_\Gamma$ are symmetric, the eigenmodes $\boldsymbol{\eta}_\alpha$ are real, orthogonal and form a basis for the liquid free surface deformations. Let us also remark that $\mathbf{S}_g$ being nonsingular, there is no zero-frequency sloshing modes.

Several examples of application of the linearized fluid–structure formulation presented here have been published during the past few years. The possibility to predict the coupling between the vibrations of the structure and the liquid sloshing using this formulation has been illustrated on test cases and on some more realistic examples [7, 19] and then validated by comparison with experimental results [18]. An application to the specific case of free structures containing liquids and a validation by comparison with some benchmark results have been published in [17]. The nonlinear effect of the hydrostatic pressure on the hydroelastic vibrations of a plate has also been analyzed in [21].

7 Conclusions and Perspectives

This paper presents a new way of deriving the linearized variational formulation for fluid–structure problems involving a deformable structure and an internal inviscid liquid with a free surface. Damping phenomena in the fluid and at the interfaces (fluid–structure and free surface) are of prime importance and are under study (first results can be found in [11]). From numerical point of view, isogeometric analysis [3] could be worthwhile of investigation for tanks of complex geometry.

References

1. Bazilevs, Y., Takizawa, K., Tezduyar, T.E.: Computational Fluid-Structure Interaction: Methods and Applications. Wiley, Chichester (2013)
2. Bermúdez, A., Rodríguez, R.: Finite element computation of the vibration modes of a fluid-solid system. Comput. Methods Appl. Mech. Eng. **119**, 355–370 (1994)
3. Cottrel, J.A., Hughes, T.J.R., Bazilevs, Y.: Isogeometric Analysis: Toward Integration of CAD and FEM. Wiley, Hoboken (2009)
4. Cruchaga, M.A., Reinoso, R.S., Storti, M.A., Celentano, D.J., Tezduyar, T.E.: Finite element computation and experimental validation of sloshing in rectangular tanks. Comput. Mech. **52**, 1301–1312 (2013)
5. Dodge, F.T.: The new "dynamic behaviour of liquids in moving containers". Southwest Research Institute, San Antonio, Texas, USA (2000)
6. El-Kamali, M., Schotté, J.S., Ohayon, R.: Three-dimensional modal analysis of sloshing under surface tension. Int. J. Numer. Methods Fluids **65**(1–3), 87–105 (2011)
7. Farhat, C., Chiu, E.K., Amsallem, D., Schotté, J.S., Ohayon, R.: On the modelling of fuel sloshing and its physical effect on flutter. AIAA J. **51**(9), 2252–2265 (2013)
8. González, J.A., Park, K.C., Lee, I., Felippa, C.A., Ohayon, R.: Partitioned vibration analysis of internal fluid-structure interaction problems. Int. J. Numer. Methods Eng. **92**, 268–300 (2012)
9. Ibrahim, R.A.: Liquid Sloshing Dynamics: Theory and Applications. Cambridge University Press, Cambridge (2005)
10. Liu, W.K., Uras, R.A.: Variational approach to fluid-structure interaction with sloshing. Nucl. Eng. Des. **106**, 69–85 (1988)
11. Miras, T., Schotté, J.S., Ohayon, R.: Liquid sloshing damping in an elastic container. J. Appl. Mech. **79**(1), 010902 (2012)
12. Morand, H.J.P., Ohayon, R.: Internal pressure effects on the vibration of partially filled elastic tanks. In: Proceedings of the World Congress on Application of Finite Element Methods to Structural Mechanics, Bournemouth (U.K.) (1975)

13. Morand, H.J.P., Ohayon, R.: Fluid-Structure Interaction: Applied Numerical Methods. Wiley, New York (1995)
14. Ohayon, R., Soize, C.: Advanced Computational Vibroacoustics – Reduced-Order Models and Uncertainty Quantification. Cambridge University Press, Cambridge (2014)
15. Ohayon, R., Soize, C.: Vibration of structures containing compressible liquids with surface tension and sloshing effects. Reduced-order model. Comput. Mech. **55**(6), 1071–1078 (2015)
16. Ryzhakov, P.B., Rossi, R., Idelsohn, S.R., Oñate, E.: A monolithic Lagrangian approach for fluid-structure interaction problems. Comput. Mech. **46**, 883–899 (2010)
17. Schotté, J.S., Ohayon, R.: Effect of gravity on a free-free elastic tank partially filled with incompressible liquid. J. Fluids Struct. **18**(2), 215–226 (2003)
18. Schotté, J.S., Ohayon, R.: Incompressible hydroelastic vibrations: finite element modelling of the elastogravity operator. Comput. Struct. **83**, 209–219 (2005)
19. Schotté, J.S., Ohayon, R.: Various modelling levels to represent internal liquid behaviour in the vibratory analysis of complex structures. Comput. Methods Appl. Mech. Eng. **198**, 1913–1925 (2009)
20. Schotté, J.S., Ohayon, R.: Linearized formulation for fluid-structure interaction: application to the linear dynamic response of a pressurized elastic structure containing a fluid with a free surface. J. Sound Vib. **332**(10), 2396–2414 (2013)
21. Schotté, J.S., Carra, S., Amabili, M., Ohayon, R.: Non linear effect of hydrostatic pressure on the hydroelastic vibrations of a plate. In: Proceedings of The Second International Conference on Dynamics, Vibration and Control. CSTAM, Beijing, China (2006)
22. Veldman, A.E.P., Gerrits, J., Luppes, R., Helder, J.A., Vreeburg, J.P.B.: The numerical simulation of liquid sloshing on board spacecraft. J. Comput. Phys. **224**, 82–99 (2007)

A Fluid–Structure Interaction Algorithm Using Radial Basis Function Interpolation Between Non-Conforming Interfaces

Simone Deparis, Davide Forti, and Alfio Quarteroni

Abstract We consider a fluid–structure interaction (FSI) problem discretized by finite elements featuring two different grids that do not necessarily agree on the interface separating the computational domain of the fluid from the one of the structure. After identifying a master domain (the structural domain) and a slave domain (the fluid domain), we build up two radial basis function (RBF) inter-grid operators, one Π_{fs} from master to slave, and the other Π_{sf} from slave to master. Then, we enforce the kinematic condition by equating the fluid velocity at the interface as the image through Π_{fs} of the temporal derivative of the structural displacement. On the other hand, the dynamic interface condition is fulfilled via a variational method where the strong form of the structural normal stress is obtained as the image through Π_{sf} of the strong form of the fluid normal stress. A numerical verification is carried out for a straight cylinder and for a patient-specific arterial bypass geometry. This new method is easy to implement and optimally accurate.

1 Introduction and Model Description

In this paper we propose a new method for the coupling of an FSI problem featuring two finite element grids that are non-conforming at the interface separating the computational fluid domain Ω^f from the computational structure domain Ω^s. This method is based on a suitable way to enforce the kinematic condition and (the strong form of) the dynamic condition via the use of two distinct RBF interpolation operators, one from Ω^f to Ω^s and the other from Ω^s to Ω^f. In our FSI model, the incompressible Navier–Stokes equations are written in ALE coordinates, whereas the linear elasticity model for the structure in a Lagrangian frame of reference. A

S. Deparis • D. Forti
EPFL – École Polytechnique Fédérale de Lausanne, Station 8, Lausanne CH-1015, Switzerland
e-mail: simone.deparis@epfl.ch; davide.forti@epfl.ch

A. Quarteroni (✉)
EPFL – École Polytechnique Fédérale de Lausanne, Station 8, Lausanne CH–1015, Switzerland
MOX - Politecnico di Milano, via Bonardi 9, Milano 20133, Italy
e-mail: alfio.quarteroni@epfl.ch

Y. Bazilevs, K. Takizawa (eds.), *Advances in Computational Fluid-Structure Interaction and Flow Simulation*, Modeling and Simulation in Science, Engineering and Technology, DOI 10.1007/978-3-319-40827-9_34

third field, the so-called geometry problem, allows to determine the displacement of the fluid domain $\mathbf{d}_f$ which defines, in turn, the ALE map. In this paper we compute $\mathbf{d}_f$ as the harmonic extension of the trace of the solid displacement $\hat{\mathbf{d}}_s$ at the reference fluid structure interface $\hat{\Gamma}$, to the fluid reference domain $\hat{\Omega}^f \subset \mathbb{R}^3$:

$$-\Delta\hat{\mathbf{d}}_f = 0 \text{ in } \hat{\Omega}^f, \quad \hat{\mathbf{d}}_f = \hat{\mathbf{d}}_s \text{ on } \hat{\Gamma}, \tag{1}$$

complemented with boundary conditions on $\partial\hat{\Omega}^f \setminus \hat{\Gamma}$. The solution of the geometry problem $\hat{\mathbf{d}}_f$ defines the ALE map $\mathscr{A}_t(\hat{\mathbf{x}}) = \hat{\mathbf{x}} + \hat{\mathbf{d}}_f(\hat{\mathbf{x}}, t)\ \forall\hat{\mathbf{x}} \in \hat{\Omega}^f$, and consequently the current fluid domain configuration at time t

$$\Omega_t^f = \{\mathbf{x} = \mathscr{A}_t(\hat{\mathbf{x}}) |\ \hat{\mathbf{x}} \in \hat{\Omega}^f\}.$$

The Navier–Stokes equations written in ALE coordinates read

$$\rho_f \left.\frac{\partial \mathbf{u}_f}{\partial t}\right|_{\hat{\mathbf{x}}} + \rho_f((\mathbf{u}_f - \mathbf{w}) \cdot \nabla)\mathbf{u}_f - \nabla \cdot \boldsymbol{\sigma}_f = 0 \text{ in } \Omega_t^f, \tag{2}$$

$$\nabla \cdot \mathbf{u}_f = 0 \text{ in } \Omega_t^f, \tag{3}$$

$$\mathbf{u}_f = \mathbf{h}_f \text{ on } \Gamma_D^f, \quad \boldsymbol{\sigma}_f \mathbf{n}_f = \mathbf{g}_f \text{ on } \Gamma_N^f \ , \ \mathbf{u}_f \circ \mathscr{A}_t = \frac{\partial \hat{\mathbf{d}}_s}{\partial t} \text{ on } \hat{\Gamma},$$

where $\left.\frac{\partial}{\partial t}\right|_{\hat{\mathbf{x}}} = \frac{\partial}{\partial t} + \mathbf{w} \cdot \nabla$ is the ALE derivative, $\mathbf{w}(\mathbf{x}) = \frac{\partial \mathscr{A}_t(\mathbf{x})}{\partial t}$ is fluid domain velocity, $\mathbf{u}_f$ and p_f are the velocity and pressure of the fluid, respectively, and $\hat{\Gamma}$ is the common interface between the fluid and the structure on the reference configuration. We denoted by ρ_f the density of the fluid and by $\boldsymbol{\sigma}_f$ the Cauchy stress tensor defined for a Newtonian fluid as $\boldsymbol{\sigma}_f = \mu_f(\nabla\mathbf{u}_f + (\nabla\mathbf{u}_f)^T) - p_f\mathbf{I}$, with $\mathbf{I}$ being the identity tensor, μ_f the dynamic viscosity of the fluid, and $\mathbf{n}_f$ the outward unit normal vector to $\partial\Omega_t^f$. The functions $\mathbf{h}_f$ and $\mathbf{g}_f$ indicate the Dirichlet and Neumann data applied at the Dirichlet and Neumann boundaries Γ_D^f and Γ_N^f, respectively, of Ω_t^f. Note that Γ_D^f, Γ_N^f, and Γ provide a disjoint partition of the boundary of Ω_t^f, i.e., their pairwise intersection is empty and $\bar{\Gamma}_D^f \cup \bar{\Gamma}_N^f \cup \bar{\Gamma} = \partial\Omega_t^f$. Although Γ_D^f, Γ_N^f and Γ^f depend on time t, we do not explicit it in the notation.

The structure dynamics is governed by the conservation of momentum law:

$$\hat{\rho}_s \frac{\partial^2 \hat{\mathbf{d}}_s}{\partial t^2} - \nabla_{\hat{\mathbf{x}}} \cdot \boldsymbol{\Pi}(\hat{\mathbf{d}}_s) = 0 \text{ in } \hat{\Omega}^s, \tag{4}$$

$$\hat{\mathbf{d}}_s = \mathbf{h}_s \text{ on } \hat{\Gamma}_D^s, \quad \boldsymbol{\Pi}(\hat{\mathbf{d}}_s)\hat{\mathbf{n}}_s = 0 \text{ on } \hat{\Gamma}_N^s \ , \quad \boldsymbol{\Pi}(\hat{\mathbf{d}}_s)\hat{\mathbf{n}}_s + \hat{\boldsymbol{\sigma}}_f \hat{\mathbf{n}}_f = 0 \text{ on } \hat{\Gamma}, \tag{5}$$

where $\hat{\Omega}^s$ denotes the reference domain occupied by the structure (the vessel wall) at rest, $\hat{\mathbf{n}}_s$ represents the outward unit normal vector to $\partial\hat{\Omega}^s$, $\hat{\boldsymbol{\sigma}}_f = (\det[\mathbf{F}])\, \mathbf{F}^{-T} \boldsymbol{\sigma}_f$, $\mathbf{F} = \mathbf{I} + \nabla_{\hat{\mathbf{x}}}\hat{\mathbf{d}}_s$ is the deformation gradient tensor, and $\boldsymbol{\Pi}(\hat{\mathbf{d}}_s)$ is the first Piola–Kirchhoff stress tensor [5]. The function $\mathbf{h}_s$ indicates the Dirichlet data applied on the Dirichlet boundary $\hat{\Gamma}^s_D$ of $\hat{\Omega}^s$, $\hat{\Gamma}^s_N$ is the Neumann boundary, and $\hat{\Gamma}^s_D$, $\hat{\Gamma}^s_N$, $\hat{\Gamma}$ provide a disjoint partition of $\partial\hat{\Omega}^s$.

The coupling between the fluid, the structure, and the geometry subproblems is ensured by imposing the geometry adherence, the continuity of the velocity (kinematic condition), and the continuity of the normal stresses at the interface. We notice that the resulting system is nonlinear due to the convective term in the fluid momentum equation and to the moving fluid domain.

In Section 2 we first introduce a discretization in space and time of both the fluid problem, the structural problem and the geometry problem. In our spatial finite element discretization we allow the fluid computational grid to be non-conforming with the structural grid at the interface. Even worse, the two interfaces could be non-conforming, a situation that arises when the two subdomains are triangulated independently, for instance, when dealing with an FSI problem in arterial vessels whose geometry is obtained from clinical images. A critical issue at the numerical level is the way the kinematic and dynamic coupling conditions are fulfilled. For readers' convenience we start in Section 2.1 by providing the complete algebraic formulation in the conforming case. Then, we treat in Section 2.2 the non-conforming case: after identifying a master domain (the structural domain) and a slave domain (the fluid one), we build up two RBF inter-grid operators: Π_{fs} from master to slave and Π_{sf} from slave to master. Then, we enforce the kinematic condition by equating the fluid velocity at the interface as the image through Π_{fs} of the temporal derivative of the structural displacement. On the other hand, the dynamic interface condition is fulfilled via a variational method where the *strong* form of the structural normal stress is obtained as the image through Π_{sf} of the *strong* form of the fluid normal stress. The new algebraic formulation differs very slightly from the conforming one. Section 3 is devoted to the numerical solution via a preconditioned Newton algorithm of the coupled algebraic system, while Section 4 addresses the numerical verification of our approach on an FSI problem for a straight flexible cylinder. Finally, in Section 5 we address the case of a patient-specific geometry of a femoropopliteal bypass. We discuss the computational efficiency of our method as well as its numerical accuracy.

2 Space and Time Discretizations

For space discretization, we consider a Galerkin finite element approximation using $\mathbb{P}_1$–$\mathbb{P}_1$ Lagrange polynomials for the representation of the fluid variables $\mathbf{u}_f$ and p_f, respectively, stabilized by the SUPG-ALE method [1, 2]. $\mathbb{P}_1$ finite elements are also used for the structure displacement $\hat{\mathbf{d}}_s$ and the harmonic extension $\hat{\mathbf{d}}_f$. We approximate the time derivative in the fluid momentum equation by means of the implicit Euler method. For the structure, we use two-steps backward-difference method

$$\mathbf{d}^{\Delta t}_{n+1} - 2\mathbf{d}^{\Delta t}_{n} + \mathbf{d}^{\Delta t}_{n-1} - \hat{\rho}_s^{-1}\Delta t^2 A_s \mathbf{d}^{\Delta t}_{n+1} = 0, \tag{6}$$

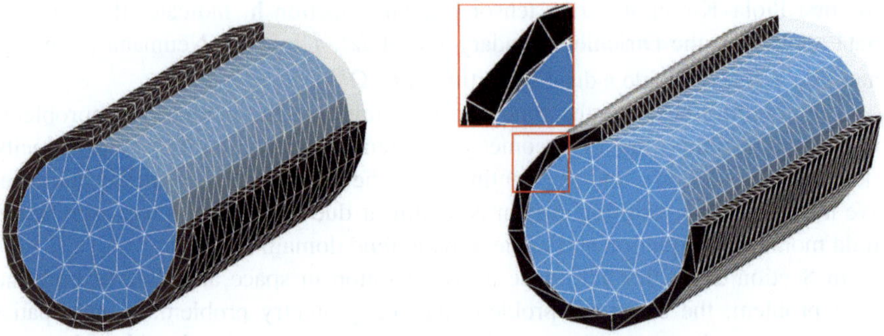

(a) Conforming meshes. (b) Non-conforming meshes.

Fig. 1 Conforming (left) and non-conforming (right) fluid–structure meshes.

where $\mathbf{d}_n^{\Delta t}$ is an approximation of $\hat{\mathbf{d}}_s$ at time $t_n = \Delta t\, n$ and A_s is the finite element stiffness matrix associated with $-\nabla_{\hat{\mathbf{x}}} \cdot \boldsymbol{\Pi}(\hat{\mathbf{d}}_s)$.

We treat the system of equations modeling the fluid–structure interaction in a monolithic fashion, i.e., we solve the FSI problem for all the unknown variables at the same time. Furthermore, we use a fully implicit scheme for which all the nonlinearities of the problem are treated implicitly.

In Sections 2.1 and 2.2 we report the fully discrete form of the nonlinear FSI system when conforming and non-conforming fluid–structure interface meshes are considered, respectively. In Fig. 1 we show an example of conforming and non-conforming fluid–structure meshes at the interface.

2.1 Conforming Fluid–Structure Interface Meshes

We assume first that at the (common) interface Γ between the fluid domain Ω_t^f and the solid domain Ω_t^s the fluid and solid meshes are the same, see Fig. 1(a). After space and time discretization, at each time step, the resulting nonlinear FSI system to be solved reads:

$$\begin{pmatrix} F(\mathbf{u}_F^{n+1}, \mathbf{d}_f^{n+1}) & 0 & I_{\Gamma^f}^T & 0 \\ 0 & S & -I_{\Gamma^s}^T & 0 \\ I_{\Gamma^f} & -I_{\Gamma^s}/\Delta t & 0 & 0 \\ 0 & -I_{\Gamma^s} & 0 & G \end{pmatrix} \begin{pmatrix} \mathbf{u}_F^{n+1} \\ \mathbf{d}_s^{n+1} \\ \boldsymbol{\lambda}^{n+1} \\ \mathbf{d}_f^{n+1} \end{pmatrix} = \begin{pmatrix} \mathbf{b}_f \\ \mathbf{b}_s \\ -I_{\Gamma^s}/\Delta t\, \mathbf{d}_s^n \\ 0 \end{pmatrix}. \tag{7}$$

In Eq. (7), we denote by $\boldsymbol{\lambda}^{n+1}$ the vector of normal stresses (in weak form) at the fluid–structure interface and $\mathbf{u}_F^{n+1} = (\mathbf{u}_f^{n+1}, p_f^{n+1})^T$. The diagonal blocks on the left-hand side of (7) account for the discretized fluid (F), structure (S), and geometry (G) problems, respectively. We remark that F is nonlinear due to the presence of the convective term and the motion of the fluid domain. The matrices I_{Γ^f} and I_{Γ^s} are the restriction of fluid and structure vectors to the interface and account for the continuity of velocities and the geometry adherence on Γ, which are imposed strongly. Their transposes account for the continuity of the normal stresses, which is imposed weakly.

2.2 Non-Conforming Fluid–Structure Interface Meshes

We consider now the case of non-conforming fluid and structure interfaces, as illustrated in Fig. 1(b). We propose a method which is able to handle both non-conforming meshes and slightly non-conforming geometries. The first case implies that the interface Γ is discretized using two different meshes depending on the side which is considered. This non-conformity is usually tackled using the mortar method [3, 9, 11]. As for the mortar method, we need to identify the so-called master and slave domains Ω_{slave} and Ω_{master}, respectively, that will play a different role. In the method proposed here, we assume the structure domain to be the master, $\hat{\Omega}^s \equiv \Omega_{master}$, while the fluid one represents the slave, $\Omega^f_t \equiv \Omega_{slave}$. In FSI problems, the normal stresses in weak form at the interface are usually computed as residuals of the fluid and structure equations. In (7) they are represented by $\boldsymbol{\lambda}$. When dealing with non-conforming meshes, they have to be represented by two different vectors, $\boldsymbol{\lambda}_f$ on the fluid side and $\boldsymbol{\lambda}_s$ on the solid one.

When using non-conforming meshes, the interface coupling conditions become more involved with respect to the conforming case since an interpolation or a projection procedure has to be performed to enable the transfer of physical information between two different grids. In our method, this procedure is carried out using the rescaled localized radial basis functions (RL-RBF) interpolant proposed in [6]. We remark that other interpolants can be used as well, however, we choose RL-RBF since it allows to consider also slightly non-conforming geometries, i.e., those for which Γ^f and Γ^s do not exactly coincide.

We define the two matrices representing the interpolation between the two sides of the interface and denote them by Π_{sf}, from Γ^f to Γ^s, and Π_{fs}, from Γ^s to Γ^f. Π_{sf} needs not to be the adjoint of Π_{fs}, neither their multiplication should yield the identity matrix. In particular, this is not the case when using RL-RBF.

To better understand the way Eq. (7) is generalized to the non-conforming case, we reformulate it using the redundant variables $\boldsymbol{\lambda}_f$ (the normal stresses in weak form from the fluid side) and $\boldsymbol{\lambda}_s$ (the normal stresses in weak form from the structure side), and set $\boldsymbol{\lambda}_f = \boldsymbol{\lambda}_s (= \boldsymbol{\lambda})$. Then, Eq. (7) can be equivalently reformulated as:

$$\begin{pmatrix} F(\mathbf{u}_F^{n+1}, \mathbf{d}_f^{n+1}) & 0 & I_{\Gamma^f}^T & 0 & 0 \\ 0 & S & 0 & -I_{\Gamma^s}^T & 0 \\ 0 & 0 & -I & I & 0 \\ I_{\Gamma^f} & -\frac{1}{\Delta t} I_{\Gamma^s} & 0 & 0 & 0 \\ 0 & -I_{\Gamma^s} & 0 & 0 & G \end{pmatrix} \begin{pmatrix} \mathbf{u}_F^{n+1} \\ \mathbf{d}_s^{n+1} \\ \boldsymbol{\lambda}_f^{n+1} \\ \boldsymbol{\lambda}_s^{n+1} \\ \mathbf{d}_f^{n+1} \end{pmatrix} = \begin{pmatrix} \mathbf{b}_f \\ \mathbf{b}_s \\ 0 \\ -\frac{1}{\Delta t} I_{\Gamma^s} \mathbf{d}_s^n \\ 0 \end{pmatrix}. \tag{8}$$

Here I is the identity matrix.

In the non-conforming case, the kinematic coupling condition becomes

$$\mathbf{u}_f^{n+1} = \Pi_{fs} \, \dot{\mathbf{d}}_s^{n+1} \quad \text{on } \Gamma^f, \tag{9}$$

which, after discretization, reads

$$\mathbf{u}_f^{n+1}|_{\Gamma^f} = \Pi_{fs} \Delta t^{-1} \left(\mathbf{d}_s^{n+1}|_{\Gamma^s} - \mathbf{d}_s^n|_{\Gamma^s} \right) \quad \text{on } \Gamma^f. \tag{10}$$

Eq. (10) replaces row 4 in Eq. (8).

We focus now on the coupling condition of the normal stresses at the fluid–structure interface. Since we are interested in interpolating quantities with the RL-RBF interpolant, we have to work with the stresses in their strong form. A simple and efficient way is to use the mass matrices of the interface on each side. In fact, denoting by $\boldsymbol{\lambda}_s^{n+1}$ the weak form of the normal stresses on Γ^s at time t^{n+1}, and by M_{Γ^s} and M_{Γ^f} the mass matrices associated with the structure and fluid sides of the interface, respectively, the discrete form of the dynamic coupling condition follows:

$$M_{\Gamma^s}^{-1} \boldsymbol{\lambda}_s^{n+1} = \Pi_{sf} M_{\Gamma^f}^{-1} \boldsymbol{\lambda}_f^{n+1} \quad \text{or} \quad \boldsymbol{\lambda}_s^{n+1} - M_{\Gamma^s} \Pi_{sf} M_{\Gamma^f}^{-1} \boldsymbol{\lambda}_f^{n+1} = 0 \quad \text{on } \Gamma^s. \tag{11}$$

This equation replaces row 3 in Eq. (8). In Eq. (11), we notice that $M_{\Gamma^f}^{-1} \boldsymbol{\lambda}_f^{n+1}$ is an approximation of the strong form of the normal stresses on Γ^f; $\Pi_{sf}(M_{\Gamma^f}^{-1} \boldsymbol{\lambda}_f^{n+1})$ is an interpolation of the normal stresses on the side of Γ^s, still in strong form, and $M_{\Gamma^s}(\Pi_{sf} M_{\Gamma^f}^{-1} \boldsymbol{\lambda}_f^{n+1})$ is again in weak form but on Γ^s. Note that order of magnitude of the entries of $\boldsymbol{\lambda}_f^{n+1}$ depend on the mesh size used to discretize Ω_t^f, that of the entries of $M_{\Gamma^s}(\Pi_{sf} M_{\Gamma^f}^{-1} \boldsymbol{\lambda}_f^{n+1})$ depend on the mesh size of $\hat{\Omega}^s$, while the order of magnitude of those of $M_{\Gamma^f}^{-1} \boldsymbol{\lambda}_f^{n+1}$ and $\Pi_{sf}(M_{\Gamma^f}^{-1} \boldsymbol{\lambda}_f^{n+1})$ are independent of the mesh sizes.

To summarize, in the non-conforming case Eq. (8) has to be replaced by:

$$\begin{pmatrix} F(\mathbf{u}_F^{n+1}, \mathbf{d}_f^{n+1}) & 0 & I_{\Gamma^f}^T & 0 & 0 \\ 0 & S & 0 & -I_{\Gamma^s}^T & 0 \\ 0 & 0 & -\Pi_{sf} M_{\Gamma^f}^{-1} & M_{\Gamma^s}^{-1} & 0 \\ I_{\Gamma^f} & -\frac{1}{\Delta t} \Pi_{fs} I_{\Gamma^s} & 0 & 0 & 0 \\ 0 & -\Pi_{fs} I_{\Gamma^s} & 0 & 0 & G \end{pmatrix} \begin{pmatrix} \mathbf{u}_F^{n+1} \\ \mathbf{d}_s^{n+1} \\ \boldsymbol{\lambda}_f^{n+1} \\ \boldsymbol{\lambda}_s^{n+1} \\ \mathbf{d}_f^{n+1} \end{pmatrix} = \begin{pmatrix} \mathbf{b}_f \\ \mathbf{b}_s \\ 0 \\ -\frac{1}{\Delta t} \Pi_{fs} I_{\Gamma^s} \mathbf{d}_s^n \\ 0 \end{pmatrix}.$$

Upon eliminating $\boldsymbol{\lambda}_s^{n+1}$ using Eq. (11), we end up with the reduced nonlinear FSI system:

$$\begin{pmatrix} F(\mathbf{u}_F^{n+1}, \mathbf{d}_f^{n+1}) & 0 & I_{\Gamma^f}^T & 0 \\ 0 & S & \boxed{-I_{\Gamma^s}^T M_{\Gamma^s} \Pi_{sf} M_{\Gamma^f}^{-1}} & 0 \\ I_{\Gamma^f} & \boxed{-\frac{1}{\Delta t}\Pi_{fs} I_{\Gamma^s}} & 0 & 0 \\ 0 & \boxed{-\Pi_{fs} I_{\Gamma^s}} & 0 & G \end{pmatrix} \begin{pmatrix} \mathbf{u}_F^{n+1} \\ \mathbf{d}_s^{n+1} \\ \boldsymbol{\lambda}_f^{n+1} \\ \mathbf{d}_f^{n+1} \end{pmatrix} = \begin{pmatrix} \mathbf{b}_f \\ \mathbf{b}_s \\ \boxed{-\frac{1}{\Delta t}\Pi_{fs} I_{\Gamma^s}} \mathbf{d}_s^n \\ 0 \end{pmatrix}, \tag{12}$$

which should be regarded as the generalization of Eq. (7) to the non-conforming case. Note that we have highlighted with a box those blocks that have actually changed.

Remark: System (12) can be interpreted as the matrix representation of a finite element formulation of the FSI in the non-conforming case that generalizes the one in the conforming case by means of suitable inter-grid interface finite element operators. Full details are given in [7].

3 Numerical Solution

We focus on solving Eq. (12) using the Newton method. The solution of (12) at time $t^n = n\,\Delta t$ is denoted by $\mathbf{X}^n = (\mathbf{u}_F^n, \mathbf{d}_s^n, \boldsymbol{\lambda}_f^n, \mathbf{d}_f^n)^T$. At each time step, we compute a sequence of approximations $\mathbf{X}_1^{n+1}, \mathbf{X}_2^{n+1}$, etc., until the numerical solution converges up to a prescribed tolerance. The generic $k+1$ iteration of the Newton method applied to (12) is described as follows. Starting from an approximation of $\mathbf{X}_k^{n+1}$, we compute the residual $\mathbf{R}_k^{n+1} = (\mathbf{r}_{F,k}^{n+1}, \mathbf{r}_{S,k}^{n+1}, \mathbf{r}_{\lambda,k}^{n+1}, \mathbf{r}_{G,k}^{n+1})^T$:

$$\mathbf{R}_k^{n+1} = \begin{pmatrix} \mathbf{b}_f \\ \mathbf{b}_s \\ -\frac{1}{\Delta t}\Pi_{fs} I_{\Gamma^s} \mathbf{d}_s^n \\ 0 \end{pmatrix} - \begin{pmatrix} F(\mathbf{u}_F^{n+1}, \mathbf{d}_f^{n+1}) & 0 & I_{\Gamma^f}^T & 0 \\ 0 & S & -I_{\Gamma^s}^T M_{\Gamma^s} \Pi_{sf} M_{\Gamma^f}^{-1} & 0 \\ I_{\Gamma^f} & -\frac{1}{\Delta t}\Pi_{fs} I_{\Gamma^s} & 0 & 0 \\ 0 & -\Pi_{fs} I_{\Gamma^s} & 0 & G \end{pmatrix} \begin{pmatrix} \mathbf{u}_{F,k}^{n+1} \\ \mathbf{d}_{s,k}^{n+1} \\ \boldsymbol{\lambda}_{f,k}^{n+1} \\ \mathbf{d}_{f,k}^{n+1} \end{pmatrix}.$$

Then, we compute the Newton correction $\delta\mathbf{X}_k^{n+1} = (\delta\mathbf{u}_{F,k}^{n+1}, \delta\mathbf{d}_{s,k}^{n+1}, \delta\boldsymbol{\lambda}_{f,k}^{n+1}, \delta\mathbf{d}_{f,k}^{n+1})^T$ by solving

$$J_{FSI}\,\delta\mathbf{X}_k^{n+1} = -\mathbf{R}_k^{n+1}, \tag{13}$$

being J_{FSI} the exact FSI Jacobian matrix. Linear system (13) is solved by the GMRES method preconditioned by FaCSI [8]. Finally, we update the solution, i.e., $\mathbf{X}_{k+1}^{n+1} = \mathbf{X}_k^{n+1} + \delta\mathbf{X}_k^{n+1}$. We stop the Newton iterations when $\|\mathbf{R}_k^{n+1}\|_\infty / \|\mathbf{R}_0^{n+1}\|_\infty \leq \epsilon$, being $\mathbf{R}_0^{n+1}$ the residual at the first Newton iteration and ϵ a given tolerance.

4 FSI in a Straight Flexible Cylinder

In this example we consider a classical benchmark problem in which the fluid geometry consists in a straight cylinder of length $L = 5$ cm and radius $R = 0.5$ cm, surrounded by a structure of constant thickness $h = 0.1$ cm. The fluid density is $\rho_f = 1.0\ g/cm^3$, the dynamic viscosity is $\mu_f = 0.03\ g/(cm\,s)$; the structure is characterized by a density $\hat{\rho}_s = 1.2\ g/cm^3$, a Young's modulus $E_s = 3 \times 10^6\ dyne/cm^2$, and a Poisson's ratio $\nu_s = 0.3$. The structure is clamped at both ends. At the fluid inflow we prescribe a constant normal stress $\boldsymbol{\sigma} \cdot \mathbf{n} = 1.33 \times 10^4\ dyne/cm^2$ for $t \leq 0.003\ s$, while a homogeneous Neumann boundary conditions is used at the fluid outflow.

In our simulations we consider 3 sets of fluid–structure meshes: in Fig. 2 we report a front view of them while in Table 1 we report their corresponding number of degrees of freedom. In Fig. 3 we show a post-processing of the solutions obtained at time $t = 0.005$ s using the three different set of meshes considered. Furthermore, in Fig. 4 we compare the solutions obtained for each set by plotting the values of the pressure and the magnitude of the solid displacements with respect to time at two specific locations. In Fig. 4(b) and Fig. 4(c) we show that the results obtained using non-conforming meshes match almost exactly those obtained in the conforming case, even when the fluid mesh is much finer than the solid one (Set 3).

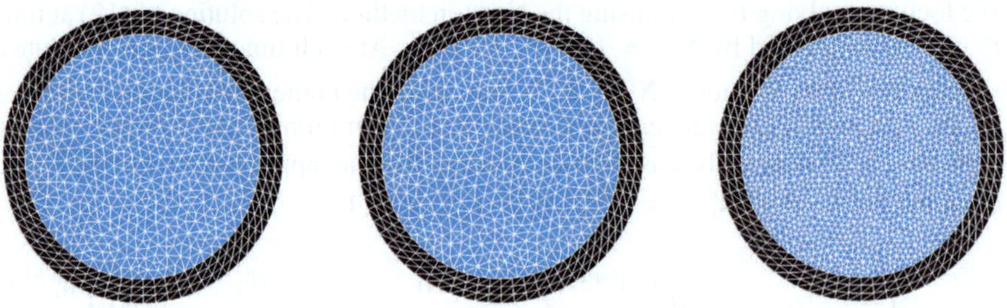

(a) Set 1. (b) Set 2. (c) Set 3.

Fig. 2 Front view of the different sets of fluid–structure meshes used. In Set 1 we consider conforming meshes. In Set 2 we keep the same meshes of Set 1 but we rotate the fluid mesh such that the fluid and structure interfaces are non-conforming. In Set 3 the structure mesh is the same of Set 1 but the fluid one is finer.

Table 1 Straight flexible tube example: number of degrees of freedom (DoF) used for each set of fluid–structure meshes considered.

Set	Fluid DoF	Structure DoF	Coupling DoF	Geometry DoF	Total
1	137'280	53'856	13'056	102'960	307'152
2	137'280	53'856	13'056	102'960	307'152
3	840'360	53'856	48'372	630'270	1'572'858

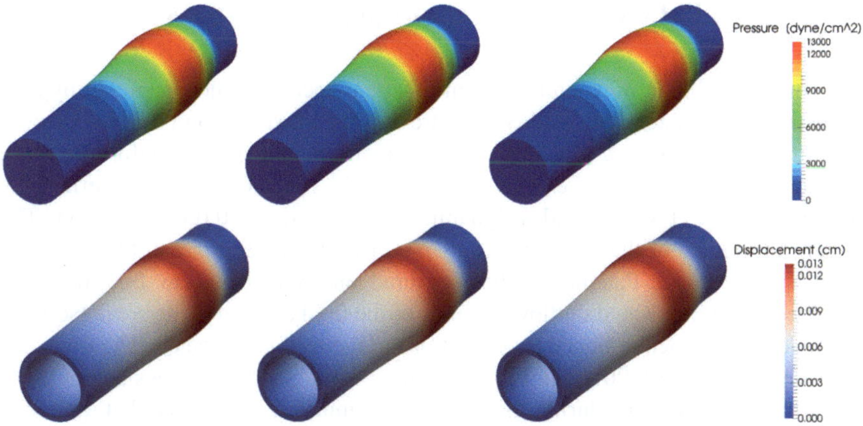

(a) Set 1. (b) Set 2. (c) Set 3.

Fig. 3 Post-processing of the results obtained at time $t = 0.005$ s. In the upper row we show the fluid pressure, while in the middle row the magnitude of the structure displacement, for the three set of meshes considered. The deformation of the fluid and structure domains is magnified by a factor 10 for visualization purposes.

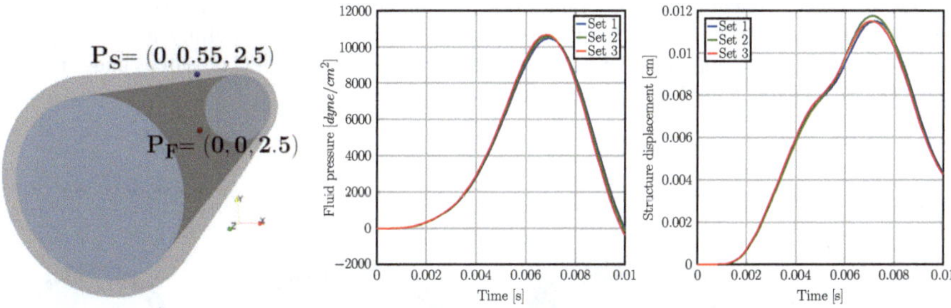

(a) Points of measure. (b) Fluid pressure at $\mathbf{P_F}$. (c) Solid displacement at $\mathbf{P_S}$.

Fig. 4 On the left we show the locations where results are taken in the fluid ($\mathbf{P}_F$) and the structure ($\mathbf{P}_S$). In the two right-most figures we report the evolution of the fluid pressure and the magnitude of the solid displacement at the point $\mathbf{P}_F$ and $\mathbf{P}_S$, respectively.

We also compared the performance of our solver when using the meshes of Set 1 and Set 2, for which the total number of degrees of freedom is the same. The average number of Newton iterations is the same and equal to 2.9, while the average number of linear solver iterations per Newton step is 20 in both cases. Finally, we observe that in the simulations performed, dealing with non-conforming meshes increases the average computational time per time step of about 15% with respect to the conforming case.

5 FSI in a Patient-Specific Femoropopliteal Bypass

We consider a blood flow simulation in a patient-specific femoropopliteal bypass. To estimate accurately the Wall-Shear Stress (WSS) distribution [10] we consider a fluid mesh featuring a boundary layer refinement that is non-conforming with respect to the one of the vessel wall, see Fig. 5. The blood is characterized by a density $\rho_f = 1\ g/cm^3$ and a dynamic viscosity $\mu_f = 0.0035\ g/(cm\, s)$. The Young's modulus of the vessel wall is $E_s = 4 \times 10^6\ dyne/cm^2$ and the Poisson's ratio is $\nu_s = 0.45$. Patient-specific measured inflow and outflow flow rates are imposed at the inflow and outflow sections of the fluid domain while we prescribe a homogeneous Dirichlet boundary conditions at the occluded branch. The time step considered is $\Delta t = 0.001\ s$. In Fig. 6 we show the WSS distributions computed at two different time steps during the heart-beat simulated. We notice that, although non-conforming meshes are used here, both the WSS magnitude and distribution are in good agreement with those reported in [4] where conforming fluid–structure meshes were adopted.

Acknowledgements The research of D. Forti was supported by the Swiss National Foundation (SNF), project No. 140184. The research of S. Deparis and A. Quarteroni was partly supported by the PASC project "Integrative HPC Framework for Coupled Cardiac Simulations". We acknowledge Prof. P. Gervasio for the fruitful discussions and suggestions. We gratefully acknowledge the Swiss National Supercomputing Centre (CSCS) for providing us the CPU resources under project ID s475. We also thank the LifeV community.

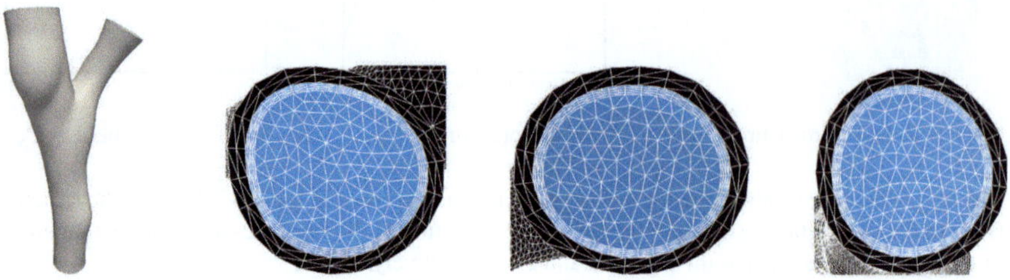

(a) Geometry. (b) Occluded branch. (c) Inflow. (d) Outflow.

Fig. 5 Bypass geometry and zoom over the inflow and outflow surfaces of the non-conforming fluid–structure meshes.

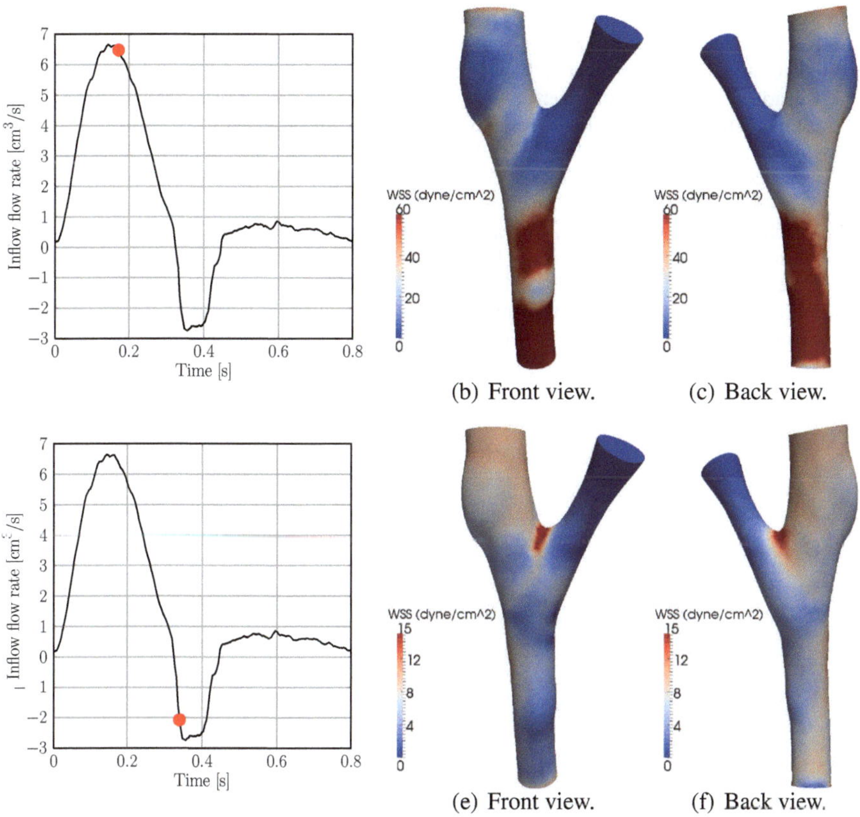

Fig. 6 WSS distributions at times $t = 0.17\ s$ (top) and at $t = 0.34\ s$ (bottom).

References

1. Bazilevs, Y., Calo, V.M., Cottrell, J.A., Hughes, T.J.R., Reali, A., Scovazzi, G.: Variational multiscale residual-based turbulence modeling for large eddy simulation of incompressible flows. Comput. Methods Appl. Mech. Eng. **197**(1–4), 173–201 (2007)
2. Bazilevs, Y., Takizawa, K., Tezduyar, T.E.: Computational Fluid–Structure Interaction: Methods and Applications. Wiley Series in Computational Mechanics. Wiley, Chichester (2013)
3. Bernardi, C., Maday, Y., Patera, A.T.: A new nonconforming approach to domain decomposition: the mortar element method. In: Brezis, H., Lions, J.L. (eds.) Nonlinear Partial Differential Equations and their Applications, pp. 13–51. Wiley, London/New York (1994)
4. Colciago, C., Deparis, S., Quarteroni, A.: Comparisons between reduced order models and full 3D models for fluid-structure interaction problems in haemodynamics. J. Comput. Appl. Math. **265**, 120–138 (2014)
5. Crosetto, P., Deparis, S., Fourestey, G., Quarteroni, A.: Parallel algorithms for fluid-structure interaction problems in haemodynamics. SIAM J. Sci. Comput. **33**(4), 1598–1622 (2011)
6. Deparis, S., Forti, D., Quarteroni, A.: A rescaled localized radial basis function interpolation on non-Cartesian and nonconforming grids. SIAM J. Sci. Comput. **36**(6), A2745–A2762 (2011/2014)

7. Deparis, S., Forti, D., Gervasio, P., Quarteroni, A.: INTERNODES: an accurate interpolation-based method for coupling the Galerkin solutions of PDEs on subdomains featuring non-conforming interfaces. Comput. Fluids (2016, in press).
8. Deparis, S., Forti, D., Grandperrin, G., Quarteroni, A.: FaCSI: a block parallel preconditioner for fluid-structure interaction in hemodynamics (2015, under review)
9. Klöppel, T., Popp, A., Küttler, U., Wall, W.A.: Fluid–structure interaction for non-conforming interfaces based on a dual mortar formulation. Comput. Methods Appl. Mech. Eng. **200**, 3111–3126 (2011)
10. Marchandise, E., Crosetto, P., Geuzaine, C., Remacle, J.F., Sauvage, E.: Quality open source mesh generation for cardiovascular flow simulations. In: Ambrosi, D., Quarteroni, A., Rozza, G. (eds.) Modeling of Physiological Flows, pp. 395–414. Springer, Milan (2012)
11. Wohlmuth, B.I.: A mortar finite element method using dual spaces for the Lagrange multiplier. SIAM J. Numer. Anal. **38**(3), 989–1012 (2000)

Elasto-Capillarity Simulations Based on the Navier–Stokes–Cahn–Hilliard Equations

E.H. van Brummelen, M. Shokrpour-Roudbari, and G.J. van Zwieten

Abstract We consider a computational model for complex-fluid–solid interaction (CSFI) based on a diffuse-interface model for the complex fluid and a hyperelastic-material model for the solid. The diffuse-interface complex-fluid model is described by the incompressible Navier–Stokes–Cahn–Hilliard (NSCH) equations with preferential-wetting boundary conditions at the fluid–solid interface. The corresponding fluid traction on the interface includes a capillary-stress contribution, and the dynamic interface condition comprises the traction exerted by the non-uniform fluid–solid surface tension. We present a weak formulation of the aggregated CSFI problem, based on an arbitrary-Lagrangian–Eulerian formulation of the NSCH equations and a proper reformulation of the complex-fluid traction and the fluid–solid surface tension. To validate the presented CSFI model, we present numerical results and conduct a comparison to experimental data for a droplet on a soft substrate.

1 Introduction

Complex fluids are fluids that consist of multiple constituents, e.g., of multiple phases of the same fluid (gas, liquid or solid) or of multiple distinct species (e.g., water and air). The interaction of such complex fluids with elastic solids leads to multitudinous intricate physical phenomena. Examples are *durotaxis*, viz., seemingly spontaneous migration of liquid droplets on solid substrates with an elasticity gradient [17], or *capillary origami*, viz., large-scale solid deformations induced by capillary forces [16]. Complex fluid–solid interaction (CFSI) is moreover of fundamental technological importance in a wide variety of applications, such as inkjet printing and additive manufacturing.

Despite significant progress in models and computational techniques for the interaction of solids and classical fluids (see [4, 19] for an overview), and for

E.H. van Brummelen (✉) • M. Shokrpour-Roudbari • G.J. van Zwieten
Department of Mechanical Engineering, Eindhoven University of Technology, P.O. Box 513, 5600 MB Eindhoven, The Netherlands
e-mail: e.h.v.brummelen@tue.nl; m.shokrpour.roudbari@tue.nl; g.j.v.zwieten@tue.nl

Y. Bazilevs, K. Takizawa (eds.), *Advances in Computational Fluid-Structure Interaction and Flow Simulation*, Modeling and Simulation in Science, Engineering and Technology, DOI 10.1007/978-3-319-40827-9_35

complex fluids separately (see, e.g., [1, 2, 10–12, 14]), CSFI has remained essentially unexplored. A notable exception is the computational CFSI model based on the Navier–Stokes–Korteweg equations in [5].

In this contribution we consider a computational model for CSFI based on a diffuse-interface complex-fluid model and a hyperelastic solid model with a Saint Venant–Kirchhoff stored-energy functional. The diffuse-interface complex-fluid model is described by the incompressible Navier–Stokes–Cahn–Hilliard (NSCH) equations. The interaction of the complex fluid with the solid substrate is represented by dynamic and kinematic interface conditions and a preferential-wetting boundary condition. The traction exerted by the complex fluid on the fluid–solid interface comprises a non-standard capillary-stress contribution, in addition to the standard pressure and viscous-stress components. The dynamic condition imposes equilibrium of this complex-fluid traction, the traction exerted by the hyperelastic solid and the traction due to the non-uniform fluid–solid surface tension. We present a weak formulation of the aggregated CSFI problem, based on an arbitrary-Lagrangian–Eulerian (ALE) formulation of the NSCH system and a suitable weak representation of the complex-fluid traction and the non-uniform fluid–solid surface tension.

To evaluate the capability of the considered CSFI model to describe elasto-capillary phenomena, we consider numerical experiments for a test case pertaining to a droplet on a soft substrate, and we present a comparison to experimental data from [18].

The remainder of this contribution is organized as follows. Section 2 presents a specification and discussion of the considered CSFI problem. In Section 3 we treat the weak formulation of the aggregated fluid–solid interaction problem. Section 4 is concerned with numerical experiments and results. Concluding remarks are presented in Section 5.

2 Problem Statement

To accommodate the complex-fluid–solid system, we consider a time interval $(0, T) \subseteq \mathbb{R}_{>0}$ and two simply connected time-dependent open subsets $\Omega_t^{\mathrm{f}} \subset \mathbb{R}^d$ $(d = 2, 3)$ and $\Omega_t^{\mathrm{s}} \subset \mathbb{R}^d$, which hold the complex-fluid and solid, respectively. The fluid–solid interface corresponds to $\Gamma_t := \partial\Omega_t^{\mathrm{f}} \cap \partial\Omega_t^{\mathrm{s}} \neq \emptyset$. We assume that the time-dependent configuration $\Omega_t := \mathrm{int}(\mathrm{cl}\,\Omega_t^{\mathrm{f}} \cup \mathrm{cl}\,\Omega_t^{\mathrm{s}})$ is the image of a time-dependent transformation $\hat{\chi}$ acting on a fixed reference domain $\hat{\Omega} := \mathrm{int}(\mathrm{cl}\,\hat{\Omega}^{\mathrm{f}} \cup \mathrm{cl}\,\hat{\Omega}^{\mathrm{s}})$ such that $\Omega_t^{\mathrm{f}} = \hat{\chi}\hat{\Omega}^{\mathrm{f}}$ and $\Omega_t^{\mathrm{s}} = \hat{\chi}\hat{\Omega}^{\mathrm{s}}$. The restrictions of $\hat{\chi}$ to $\hat{\Omega}^{\mathrm{f}}$ and $\hat{\Omega}^{\mathrm{s}}$ are denoted by $\hat{\chi}^{\mathrm{f}}$ and $\hat{\chi}^{\mathrm{s}}$, respectively. The reference domains $\hat{\Omega}^{\mathrm{f}} := \Omega_0^{\mathrm{f}}$ and $\hat{\Omega}^{\mathrm{s}} := \Omega_0^{\mathrm{s}}$ are identified with the initial configurations.

2.1 Navier–Stokes–Cahn–Hilliard Complex-Fluid Model

We consider a complex fluid composed of two immiscible incompressible constituents, separated by a thin diffuse interface. The behavior of the complex fluid is described by the NSCH equations. The two species are identified by an order parameter $\varphi : \Omega_t^{\mathrm{f}} \to [-1, 1]$. Typically, φ is selected as either volume fraction [1, 11, 12] or mass fraction [10, 14], such that $\varphi = 1$ (resp. $\varphi = -1$) pertains to a pure species-1 (resp. species-2) composition of the fluid, and $\varphi \in (-1, 1)$ indicates a mixture. Depending on the definition of the phase indicator φ as mass or volume fraction, and the definition of mixture velocity as mass-averaged or volume-averaged species velocity, various forms of the NSCH equations can be derived. In mass-averaged-velocity formulations, the mixture is generally quasi-incompressible. In volume-averaged-velocity formulations, the mixture is incompressible. We select φ as volume fraction and consider a volume-averaged-velocity formulation. The behavior of the complex fluid is described by [2, 12]:

$$\left.\begin{aligned} \partial_t(\rho u) + \nabla \cdot (\rho u \otimes u) + \nabla p - \nabla \cdot \tau + \tilde{\sigma}\epsilon \nabla \cdot (\nabla\varphi \otimes \nabla\varphi) &= 0 \\ \nabla \cdot u &= 0 \\ \partial_t \varphi + \nabla \cdot (\varphi u) - \gamma \Delta \mu &= 0 \\ \mu + \tilde{\sigma}\epsilon \Delta\varphi - \tilde{\sigma}\epsilon^{-1} W'(\varphi) &= 0 \end{aligned}\right\} \text{ in } \Omega_t^{\mathrm{f}} \qquad (1)$$

with $\rho := \rho(\varphi) = \rho_1(1+\varphi)/2 + \rho_2(1-\varphi)/2$ as mixture density, $u : \Omega_t^{\mathrm{f}} \to \mathbb{R}^d$ as volume-averaged mixture velocity, $p = p : \Omega_t^{\mathrm{f}} \to \mathbb{R}$ as pressure, $\tau = \nu \nabla^s u$ as viscous-stress tensor and $\mu : \Omega_t^{\mathrm{f}} \to \mathbb{R}$ as chemical potential. The mixture viscosity is defined as $\nu := \nu(\varphi) = \nu_1(1+\varphi)/2 + \nu_2(1-\varphi)/2$. The parameter $\tilde{\sigma}$ is related to the fluid–fluid surface tension σ by $2\sqrt{2}\,\tilde{\sigma} = 3\,\sigma$, and $\gamma > 0$ designates mobility. The energy density associated with mixing of the constituents is represented by the standard double-well potential $W(\varphi) = \frac{1}{4}(\varphi^2 - 1)^2$. The parameter $\epsilon > 0$ controls the thickness of the diffuse interface between the fluid constituents.

Suitable initial conditions for (1) are provided by a specification of the initial phase distribution and the initial velocity, according to $\varphi(0, \cdot) = \varphi_0$ and $u(0, \cdot) = u_0$, with $\varphi_0 : \Omega_0^{\mathrm{f}} \to [-1, 1]$ and $u_0 : \Omega_0^{\mathrm{f}} \to \mathbb{R}^d$ exogenous data. Equations (1_1), (1_3), and (1_4) are typically furnished with Dirichlet or Neumann boundary conditions:

$$\begin{aligned} u &= g_{\mathrm{D}}^u \quad \text{on } \Gamma_{\mathrm{D}}^u & -pn + \tau n - \tilde{\sigma}\epsilon \partial_n \varphi \nabla\varphi &= g_{\mathrm{N}}^u \quad \text{on } \Gamma_{\mathrm{N}}^u \\ \varphi &= g_{\mathrm{D}}^{\varphi} \quad \text{on } \Gamma_{\mathrm{D}}^{\varphi} & -\tilde{\sigma}\epsilon \partial_n \varphi &= g_{\mathrm{N}}^{\varphi} \quad \text{on } \Gamma_{\mathrm{N}}^{\varphi} \\ \mu &= g_{\mathrm{D}}^{\mu} \quad \text{on } \Gamma_{\mathrm{D}}^{\mu} & \gamma \partial_n \mu &= g_{\mathrm{N}}^{\mu} \quad \text{on } \Gamma_{\mathrm{N}}^{\mu} \end{aligned} \qquad (2)$$

with n the exterior unit normal vector to $\partial\Omega_t^{\mathrm{f}}$. The right-hand sides in (2) correspond to exogenous data. It generally holds that $\Gamma_{\mathrm{D}}^{(\cdot)} \cap \Gamma_{\mathrm{N}}^{(\cdot)} = \emptyset$. The Neumann condition in (2_1) provides a specification of the fluid traction on the boundary Γ_{N}^u. If $\Gamma_{\mathrm{N}}^{\mu}$ corresponds to a material boundary, homogeneous data $g_{\mathrm{N}}^{\mu} = 0$ provide phase conservation at the boundary. Indeed, from (1_3), the Neumann condition in (2_3) and the Reynolds transport theorem it follows that:

$$\frac{d}{dt}\int_{\Omega_t^{\mathrm{f}}}\varphi = \int_{\partial\Omega_t^{\mathrm{f}}\setminus\Gamma_{\mathrm{N}}^{\mu}}\left(\gamma\partial_n\mu - \varphi(u_n - w_n)\right) - \int_{\Gamma_{\mathrm{N}}^{\mu}}\varphi(u_n - w_n) + \int_{\Gamma_{\mathrm{N}}^{\mu}} g_{\mathrm{N}}^{\mu} \tag{3}$$

with u_n and w_n the normal velocities of the fluid and of the boundary, respectively. Material boundaries satisfy $u_n = w_n$ and, accordingly, the penultimate term in (3) vanishes. Therefore, the contribution of $\Gamma_{\mathrm{N}}^{\mu}$ to production of φ vanishes if $g_{\mathrm{N}}^{\mu} = 0$. An important alternative to (2_2) is the nonlinear Robin-type condition

$$\tilde{\sigma}\epsilon\partial_n\varphi + \sigma_{\mathrm{W}}'(\varphi) = 0 \quad \text{on}\,\Gamma_{\mathrm{W}} \tag{4}$$

with $\sigma_{\mathrm{W}}(\varphi) = \frac{1}{4}(\varphi^3 - 3\varphi)(\sigma_2 - \sigma_1) + \frac{1}{2}(\sigma_1 + \sigma_2)$ the surface tension of the complex-fluid–solid interface, and $\sigma_1 > 0$ and $\sigma_2 > 0$ the fluid–solid surface tensions of species 1 and 2, respectively; see [13]. Note that $\sigma_{\mathrm{W}}(\cdot)$ provides an interpolation of the pure-species fluid–solid surface tensions, i.e., $\sigma_{\mathrm{W}}(1) = \sigma_1$ and $\sigma_{\mathrm{W}}(-1) = \sigma_2$. Equation (4) describes preferential wetting of Γ_{W} by the two fluid components. In particular, the angle $\theta_{\mathrm{s}} = \arccos((\sigma_2 - \sigma_1)/\sigma)$ corresponds to the static contact angle between the diffuse interface and Γ_{W} (interior to fluid 1). Interaction of the complex fluid (1) with a solid substrate is modeled by Dirichlet condition (2_1), (homogeneous) Neumann condition (2_3), and preferential-wetting condition (4).

2.2 *Hyperelastic Saint Venant–Kirchhoff Solid Model*

We consider a hyperelastic solid with Saint Venant–Kirchhoff stress-strain relation. Denoting the initial density of the solid by $\hat{\rho} := \rho_0 : \hat{\Omega}^{\mathrm{s}} \to \mathbb{R}_{>0}$, the solid deformation $\hat{\chi}^{\mathrm{s}} : \hat{\Omega}^{\mathrm{s}} \to \Omega_t^{\mathrm{s}}$ satisfies the equation of motion:

$$\hat{\rho}\partial_t^2\hat{\chi}^{\mathrm{s}} - \hat{\nabla}\cdot\hat{P} = 0 \quad \text{in}\,\hat{\Omega}^{\mathrm{s}} \tag{5}$$

with $\hat{P}$ the first Piola–Kirchhoff stress tensor and $\hat{\nabla}\cdot$ the divergence operator in the reference configuration. For hyperelastic materials, $\hat{\nabla}\cdot\hat{P}$ is the vector-valued function such that $-\int_{\hat{\Omega}^{\mathrm{s}}}\hat{x}\cdot(\hat{\nabla}\cdot\hat{P}) = \mathscr{W}'(\hat{\chi}^{\mathrm{s}};\hat{x})$ for all $\hat{x} \in C_0^{\infty}(\hat{\Omega}^{\mathrm{s}},\mathbb{R}^d)$, with $\mathscr{W}$ the stored-energy functional, $\mathscr{W}'(\hat{\chi}^{\mathrm{s}};\cdot)$ its Fréchet derivative at $\hat{\chi}^{\mathrm{s}}$, and $C_0^{\infty}(\hat{\Omega}^{\mathrm{s}},\mathbb{R}^d)$ the class of $\mathbb{R}^d$-valued smooth functions with compact support in $\hat{\Omega}^{\mathrm{s}}$. Denoting by $F := F(\hat{\chi}^{\mathrm{s}})$ the deformation tensor and by $E := \frac{1}{2}(F^TF - I)$ the Green–Lagrange strain tensor, the Saint Venant–Kirchhoff relation specifies the strain-energy density associated with $\hat{\chi}^{\mathrm{s}}$ as $\frac{1}{2}\lambda_{\mathrm{L}}(\operatorname{tr}E)^2 + \mu_{\mathrm{L}}(\operatorname{tr}E^2)$ with λ_{L} and μ_{L} the Lamé parameters.

The identification of the reference configuration and the initial configuration yields the initial condition $\hat{\chi}_0^{\mathrm{s}} = \mathrm{Id}$. Equation (5) is generally furnished with Dirichlet or Neumann conditions:

$$\hat{\chi}^{\mathrm{s}} = g_{\mathrm{D}}^{\hat{\chi}} \quad \text{on } \hat{\Gamma}_{\mathrm{D}}^{\hat{\chi}} \qquad \hat{P}\hat{n} = g_{\mathrm{N}}^{\hat{\chi}} \quad \text{on } \hat{\Gamma}_{\mathrm{N}}^{\hat{\chi}} \tag{6}$$

with $g_{\mathrm{D}}^{\hat{\chi}}$ and $g_{\mathrm{N}}^{\hat{\chi}}$ deformation and traction data on $\hat{\Gamma}_{\mathrm{D}}^{\hat{\chi}}$ and $\hat{\Gamma}_{\mathrm{N}}^{\hat{\chi}}$, respectively.

2.3 Interface Conditions

The complex fluid (1) and solid (5) are interconnected at the interface by kinematic and dynamic interface conditions. The kinematic condition identifies the mixture velocity and the structural velocity at the interface. This condition can be interpreted as a Dirichlet boundary condition for fluid velocity in accordance with (2_1):

$$u = g_{\mathrm{D}}^{u} := \partial_t \hat{\chi}^{\mathrm{s}} \circ \hat{\chi}^{-1} \text{on } \Gamma_t \subseteq \Gamma_{\mathrm{D}}^{u} \tag{7}$$

Kinematic condition (7) constitutes a partial solid-wall condition for (1). The condition is complemented by a homogeneous Neumann condition (2_3) to impose conservation of phase, and wettability boundary condition (4).

The dynamic condition imposes equilibrium of the fluid and solid tractions and the traction exerted on the interface by the fluid–solid surface tension. The traction due to the fluid–solid surface tension is given by the Young–Laplace relation for non-uniform surface tension according to $\Sigma^{\mathrm{fs}} = \sigma_{\mathrm{W}}(\varphi)\kappa\, n + \nabla_{\Gamma}\sigma_{\mathrm{W}}(\varphi)$, with κ as the additive curvature of Γ_t and $\nabla_{\Gamma}(\cdot)$ the tangential gradient on Γ_t; see, e.g., [9]. We adopt the convention that curvature is negative if the osculating circle in the normal plane is located in the fluid domain. The complex fluid in (1) exerts traction $\Sigma^{\mathrm{f}} := pn - \tau n + \tilde{\sigma}\epsilon\partial_n\varphi\nabla\varphi$ on the interface; cf. (2_1). Note the capillary-stress contribution, $\tilde{\sigma}\epsilon\partial_n\varphi\nabla\varphi$, to the fluid traction. The traction exerted by the solid (5) is $\hat{\Sigma}^{\mathrm{s}} := -\hat{P}\hat{n}$ with $\hat{n}$ the exterior unit normal vector to $\partial\hat{\Omega}^{\mathrm{s}}$; cf. (6). To account for the fact that fluid traction and surface-tension traction are expressed in the current configuration and solid traction is expressed in the reference configuration, we consider the dynamic condition in distributional form:

$$\int_{\hat{\Gamma}} \hat{v}\cdot\hat{\Sigma}^{\mathrm{s}}\,\mathrm{d}\hat{s} = -\int_{\Gamma_t} (\hat{v}\circ\hat{\chi}^{-1})\cdot\left(\Sigma^{\mathrm{f}} + \Sigma^{\mathrm{fs}}\right)\mathrm{d}s \quad \forall \hat{v} \in C_0^{\infty}(\hat{\Gamma}) \tag{8}$$

with $\mathrm{d}\hat{s}$ and $\mathrm{d}s$ the surface measures carried by $\hat{\Gamma}$ and Γ_t, respectively. A precise interpretation of (8) based on weak traction evaluation is presented in Section 3.2.

3 Weak Formulation

In this section we present a consistent weak formulation of the fluid–solid interaction problem in Section 2. We first consider a weak ALE formulation of the NSCH system (1) in Section 3.1. Section 3.2 presents a weak formulation of aggregated FSI problem, including a weak formulation of the solid subsystem (5) and an appropriate weak formulation of the traction exerted by the fluid on the solid at the interface in conformity with the dynamic condition.

3.1 ALE Formulation of NSCH Equations

To accommodate the motion of the fluid domain, we consider a weak formulation of (1) in ALE form. The weak formulation is set in the current configuration. The deformation of the fluid domain, $\hat{\chi}^{\mathrm{f}}$, induces domain velocity $\hat{w} := \partial_t \hat{\chi}^{\mathrm{f}}$. To derive the ALE formulation, we note that

$$\int_{\Omega_t^{\mathrm{f}}} z\partial_t\psi = \frac{d}{dt}\int_{\Omega_t^{\mathrm{f}}} z\psi - \int_{\Omega_t^{\mathrm{f}}} \nabla\cdot(\psi w z) - \int_{\Omega_t^{\mathrm{f}}} \psi\,\partial_t z = \frac{d}{dt}\int_{\Omega_t^{\mathrm{f}}} z\psi - \int_{\Omega_t^{\mathrm{f}}} z\nabla\cdot(\psi w) \quad (9)$$

for all $\hat{z} \in C^\infty(\hat{\Omega}^{\mathrm{f}})$ and $\psi \in C^\infty(\Omega_t^{\mathrm{f}})$, with $z = \hat{z} \circ \hat{\chi}^{-1}$ and $w = \hat{w} \circ \hat{\chi}^{-1}$. The identities in (9) follow from the transport theorem and $\partial_t(\hat{z} \circ \hat{\chi}^{-1}) = -w \cdot \nabla z$. From (9) it follows that (1_1) and (1_3) subject to (2) can be recast into the weak ALE form:

$$\begin{aligned} d_t\langle \rho u, v\rangle + \mathscr{A}_{\mathrm{N}}(u, w, \varphi; v) + \mathscr{B}(p, v) &= \mathscr{L}_{\mathrm{N}}(u, w, \varphi, p; v) \qquad \forall \hat{v} \in C^\infty(\hat{\Omega}^{\mathrm{f}}, \mathbb{R}^d) \\ d_t\langle \varphi, z\rangle + \mathscr{A}_{\mathrm{C}}(u, w, \varphi, \mu; z) &= \mathscr{L}_{\mathrm{C}}(u, w, \varphi, \mu; z) \qquad \forall \hat{z} \in C^\infty(\hat{\Omega}^{\mathrm{f}}) \end{aligned} \quad (10)$$

with $\langle \rho u, v\rangle = \int_{\Omega_t^{\mathrm{f}}} v \cdot \rho u$ and $\langle \varphi, z\rangle = \int_{\Omega_t^{\mathrm{f}}} z\varphi$, and

$$\begin{aligned} \mathscr{A}_{\mathrm{N}}(u, w, \varphi; v) &= \int_{\Omega_t^{\mathrm{f}}} \nabla v : \big(\tau - \rho u \otimes (u - w) - \tilde{\sigma}\epsilon\nabla\varphi \otimes \nabla\varphi\big) \\ \mathscr{L}_{\mathrm{N}}(u, w, \varphi, p; v) &= \int_{\Gamma_{\mathrm{N}}^u} v \cdot g_{\mathrm{N}}^u - \int_{\partial\Omega_t^{\mathrm{f}}\setminus\Gamma_{\mathrm{N}}^u} v \cdot \Sigma^{\mathrm{f}} - \int_{\partial\Omega_t^{\mathrm{f}}} v \cdot \rho u(u_n - w_n) \\ \mathscr{A}_{\mathrm{C}}(u, w, \varphi, \mu; z) &= \int_{\Omega_t^{\mathrm{f}}} \nabla z \cdot \big(\gamma\nabla\mu - \varphi(u - w)\big) \\ \mathscr{L}_{\mathrm{C}}(u, w, \varphi, \mu; z) &= \int_{\Gamma_{\mathrm{N}}^\mu} z g_{\mathrm{N}}^\mu + \int_{\partial\Omega_t^{\mathrm{f}}\setminus\Gamma_{\mathrm{N}}^\mu} z\gamma\partial_n\mu - \int_{\partial\Omega_t^{\mathrm{f}}} z\varphi(u_n - w_n) \\ \mathscr{B}(p, v) &= \int_{\Omega_t^{\mathrm{f}}} -p\,\nabla\cdot v \end{aligned} \quad (11)$$

From (1_2) and (1_4), the Neumann condition in (2_2) and the wetting condition (4), we moreover infer:

$$\begin{aligned} \mathscr{B}(q, u) &= 0 \qquad \forall q \in C^\infty(\Omega_t^{\mathrm{f}}) \\ \mathscr{A}_{\mathrm{P}}(\varphi; \mu, y) &= \mathscr{L}_{\mathrm{P}}(\varphi, y) \qquad \forall y \in C^\infty(\Omega_t^{\mathrm{f}}) \end{aligned} \quad (12)$$

with

$$\begin{aligned} \mathscr{A}_{\mathrm{P}}(\varphi; \mu, y) &= \int_{\Omega_t^{\mathrm{f}}} y(\mu - \tilde{\sigma}\epsilon^{-1}W'(\varphi)) - \int_{\Omega_t^{\mathrm{f}}} \nabla y \cdot \tilde{\sigma}\epsilon\nabla\varphi - \int_{\Gamma_{\mathrm{W}}} y\sigma_{\mathrm{W}}'(\varphi) \\ \mathscr{L}_{\mathrm{P}}(\varphi, y) &= \int_{\Gamma_{\mathrm{N}}^\varphi} y g_{\mathrm{N}}^\varphi - \int_{\partial\Omega_t^{\mathrm{f}}\setminus(\Gamma_{\mathrm{W}}\cup\Gamma_{\mathrm{N}}^\varphi)} y\tilde{\sigma}\epsilon\partial_n\varphi \end{aligned} \quad (13)$$

It is important to note that the fluid–solid interface satisfies $\Gamma_t \subseteq \Gamma_{\mathrm{W}} \cap \Gamma_{\mathrm{N}}^\mu \cap \Gamma_{\mathrm{D}}^u$.

The configuration of the fluid domain, $\hat{\chi}^{\mathrm{f}}$, can be constructed in various manners. We select $\hat{\chi}^{\mathrm{f}} = h_{\hat{\chi}^{\mathrm{s}}|_{\hat{\Gamma}}}$ as the harmonic extension of the trace of the solid displacement on the interface onto $\hat{\Omega}^{\mathrm{f}}$. Accordingly, it holds that $\hat{w} = \partial_t h_{\hat{\chi}^{\mathrm{s}}|_{\hat{\Gamma}}}$.

3.2 Aggregated Fluid–Solid Interaction Problem

From the equation of motion of the solid in (5), we infer the weak formulation:

$$d_t^2\langle\hat{\rho}\hat{\chi}^{\mathrm{s}},\hat{x}\rangle + \mathscr{W}'(\hat{\chi}^{\mathrm{s}};\hat{x}) = \int_{\hat{\Gamma}_{\mathrm{N}}^{\hat{\chi}}} \hat{x}\cdot g_{\mathrm{N}}^{\hat{\chi}} - \int_{\partial\hat{\Omega}^{\mathrm{s}}\setminus(\hat{\Gamma}\cup\hat{\Gamma}_{\mathrm{N}}^{\hat{\chi}})} \hat{x}\cdot\hat{\Sigma}^s - \int_{\hat{\Gamma}} \hat{x}\cdot\hat{\Sigma}^s \tag{14}$$

for all $\hat{x} \in C^{\infty}(\hat{\Omega}^{\mathrm{s}},\mathbb{R}^d)$. The ultimate term in (14) constitutes the solid traction on the interface. The dynamic condition imposes that this term coincides with the right-hand side of (8). Noting that $\Gamma_t \subseteq \partial\Omega_t^{\mathrm{f}} \setminus \Gamma_{\mathrm{N}}^u$, equations (10)–(11) convey that the fluid-traction contribution can be expressed as:

$$-\int_{\Gamma_t} x\cdot\Sigma^{\mathrm{f}} = \int_{\Gamma_t} \ell_x\cdot\rho u(u_n - w_n) + d_t\langle\rho u,\ell_x\rangle + \mathscr{A}_{\mathrm{N}}(u,w,\varphi;\ell_x) + \mathscr{B}(p,\ell_x) \tag{15}$$

where ℓ_x represents an appropriate lifting of x, viz. any suitable function $\Omega_t^{\mathrm{f}} \to \mathbb{R}^d$ such that $\ell_x|_{\Gamma_t} = x$ and ℓ_x vanishes on $\partial\Omega_t^{\mathrm{f}} \setminus \Gamma_t$. The right member of (15) provides a weak formulation of the traction functional in the left member of (15), in the sense that the identity (15) holds for all solutions of (1) for which the left-hand side of (15) is defined, but the right-hand side is defined for a larger class of solutions to (1) with weaker regularity; see also [15, 21, 22]. The contribution of the fluid–solid surface tension in the right-member of (8) can be reformulated as:

$$-\int_{\Gamma_t} x\cdot\big(\sigma_{\mathrm{W}}(\varphi)\kappa n + \nabla_{\Gamma}\sigma_{\mathrm{W}}(\varphi)\big) = \mathscr{C}(\hat{\chi}^{\mathrm{s}},\varphi;x) + \int_{\partial\Gamma_t} \upsilon\cdot\big(\sigma_{\mathrm{w}}(\varphi)\,x\big) \tag{16}$$

with

$$\mathscr{C}(\hat{\chi}^{\mathrm{s}},\varphi;x) = \int_{\Gamma_t} \sigma_{\mathrm{w}}(\varphi)\,\nabla_{\Gamma}\mathrm{Id}_{\Gamma_t} : \nabla_{\Gamma}x \tag{17}$$

and Id_{Γ_t} the identity on Γ_t and υ the exterior unit normal vector to $\partial\Gamma_t$ in the tangent bundle of Γ_t; see [3, 7]. Let us note that the right member of (17) depends implicitly on the solid deformation $\hat{\chi}^{\mathrm{s}}$ via the shape of the interface Γ_t. The second term in the right member of (16) cannot generally be bounded in weak formulations, and it must vanish by virtue of boundary conditions on $\hat{\chi}^{\mathrm{s}}$ or x.

To provide a setting for the weak formulation of the fluid–solid interaction problem, let $L^2(\omega)$ denote the class of square-integrable functions on any $\omega \subset \mathbb{R}^d$, $H^1(\omega)$ the Sobolev space of functions in $L^2(\omega)$ with weak derivatives in $L^2(\omega)$, and

$H^1_{0,\Xi}(\omega)$ the subspace of functions that vanish on $\Xi \subseteq \partial\Omega$. For a vector space $X(\omega)$ of scalar-valued functions, $X(\omega, \mathbb{R}^d)$ is the extension to the corresponding vector space of $\mathbb{R}^d$-valued functions. Given a vector space V and a time interval $(0, T)$, $W(0, T; V)$ represents a (suitable) class of functions from $(0, T)$ into $X(\omega)$.

We collect the ambient spaces for the fluid and solid variables into[1]:

$$V := H^1_{0,\Gamma^u_{\rm D}}(\Omega^{\rm f}_t, \mathbb{R}^d) \times L^2(\Omega^{\rm f}_t) \times H^1_{0,\Gamma^\varphi_{\rm D}}(\Omega^{\rm f}_t) \times H^1_{0,\Gamma^\mu_{\rm D}}(\Omega^{\rm f}_t) \times H^1_{0,\Gamma^{\hat\chi}_{\rm D}}(\hat\Omega^{\rm s}, \mathbb{R}^d) \quad (18)$$

For conciseness, we assume $g^u_{\rm D}|_{\Gamma^u_{\rm D}\backslash\Gamma_t} = 0$, $g^\varphi_{\rm D} = 0$, $g^\mu_{\rm D} = 0$, and $g^{\hat\chi}_{\rm D} = \mathrm{Id}$. The aggregated fluid–solid interaction problem can then be condensed into:

Find $(u, p, \varphi, \mu, \hat\chi^{\rm s} - \mathrm{Id}) \in W(0, T; V)$ *such that almost everywhere in* $(0, T)$:

$$\begin{aligned} & d_t\langle\rho(u + \ell_{w|\Gamma_t}), v + \ell_{x|\Gamma_t}\rangle + \mathscr{A}_{\rm N}\big(u + \ell_{w|\Gamma_t}, w, \varphi; v + \ell_{x|\Gamma_t}\big) + \mathscr{B}(p, v + \ell_{x|\Gamma_t}) \\ &\quad + \mathscr{B}(q, u + \ell_{w|\Gamma_t}) + d_t\langle\varphi, z\rangle + \mathscr{A}_{\rm C}(u + \ell_{w|\Gamma_t}, w, \varphi, \mu; z) + \mathscr{A}_{\rm P}(\varphi; \mu, y) \\ & + d^2_t\langle\hat\rho\hat\chi^{\rm s}, \hat x\rangle + \mathscr{W}'(\hat\chi^{\rm s}, \hat x) + \mathscr{C}(\hat\chi^{\rm s}, \varphi; x) = \mathscr{L}_{\rm A}(v, z, y, \hat x) \quad \forall (v, q, z, y, \hat x) \in V \end{aligned} \quad (19)$$

with $w = \partial_t h_{\hat\chi^{\rm s}|_{\hat\Gamma}} \circ \hat\chi^{-1}$ and $x = \hat x \circ \hat\chi^{-1}$, and the aggregated linear form:

$$\mathscr{L}_{\rm A}(v, z, y, \hat x) = \int_{\Gamma^u_{\rm N}} v \cdot g^u_{\rm N} + \int_{\Gamma^\mu_{\rm N}} z\, g^\mu_{\rm N} + \int_{\Gamma^\varphi_{\rm N}} y\, g^\varphi_{\rm N} + \int_{\Gamma^{\hat\chi}_{\rm N}} \hat x \cdot g^{\hat\chi}_{\rm N}$$

It is to be noted that $\hat\chi^{\rm s} - \mathrm{Id}$ represents solid displacement. Furthermore, by virtue of $(v + \ell_{x|\Gamma_t})|_{\Gamma_t} = \hat x|_{\hat\Gamma} \circ \hat\chi^{-1}$ and $(u + \ell_{w|\Gamma_t})|_{\Gamma_t} = \partial_t\hat\chi|_{\hat\Gamma} \circ \hat\chi^{-1}$, the test spaces for the equations of motion of the fluid and the solid and the trial spaces for the fluid and solid velocity in (19) are essentially continuous across the interface.

4 Numerical Experiments

To evaluate the predictive capabilities of the presented CFSI model, we consider numerical approximations of (19) for the experimental setup in [18]. The test case concerns a 13.8 *p*l droplet on a soft substrate; see Fig. 1 (*left*). We characterize the substrate by a nearly incompressible solid with Saint Venant–Kirchhoff constitutive behavior, with Lamé parameters $\mu_{\rm L} = \tilde E/(2 + 2\tilde\nu)$ and $\lambda_{\rm L} = \tilde\nu\tilde E/(1 + \tilde\nu)(1 - 2\tilde\nu)$, and Young's modulus $\tilde E = 3\,k\,\mathrm{Pa}$ and Poisson ratio $\tilde\nu = 0.499$. The surface tension of the interface between the droplet (fluid 1) and ambient fluid (fluid 2) is $\sigma =$

[1]The admissible solid deformations must in fact satisfy auxiliary conditions at the interface to ensure that the surface-tension contributions are well-defined. Detailed treatment of this aspect is beyond the scope of this work.

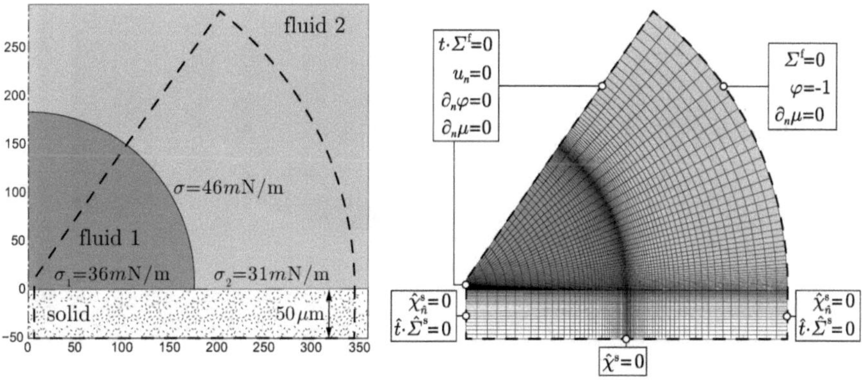

Fig. 1 Illustration of the considered experimental configuration (*left*) and the corresponding computational setup (*right*).

$46\,m\,\mathrm{N/m}$. The fluid/solid surface tension of fluid 1 (resp. fluid 2) is $\sigma_1 = 36\,m\,\mathrm{N/m}$ (resp. $\sigma_2 = 31\,m\,\mathrm{N/m}$). The diffuse-interface thickness is set to $\epsilon = 2\,\mu\mathrm{m}$. Our interest is restricted to steady solutions and, hence, ρ_1, ρ_2, $\hat{\rho}$, ν, and γ are essentially irrelevant. For completeness, we mention that we select matched fluid densities $\rho = \rho_1 = \rho_2 = 1.26\,pg/(\mu\mathrm{m})^3$, matched fluid viscosities $\nu = \nu_1 = \nu_2 = 1412\,m\,\mathrm{Pa\,s}$, solid density $\hat{\rho} = 12.6\,pg/(\mu\mathrm{m})^3$, and mobility $\gamma = 0.01\,(\mu\mathrm{m})^3\,\mu\mathrm{s}/pg$. We refer to Fig. 1 for further details of the experimental configuration.

We incorporate the rotational symmetry of the experimental setup in the discrete approximation of (19). The considered approximation spaces are based on a locally refined mesh, adapted to the diffuse interface; see Fig. 1 (*right*). We apply Raviart–Thomas compatible B-spline approximations for velocity (u) and pressure (p) with order $((3,2),(2,3))$ and 2, respectively; see [6, 8]. The order parameter (φ) and chemical potential (μ) are approximated by means of quadratic B-splines. The solid deformation ($\hat{\chi}^{\mathrm{s}}$) and the deformation of the fluid domain ($\hat{\chi}^{\mathrm{f}}$) are approximated with quadratic B-splines as well. Let us note that by virtue of the C^1-continuity of the solid deformation, the interface Γ_t corresponds to a C^1 manifold. The temporal discretization of (19) is based on backward Euler approximation of the time derivatives, with time step $0.5\,ms$. In each time step, the aggregated fluid–solid interaction problem is solved by means of subiteration with underrelaxation; see, for instance, [20].

Figure 2 (*left*) presents a comparison of the computed interface configuration, Γ_t, at $t \in \{0, 0.5, 1, 2, 4, 8, 16\}\,ms$ and experimental data from [18]. At $t = 16\,ms$, the interface has essentially reached its equilibrium deformation. The surface tension of the fluid–fluid interface yields a localized load on the fluid–solid interface near the contact line, resulting in a kink in the surface deformation of the soft substrate. In addition, the fluid–fluid surface tension leads to an increased pressure in the droplet relative to the ambient pressure (see also Fig. 2 (*right*)), viz. Laplace pressure, and a corresponding depression of the substrate. Comparison of the experimental and

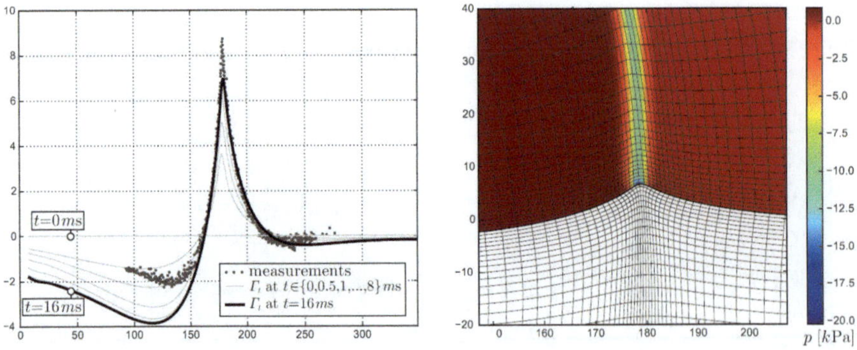

Fig. 2 Comparison of the computed fluid–solid interface configuration Γ_t at $t \in \{0, 0.5, 1, 2, 4, 8\}\,ms$ (*grey*) and at $t = 16\,ms$ (*black*) and rendering of experimental results from [18] (*left*), and magnification of the contact-line region at $t = 16\,ms$ with deformed fluid and solid meshes and computed pressure distribution (*right*).

computed results conveys that the fluid–solid interface elevation at the contact line is underestimated by approximately 25%. The underestimation can be attributed to the regularizing effect of the diffuse interface. It is anticipated that further reduction of the diffuse-interface thickness (ϵ) and corresponding refinement of the mesh leads to an increase in the fluid–solid interface elevation at the contact line. The indentation of the substrate below the droplet is noticeably overestimated. In this regard, it is to be mentioned that on account of the nearly incompressible behavior of the solid, its volume at $t = 16\,ms$ has decreased by only 0.16% relative to the initial volume.

Figure 2 (*right*) presents a magnification of the contact-line region at $t = 16\,ms$ with the fluid and solid meshes in the actual configuration and the computed pressure distribution in the complex fluid. It is noteworthy that the pressure in the diffuse interface exhibits a localized minimum at the contact line. The pressure in the droplet is virtually uniform with value $p \approx 520\,\text{Pa}$, which is close to the theoretical Laplace pressure $2\sigma/R$ in a droplet on a rigid substrate with radius $R = 178\,\mu\text{m}$ and surface tension $\sigma = 46\,m\,\text{N/m}$.

5 Conclusion

We presented a model for the interaction of a complex fluid with an elastic solid, in which the complex fluid is represented by the NSCH equations and the solid is characterized by a hyperelastic material with a Saint Venant–Kirchhoff stored-energy functional. The interaction between the fluid and the solid at their mutual interface is described by a preferential-wetting condition in addition to the usual kinematic and dynamic interface conditions. The fluid traction on the fluid–solid interface comprises a non-standard capillary-stress contribution, and the dynamic condition contains a contribution from the non-uniform fluid–solid surface tension. A weak formulation of the CFSI problem was presented, based on an ALE formulation of the NSCH system and a suitable reformulation of the complex-fluid traction and the fluid–solid surface-tension traction.

Numerical results were presented for a stationary droplet on a soft solid substrate, based on finite-element approximation of the weak formulation of the aggregated CFSI problem. Comparison of the computed results with experimental data for the considered test case exhibited very good agreement in the contact-line region. The substrate depression below the droplet was noticeably overestimated relative to the experimental data. In view of the close agreement between the computed pressure in the droplet and the theoretical Laplace pressure, it appears that the discrepancy between the computed and observed depression is to be attributed to corresponding differences in the constitutive behavior of the solid substrate. The overall good agreement between the computed and experimental data indicates the potential of computational CFSI models based on the NSCH equations to predict elasto-capillary phenomena.

Acknowledgements This work is supported by NanoNextNL, a micro and nanotechnology consortium of the Government of the Netherlands and 130 partners.

References

1. Abels, H., Garcke, H., Grün, G.: Thermodynamically consistent, frame indifferent diffuse interface models for incompressible two-phase flows with different densities. Math. Models Methods Appl. Sci. **22**, 1150013 (2012)
2. Aland, S., Voigt, A.: Benchmark computations of diffuse interface models for two-dimensional bubble dynamics. Int. J. Numer. Meth. Fluids **69**, 747–761 (2012)
3. Bänsch, E.: Finite element discretization of the Navier–Stokes equations with a free capillary surface. Numer. Math. **88**, 203–235 (2001)
4. Bazilevs, Y., Takizawa, K., Tezduyar, T.E.: Computational Fluid-Structure Interaction: Methods and Applications. Wiley, Chichester (2013)
5. Bueno, J., Bona-Casas, C., Bazilevs, Y., Gomez, H.: Interaction of complex fluids and solids: theory, algorithms and application to phase-change-driven implosion. Comput. Mech. **55**, 1105–1118 (2015)
6. Buffa, A., de Falco, C., Sangalli, G.: Isogeometric analysis: stable elements for the 2D Stokes equation. Int. J. Numer. Meth. Fluids **65**, 1407–1422 (2011)
7. Dziuk, G.: An algorithm for evolutionary surfaces. Numer. Math. **58**, 603–611 (1991)
8. Evans, J.A., Hughes, T.J.R.: Isogeometric divergence-conforming B-splines for the steady Navier-Stokes equations. Math. Models Methods Appl. Sci. **23**, 1421–1478 (2012)
9. Gouin, H.: Interfaces endowed with nonconstant surface energies revisited with the d'Alembert–Lagrange principle. Math. Mech. Complex Syst. **2**, 23–43 (2014)
10. Guo, Z., Lin, P., Lowengrub, J.S.: A numerical method for the quasi-incompressible Cahn–Hilliard–Navier–Stokes equations for variable density flows with a discrete energy law. J. Comput. Phys. **276**, 486–507, 11 (2014)
11. Hohenberg, P.C., Halperin, B.I.: Theory of dynamic critical phenomena. Rev. Mod. Phys. **49**, 435–479 (1977)
12. Jacqmin, D.: Calculation of two-phase Navier–Stokes flows using phase-field modeling. J. Comput. Phys. **155**, 96–127, 10 (1999)

13. Jacqmin, D.: Contact-line dynamics of a diffuse fluid interface. J. Fluid Mech. **402**, 57–88 (2000)
14. Lowengrub, J., Truskinovsky, L.: Quasi-incompressible Cahn–Hilliard fluids and topological transitions. Proc. Math. Phys. Eng. Sci. **454**, 2617–2654 (1998)
15. Melbø, H., Kvamsdal, T.: Goal oriented error estimators for Stokes equations based on variationally consistent postprocessing. Comput. Methods Appl. Mech. Eng. **192**, 613–633 (2003)
16. Py, C., Reverdy, P., Doppler, L., Bico, J., Roman, B., Baroud, C.N.: Capillary origami: spontaneous wrapping of a droplet with an elastic sheet. Phys. Rev. Lett. **98**, 156103 (2007)
17. Style, R.W., et al.: Patterning droplets with durotaxis. PNAS **110**(31), 12541–12544 (2013)
18. Style, R.W., Boltyanskiy, R., Che, Y., Wettlaufer, J.S., Wilen, L.A., Dufresne, E.R.: Universal deformation of soft substrates near a contact line and the direct measurement of solid surface stresses. Phys. Rev. Lett. **110**, 066103 (2013)
19. Tezduyar, T.E., Takizawa, K., Moorman, C., Wright, S., Christopher, J.: Space–time finite element computation of complex fluid–structure interactions. Int. J. Numer. Meth. Fluids **64**, 1201–1218 (2010)
20. van Brummelen, E.H.: Partitioned iterative solution methods for fluid-structure interaction. Int. J. Numer. Meth. Fluids **65**, 3–27 (2011)
21. van Brummelen, E.H., van der Zee, K.G., Garg, V.V., Prudhomme, S.: Flux evaluation in primal and dual boundary-coupled problems. J. Appl. Mech. **79**, 010904-8 (2012)
22. van der Zee, K.G., van Brummelen, E.H., Akkerman, I., de Borst, R.: Goal-oriented error estimation and adaptivity for fluid-structure interaction using exact linearized adjoints. Comput. Methods Appl. Mech. Eng. **200**, 2738–2757 (2011)

Fluid–Structure Interaction Modeling and Isogeometric Analysis of a Hydraulic Arresting Gear at Full Scale

Ming-Chen Hsu, Chenglong Wang, Michael C.H. Wu, Fei Xu, and Yuri Bazilevs

Abstract Fluid–structure interaction (FSI) analysis of a full-scale hydraulic arresting gear used to retard the forward motion of an aircraft landing on an aircraft-carrier deck is performed. The simulations make use of the recently developed core and special-purpose FSI techniques for other problem classes, specialized to the present application. A recently proposed interactive geometry modeling and parametric design platform for isogeometric analysis (IGA) is directly employed to create the arresting gear model, and illustrates a natural application of IGA to this problem class. The fluid mechanics and FSI simulation results are reported in terms of the arresting gear rotor loads and blade structural deformation and vibration. Excellent agreement is achieved with the experimental results for the arresting gear design simulated in this work.

1 Introduction

Military aircraft, during landing on the deck of an aircraft carrier, ejects a "hook" that engages a wire connected to a tape drum. The resultant tape-drum angular momentum is transferred to the rotor inside a hydraulic energy absorber (or a hydraulic arresting gear). The rotor, which is a steel structure several feet in diameter, accelerates rapidly, reaching speeds of 800 rpm. The rotor acceleration is then arrested by the drag forces coming from the surrounding water inside the arresting gear. This, in turn, puts the wire in tension and rapidly slows the aircraft forward motion. The rotor speed and blade topology, geometry, and structural design play a critical role in the performance of the device, both in its function

M.-C. Hsu • C. Wang • M.C.H. Wu • F. Xu
Department of Mechanical Engineering, Iowa State University, 2025 Black Engineering, Ames, IA 50011, USA

Y. Bazilevs (✉)
Department of Structural Engineering, University of California, San Diego, 9500 Gilman Drive, La Jolla, CA 92093, USA
e-mail: yuri@ucsd.edu

Y. Bazilevs, K. Takizawa (eds.), *Advances in Computational Fluid-Structure Interaction and Flow Simulation*, Modeling and Simulation in Science, Engineering and Technology, DOI 10.1007/978-3-319-40827-9_36

to arrest the motion of landing aircraft, as well as in its ability to withstand the internal hydrodynamic loads and perform multiple consecutive aircraft arrests without failure. As a result, accurate prediction of rotor loads and the structure response to these loads is important, requiring advanced modeling and simulation, which we undertake in this work using isogeometric analysis (IGA) [1, 2] and fluid–structure interaction (FSI) [3].

The paper is outlined as follows. In Section 2, we describe the geometry of the Virginia Tech (VT) arresting gear design [4], which belongs to the Model 64 [5] energy absorber system. We describe a novel technique for IGA analysis-suitable geometry construction of the arresting gear design. We make use of a recently proposed interactive geometry modeling and parametric design platform [6], which is based on the Rhino 3D CAD software [7] with an embedded visual programming tool Grasshopper [8]. Rhino 3D gives the user access to complex geometry modeling functionality with objects such as NURBS and T-splines, while Grasshopper is employed for the generative algorithm approach to arresting gear geometric design. In Section 3, we present the governing equations involved in the FSI model and summarize the numerical formulations employed. In Section 4, we present the results of standalone fluid and structural mechanics, and FSI analyses of the VT arresting gear at full scale.

2 Geometry Modeling and Meshing for the Arresting Gear FSI Analysis

In this work we simulate the Virginia Tech (VT) arresting gear design described in [4] and shown in Figure 1 (Left). We consider a full-scale model with slightly simplified geometry, but with all the important structural components represented. The VT model has experimental data available for hydrodynamic loads acting on the rotor operating at speeds ranging from 200 rpm to 800 rpm, which are typical rotor speeds during the aircraft arrest. The availability of experimental data enables one to perform methods validation at full scale, and to assess the computational effort needed for this challenging problem class.

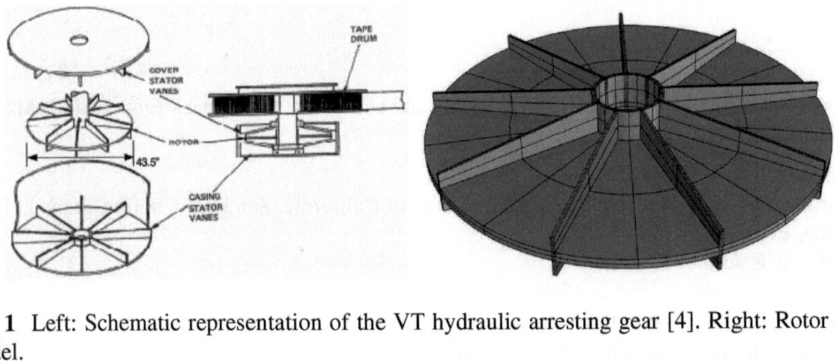

Fig. 1 Left: Schematic representation of the VT hydraulic arresting gear [4]. Right: Rotor solid model.

The VT arresting gear design includes two main parts: A rotating turbine (rotor) and a stationary reel (stator). The rotor diameter is 43.5 in. There are nine blades on the rotor side and eight vanes on the stator side, with a small gap present between the rotor blades and stator vanes. The rotor cross-section geometry is shown in Figure 2 (Left), while the key rotor geometric parameters and dimensions are summarized in table of Figure 2 (Right).

Using this data input, the arresting gear fluid-mechanics domain geometry is created with the help of the interactive geometry modeling and parametric design platform described in [6]. The generative algorithm employed for the rotor design is depicted in Figure 3. The algorithm, which is implemented using the visual programming interface Grasshopper, takes the rotor parametric input and, using the existing Rhino 3D functionality, constructs the underlying NURBS model of the arresting gear geometry. The resulting rotor geometric model is shown in Figure 1 (Right). The stator geometry is constructed in an analogous fashion. Figure 4 shows the complete arresting gear geometry, as well as a view of the model interior

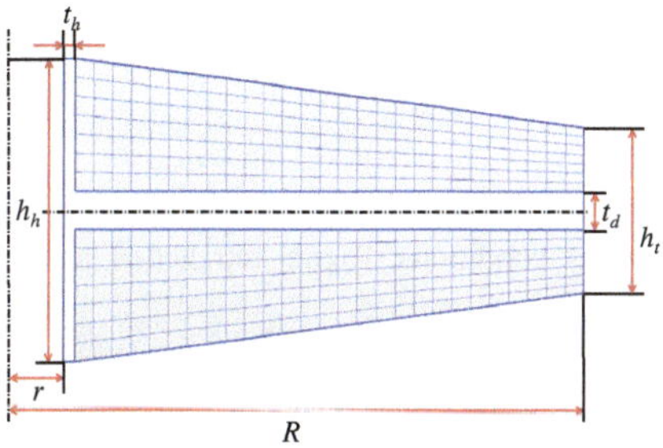

Parameters	Name	Unit (in)
Hub thickness	t_h	0.4
Hub height	h_h	7.96
Inner radius	r	3.48
Outer radius	R	21.75
Disc thickness	t_d	1.0
Tip height	h_t	4.35
Blade thickness	t_b	0.348

Fig. 2 Left: Rotor cross-section with dimensions. Right: Arresting gear rotor dimensions.

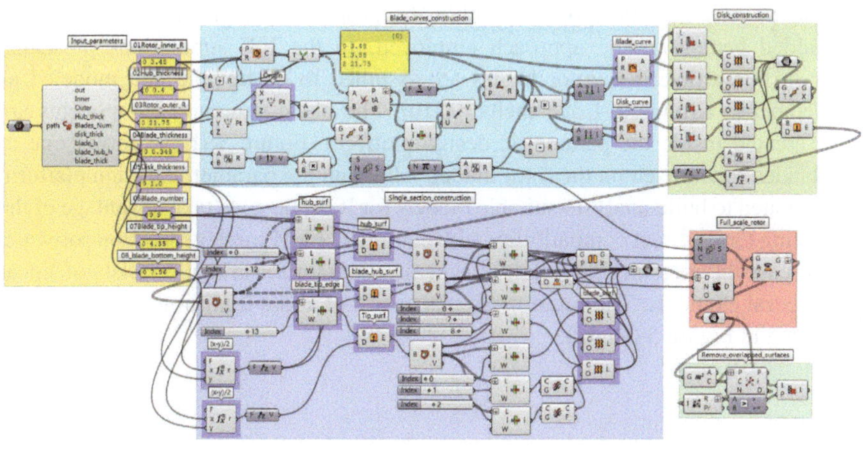

Fig. 3 Generative algorithm for parametric geometry design of the arresting gear rotor expressed in a visual programming interface Grasshopper.

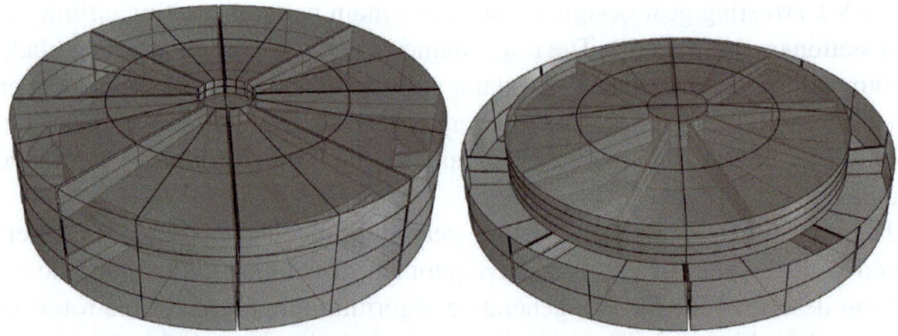

Fig. 4 Left: NURBS-based IGA model of the VT arresting gear, including rotor, stator, and outer casing. Right: NURBS-based IGA model of the VT arresting gear with a zoom on the sliding interface between rotor and stator domains.

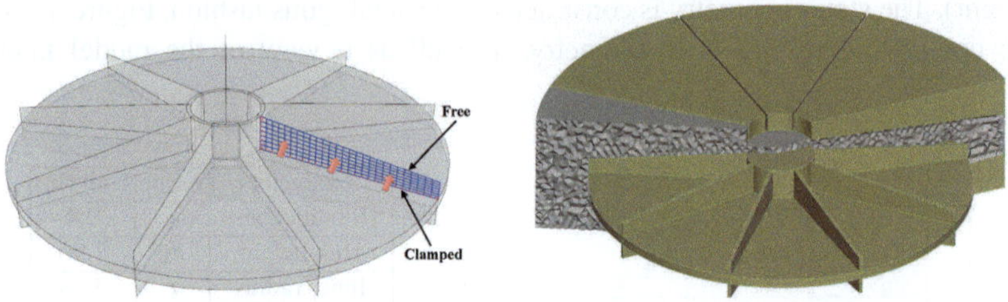

Fig. 5 Left: Rotor blade with clamped boundary conditions. Right: Tetrahedral mesh of the arresting gear fluid-mechanics domain.

showing the sliding interface design. The sliding interface is needed for the fluid mechanics part of the FSI problem to account for the relative motion between the rotor and stator components. It is interesting to note that such objects as sliding interfaces may also be modeled within the proposed design platform, which is an added benefit of the approach employed. In the FSI simulations presented later in this article, the rotor blades are assumed to be flexible and modeled as isogeometric Kirchhoff–Love shells [9, 10]. The blade structural components are also created inside the parametric design platform. Figure 5 (Left) shows one of the blades superposed on the rotor model. Although the parametric design platform was created to build analysis-suitable models for IGA, here we take advantage of the modeling tool's direct compatibility with automatic FEM mesh generation software, and create the arresting gear fluid-mechanics domain mesh that consists of linear tetrahedral elements. The finest mesh makes use of about 2 million tetrahedra, and is shown in Figure 5 (Right).

3 Governing Equations and Numerical Methods

The hydrodynamics of the arresting gear is governed by the Navier–Stokes equations of incompressible flows, which are posed on a moving spatial domain and written in the Arbitrary Lagrangian–Eulerian (ALE) frame [11] as follows:

$$\rho_1 \left(\left. \frac{\partial \mathbf{u}}{\partial t} \right|_{\hat{x}} + (\mathbf{u} - \hat{\mathbf{u}}) \cdot \nabla \mathbf{u} - \mathbf{f}_1 \right) - \nabla \cdot \boldsymbol{\sigma}_1 = \mathbf{0}, \tag{1}$$

$$\nabla \cdot \mathbf{u} = 0. \tag{2}$$

Here ρ_1 is the fluid density, $\mathbf{u}$ is the velocity, $\mathbf{f}_1$ is the body force per unit mass, and $\hat{\mathbf{u}}$ is the velocity of the fluid mechanics domain. The Cauchy stress, $\boldsymbol{\sigma}_1$, is given by

$$\boldsymbol{\sigma}_1 (\mathbf{u}, p) = -p\mathbf{I} + 2\mu\varepsilon (\mathbf{u}), \tag{3}$$

where p is the pressure, $\mathbf{I}$ is the identity tensor, μ is the dynamic viscosity, and $\varepsilon (\mathbf{u})$ is the strain-rate tensor defined as

$$\varepsilon (\mathbf{u}) = \frac{\nabla \mathbf{u} + \nabla \mathbf{u}^T}{2}. \tag{4}$$

The time derivative in Eq. (1) is taken with respect to the fixed referential-domain coordinates $\hat{\mathbf{x}}$. All space derivatives are taken with respect to spatial coordinates of the current configuration $\mathbf{x}$.

The governing equations of structural mechanics are written in the Lagrangian frame [12] and consist of the local balance of linear momentum:

$$\rho_2 \left(\frac{\mathrm{d}^2 \mathbf{y}}{\mathrm{d}t^2} - \mathbf{f}_2 \right) - \nabla \cdot \boldsymbol{\sigma}_2 = \mathbf{0}. \tag{5}$$

Here ρ_2 is the structural mass density, $\mathbf{f}_2$ is the body force per unit mass, $\boldsymbol{\sigma}_2$ is the Cauchy stress, and $\mathbf{y}$ is the unknown structural displacement. The time derivative in Eq. (5) is taken with respect to the fixed material coordinates of the structure reference configuration.

Compatibility of the kinematics and tractions is enforced at the fluid–structure interface, namely

$$\mathbf{u} - \frac{\mathrm{d}\mathbf{y}}{\mathrm{d}t} = \mathbf{0}, \tag{6}$$

$$\boldsymbol{\sigma}_1 \mathbf{n}_1 + \boldsymbol{\sigma}_2 \mathbf{n}_2 = \mathbf{0}, \tag{7}$$

where $\mathbf{n}_1$ and $\mathbf{n}_2$ are the unit outward normal vectors to the fluid and structural mechanics domains, respectively.

To discretize the arresting gear hydrodynamics, the ALE–VMS method [13, 14] and weakly enforced essential boundary conditions [15–17] are employed. The former is an extension of the residual-based variational multiscale large eddy simulation turbulence model [18] to moving domains using the ALE technique, while the latter acts as a "near-wall model" in that it relaxes boundary-layer resolution requirements to achieve good accuracy of fluid solution and loads prediction on meshes without excessive boundary-layer refinement [19–24].

In the arresting gear design the stator is located in close proximity of the rotor, leaving only a small gap as the rotor blades pass the stator vanes during the device operation. To capture the complex dynamics of arresting gear rotor–stator interaction, the sliding-interface technique from [25, 26] is employed.

The structural mechanics of rotor blades and stator vanes is modeled using Kirchhoff–Love shells [9, 10]. These are discretized using IGA based on NURBS [1, 2] and make use of only displacement degrees of freedom. Using rotation-free IGA shells to model the blades presents a good combination of efficiency, accuracy, and robustness.

The coupled FSI problem is formulated using an augmented Lagrangian approach for FSI, which was originally proposed in [27] to handle boundary-fitted mesh computations with non-matching fluid–structure interface discretizations. While at the fluid–structure interface the fluid mechanics mesh follows the motion of the blades, the outer boundaries of the rotor subdomain are restricted to only undergo rigid rotation. This choice of domain motion preserves the geometry of the sliding interface.

The rest of the mesh motion is obtained by solving the equations of elastostatics with Jacobian-based stiffening [28–33]. The generalized-α method [34–36] is employed to advance to FSI equations in time, while block-iterative coupling strategy [3, 33, 37, 38] is used to solve the coupled FSI system at each time step.

4 Simulations of the VT Arresting Gear Model

4.1 Mesh Convergence Study for the Fluid Mechanics Simulation

We first perform standalone fluid mechanics computations assuming the structure is rigid and rotor speed is prescribed. Four cases corresponding to rotor speeds ranging from 200 rpm to 800 rpm at 200 rpm intervals are computed. Furthermore, mesh refinement in each case is performed to assess the mesh resolution requirements in achieving the desired level of accuracy in the rotor hydrodynamic torque prediction. Table 1 shows the element and time-step sizes employed in the mesh convergence study. For this study the simulations are started impulsively and continued until a statistically stationary value of the rotor torque is achieved. The results of the mesh refinement study are shown in Table 1 and Figure 6. In all cases, the converged

Table 1 Mesh statistics for the refinement study.

Torque (ft lb)	Mesh 1	Mesh 2	Mesh 3	Mesh 4	Experiment
Mesh size (in)	3.0	2.0	1.0	0.5	N/A
Time-step size (s)	7.5×10^{-4}	5.0×10^{-4}	2.5×10^{-4}	1.25×10^{-4}	N/A
Number of elements	16,395	49,644	314,462	2,008,047	N/A
200 rpm	4,690	6,294	6,215	6,116	6,196
400 rpm	17,734	23,714	26,559	24,730	24,782
600 rpm	37,726	50,963	59,948	56,924	55,760
800 rpm	63,991	87,302	107,043	104,733	99,128

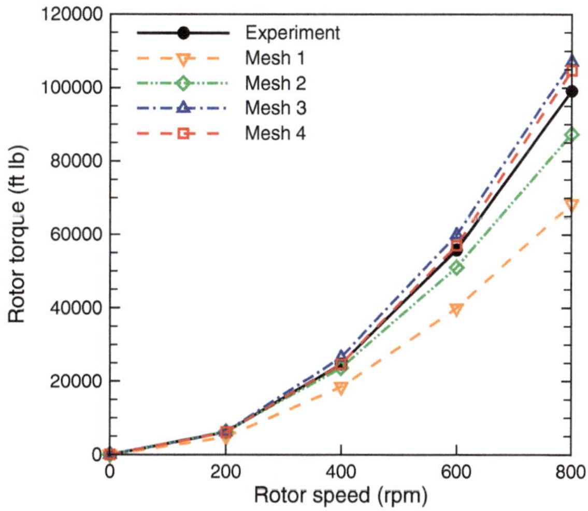

Fig. 6 Rotor-torque values obtained in the mesh refinement study of the hydraulic arresting gear. Excellent mesh convergence results are achieved for all rotor speeds. Note that the mesh corresponding to the blue curve gives reasonably accurate results while consisting of only slightly over 300,000 elements.

results are in very good agreement with the experimental torque data. Note that, a coarse mesh of around 300,000 elements is already capable of producing accurate rotor loads, especially for lower-rpm cases.

4.2 *Ramp-Up Simulation*

In this section we present a standalone fluid mechanics arresting gear ramp-up simulation, which is representative of the device operation during the aircraft arrest. For this, we employ the second finest mesh in Table 1, and prescribe time-dependent rotor speed. We assume the rotor angular acceleration decreases from 670.2 rad/s^2 to zero within 0.25 s, at which point the rotor reaches the terminal speed of 800 rpm. The resulting rotor torque time history is shown in Figure 7. Within 0.25 s, the torque rapidly climbs to a statistically stationary value that is consistent with the impulsive-start simulation results presented in Table 1 and Figure 6.

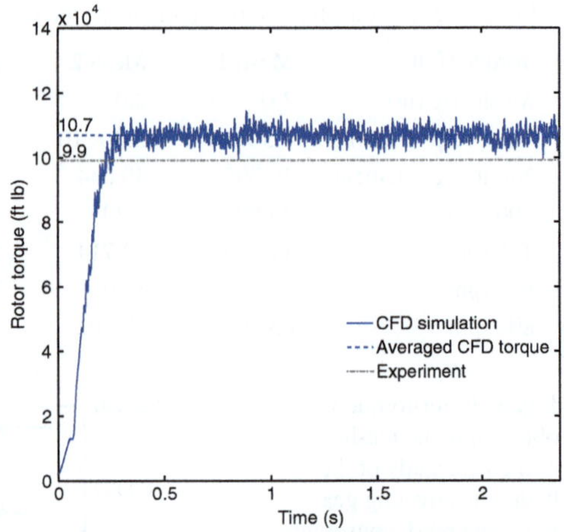

Fig. 7 Time history of the rotor hydrodynamic torque for the terminal rotor speed of 800 rpm. Experimentally measured torque value at constant rotor speed of 800 rpm is plotted for comparison.

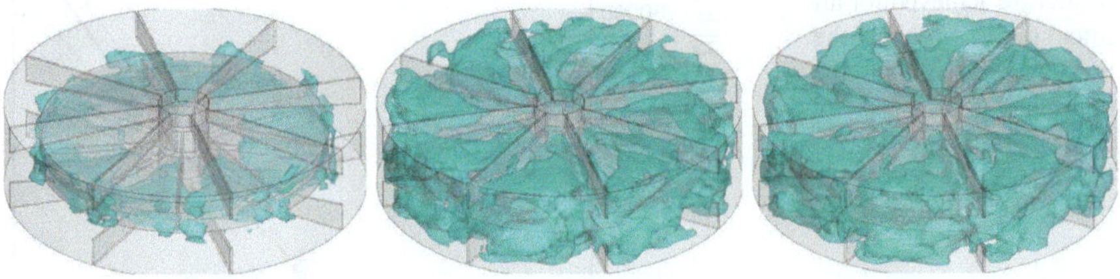

(a) $t = 0.0875$ s. (b) $t = 0.25$ s. (c) $t = 0.45$ s.

Fig. 8 Isosurfaces of water speed for the ramp-up simulation at different time instances showing the development of complex, recirculating turbulent flow.

Figure 8 shows the flow speed isosurfaces at three different time instances as the rotor spins inside the arresting gear. The flow quickly transitions to turbulence and appears to be fully developed and statistically stationary soon after the terminal rotor speed is achieved. The full turbulent cascade, from large vortices to small eddies, is present inside the arresting gear, which is causing additional resistance to rotor motion. Instantaneous pressure contours on the rotor are shown in Figure 9. Note that the pressure distribution is not symmetric, which is in part due to the unsteady, turbulent nature of the flow, and in part due to the asymmetry of the arresting gear geometry stemming from the different number of rotor blades and stator vanes employed. Additionally, although the flow is unsteady and pressure fluctuates significantly, maximum pressure loads act somewhere between the blade midspan and tip.

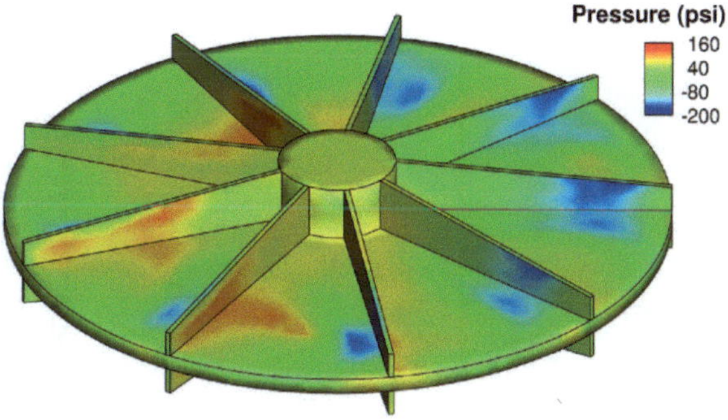

Fig. 9 Pressure (in psi) on the rotor surface at a time instant after the terminal rotor speed is reached. Note that the pressure solution is completely nonsymmetric, which is a consequence of the unsteady, turbulent nature of the flow, and the asymmetry of the arresting gear geometry stemming from the different number of rotor blades and stator vanes employed.

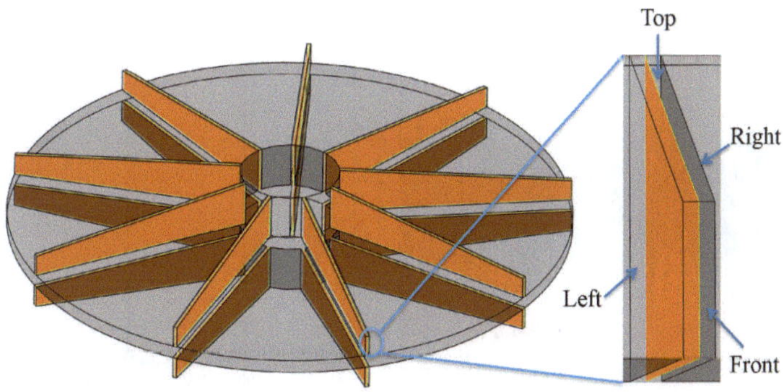

Fig. 10 Fluid mechanics domain built with an assumption of a solid blade superposed with the blade structural shell midsurface. This discretization choice creates a geometric and parametric mismatch between the fluid and structural meshes at their interface. However, such situations are naturally handled in the FSI framework employed in this study.

4.3 FSI Simulation

In this section we present FSI simulation results for the VT arresting gear model. In the model the casing and middle plate are assumed to be rigid, while the rotor blades and stator vanes are assumed to be flexible, and modeled as rotation-free Kirchhoff–Love shells. Because the fluid mechanics mesh is built with an assumption of a solid blade, while a shell model is used for the blade structure, the fluid and structural meshes are mismatched both geometrically and parametrically at their interface (see Figure 10). However, the FSI framework employed here is naturally suited for such situations [27].

We perform FSI simulations using ramp-up conditions described in Section 4.2, which lead to the terminal rotor speed of 800 rpm. We test two cases corresponding

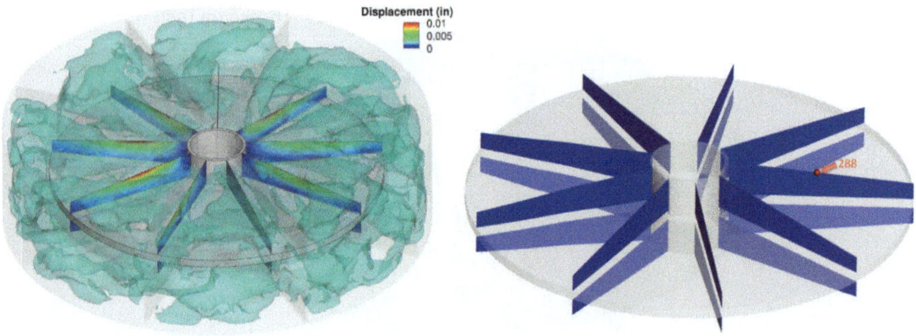

Fig. 11 Left: Rotor blade deflection contours superposed with the flow speed isosurfaces for Case 1. Right: One spatial location on the blade midsurface where the displacement time history is analyzed.

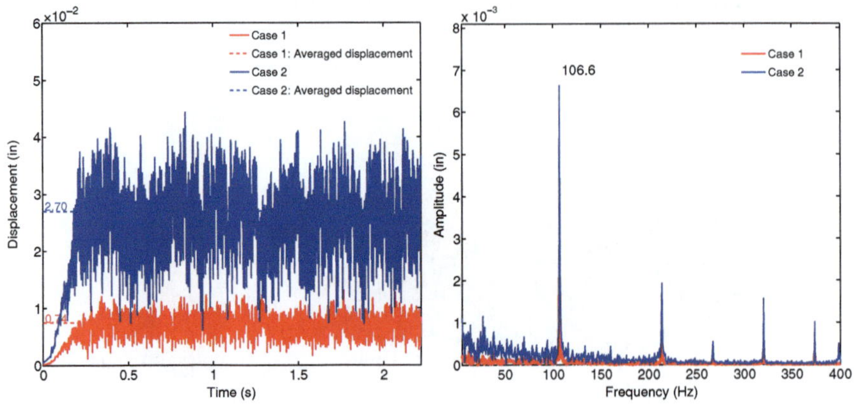

(a)Time domain plot. (b)Frequency domain plot.

Fig. 12 Blade displacement magnitude at location No. 288.

to the blade thickness of 0.348 in (Case 1) and 0.226 in (Case 2). Figure 11 (Left) shows instantaneous blade-structure deformed configuration and the surrounding turbulent flow field for Case 1.

We extract the displacement history for one spatial location, labeled No. 288, and shown in Figure 11 (Right). The displacement magnitude at both locations is shown, in the time and frequency domains, in Figure 12. Although the thinner Case 2 exhibits displacement levels that are 2.5 times larger than Case 1, the blade displacement magnitudes are relatively small compared to the device length scale in both cases. The frequency of rotor-stator interaction (i.e., the frequency at which the rotor blade passes the stator vane) is 106.67 Hz. That frequency is clearly visible in the frequency-domain displacement plots. Several harmonics of that frequency are also excited in the blade response, which is especially pronounced for the thinner blade at location No. 288. Contours of the blade displacement and MIPE are shown in Figures 13 and 14.

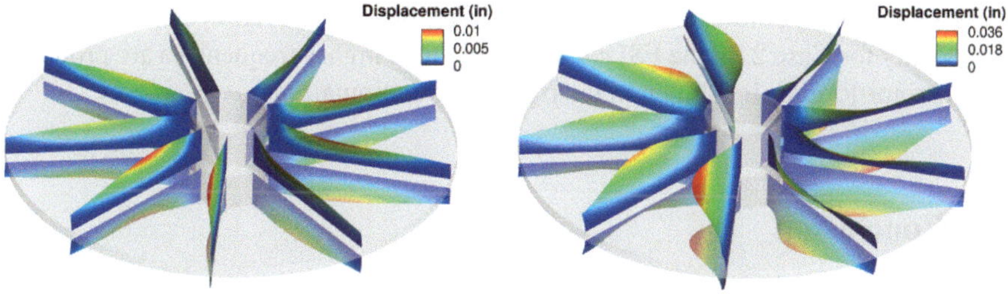

(a) Case 1. (b) Case 2.

Fig. 13 Contours of blade displacement magnitude.

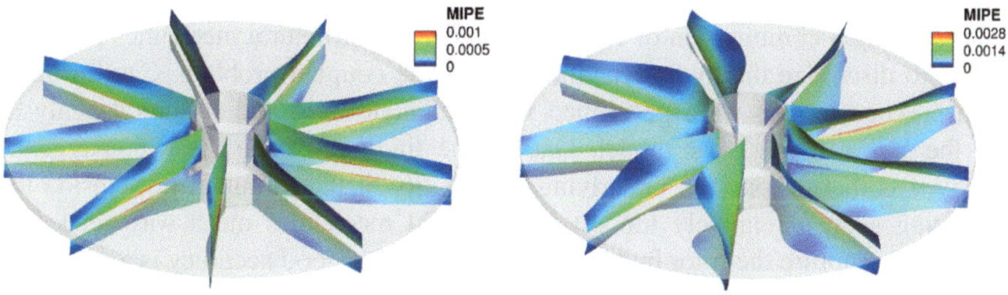

(a) Case 1. (b) Case 2.

Fig. 14 Contours of MIPE.

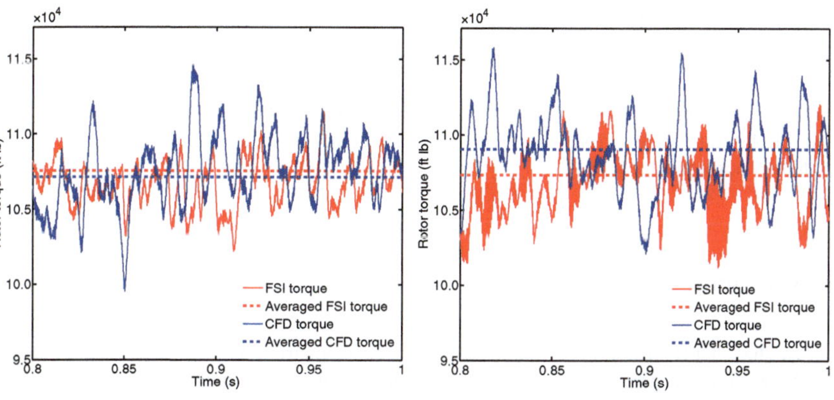

(a) Case 1. (b) Case 2.

Fig. 15 Rotor-torque time history for FSI and rigid-rotor simulations, the latter denoted by "CFD torque."

The rotor-torque time history, including comparison between the FSI and rigid-rotor simulations, is shown in Figure 15. Due to the relatively small blade deformation, the mean values of the torque are similar for the FSI and rigid-rotor simulations. As expected, the thinner blade gives a more pronounced FSI effect. The frequency content, however, is quite different for the FSI and rigid-rotor simulations, especially for Case 2. In the FSI computation much higher frequencies are produced in the rotor-torque signal than in the rigid-rotor computation.

5 Conclusion

In this work we developed a "pipeline" for geometry modeling and predictive FSI simulation of hydraulic arresting gears at full scale. A parametric modeling platform recently proposed in [6] was adapted to generate analysis-suitable IGA models for this complex system. The FSI simulations were carried out for the VT arresting gear model using a combination of IGA to discretize the structural mechanics part, and FEM to discretize the hydrodynamics part of the coupled problem. Careful mesh convergence studies were performed for standalone fluid and structural analyses of the VT arresting gear model. One of the findings of the mesh convergence study was that despite the underlying complexity of the turbulent flow inside the arresting gear, a relatively modest size of fluid mechanics mesh was needed to accurately capture the rotor hydrodynamic loads. This good accuracy is attributable to the use of the ALE-VMS technique with weakly enforced boundary conditions as the underlying numerical methodology. The FSI simulations produced rotor blade displacements that are relatively low compared to the device length scale. However, the vibrational response predicted the presence of multiple highly pronounced frequencies, especially for the thinner blade design. This finding is intriguing and needs to be investigated in the future work.

Acknowledgements This work was supported by NAVAIR, Program Manager Dr. Nam Phan, and ARO grant No. W911NF-14-1-0296, Program Manager Dr. Joseph Myers. This support is gratefully acknowledged.

References

1. Hughes, T.J.R., Cottrell, J.A., Bazilevs, Y.: Isogeometric analysis: CAD, finite elements, NURBS, exact geometry, and mesh refinement. Comput. Methods Appl. Mech. Eng. **194**, 4135–4195 (2005)
2. Cottrell, J.A., Hughes, T.J.R., Bazilevs, Y.: Isogeometric Analysis: Toward Integration of CAD and FEA. Wiley, Chichester (2009)
3. Bazilevs, Y., Takizawa, K., Tezduyar, T.E.: Computational Fluid–Structure Interaction: Methods and Applications. Wiley, Chichester (2013)
4. Chiu, Y.-T.: Computational fluid dynamics simulations of hydraulic energy absorber. Master's thesis, Virginia Polytechnic Institute and State University (1999)
5. Parker, R.V.: Arrestment considerations for the space shuttle. In: The Space Congress Proceedings, Wilmington, Delaware (1971)
6. Hsu, M.-C., Wang, C., Herrema, A.G., Schillinger, D., Ghoshal, A., Bazilevs, Y.: An interactive geometry modeling and parametric design platform for isogeometric analysis. Comput. Math. Appl. (2015). http://dx.doi.org/10.1016/j.camwa.2015.04.002
7. Rhinoceros. http://www.rhino3d.com/ (2015)

8. Grasshopper. http://www.grasshopper3d.com/ (2015)
9. Kiendl, J., Bletzinger, K.-U., Linhard, J., Wüchner, R.: Isogeometric shell analysis with Kirchhoff–Love elements. Comput. Methods Appl. Mech. Eng. **198**, 3902–3914 (2009)
10. Kiendl, J., Bazilevs, Y., Hsu, M.-C., Wüchner, R., Bletzinger, K.-U.: The bending strip method for isogeometric analysis of Kirchhoff–Love shell structures comprised of multiple patches. Comput. Methods Appl. Mech. Eng. **199**, 2403–2416 (2010)
11. Hughes, T.J.R., Liu, W.K., Zimmermann, T.K.: Lagrangian–Eulerian finite element formulation for incompressible viscous flows. Comput. Methods Appl. Mech. Eng. **29**, 329–349 (1981)
12. Belytschko, T., Liu, W.K., Moran, B.: Nonlinear Finite Elements for Continua and Structures. Wiley, Chichester (2000)
13. Takizawa, K., Bazilevs, Y., Tezduyar, T.E.: Space–time and ALE-VMS techniques for patient-specific cardiovascular fluid–structure interaction modeling. Arch. Comput. Methods Eng. **19**, 171–225 (2012)
14. Bazilevs, Y., Hsu, M.-C., Takizawa, K., Tezduyar, T.E.: ALE-VMS and ST-VMS methods for computer modeling of wind-turbine rotor aerodynamics and fluid–structure interaction. Math. Mod. Methods Appl. Sci. **22**(supp02), 1230002 (2012)
15. Bazilevs,Y., Hughes, T.J.R.: Weak imposition of Dirichlet boundary conditions in fluid mechanics. Comput. Fluids **36**, 12–26 (2007)
16. Bazilevs, Y., Michler, C., Calo, V.M., Hughes, T.J.R.: Weak Dirichlet boundary conditions for wall-bounded turbulent flows. Comput. Methods Appl. Mech. Eng. **196**, 4853–4862 (2007)
17. Bazilevs, Y., Michler, C., Calo, V.M., Hughes, T.J.R.: Isogeometric variational multiscale modeling of wall-bounded turbulent flows with weakly enforced boundary conditions on unstretched meshes.Comput. Methods Appl. Mech. Eng. **199**, 780–790 (2010)
18. Bazilevs, Y., Calo, V.M., Cottrel, J.A., Hughes, T.J.R., Reali, A., Scovazzi, G.: Variational multiscale residual-based turbulence modeling for large eddy simulation of incompressible flows. Comput. Methods Appl. Mech. Eng. **197**, 173–201 (2007)
19. Akkerman, I., Bazilevs, Y., Kees, C.E., Farthing, M.W.: Isogeometric analysis of free-surface flow. J. Comput. Phys. **230**, 4137–4152 (2011)
20. Kees, C.E., Akkerman, I., Farthing, M.W., Bazilevs, Y.: A conservative level set method suitable for variable-order approximations and unstructured meshes. J. Comput. Phys. **230**, 4536–4558 (2011)
21. Akkerman, I., Bazilevs, Y., Benson, D.J., Farthing, M.W., Kees, C.E.: Free-surface flow and fluid–object interaction modeling with emphasis on ship hydrodynamics. J. Appl. Mech. **79**, 010905 (2012)
22. Hsu, M.-C., Akkerman, I., Bazilevs, Y.: Wind turbine aerodynamics using ALE–VMS: validation and the role of weakly enforced boundary conditions. Comput. Mech. **50**, 499–511 (2012). doi:10.1007/s00466-012-0686-x
23. Akkerman, I., Dunaway, J., Kvandal, J., Spinks, J., Bazilevs, Y.: Toward free-surface modeling of planing vessels: simulation of the Fridsma hull using ALE-VMS. Comput. Mech. **50**, 719–727 (2012)
24. Hsu, M.-C., Akkerman, I., Bazilevs, Y.: Finite element simulation of wind turbine aerodynamics: validation study using NREL Phase VI experiment. Wind Energy **17**, 461–481 (2014)
25. Bazilevs, Y., Hughes, T.J.R.: NURBS-based isogeometric analysis for the computation of flows about rotating components. Comput. Mech. **43**, 143–150 (2008)
26. Hsu, M.-C., Bazilevs, Y.: Fluid–structure interaction modeling of wind turbines: simulating the full machine. Comput. Mech. **50**, 821–833 (2012)
27. Bazilevs, Y., Hsu, M.-C., Scott, M.A.: Isogeometric fluid–structure interaction analysis with emphasis on non-matching discretizations, and with application to wind turbines. Comput. Methods Appl. Mech. Eng. **249–252**, 28–41 (2012)
28. Tezduyar, T.E., Behr, M., Mittal, S., Johnson, A.A.: Computation of unsteady incompressible flows with the finite element methods – space–time formulations, iterative strategies and massively parallel implementations. In: New Methods in Transient Analysis, PVP-Vol.246/AMD-Vol.143, pp. 7–24. ASME, New York (1992)

29. Tezduyar, T., Aliabadi, S., Behr, M., Johnson, A., Mittal, S.: Parallel finite-element computation of 3D flows. Computer **26**(10), 27–36 (1993)
30. Johnson, A.A., Tezduyar, T.E.: Mesh update strategies in parallel finite element computations of flow problems with moving boundaries and interfaces. Comput. Methods Appl. Mech. Eng. **119**, 73–94 (1994)
31. Tezduyar, T.E.: Finite element methods for flow problems with moving boundaries and interfaces. Arch. Comput. Methods Eng. **8**, 83–130 (2001)
32. Stein, K., Tezduyar, T., Benney, R.: Mesh moving techniques for fluid–structure interactions with large displacements. J. Appl. Mech. **70**, 58–63 (2003)
33. Tezduyar, T.E., Sathe, S.: Modeling of fluid–structure interactions with the space–time finite elements: Solution techniques. Int. J. Numer. Methods Fluids **54**, 855–900 (2007)
34. Bazilevs, Y., Calo, V.M., Hughes, T.J.R., Zhang, Y.: Isogeometric fluid–structure interaction: theory, algorithms, and computations. Comput. Mech. **43**, 3–37 (2008)
35. Chung, J., Hulbert, G.M.: A time integration algorithm for structural dynamics with improved numerical dissipation: the generalized-α method. J. Appl. Mech. **60**, 371–75 (1993)
36. Jansen, K.E., Whiting, C.H., Hulbert, G.M.: A generalized-α method for integrating the filtered Navier-Stokes equations with a stabilized finite element method. Comput. Methods Appl. Mech. Eng. **190**, 305–319 (2000)
37. Tezduyar, T.E., Sathe, S., Stein, K.: Solution techniques for the fully-discretized equations in computation of fluid–structure interactions with the space–time formulations. Comput. Methods Appl. Mech. Eng. **195**, 5743–5753 (2006)
38. Tezduyar, T.E., Sathe, S., Keedy, R., Stein, K.: Space–time finite element techniques for computation of fluid–structure interactions. Comput. Methods Appl. Mech. Eng. **195**, 2002–2027 (2006)

Finite-Element/Boundary-Element Coupling for Inflatables: Effective Contact Resolution

T.M. van Opstal

Abstract The finite-element/boundary-element technique for inflatable structures has been utilized successfully for airbag simulations. Its virtues in resolving the complex contact problem in such cases have been demonstrated in 2D in van Opstal et al. (Comput. Mech. 50(6):779–788, 2012) and theoretically motivated in 3D in van Opstal et al. (Comput. Mech. Comput. Methods Appl. Mech. Eng. 284:637–663, 2015). In this contribution, this is extended to a 3D test case, demonstrating the merits of this technique for real-world applications. Secondly, it is shown how this technique naturally fits into and greatly profits from an isogeometric setting.

1 Introduction

A fluid–structure interaction technique for the simulation of inflatable structures is introduced. The presented finite-element/boundary-element (FEBE) technique couples an isogeometric finite-element discretization of a flexible shell structure with an isogeometric boundary-element discretization of a Stokes fluid. This technique was introduced in [18] for planar problems and extended to 3D in [19]. One of the marked advantages of this approach is the contact mechanism, lubrication, which is an inherent attribute of the flow model. Its role in preventing contact was theoretically substantiated, but only demonstrated in the planar setting. In the present work, we demonstrate the effectiveness of the contact mechanism in the isogeometric and three-dimensional setting, discussing various aspects pertaining to the accurate resolution of the traction responsible for contact prevention.

An extensive body of literature already covers the interface of the fields of fluid–structure interaction and contact mechanics. If the problem under consideration requires that the gap between two contact surfaces may vanish, an interface tracking technique may be applied, sometimes locally [5, 15], e.g., [3, 7, 21]. If, however, a problem allows for a finite gap to be maintained between the contacting surfaces,

T.M. van Opstal (✉)
Department of Mathematical Sciences, Norwegian University of Science and Technology, 7491 Trondheim, Norway
e-mail: timo.vanopstal@math.ntnu.no

Y. Bazilevs, K. Takizawa (eds.), *Advances in Computational Fluid-Structure Interaction and Flow Simulation*, Modeling and Simulation in Science, Engineering and Technology, DOI 10.1007/978-3-319-40827-9_37

arbitrary Lagrangian–Eulerian [4] and space-time [14] techniques can be used to compute problems as challenging as the disreefing of parachute clusters [11–13] and 1000 spheres falling through a tube [6]. In this work a finite (but arbitrarily small) gap will likewise be maintained in contact regions.

The target application for the current approach is inflatable structures. These typically undergo large deformations with ubiquitous self-contact during the inflation process, which often starts from a complex, folded initial configuration. Correct simulation is contingent to the resolution of every single contact mode throughout the process. To tackle this type of problem, contact treatment based on the lubrication effect inherent to viscous flow is highly advantageous:

1. it avoids explicit contact detection, which becomes prohibitively expensive and non-robust for complex geometries such as folded inflatables.
2. accuracy is not impaired by artificial terms added to the model for soft contact treatment, such as in, e.g., [17].

Correct simulation of the contact mechanics is contiguous to the accurate resolution of the lubrication tractions, which increase singularly as the contact gap closes. Thus, after recapitulating the mathematical model in §2, accurate approximation and solution through isogeometric analysis and adaptive quadrature schemes are covered in §3. The resulting methodology is demonstrated in a numerical test case, a deflating balloon, in §4. Finally, conclusions are drawn in §5.

2 Mathematical Model of an Inflatable Structure

The governing equations of the fluid–structure interaction are elaborated in [19]. In this section, this model is recapitulated concisely, and we refer to [19] for full details. An inflatable structure occupies a boundary segment Γ_t at time t, which, together with a fixed inflow boundary segment $\Gamma_{\rm in}$ encloses the interior domain Ω_t of the inflatable. This is schematized in fig. 1. The configuration $\theta : (0, T)\times\underline{\Gamma} \mapsto \Gamma_t$ maps a material point in the reference manifold $x \in \underline{\Gamma}$ to its position on the current manifold $\theta(t, x) \in \Gamma_t$, and is sometimes abbreviated as $\theta_t(\cdot) := \theta(t, \cdot)$. Throughout, entities related to the reference manifold are underlined.

The flow on both the interior and exterior domains, resp. Ω_t and $\mathbb{R}^3 - \overline{\Omega_t}$, are governed by the boundary integral formulation introduced in §2.1. The motion of the inflatable structure itself is governed by a shell formulation, cf. §2.2. Finally, the coupling of these two subsystems, as well as a condition enforcing compatibility between fluid and structure solutions, is treated in §2.3.

2.1 Boundary Integral Formulation of the Fluid

The flow interior and exterior to the flexible structure is described by Stokes flow, supplemented by Dirichlet conditions g on the boundary $\partial\Omega_t$ and suitable radiation conditions in the far-field. It will be more convenient to pull back the formulation

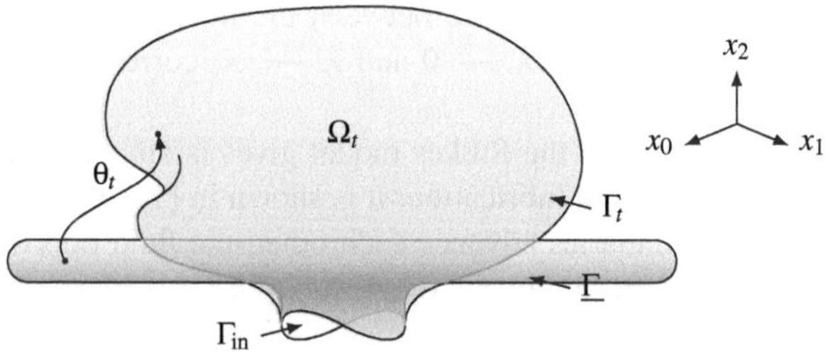

Fig. 1 Schematic geometry of an inflatable structure for the case $d = 3$. The reference and current domains are overlayed, as they coincide over Γ_{in}.

to the reference configuration, denoting the boundary condition by $\underline{g} = g \circ \partial\Omega_t$ and the traction jump across $\partial\Omega_t$ due to the interior and exterior flows by $\underline{\tau}_{\text{j}}$. The weak formulation of the direct boundary integral equation becomes

given $\underline{g} \in H^{1/2}(\underline{\partial\Omega})$, *find* $(\underline{\tau}_{\text{j}}, \zeta) \in H^{-1/2}(\underline{\partial\Omega}) \times \mathbb{R}$ *such that*

$$a_{\text{ft}}(\underline{\tau}_{\text{j}}, \underline{\psi}) + b_t(\underline{\tau}_{\text{j}}, \upsilon) + b_t(\underline{\psi}, \zeta) = F_{\text{ft}}(\underline{g}, \underline{\psi}) \qquad \forall(\underline{\psi}, \upsilon) \in H^{-1/2}(\partial\Omega_t) \times \mathbb{R} \tag{1}$$

with bilinear forms

$$\begin{aligned} a_{\text{ft}}(\phi, \psi) &:= \langle V_t\phi, \psi\rangle_{H^{1/2}(\underline{\partial\Omega})}, \\ b_t(\psi, \upsilon) &:= \upsilon\langle \psi, J_t n \circ \theta_t\rangle_{H^{1/2}(\underline{\Gamma})} + \upsilon\langle \psi, J_t n \circ \theta_t\rangle_{H^{1/2}(\Gamma_{\text{in}})}, \\ F_{\text{ft}}(\phi, \psi) &:= \langle (1+\lambda)J_t\phi/2 + (1-\lambda)K_t\phi, \psi\rangle_{H^{1/2}(\underline{\partial\Omega})}, \end{aligned}$$

and boundary integral operators

$$\begin{aligned} (V_t\psi)_i &:= J_t \frac{1}{8\pi} \oint_{\underline{\partial\Omega}} \left(\delta_{ik}\frac{1}{r} + \frac{(x_i - y_i)(x_k - y_k)}{r^3} \right) \circ (\theta_t \times \theta_t)\psi_k(y) J_t\, ds(y), \\ (K_t\phi)_i &:= J_t \frac{3}{4\pi} \oint_{\underline{\partial\Omega}} \left(\frac{(x_i - y_i)(x_k - y_k)(x_j - y_j)n_j(y)}{r^5} \right) \circ (\theta_t \times \theta_t)\phi_k(y) J_t\, ds(y), \end{aligned}$$

with $i, k \in \{0, 1, 2\}$, $r = |x - y|$ the Euclidean distance and

$$J_t := ((\underline{t}^0 \cdot \nabla_{\underline{\Gamma}}\theta_t) \times (\underline{t}^1 \cdot \nabla_{\underline{\Gamma}}\theta_t)) \cdot \underline{n} \tag{2}$$

the determinant of the Jacobian of the map θ_t. In the above, $\underline{n}$ and $\underline{t}^\alpha$, $\alpha \in \{0, 1\}$ denote the normal and tangent vectors to $\underline{\partial\Omega}$. Furthermore, the composition with $(\theta_t \times \theta_t)$ serves to transport the kernel of the integral operator to the reference manifold.

In (1), the Lagrange multiplier ζ constrains the kernel of the boundary integral operator V_t. This kernel is related to the undefined average pressure level p_0, similar to the Stokes PDE with pure Dirichlet boundary conditions. Furthermore,

$0 \le \lambda \le \infty$ denotes the viscosity ratio between the fluids interior and exterior to $\partial\Omega$, where the respective limits $\lambda \to 0$ and $\lambda \to \infty$ correspond to the separate interior and exterior problems.

One of the advantages that the Stokes model gives is an automatic mechanism for contact prevention, namely lubrication. It is shown in [16, Thms. 33,40] that, as two smooth material boundaries advance each other, the fluid opposes this motion with a repulsive traction of singular strength of $O(h^{-3})$, where h is the distance between the boundaries. Thus, the fluid formulation has a built-in contact prevention mechanism which avoids the highly nontrivial explicit treatment of self-contact which characterizes folded inflatable structures.

We conclude this section with a number of remarks:

1. The boundary integral formulation (1) gives a direct relation between the Dirichlet and Neumann data at the boundary, mapping velocities imposed by the structure to the tractions imposed on the structure, which is precisely the relation typically required in a fluid–structure interaction problem. No meshing and approximation of the interior domain is thus required, although the solution in the interior can be reconstructed *a posteriori* if desired.
2. Contrary to a discretization based on a volumetric formulation, it is not more expensive to treat the flow in the exterior domain concurrently, it simply amounts to adjusting the ratio λ. Moreover, when the conditions in both fluids are the same, $\lambda = 1$ and we see that the problem actually becomes cheaper to approximate, as the dual layer operator K_t need not be assembled.

2.2 *Parametrization-Free Kirchhoff–Love Structure*

We now turn to the equation governing the configuration θ, which is assumed to be a member of a Bochner space. A Bochner space $L^2(0,T;Y)$ contains functions $f : [0,T] \to Y$, such that $\int_0^T \|f(t,\cdot)\|_Y^2\, dt < \infty$. First and second time derivatives are denoted $\dot{(\cdot)}$ and $\ddot{(\cdot)}$, respectively. We set $X := H^2(\underline{\Gamma}) \cap H_0^1(\underline{\Gamma})$ and write the structure problem as

given $(\chi_0, \chi_1) \in H^2(\underline{\Gamma}) \times L^2(\underline{\Gamma})$ *and* $F \in L^2(0,T;L^2(\underline{\Gamma}))$, *find* $\theta \in (\chi_0 + \{\theta \in L^2(0,T;X) : \dot{\theta} \in L^2(0,T;L^2(\underline{\Gamma}))\})$ *such that*

$$\langle \ddot{\theta}_t, \rho \rangle_X + W'(\theta_t;\rho) = F(t;\rho) \qquad a.e.\ t \in (0,T),\ \forall \rho \in X, \tag{3a}$$

$$\theta_0 = \chi_0, \tag{3b}$$

$$\dot{\theta}_0 = \chi_1. \tag{3c}$$

In this problem, F is an external load, χ_0 and χ_1 are the initial position and velocity, and W is the internal energy of a Kirchhoff–Love shell, given in terms of a configuration θ as

$$W(\theta) = \frac{1}{2}\int_{\underline{\Gamma}} \Xi^{ijkl}\left(\varepsilon_{ij}(\theta)\varepsilon_{kl}(\theta) + \epsilon^2 \kappa_{ij}(\theta)\kappa_{kl}(\theta)\right) ds, \tag{4}$$

where

$$\varepsilon_{ij}(\theta) := \tfrac{1}{2}(\underline{\Pi}_{ij} - \theta_{m;i}\theta_{m;j}),$$
$$\kappa_{ij}(\theta) := \underline{\Pi}_{mi}\underline{n}_{m;j} - \theta_{m;i}n_{m;j},$$

are the components of the membrane and bending strains; and

$$\Xi^{ijkl} := \upsilon\delta_{ij}\delta_{kl} + \tfrac{1}{2}(1-\upsilon)(\delta_{ik}\delta_{jl} + \delta_{il}\delta_{jk}),$$

is the constitutive tensor of the Saint Venant–Kirchhoff material law, which is especially suited to the anticipated small strains and large rotations [1]. The tensor $\underline{\Pi}_{ij} := \delta_{ij} - \underline{n}_i\underline{n}_j$ denotes the projection into the tangent space of $\underline{\Gamma}$ and $f_{i;j}$ denotes the jth component of the gamma gradient of component i of a vector function f.

The structural model introduces two model parameters, ϵ and υ, which represent the flexural rigidity and Poisson's ratio, respectively.

This form of the Kirchhoff–Love energy has been dubbed *parametrization-free* in [19] as it is formulated on the stress-free state $\underline{\Gamma}$, without reference to some parametrization of this surface. The main advantage is conceptual: eliminating the superfluous and arbitrary parametrization from the formulation. Implementation aspects for this formulation are treated in [19].

2.3 Transmission Conditions and Full Problem

Standard Dirichlet–Neumann coupling is employed between the fluid and the structure. The kinematic condition imposes a velocity on the fluid at the interface and reads

$$\underline{g} = \begin{cases} \dot{\theta}_t & \text{on } \underline{\Gamma}, \\ qn & \text{on } \Gamma_{\text{in}}, \end{cases} \tag{5a}$$

with $q : \Gamma_{\text{in}} \to \mathbb{R}$ the normal inflow velocity. The dynamic condition imposes continuity of tractions at the fluid–structure interface

$$F(t;\cdot) = -\varpi\,\langle J_t\underline{\tau}_{\text{j}},\cdot\rangle. \tag{5b}$$

with J_t the determinant of θ_t, according to (2). The model parameter ϖ can be interpreted as a *coupling strength* and arises in the nondimensionalization [19] by agglomeration of several parameters (such as the Young's modulus) from the dimensional models.

In addition to the coupling conditions, an auxiliary condition is to be satisfied, connected to the incompressibility of the fluid encapsulated by the structure. The configuration provided by the structural equation of motion (3) has to conserve the volume, i.e., $Q_t(\theta_t) = 0$ with

$$Q_t(\theta_t) = \frac{1}{3}\,(n, x)_{\partial\Omega_t} - \left(\frac{1}{3}\,(n, x)_{\underline{\partial\Omega}} - \int_0^t \int_{\Gamma_{\text{in}}} q\,ds\,dt\right), \tag{6}$$

where the first term is the current volume, the second the initial volume, and the last the total influx until time t. This auxiliary constraint is imposed by the Lagrange multiplier method, and it is proven in [19] that this Lagrange multiplier can be seen as the total excess pressure p_0 required to uniformly expand the inflatable structure to the correct volume. This constant pressure coincides with the undetermined mode constrained from the fluid subproblem (1).

The above discussion leads to the aggregated FSI model, composed of the weak form of the interior and exterior fluid boundary integral equations (1), the Lagrangian equation of motion for the structure (3) augmented with the compatibility condition and coupled through the transmission conditions (5):

given $q \in L^2(0, T; L^2(\Gamma_{\text{in}}))$ *and* $(\chi_0, \chi_1) \in X \times L^2(\underline{\Gamma})$, *find* $(\theta_t, \underline{\tau}_{\text{j}}, p_0, \zeta) \in (\chi_0 + X) \times H^{-1/2}(\underline{\partial\Omega}) \times \mathbb{R} \times \mathbb{R}$ *such that*

$$\begin{aligned}
&\langle \ddot{\theta}_t, \rho \rangle_X + W'(\theta_t; \rho) + p_0 Q_t'(\theta_t; \rho) + \mu Q_t(\theta_t) + \varpi \langle J_t \underline{\tau}_{\text{j}}, \rho \rangle_X \\
&\qquad + a_{\text{ft}}(\underline{\tau}_{\text{j}}, \underline{\psi}) + b_t(\underline{\tau}_{\text{j}}, \upsilon) + b_t(\underline{\psi}, \zeta) - F_{\text{ft}}(\dot{\theta}_t, \underline{\psi}) = F_{\text{ft}}(q(t, \cdot)n, \underline{\psi}) \\
&\qquad\qquad a.e.\ t \in (0, T),\ \forall (\rho, \underline{\psi}, \mu, \upsilon) \in X \times H^{-1/2}(\underline{\partial\Omega}) \times \mathbb{R} \times \mathbb{R},
\end{aligned} \tag{7a}$$

$$\theta_0 = \chi_0, \tag{7b}$$

$$\dot{\theta}_0 = \chi_1, \tag{7c}$$

where we extend q and $\dot{\theta}_t$ by 0 on $\underline{\Gamma}$ and Γ_{in}, respectively.

3 Approximation and Solution

In treating the discretization and solution of the mathematical model (7), emphasis is put on lubrication and contact mechanics. It was mentioned in §2.1 that the magnitude of the traction forces grow as $O(h^{-3})$ with h the gap size between smooth surfaces in contact. Thus, an isogeometric matching-mesh discretization, which provides the required smoothness and geometrical exactness, is presented in §3.1. As the gap size h vanishes, local contributions to the boundary integrals become more singular on account of the $1/r$ and $1/r^2$ dependence of the kernels of V_t and K_t. The accurate evaluation of integrals in such circumstances is treated in §3.2. Finally, as the fluid–structure coupling is anticipated to be strong when contact occurs, we outline the partitioned iterative procedure for resolving the coupled system in §3.3.

3.1 Subdivision Surface Approximation Spaces

Two considerations have lead to the choice of approximation space for this problem. Firstly, a C^1-continuous basis is desired to have a *rotation free* discretization of the shell structure, and facilitates contact treatment through the lubrication effect. The theoretical results depend on a smooth representation of the fluid boundary. In that case, there is a time at which the gap width is much smaller than the diameter of the contact surface, and the fluid response resembles lubrication. Secondly, an unstructured basis is required, to be able to represent complex geometries, such as folded inflatables; and to perform local refinement where it is required to resolve fine geometrical features and near-singular tractions in near contact.

One of the natural candidates is subdivision surfaces with the Catmull–Clark subdivision scheme [2]. Recently, (truncated) hierarchical refinement has been presented for this basis in [20, 22]. One of the freely available CAGD packages in which subdivision surface geometries can be designed is Blender™, which has been used in this context.

Catmull–Clark subdivision surfaces can be seen as an extension of cubic B-splines to unstructured grids composed of quadrilaterals. Defining the valence of a vertex as the number of edges terminating there, we observe that interior vertices of a Cartesian mesh are of valency four. An interior vertex of valency other than four is called an *extraordinary vertex* (EV). To extend the definition of the spline basis near an EV, a subdivision mask is introduced, in this case the Catmull–Clark mask, which preserves C^1-continuity almost everywhere. As such a mask only provides an implicit definition of the basis functions at specific points, Stam's approach [10] is used to evaluate the basis functions and their gradients in numerical quadrature. This approach involves l levels of virtual refinement, such that the quadrature point lies in a sub-element away from the EV, and can be evaluated as a regular basis function.

As the domains of the fluid and structure subproblems coincide, it is convenient to use the discretization of the structure for the fluid as well. Thus, this approach can be seen as a *matching mesh* approach, such that any geometrical approximation is avoided. The advantages are twofold. Firstly, errors due to geometrical approximation, which are significant in the BEM [9], are now entirely eliminated. One can imagine especially in the pervading near-contact modes, that inaccuracies in the tractions due to geometrical approximation may easily become disastrous, due to the $O(h^{-3})$ relation to the gap size h. Secondly, we completely eliminate the volumetric meshing step, which is otherwise a formidable challenge in the light of the complex initial geometries and large deformations characterizing inflatable structures.

3.2 Boundary-Element Approximation of the Fluid

To evaluate the singular integrals on product domains in the boundary-element problem for the fluid, the regularizing transformations described in [8] are used. The convergence rates for standard Gauss quadrature rules are known for this set of transformations and class of integrands, cf. [8]. This suggests an adaptive quadrature scheme in which the quadrature order is increased until some heuristic criterion is met. Let $F_{\kappa,q}$ be the approximation of the integral over element κ with quadrature order q. In this work, the criterion $|F_{\kappa,q-1} - F_{\kappa,q}| < 10^{-7}$ is used.

It is observed that this adaptive procedure converges and that the amount of evaluations requiring an order q decreases rapidly with q, cf. fig. 2: selection of a scheme of $q > 10$ is approximately 10^4 times less likely than selection of the scheme $q = 3$. In fig. 2 it can also be seen how many levels would be required if an order q scheme would be evaluated near an extraordinary vertex. Such an evaluation would require the $(l-1)^{\text{th}}$ power of the subdivision matrix, which can be cached once per level. Observe that the number of required levels grows only very slowly and that in this case 8 such levels are required.

3.3 Partitioned Iterative Solution

Monolithic solution of a discretization of (7) is untractable, as this requires linearization of the coupling terms. In fact, this leads to hypersingular integrals, i.e., integrands with singularities stronger than those in (7). Despite the strongly coupled nature of this problem due to lubrication, a partitioned scheme is therefore preferred.

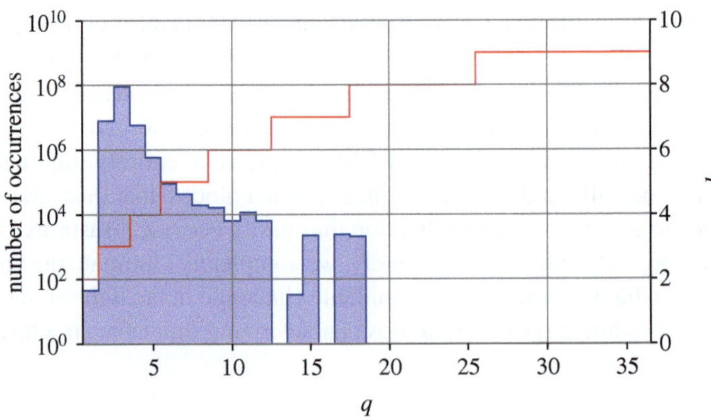

Fig. 2 The blue histogram shows the number of occurrences of a quadrature order q from a sample of 10^9 element evaluations in the simulation of §4. The red line shows the required number of levels l in Stam's algorithm to evaluate an order q Gauss rule.

Thus, the fluid and structure subproblems, the $(\underline{\psi}, \upsilon)$ and (ρ, μ) terms of (7), respectively, are solved separately. The required damping for stability of the coupled problem is introduced through the dissipative implicit Euler time-integration scheme applied to the structure.

A subiteration step within our partitioned procedure starts with a linear extrapolation of the initial data (provided by the solution at the previous time interval), which serves as a first approximation of the new coupled solution. Within a fluid–structure subiteration, a structural solve is performed first, to ensure compatibility of the fluid boundary data. The subiteration is considered converged if (1) the norm of the structure residual is below the tolerance before a Newton solve is performed and, in addition, (2) the norm of the fluid update is below that same tolerance.

4 Deflation of a Balloon

The prime objective of the numerical experiment presented here is to assess the accuracy of the solution method to resolve the complex self-contact that often arises in the simulation of inflatable structures. To this end, as in [18], an inverted problem is considered, namely the deflation of a balloon.

The initial geometry $\underline{\partial\Omega}$ contains 320 elements in Γ_{in} and 832 elements in $\underline{\Gamma}$. Both these subsets contain 4 extraordinary vertices of valence 3. A small random perturbation is applied to $\underline{\Gamma}$ away from $\partial\underline{\Gamma}$ so that $\underline{\partial\Omega}$ remains connected. The random perturbations cause the structure to wrinkle instead of contract uniformly, rendering the structural response closer to reality. The initial configuration is plotted in fig. 3.

The model parameters are $\upsilon = 0$; $\epsilon = 5.77\ 10^{-4}$ corresponding to a flexible shell; $\lambda = 1$ corresponding to identical fluids in the interior and exterior; and

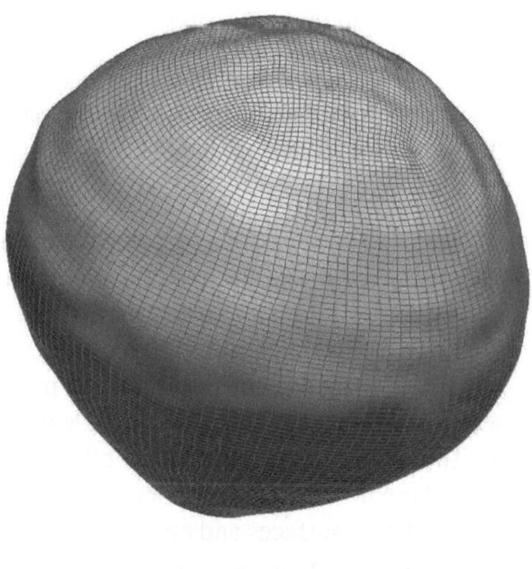

Fig. 3 Initial configuration for the deflation of a balloon.

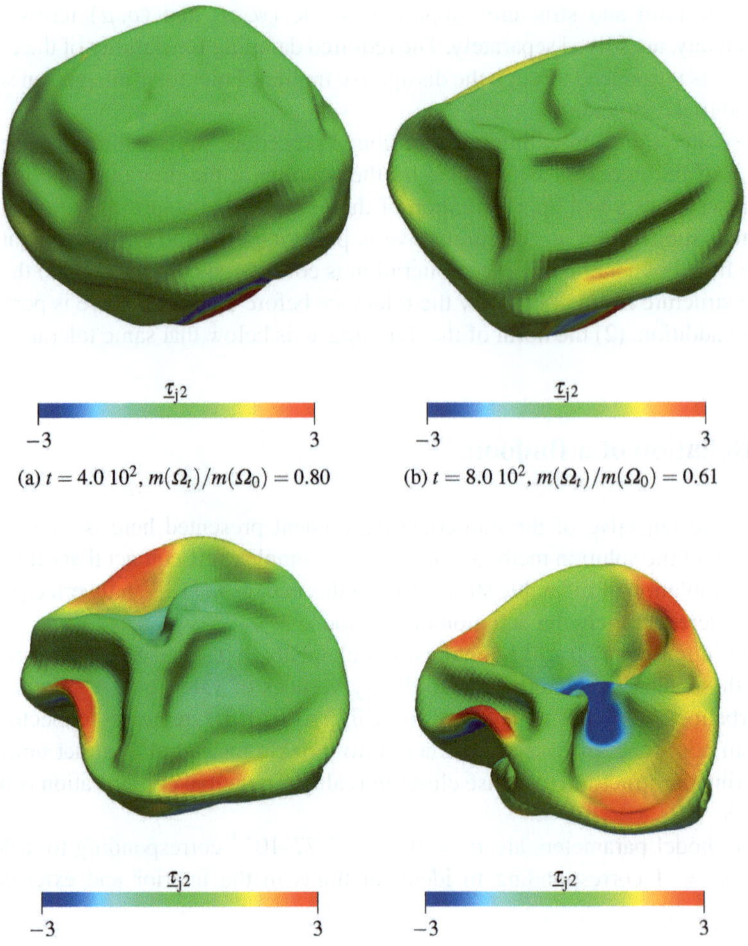

(a) $t = 4.0\ 10^2$, $m(\Omega_t)/m(\Omega_0) = 0.80$

(b) $t = 8.0\ 10^2$, $m(\Omega_t)/m(\Omega_0) = 0.61$

(c) $t = 1.20\ 10^3$, $m(\Omega_t)/m(\Omega_0) = 0.41$

(d) $t = 1.31\ 10^3$, $m(\Omega_t)/m(\Omega_0) = 0.26$

Fig. 4 Snapshots of the configuration and fluid traction at different time levels during the deflation process. The color coding corresponds to the vertical component of the traction, $\underline{\tau}_{j2}$ and $m(A)$ denotes the volume of set A.

$\varpi = 10^{-5}$. The outflow q is constant in time and has a sine-shaped profile with a magnitude such that the volume would vanish at $T = 2^{12}$. The time step size is $\tau = 4$ and the tolerance in the partitioned solver TOL is set to 10^{-6}, settings for which only 1–2 subiterations are required at each time level.

In fig. 4, snapshots of the deflation process are shown. Recall that the plotted traction component $\underline{\tau}_{j2}$ does not include the contribution of the total excess pressure $p_0 n$. The action of the lubrication effect can clearly be observed along horizontal folds, where the vertical component is positive (red) on upper surfaces and negative (blue) on lower surfaces and close to zero elsewhere in fig. 4a–4c. Thus, the collapse of the structure under the action of the total excess pressure $p_0 n$, which is uniform over Γ_t, is observed to be counteracted locally by $\underline{\tau}_j$ to prevent contact.

As the limits of the color bars are identical, it is clear that the magnitude of these lubrication forces increases as the deflation proceeds. At the same time, the folds in the fabric become gradually sharper. The deflation process is terminated at $t = 1.31\ 10^3$, cf. fig. 4(d). As can be seen from this snapshot, a large downward traction is exerted on the part of the structure near Γ_{in}, where the outflow condition (5a) is prescribed. The outflux q does not correspond to the velocity of Γ_{in}. The segments therefore seem to recede and no lubrication force is instigated. Ultimately, this self-intersection is therefore not due to insufficient resolution of the fluid response, but rather to the artificial problem setup where a structure is deflated instead of inflated.

5 Conclusions

In this chapter we have presented an FEBE method for the simulation of inflatable structures. A Kirchhoff–Love shell with low flexural rigidity, discretized with the finite-element method (using a Catmull–Clark subdivision surface basis) is coupled with a boundary-element method discretization of both the interior and exterior fluid. It is advantageous to let the fluid inherit the structural mesh, as this leads to a geometrically exact matched discretization at the fluid–structure interface, which in turn enables accurate approximation of contact forces. The coupled system may be advanced in a partitioned iterative fashion.

We have furthermore assessed the capabilities of this method in the presence of multiple modes of self-contact, by considering a deflation problem, in which the fluid response is anticipated to be dominated by lubrication forces. It was observed that the computed traction forces effectively prevent self-intersection of the structure, although very large deformations are sustained.

References

1. Bischoff, M., Wall, W.A., Bletzinger, K.-U., Ramm, E.: Models and finite elements for thin-walled structures. In: Encyclopedia of Computational Mechanics, vol. 2: Structures, chap. 3, pp. 59–137. Wiley, Chichester (2004)
2. Catmull, E., Clark, J.: Recursively generated b-spline surfaces on arbitrary topological meshes. Comput. Aided Des. **10**(6), 350–355 (1978)
3. Hsu, M.-C., Kamensky, D., Bazilevs, Y., Sacks, M.S., Hughes, T.J.R.: Fluid–structure interaction analysis of bioprosthetic heart valves: significance of arterial wall deformation. Comput. Mech. **54**, 1055–1071 (2014)
4. Hughes, T.J.R., Liu, W.K., Zimmermann, T.K.: Lagrangian-Eulerian finite element formulation for incompressible viscous flows. Comput. Methods Appl. Mech. Eng. **29**, 329–349 (1981)
5. Johnson, A.A., Tezduyar, T.E.: Mesh update strategies in parallel finite element computations of flow problems with moving boundaries and interfaces. Comput. Methods Appl. Mech. Eng. **119**, 73–94 (1994)
6. Johnson, A.A., Tezduyar, T.E.: Advanced mesh generation and update methods for 3d flow simulations. Comput. Mech. **23**, 130–143 (1999)

7. Kamensky, D., Hsu, M.-C., Schillinger, D., Evans, J.A., Aggarwal, A., Bazilevs, Y., Sacks, M.S., Hughes, T.J.R.: An immersogeometric variational framework for fluid–structure interaction: application to bioprosthetic heart valves. Comput. Methods Appl. Mech. Eng. **284**, 1005–1053 (2015)
8. Sauter, S.A., Schwab, C.: Boundary Element Methods. Springer Series in Computational Mechanics. Springer, New York (2011)
9. Scott, M.A., Simpson, R.N., Evans, J.A., Lipton, S., Bordas, S.P.A., Hughes, T.J.R., Sederberg, T.W.: Isogeometric boundary element analysis using unstructured T-splines. Comput. Methods Appl. Mech. Eng. **254**, 197–221 (2013)
10. Stam, J.: Exact evaluation of Catmull-Clark subdivision surfaces at arbitrary parameter values. In: Proceedings of the 25th Annual Conference on Computer Graphics and Interactive Techniques, pp. 295–404. ACM, New York (1998)
11. Takizawa, K., Spielman, T., Tezduyar, T.E.: Space–time FSI modeling and dynamical analysis of spacecraft parachutes and parachute clusters. Comput. Mech. **48**, 345–364 (2011)
12. Takizawa, K., Fritze, M., Montes, D., Spielman, T., Tezduyar, T.E.: Fluid-structure interaction modeling of ringsail parachutes with disreefing and modified geometric porosity. Comput. Mech. **50**, 835–854 (2012)
13. Takizawa, K., Tezduyar, T.E., Boswell, C., Kolesar, R., Montel, K.: FSI modeling of the reefed stages and disreefing of the Orion spacecraft parachutes. Comput. Mech. **54**, 1203–1220 (2014)
14. Tezduyar, T.E.: Stabilized finite element formulations for incompressible flow computations. Adv. Appl. Mech. **38**, 1–44 (1992)
15. Tezduyar, T.E., Takizawa, K., Moorman, C., Wright, S., Christopher, J.: Space-time finite element computation of complex fluid-structure interactions. Int. J. Numer. Methods Fluids **64**, 1201–1218 (2010)
16. van Opstal, T.M.: Numerical methods for inflatables with multiscale geometries. Ph.D. thesis, Eindhoven University of Technology (2013)
17. van Opstal, T.M., van Brummelen, E.H.: A finite-element/boundary-element method for large-displacement fluid-structure interaction with potential flow. Comput. Methods Appl. Mech. Eng. **266**, 57–69 (2013). http://dx.doi.org/10.1016/j.cma.2013.07.0092013
18. van Opstal, T.M., van Brummelen, E.H., de Borst, R., Lewis, M.R.: A finite-element/boundary-element method for large-displacement fluid-structure interaction. Comput. Mech. **50**(6), 779–788 (2012)
19. van Opstal, T.M., van Brummelen, E.H., van Zwieten, G.J.: A finite-element/boundary-element method for three-dimensional, large-displacement fluid-structure interaction. Comput. Methods Appl. Mech. Eng. **284**, 637–663 (2015)
20. Wei, X., Zhang, Y., Hughes, T.J.R., Scott, M.A.: Truncated hierarchical Catmull–Clark subdivision with local refinement. Comput. Methods Appl. Mech. Eng. **291**, 1–20 (2015)
21. Wick, T.: Flapping and contact FSI computations with the fluid–solid interface-tracking/interface-capturing technique and mesh adaptivity. Comput. Mech. **53**(1), 29–43 (2014)
22. Zore, U., Jüttler, B., Kosinka, J.: On the linear independence of (truncated) hierarchical subdivision splines. Geometry+ Simulation Report, 17 (2014)

Recent Advances in Fluid–Structure Interaction Simulations of Wind Turbines

A. Korobenko, X. Deng, J. Yan, and Y. Bazilevs

Abstract In this chapter the numerical challenges of simulating aerodynamics and fluid–structure interaction (FSI) of wind turbines are summarized, and the recently developed computational methods that address these challenges are presented. Several wind-turbine computations at full scale and with full complexity of the geometry and material composition are presented, which illustrate the accuracy, robustness, and general applicability of the methods developed for this problem class.

1 Introduction

Wind energy is one of the fastest growing power production sectors with a significant remaining cost reduction potential. Nowadays the wind energy cost is comparable to conventional power sources such as coal, natural gas, and nuclear energy. Further cost reduction is possible through advanced research and developments enhanced with high-precision predictive modeling methods and tools. It is our belief that advanced computational analysis of wind turbines, including time-dependent, full-scale fluid–structure interaction (FSI) simulation needs to be incorporated into the wind-turbine design process to improve their performance under a wide range of operating conditions and to prevent failure of the main turbine components. During wind-turbine operation, or at the stage of experimental testing, advanced data-driven simulations can serve as a valuable tool to produce high-fidelity outputs for quantities of interest for which measurements are not readily available.

Accurate and robust full-machine wind-turbine FSI simulations, however, engender several significant challenges when it comes to modeling of the aerodynamics. In the near-tip region of the offshore horizontal-axis wind-turbine (HAWT) blades the flow Reynolds number is $O(10^7)$, which results in fully turbulent, wall-bounded flow. In order to accurately predict the blade aerodynamic loads in this regime, the numerical formulation must be stable and sufficiently accurate in the presence of

A. Korobenko • X. Deng • J. Yan • Y. Bazilevs (✉)
University of California San Diego, La Jolla, CA 92093, USA
e-mail: yuri@ucsd.edu

Y. Bazilevs, K. Takizawa (eds.), *Advances in Computational Fluid-Structure Interaction and Flow Simulation*, Modeling and Simulation in Science, Engineering and Technology, DOI 10.1007/978-3-319-40827-9_38

thin, transitional turbulent boundary layers. In a case of vertical-axis wind turbines (VAWT) the aerodynamic complexity of a problem is further increased. The wind-turbine blades experience rapid and large variations in the angle of attack resulting in an air flow that is constantly switching from being fully attached to being fully separated, even under steady wind and rotor speeds. Another challenge is to represent how the turbulent flow features generated by the upstream blades affect the aerodynamics of the downstream blades. In addition, as the flexibility in VAWTs does not come from the blades, which are practically rigid (although blades deform at high rotational speeds), but rather from the tower itself, and its connection to the rotor and ground, the challenge remains to be able to simulate a spinning rotor that is mounted on a flexible tower. These computational challenges are addressed in this book chapter.

In what follows, the book chapter presents computational techniques for advanced FSI simulations of both horizontal- and vertical-axis wind-turbine designs, ranging from a few kW to multi-MW systems. In Section 2 we briefly recall the aerodynamics and structural mechanics modeling approaches and novel mesh moving techniques for objects in relative motions. In Section 3 we present an FSI computation of a 5 MW offshore wind-turbine undergoing yawing motion, and an FSI computation of a well-known VAWT design. We conclude the section with an aerodynamic simulation of a 5 MW offshore wind-turbine rotor in an atmospheric boundary layer (ABL). In Section 4 we draw conclusions.

2 Methods for Modeling and Simulations

2.1 *Governing Equations at the Continuum Level*

To perform the HAWT and VAWT simulations, we adopt the FSI framework developed in [9]. The wind-turbine aerodynamics is governed by the Navier–Stokes equations of incompressible flows. The incompressible-flow assumption is valid for the present application because the Mach number is low. The Navier–Stokes equations are posed on a moving spatial domain, and are written in the arbitrary Lagrangian–Eulerian (ALE) frame [20] as follows:

$$\rho_1 \left(\left. \frac{\partial \mathbf{u}}{\partial t} \right|_{\hat{x}} + (\mathbf{u} - \hat{\mathbf{u}}) \cdot \nabla \mathbf{u} - \mathbf{f}_1 \right) - \nabla \cdot \boldsymbol{\sigma}_1 = \mathbf{0}, \tag{1}$$

$$\nabla \cdot \mathbf{u} = 0, \tag{2}$$

where ρ_1 is the fluid density, $\mathbf{f}_1$ is the external force per unit mass, $\mathbf{u}$ and $\hat{\mathbf{u}}$ are velocities of the fluid and fluid mechanics domain, respectively. The stress tensor $\boldsymbol{\sigma}_1$ is defined as

$$\boldsymbol{\sigma}_1 (\mathbf{u}, p) = -p\mathbf{I} + 2\mu \boldsymbol{\varepsilon} (\mathbf{u}), \tag{3}$$

where p is the pressure, $\mathbf{I}$ is the identity tensor, μ is the dynamic viscosity, and $\boldsymbol{\varepsilon}(\mathbf{u})$ is the strain-rate tensor given by

$$\boldsymbol{\varepsilon}(\mathbf{u}) = \frac{1}{2}\left(\nabla\mathbf{u} + \nabla\mathbf{u}^T\right). \tag{4}$$

In Eq. (1), $\Big|_{\hat{x}}$ denotes the time derivative taken with respect to a fixed referential domain spatial coordinates $\hat{\mathbf{x}}$. The spatial derivatives in the above equations are taken with respect to the spatial coordinates $\mathbf{x}$ of the current configuration.

The governing equations of structural mechanics written in the Lagrangian frame consist of the local balance of linear momentum, and are given by

$$\rho_2\left(\frac{\mathrm{d}^2\mathbf{y}}{\mathrm{d}t^2} - \mathbf{f}_2\right) - \nabla\cdot\boldsymbol{\sigma}_2 = \mathbf{0}, \tag{5}$$

where ρ_2 is the structural density, $\mathbf{f}_2$ is the body force per unit mass, $\boldsymbol{\sigma}_2$ is the structural Cauchy stress, and $\mathbf{y}$ is the unknown structural displacement vector.

At the fluid–structure interface, compatibility of the kinematics and tractions is enforced, namely

$$\mathbf{u} - \frac{\mathrm{d}\mathbf{y}}{\mathrm{d}t} = \mathbf{0}, \tag{6}$$

$$\boldsymbol{\sigma}_1\mathbf{n}_1 + \boldsymbol{\sigma}_2\mathbf{n}_2 = \mathbf{0}, \tag{7}$$

where $\mathbf{n}_1$ and $\mathbf{n}_2$ are the unit outward normal vectors to the fluid and structural mechanics domain at their interface. Note that $\mathbf{n}_1 = -\mathbf{n}_2$.

The above equations constitute the basic formulation of the FSI problem at the continuous level. The aerodynamics formulation makes use of the FEM-based ALE–VMS approach [8, 20, 32] augmented with weakly enforced boundary conditions [2, 4]. The former acts as a turbulence model, while the latter relaxes the mesh size requirements in the boundary layer without sacrificing the solution accuracy. The ALE–VMS technique was successfully applied to several challenging problems, including full-scale 3D aerodynamics and FSI of horizontal- and vertical-axis wind turbines [5, 6, 8, 11, 11, 24, 25], FSI of compliant hydrofoils [1], patient-specific cardiovascular FSI [32], and FSI of bioprosthetic heart valves [19].

The structural mechanics of wind-turbine blades is modeled using a combination of a rotation-free multilayer composite Kirchhoff–Love shell [6, 13, 23, 25] and beam/cable [27] formulations. Both are discretized using IGA [17, 21] based on non-uniform rational B-splines (NURBS) [26]. This approach gives a good combination of accuracy due to the higher-order and higher-continuity representation of the geometry and solution, and efficiency due to the lack of rotational degrees of freedom in the formulation.

For a variety of discretization options, FSI coupling strategies, and applications to a large class of problems in engineering the reader is referred to the recent book on computational FSI [10].

2.2 *Special Techniques for Handling Nonmatching Interfaces*

In order to simulate complicated FSI scenarios, such as rotor yawing for HAWTs, or even basic operation for VAWTs, additional computational techniques are required. In the case of HAWT rotor yawing motion, the entire gearbox undergoes rotation parallel to the tower axis, and this rotation must be transferred to the rotor and hub without interfering with the rotor spinning motion. In the case of basic VAWT operation, the air flow spins the rotor, which is connected to a flexible tower with struts. Furthermore, the moving-mesh aerodynamics formulation for this expanded problem class can no longer have a fixed sliding interface. For example, in the case of the rotor yawing motion, in order to keep the good quality of the aerodynamics mesh and prevent the rotor blades from crossing the boundary of the rotor cylindrical domain, it is preferred that the sliding interface follows the motion of the gearbox, while accommodating the spinning rotor and deforming blades.

To accommodate the spinning motion of the rotor superposed on the global elastic deformation of the structure, and to maintain a moving-mesh discretization with good boundary-layer resolution critical for aerodynamics accuracy, the sliding interface technique [3, 18, 24] is "upgraded" to handle more complex structural motions [11, 14]. While at the fluid–structure interface the fluid mechanics mesh follows the rotor motion, the outer boundary of the cylindrical domain that encloses the rotor is only allowed to move as a rigid object. The rigid-body motion part is extracted from the rotor structural mechanics solution (see, e.g., [33]), and is applied directly to the outer boundary of the cylindrical domain enclosing the rotor. The inner boundary of the domain that encloses the cylindrical subdomain also moves as a rigid object. It follows the motion of the cylindrical subdomain, but with the spinning component of the motion removed. The fluid mechanics mesh motion in the interior of the two subdomains is governed by the equations of elastostatics with Jacobian-based stiffening [22, 28, 38, 40–42] to preserve the aerodynamic mesh quality. Continuity of the kinematic and traction variables across the nonmatching sliding interface is enforced weakly.

We note that in application of the FEM to flows with moving mechanical components, alternatively to the sliding interface approach, the Shear–Slip Mesh Update Method [15, 16, 43] and its more general versions [38, 39] may also be used to handle objects in relative motion. A recently developed set of space-time (ST) methods can serve as a third alternative in dealing with objects in relative motion. The components of this set include the ST/NURBS Mesh Update Method [34, 36, 37], ST interface tracking with topology change [29], and ST computation technique with continuous representation in time [35].

3 Numerical Examples

3.1 FSI Simulation of an Offshore Wind Turbine with Yawing

We first show an FSI simulations of a 5 MW offshore wind turbine undergoing yawing motion [14]. The wind-turbine rotor is positioned at 80 m above ground and is tilted by 5° to avoid the blade hitting the tower as the rotor spins. (Another way to have sufficient tower clearance is to "prebend" the rotor blades. See [7] for details.) Furthermore, the wind-turbine rotor plane is initially placed at 15° relative to the wind direction. A fixed yawing rotational speed is applied to the gearbox to slowly turn the rotor into the wind at 0.03 rad/s. The inflow wind speed is set to 11.4 m/s. The initial rotor speed is set to 12.1 rpm, and the rotor is allowed to spin freely during the prescribed yawing motion.

The structural mechanics mesh of the full turbine has 13,273 quadratic NURBS shell elements and two quadratic NURBS beam elements. The aerodynamics mesh has a total of 5,458,185 linear elements. Triangular prisms are employed in the blade boundary layers, and tetrahedral elements are used elsewhere in the aerodynamics domain. The size of the first boundary-layer element in the wall-normal direction is 1 cm. The time step of 0.0001 s is employed in the computation.

Snapshots of the vorticity isosurfaces colored by flow speed are shown in Figure 1. Figure 2 shows the time history of the aerodynamic torque axial component (i.e., the component in the direction of the vector aligned with the axis of rotation). As the rotor turns into the wind, the value is increasing, as expected, and, as the rotor passed the point of maximum power production, the aerodynamic torque value

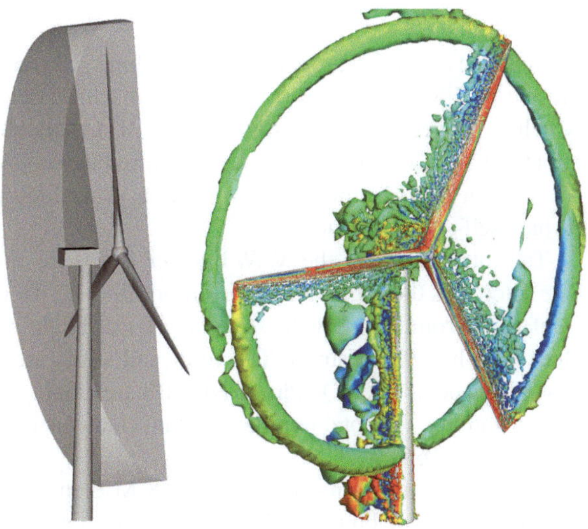

Fig. 1 Left: NURBS-based IGA structural model of a HAWT showing a portion of the sliding interface. Right: Snapshot of vorticity colored by air speed as the rotor spins and yaws.

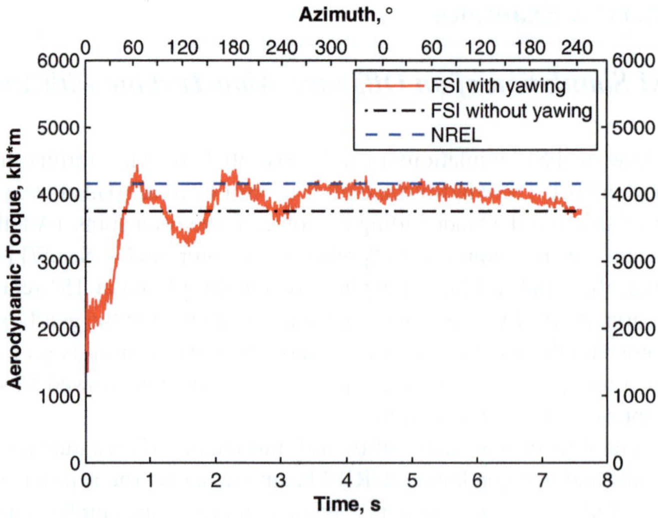

Fig. 2 Time history of the aerodynamic torque for the FSI simulation with yawing.

starts to decrease. The level of the computed aerodynamic torque is consistent with the earlier simulations for this wind turbine operating under similar wind- and rotor-speed conditions (see, e.g., [5, 6, 9, 30, 31, 36]).

3.2 FSI Simulation of a Windspire VAWT

To illustrate an application of the wind-turbine FSI framework to VAWTs, we briefly present an FSI simulation of a 1.2 kW machine designed by Windspire Energy, which is a three-bladed, medium-solidity Darrieus turbine. For more information about the VAWT model the reader is referred to a companion paper [11], where the focus is on VAWT simulation, validation using field-test data, and assessment of turbine self-starting issues.

The total height of the VAWT tower is 9.0 m, and the rotor height is 6.0 m. The rotor uses the DU06W200 airfoil profile with the chord length of 0.127 m, and is of the Giromill type with straight vertical blade sections attached to the main shaft with horizontal struts. The rotor and struts are made of aluminum, and the tower is made of steel. Quadratic NURBS are employed for both the beam and shell discretizations. The total number of beam elements is 116, and total number of shell elements is 7,029.

The aerodynamics mesh has about 8 M elements, which are linear triangular prisms in the blade boundary layers, and linear tetrahedra elsewhere. The boundary-layer mesh is constructed using 18 layers of elements, with the size of the first element in the wall-normal direction of 0.0003 m, and growth ratio of 1.1.

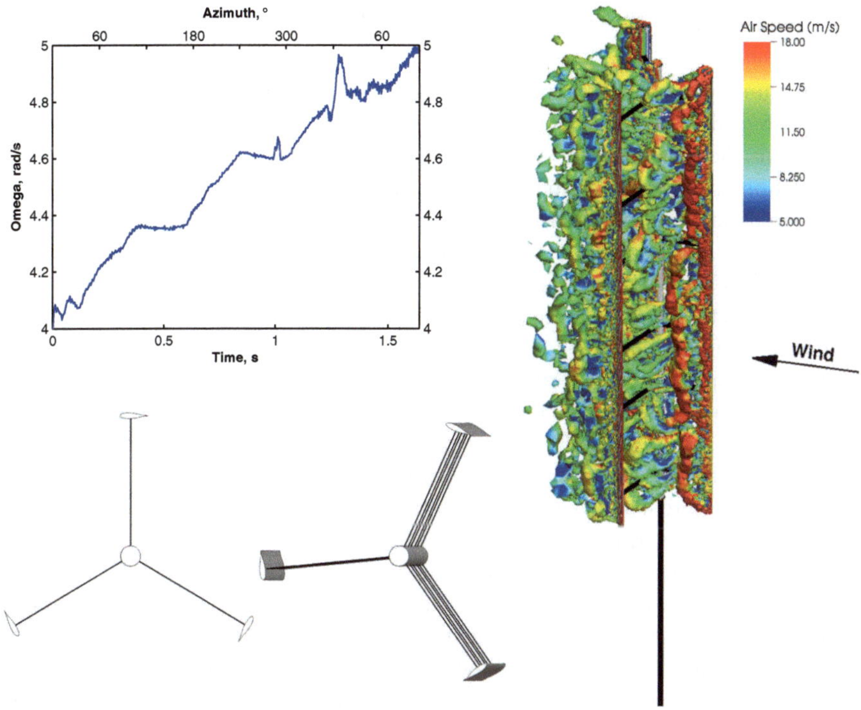

Fig. 3 FSI simulation of a Windspire VAWT. Top-Left: Time history of the rotor speed. Bottom-Left: Top view of the wind-turbine structure in the reference and current configuration illustrating the range of the tower-tip displacement during operation. Right: Vorticity colored by air speed illustrating the air-flow complexity.

We fix the inflow wind speed at 11.4 m/s, and consider the initial rotor speed of 4 rad/s. The VAWT is allowed to spin freely and accelerate under the action of incoming wind. The time step in the computation is set to 2.0×10^{-5} s.

Figure 3 shows the snapshots of vorticity colored by velocity magnitude. The flow is fully turbulent and requires relatively fine boundary-layer resolution for accurate aerodynamic load prediction. Also, Figure 3 shows the time history of rotor speed. A gradual increase in the rotor speed is observed, suggesting a possibility that the rotor may eventually reach the speed necessary for efficient operation. Finally, Figure 3 shows the snapshots of the top view of the VAWT structure in the reference and current configurations. The tip displacement amplitude is predicted to be about 12 cm, which we find reasonable given the tower height of 9 m, and one of the VAWT design objectives being that the structure is not too flexible.

3.3 Effect of Atmospheric Boundary Layer

In the numerical examples presented so far the simulations were carried out under simplified constant uniform wind speed conditions. In this section we investigate the influence of ABL on the turbine performance. The ALE–VMS formulation for stratified turbulent incompressible flows is adopted to perform a pure aerodynamic simulation of a full-scale 5 MW offshore wind-turbine rotor in an ABL with temperature stratification. The formulation couples the Navier–Stokes equations of incompressible flows with the Boussinesq approximation, and a scalar advection–diffusion equation for the temperature field. More details on the numerical formulation can be found in [12].

The rotor speed is set to 9 rpm, which gives the optimal tip-speed ratio for 8 m/s wind speed, and which is also the geostrophic wind speed used in the computation. The time-step size of 2.0×10^{-4} s is employed. The flow is initialized using the LES data interpolated to the interior of the rotor mesh, and the computation is started impulsively. Figure 4 shows vorticity isosurfaces at $t = 8.5$ s. Due to the presence of the inversion layer, tip vortices travel with different speeds, faster near the top and slower near the bottom of the domain. As a result, the perfect helical pattern of tip vortices, which is expected in the case of uniform flow, is no longer present. As the rotor turns and blades travel in and out of the inversion layer, they introduce a certain amount of mixing in the flow (see Figure 4). This inversion-layer mixing propagates downstream and gives a complex and largely unstudied wake behavior. The rotor-thrust time history is shown in Figure 5, where thrust is plotted for each individual blade. The presence of ABL produces an 18% fluctuation in the thrust load during the cycle. This indicates that stratification strongly affects the transient

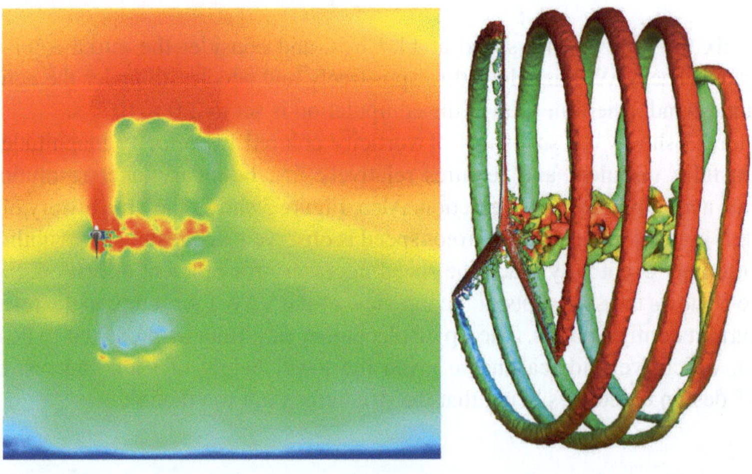

Fig. 4 Wind-turbine rotor in an ABL. Left: Flow speed (in m/s) contours on a 2D slice of the problem domain showing mixing of the inversion layer in the rotor wake. Right: Vorticity isosurfaces colored by flow speed (in m/s) at $t = 8.5$ s.

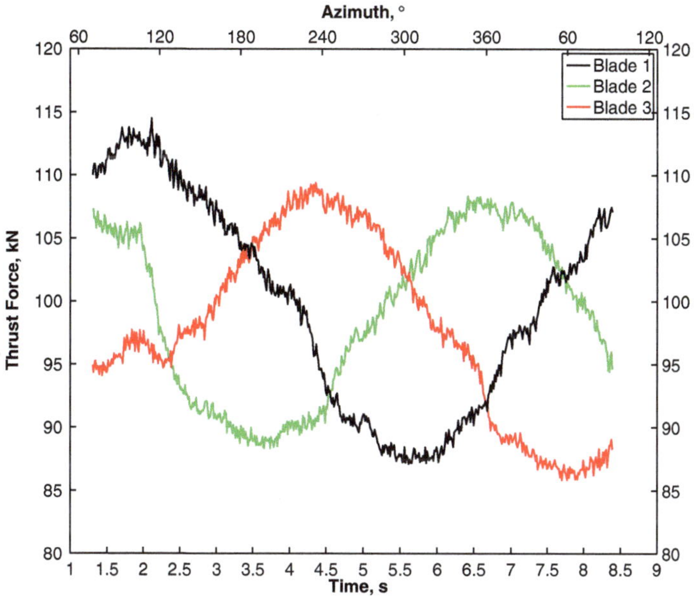

Fig. 5 Wind-turbine rotor in an ABL. Time history of the thrust force acting on each blade.

rotor loads, and should be considered in the structural and FSI analyses of offshore wind-turbine blades, especially for the prediction of fatigue damage and remaining useful life of these structures.

4 Conclusions

In this book chapter we targeted more advanced FSI simulations of wind turbines, such as rotor yawing for HAWTs, and full-machine FSI of VAWTs. Structural models of wind-turbine designs were constructed and discretized using the recently proposed IGA-based rotation-free shell and beam formulations. This approach presents a good combination of accuracy due to the structural geometry representation using smooth, higher-order functions, and efficiency due to the fact that only displacement degrees of freedom are employed in the formulation. The ALE–VMS technique for aerodynamics modeling was augmented with an improved version of the sliding interface formulation, which allows the interface to move in space as a rigid object and accommodates the global turbine deflections in addition to the rotor spinning motion.

Using advanced FSI techniques developed, we were able to simulate a large-scale 5 MW HAWT undergoing yawing motion. We also present FSI simulations of full-scale Darrieus VAWT. The numerical examples presented in this book chapter illustrate the successful application of the proposed techniques to the FSI simulation of wind turbines at full scale. Finally, in the last example, we simulated a full-scale 5 MW offshore wind-turbine rotor in a thermally stratified ABL and observed significant variations in the rotor-blade thrust during the rotation cycle. In the near future we plan to combine the newly developed ALE–VMS formulation of stratified flows with our full-machine FSI framework to simulate wind turbines operating in an ABL. These simulations will likely produce more realistic structural mechanics response of the blades, and will give us better understanding of the turbulent wake structures created downstream the turbine.

Acknowledgements This work was supported by the AFOSR Award FA9550-12-1-0005, the AFOSR Award FA9550-12-1-0046, the NSF Award CBET-1306869, and the NSF CAREER Award 1055091. This support is gratefully acknowledged.

References

1. Augier, B., Yan, J., Korobenko, A., Czarnowski, J., Ketterman, G., Bazilevs, Y.: Experimental and numerical FSI study of compliant hydrofoils. Comput. Mech. **55**, 1079–1090 (2015)
2. Bazilevs, Y., Hughes, T.J.R.: Weak imposition of Dirichlet boundary conditions in fluid mechanics. Comput. Fluids **36**, 12–26 (2007)
3. Bazilevs, Y., Hughes, T.J.R.: NURBS-based isogeometric analysis for the computation of flows about rotating components. Comput. Mech. **43**, 143–150 (2008)
4. Bazilevs, Y., Michler, C., Calo, V.M., Hughes, T.J.R.: Isogeometric variational multiscale modeling of wall-bounded turbulent flows with weakly enforced boundary conditions on unstretched meshes. Comput. Methods Appl. Mech. Eng. **199**, 780–790 (2010)
5. Bazilevs, Y., Hsu, M.-C., Akkerman, I., Wright, S., Takizawa, K., Henicke, B., Spielman, T., Tezduyar, T.E.: 3D simulation of wind turbine rotors at full scale. Part I: geometry modeling and aerodynamics. Int. J. Numer. Methods Fluids **65**, 207–235 (2011)
6. Bazilevs, Y., Hsu, M.-C., Kiendl, J., Wüchner, R., Bletzinger, K.-U.: 3D simulation of wind turbine rotors at full scale. Part II: fluid–structure interaction modeling with composite blades. Int. J. Numer. Methods Fluids **65**, 236–253 (2011)
7. Bazilevs, Y., Hsu, M.-C., Kiendl, J., Benson, D.J.: A computational procedure for pre-bending of wind turbine blades. Int. J. Numer. Methods Eng. **89**, 323–336 (2012)
8. Bazilevs, Y., Hsu, M.-C., Takizawa, K., Tezduyar, T.E.: ALE-VMS and ST-VMS methods for computer modeling of wind-turbine rotor aerodynamics and fluid–structure interaction. Math. Mod. Methods Appl. Sci. **22**(supp02), 1230002 (2012)
9. Bazilevs, Y., Hsu, M.-C., Scott, M.A.: Isogeometric fluid–structure interaction analysis with emphasis on non-matching discretizations, and with application to wind turbines. Comput. Methods Appl. Mech. Eng. **249–252**, 28–41 (2012)
10. Bazilevs, Y., Takizawa, K., Tezduyar, T.E.: Computational Fluid–Structure Interaction: Methods and Applications. Wiley, Chichester (2013)
11. Bazilevs, Y., Korobenko, A., Deng, X., Yan, J., Kinzel, M., Dabiri, J.O.: FSI modeling of vertical-axis wind turbines. J. Appl. Mech. **81**, 081006 (2014)
12. Bazilevs, Y., Korobenko, A., Yan, J., Pal, A., Gohari, S.M.I., Sarkar, S.: ALE–VMS formulation for stratified turbulent incompressible flows with applications. Math. Models Methods Appl. Sci. **25**, 1540011 (2015)
13. Bazilevs, Y., Deng, X., Korobenko, A., Lanza Di Scalea, F., Todd, M.D., Taylor, S.G.: Isogeometric fatigue damage prediction in large-scale composite structures driven by dynamic sensor data. J. Appl. Mech. **82**, 091008 (2015)

14. Bazilevs, Y., Korobenko, A., Deng, X., Yan, J.: Novel structural modeling and mesh moving techniques for advanced FSI simulation of wind turbines. Int. J. Numer. Methods Eng. **102**, 766–783 (2015)
15. Behr, M., Tezduyar, T.: The shear-slip mesh update method. Comput. Methods Appl. Mech. Eng. **174**, 261–274 (1999)
16. Behr, M., Tezduyar, T.: Shear-slip mesh update in 3D computation of complex flow problems with rotating mechanical components. Comput. Methods Appl. Mech. Eng. **190**, 3189–3200 (2001)
17. Cottrell, J.A., Hughes, T.J.R., Bazilevs, Y.: Isogeometric Analysis. Toward Integration of CAD and FEA. Wiley, Chichester (2009)
18. Hsu, M.-C., Akkerman, I., Bazilevs, Y.: Finite element simulation of wind turbine aerodynamics: validation study using NREL Phase VI experiment. Wind Energy **17**, 461–481 (2014)
19. Hsu, M.-C., Kamensky, D., Bazilevs, Y., Sacks, M.S., Hughes, T.J.R.: Fluid–structure interaction analysis of bioprosthetic heart valves: significance of arterial wall deformation. Comput. Mech. **54**, 1055–1071 (2014)
20. Hughes, T.J.R., Liu, W.K., Zimmermann, T.K.: Lagrangian–Eulerian finite element formulation for incompressible viscous flows. Comput. Methods Appl. Mech. Eng. **29**, 329–349 (1981)
21. Hughes, T.J.R., Cottrell, J.A., Bazilevs, Y.: Isogeometric analysis: CAD, finite elements, NURBS, exact geometry, and mesh refinement. Comput. Methods Appl. Mech. Eng. **194**, 4135–4195 (2005)
22. Johnson, A.A., Tezduyar, T.E.: Mesh update strategies in parallel finite element computations of flow problems with moving boundaries and interfaces. Comput. Methods Appl. Mech. Eng. **119**, 73–94 (1994)
23. Kiendl, J., Bazilevs, Y., Hsu, M.-C., Wüchner, R., Bletzinger, K.-U.: The bending strip method for isogeometric analysis of Kirchhoff–Love shell structures comprised of multiple patches. Comput. Methods Appl. Mech. Eng. **199**, 2403–2416 (2010)
24. Korobenko, A., Hsu, M.-C., Akkerman, I., Bazilevs, Y.: Aerodynamic simulation of vertical-axis wind turbines. J. Appl. Mech. **81**, 021011 (2013)
25. Korobenko, A., Hsu, M.-C., Akkerman, I., Tippmann, J., Bazilevs, Y.: Structural mechanics modeling and FSI simulation of wind turbines. Math. Mod. Methods Appl. Sci. **23**, 249–272 (2013)
26. Piegl, L., Tiller, W.: The NURBS Book (Monographs in Visual Communication), 2nd edn. Springer, New York (1997)
27. Raknes, S.B., Deng, X., Bazilevs, Y., Benson, D.J., Mathisen, K.M., Kvamsdal, T.: Isogeometric rotation-free bending-stabilized cables: statics, dynamics, bending strips and coupling with shells. Comput. Methods Appl. Mech. Eng. **263**, 127–143 (2013)
28. Stein, K., Tezduyar, T., Benney, R.: Mesh moving techniques for fluid–structure interactions with large displacements. J. Appl. Mech. **70**, 58–63 (2003)
29. Takizawa, K.: Computational engineering analysis with the new-generation space–time methods. Comput. Mech. **54**, 193–211 (2014)
30. Takizawa, K., Henicke, B., Montes, D., Tezduyar, T.E., Hsu, M.-C., Bazilevs, Y.: Numerical-performance studies for the stabilized space–time computation of wind-turbine rotor aerodynamics. Comput. Mech. **48**, 647–657 (2011)
31. Takizawa, K., Henicke, B., Tezduyar, T.E., Hsu, M.-C., Bazilevs, Y.: Stabilized space–time computation of wind-turbine rotor aerodynamics. Comput. Mech. **48**, 333–344 (2011)
32. Takizawa, K., Bazilevs, Y., Tezduyar, T.E.: Space–time and ALE-VMS techniques for patient-specific cardiovascular fluid–structure interaction modeling. Arch. Comput. Methods Eng. **19**, 171–225 (2012)
33. Takizawa, K., Tezduyar, T.E., Boben, J., Kostov, N., Boswell, C., Buscher, A.: Fluid–structure interaction modeling of clusters of spacecraft parachutes with modified geometric porosity. Comput. Mech. **52**, 1351–1364 (2013)
34. Takizawa, K., Bazilevs, Y., Tezduyar, T.E., Hsu, M.-C., Øiseth, O., Mathisen, K.M., Kostov, N., McIntyre, S.: Engineering analysis and design with ALE-VMS and space–time methods. Arch. Comput. Methods Eng. **21**, 481–508 (2014)

35. Takizawa, K., Tezduyar, T.E., Kostov, N.: Sequentially-coupled space–time FSI analysis of bio-inspired flapping-wing aerodynamics of an MAV. Comput. Mech. **54**, 213–233 (2014)
36. Takizawa, K., Tezduyar, T.E., McIntyre, S., Kostov, N., Kolesar, R., Habluetzel, C.: Space–time VMS computation of wind-turbine rotor and tower aerodynamics. Comput. Mech. **53**, 1–15 (2014)
37. Takizawa, K., Bazilevs, Y., Tezduyar, T.E., Long, C.C., Marsden, A.L., Schjodt, K.: ST and ALE-VMS methods for patient-specific cardiovascular fluid mechanics modeling. Math. Mod. Methods Appl. Sci. **24**, 2437–2486 (2014)
38. Tezduyar, T.E.: Finite element methods for flow problems with moving boundaries and interfaces. Arch. Comput. Methods Eng. **8**, 83–130 (2001)
39. Tezduyar, T.E.: Finite elements in fluids: special methods and enhanced solution techniques. Comput. Fluids **36**, 207–223 (2007)
40. Tezduyar, T.E., Sathe, S.: Modeling of fluid–structure interactions with the space–time finite elements: solution techniques. Int. J. Numer. Methods Fluids **54**, 855–900 (2007)
41. Tezduyar, T.E., Behr, M., Mittal, S., Johnson, A.A.: Computation of unsteady incompressible flows with the finite element methods – space–time formulations, iterative strategies and massively parallel implementations. In: New Methods in Transient Analysis, PVP-Vol.246/AMD-Vol. 143, pp. 7–24. ASME, New York (1992)
42. Tezduyar, T., Aliabadi, S., Behr, M., Johnson, A., Mittal, S.: Parallel finite-element computation of 3D flows. Computer **26**(10), 27–36 (1993)
43. Tezduyar, T., Aliabadi, S., Behr, M., Johnson, A., Kalro, V., Litke, M.: Flow simulation and high performance computing. Comput. Mech. **18**, 397–412 (1996)

MIX
Papier aus verantwortungsvollen Quellen
Paper from responsible sources
FSC® C105338
FSC
www.fsc.org

If you have any concerns about our products,
you can contact us on
ProductSafety@springernature.com

In case Publisher is established outside the EU,
the EU authorized representative is:
Springer Nature Customer Service Center GmbH
Europaplatz 3, 69115 Heidelberg, Germany

Printed by Libri Plureos GmbH
in Hamburg, Germany